STUDENT SOLUTIONS MANUAL

CINDY TRIMBLE & ASSOCIATES

INTERMEDIATE ALGEBRA

SIXTH EDITION

John Tobey
North Shore Community College

Jeffrey Slater
North Shore Community College

Prentice Hall
is an imprint of

Reproduced by Pearson Prentice Hall from electronic files supplied by the author.

Publishing as Prentice Hall, Upper Saddle River, NJ 07458.

ISBN-13: 978-0-321-57838-9
ISBN-10: 0-321-57838-4

1 2 3 4 5 6 BRR 12 11 10 09 08

Prentice Hall
is an imprint of

www.pearsonhighered.com

Contents

Chapter 1

1.1 Exercises

1. Integers include negative numbers. Whole numbers are nonnegative integers.

3. A terminating decimal is a decimal that comes to an end.

5. 13,001 is a natural number: Natural, Whole, Integer, Rational, Real

7. –42 is an integer: Integer, Rational, Real

9. –6.1313 is a terminating decimal: Rational, Real

11. $-\frac{8}{7}$ is a quotient of integers: Rational, Real

13. $\frac{\pi}{5}$, π is a decimal that is nonterminating and nonrepeating: Irrational, Real

15. 0.79 is a terminating decimal: Rational, Real

17. 7.040040004... is a nonterminating and nonrepeating decimal: Irrational, Real

19. $-25, -\frac{28}{7}$

21. $-\frac{28}{7}, -25, -\frac{18}{5}, -\pi, -0.763, -0.333...$

23. $\frac{1}{10}, \frac{2}{7}, 9, \frac{283}{5}, 52.8$

25. $-\frac{18}{5}, -0.763, -0.333...$

27. 1, 2, 3, 4, 5, 6, 7

29. $8 + (-5) = (-5) + 8$
Commutative property of addition

31. $\frac{1}{14} \cdot 14 = 1$
Inverse property of multiplication

33. $5.6 + 0 = 5.6$
Identity property of addition

35. $-\frac{1}{6} \cdot \frac{5}{9} = \frac{5}{9} \cdot \left(-\frac{1}{6}\right)$
Commutative property of multiplication

37. $\frac{3}{7} \cdot 1 = \frac{3}{7}$
Identity property of multiplication

39. $-6(4.5 + 3) = -6(4.5) + (-6)(3)$
Distributive property of multiplication over addition

41. $2.4 + \left(3 + \sqrt{5}\right) = (2.4 + 3) + \sqrt{5}$
Associative property of addition

43. $4 \cdot (3 \cdot x) = (4 \cdot 3) \cdot x$
Associative property of multiplication

45. $-\sqrt{3}\left(5 \cdot \frac{1}{2}\right) = \left(-\sqrt{3} \cdot 5\right) \cdot \frac{1}{2}$
Associative property of multiplication

47. 12 is the additive inverse of –12.

49. $-\frac{3}{5}$ is the multiplicative inverse of $-\frac{5}{3}$.

51. **a.** $\frac{3}{8} = 0.375 = 37.5\%$
37.5% of subspecies of tiger have become extinct.

    b. $\frac{5}{8} = 0.625 = 62.5\%$
62.5% of subspecies of tiger are endangered.

    c. None

53. There are several possibilities. The second and third numbers increase by \$100. \$500 does not follow this pattern. The fourth through the eleventh numbers follow a doubling pattern. \$125,000 does not follow this pattern.

55. No. When the winnings reach \$15,625, the next smallest winning amount would be one half of that amount. This next amount would be \$7812.50 which is not a whole number.

57. 154(0.4) = 61.6; about 62 people per year are killed in house fires.

Quick Quiz 1.1

1. The irrational numbers are numbers whose decimal forms are nonterminating and nonrepeating: -2π, $\sqrt{7}$

2. The negative real numbers are -2π, $-\frac{3}{11}$, $-0.5333...$

3. 3(−3.56 + 9) = 3(−3.56) + 3(9) illustrates the Distributive Property of Multiplication over Addition.

4. Answers may vary. Possible solution: (4)(3.5 + 9.3) = (4)(9.3 + 3.5) illustrates the Commutative property of addition. Then (4)(9.3 + 3.5) = (9.3 + 3.5)(4) illustrates the Commutative property of multiplication.

1.2 Exercises

1. To add two real numbers with the same sign, add their absolute values. The sum takes the common sign. To add two real numbers with different signs, find the difference of their absolute values. The answer takes the sign of the number with the larger absolute value.

3. $\left|-\frac{2}{3}\right|=\frac{2}{3}$

5. $|8.3| = 8.3$

7. $|9 - 14| = |-5| = 5$

9. $|-b| = b$

11. $-6 + (-12) = -18$

13. $-3 - (-5) = -3 + 5 = 2$

15. $5\left(-\frac{1}{3}\right)=-\frac{5}{3}=-1\frac{2}{3}$

17. $\frac{-18}{-2}=9$

19. $(-0.3)(0.1) = -0.03$

21. $-4.9 + 10.5 = 5.6$

23. $-\frac{5}{12}+\frac{7}{18}=-\frac{15}{36}+\frac{14}{36}=-\frac{1}{36}$

25. $(-2.4) \div (6) = -0.4$

27. $\left(-\frac{2}{3}\right)\left(-\frac{7}{4}\right)=\frac{2\cdot 7}{3\cdot 4}=\frac{7}{3\cdot 2}=\frac{7}{6}=1\frac{1}{6}$

29. $(-4)\left(\frac{1}{2}\right)+(-7)(-3)=-2+21=19$

31. $12 + (-12) = 0$

33. $\frac{-5}{0}$ Undefined

35. $\frac{0}{-3}=0$

37. $\frac{3-3}{-8}=\frac{0}{-8}=0$

39. $\frac{-7+(-7)}{-14}=\frac{-14}{-14}=1$

41. $\frac{17}{18}+\left(-\frac{5}{6}\right)=\frac{17}{18}+\left(-\frac{15}{18}\right)=\frac{2}{18}=\frac{1}{9}$

43. $$\begin{aligned}\frac{5}{6}\div\left(-\frac{3}{4}\right)&=\frac{5}{6}\times\left(-\frac{4}{3}\right)\\&=-\frac{20}{18}\\&=-\frac{5\cdot 2\cdot 2}{3\cdot 3\cdot 2}\\&=-\frac{10}{9}\\&=-1\frac{1}{9}\end{aligned}$$

45. $$\begin{aligned}5+6-(-3)-8+4-3&=5+6+3+(-8)+4+(-3)\\&=11+3+(-8)+4+(-3)\\&=14+(-8)+4+(-3)\\&=6+4+(-3)\\&=10+(-3)\\&=7\end{aligned}$$

47. $\frac{9(-3)+7}{3-7}=\frac{-27+7}{3-7}=\frac{-20}{-4}=5$

49. $6(-2)+3-5(-3)-4=-12+3-(-15)-4$
$=-9+15+(-4)$
$=6+(-4)$
$=2$

51. $15+20\div 2-4(3)=15+10-4(3)$
$=15+10-12$
$=25-12$
$=13$

53. $\frac{6-2(7)}{5-6}=\frac{6-14}{-1}=\frac{-8}{-1}=8$

55. $\frac{1+49\div(-7)-(-3)}{-1-2}=\frac{1+(-7)+3}{-3}$
$=\frac{-6+3}{-3}$
$=\frac{-3}{-3}$
$=1$

57. $4\left(-\frac{1}{2}\right)+\frac{2}{3}(9)=-2+6=4$

59. $\frac{1.63482-2.48561}{(16.05436)(0.07814)}=\frac{-0.85079}{1.25448769}$
≈ -0.678197169

61. One or three of the quantities a, b, c must be negative. In other words, when multiplying an odd number of negative numbers, the answer will be negative.

Cumulative Review

63. 5 + 17 = 17 + 5
Commutative property of addition

64. 4 · (3 · 6) = (4 · 3) · 6
Associative property of multiplication

65. $-\frac{1}{2}\pi$, $\sqrt{3}$ are irrational numbers.

66. $-16, 0, \frac{19}{2}, 9.36, 10.\overline{5}$ are rational numbers.

Quick Quiz 1.2

1. −8 + 3(−4) = −8 − 12 = −20

2. 5.6 − (−3.8) = 5.6 + 3.8 = 9.4

3. $-12+15\div(-5)+3(6)=-12+(-3)+3(6)$
$=-12-3+18$
$=-15+18$
$=3$

4. Answers may vary. Possible solution: Simplify the numerator and denominator separately. Then divide.
$\frac{3(-2)+8}{5-9}=\frac{-6+8}{-4}=\frac{2}{-4}=-\frac{1}{2}$

1.3 Exercises

1. The base is a and the exponent is 3.

3. Positive

5. −11 and 11, (−11)(−11) = 121
(+11)(+11) = 121

7. $9\cdot 9\cdot 9\cdot 9=9^4$

9. $(-6)(-6)(-6)(-6)(-6)=(-6)^5$

11. $x\cdot x\cdot x\cdot x\cdot x\cdot x\cdot y\cdot y\cdot y=x^6y^3$

13. $2^5=2\cdot 2\cdot 2\cdot 2\cdot 2=32$

15. $(-5)^2=(-5)(-5)=25$

17. $-6^2=-(6)(6)=-36$

19. $-1^4=-(1)(1)(1)(1)=-1$

21. $\left(\frac{2}{3}\right)^2=\left(\frac{2}{3}\right)\left(\frac{2}{3}\right)=\frac{4}{9}$

23. $\left(-\frac{1}{4}\right)^4=\left(-\frac{1}{4}\right)\left(-\frac{1}{4}\right)\left(-\frac{1}{4}\right)\left(-\frac{1}{4}\right)=\frac{1}{256}$

25. $(0.7)^2=(0.7)(0.7)=0.49$

27. $(0.04)^3=(0.04)(0.04)(0.04)=0.000064$

29. $\sqrt{81}=9$ because $9^2=81$.

31. $-\sqrt{16}=-4$

33. $\sqrt{\frac{4}{9}}=\frac{\sqrt{4}}{\sqrt{9}}=\frac{2}{3}$

35. $\sqrt{0.09}=0.3$

37. $\sqrt{9+7}=\sqrt{16}=4$

39. $\sqrt{3599+1}=\sqrt{3600}=60$

41. $\sqrt{\frac{5}{36}+\frac{31}{36}}=\sqrt{\frac{36}{36}}=\sqrt{1}=1$

43. $\sqrt{-36}$ not a real number

45. $-\sqrt{-0.36}$ not a real number

47. $5(3-9)+7=5(-6)+7=-30+7=-23$

49. $-15\div 3+7(-4)=-5-28=-33$

51. $(-2)(-10)+(-4)^2=20+16=36$

53.
$$\begin{aligned}(5+2-8)^3-(-7)&=(7-8)^3+7\\&=(-1)^3+7\\&=-1+7\\&=6\end{aligned}$$

55. $(-2)^3+(-4)^2-3=-8+16-3=8-3=5$

57. $-5^2+3(1-8)=-25+3(-7)=-25+(-21)=-46$

59. $5[(1.2-0.4)-0.8]=5(0.8-0.8)=5(0)=0$

61. $4(-6)-3^2+\sqrt{25}=-24-9+5=-33+5=-28$

63. $\frac{7+2(-4)+5}{8-6}=\frac{7+(-8)+5}{2}=\frac{-1+5}{2}=\frac{4}{2}=2$

65. $\frac{-3(2^3-1)}{3-10}=\frac{-3(8-1)}{-7}=\frac{-3(7)}{-7}=\frac{-21}{-7}=3$

67.
$$\begin{aligned}\frac{|2^2-5|-3^2}{-5+3}&=\frac{|4-5|-9}{-2}\\&=\frac{|-1|-9}{-2}\\&=\frac{1-9}{-2}\\&=\frac{-8}{-2}\\&=4\end{aligned}$$

69.
$$\begin{aligned}\frac{\sqrt{(-5)^2-3+14}}{|19-6+3-25|}&=\frac{\sqrt{25-3+14}}{|13+3-25|}\\&=\frac{\sqrt{22+14}}{|16-25|}\\&=\frac{\sqrt{36}}{|-9|}\\&=\frac{6}{9}\\&=\frac{2}{3}\end{aligned}$$

71. $\frac{\sqrt{6^2-3^2-2}}{(-3)^2-4}=\frac{\sqrt{36-9-2}}{9-4}=\frac{\sqrt{25}}{5}=\frac{5}{5}=1$

73. $(5.986)^5\approx 7685.702373$

75. $2^8-2^6=256-64=192$
There are 192 more results.

Cumulative Review

77. $a\cdot\frac{1}{a}=1$
Inverse property of multiplication

78. $b+(-b)=0$
Inverse property of addition

79. $\frac{40{,}000-5000}{5000}=\frac{35{,}000}{5000}=7$
There is an increase of 700%.

80. $\frac{81{,}000}{27{,}000}=3$
The pressure is three times greater.

81. $1+0.1+1.1(0.1)=1.21$
His income had an increase of 21%.

82. last year: $n \to 0.9n$
this year: $0.9n \to 0.8(0.09n) = 0.72n$
After the two years the number of employees is 72% of the original number of employees, a 28% decrease.

Quick Quiz 1.3

1. $\left(\frac{2}{3}\right)^5 = \left(\frac{2}{3}\right)\left(\frac{2}{3}\right)\left(\frac{2}{3}\right)\left(\frac{2}{3}\right)\left(\frac{2}{3}\right) = \frac{32}{243}$

2. $\frac{8+3(-2)+4}{9-12} = \frac{8+(-6)+4}{9-12} = \frac{6}{-3} = -2$

3. $$\begin{aligned}&(8-10)^3 - 30 \div (-3) + \sqrt{17+8}\\ &= (-2)^3 - 30 \div (-3) + \sqrt{25}\\ &= -8 - 30 \div (-3) + 5\\ &= -8 - (-10) + 5\\ &= -8 + 10 + 5\\ &= 7\end{aligned}$$

4. Answers may vary. Possible solution:
First simplify the numerator by:
$$\begin{aligned}\sqrt{(-3)^3 - 6(-2) + 15} &= \sqrt{-27 - 6(-2) + 15}\\ &= \sqrt{-27 + 12 + 15}\\ &= \sqrt{0}\\ &= 0\end{aligned}$$
Then evaluate the denominator by:
$|3 - 5| = |-2| = 2$
Since the numerator is 0 (and the denominator is not), it evaluates to 0.

How Am I Doing? Sections 1.1–1.3

1. π, $\sqrt{7}$ are irrational real numbers.

2. $\sqrt{9}, -5, 3, \frac{6}{2}, 0$ are integers.

3. $\sqrt{3}$ belongs to the real number set and the irrational number set.

4. $(x + y) + z = x + (y + z)$
The associative property of addition

5. $12\left(\frac{1}{12}\right) = 1$
The inverse property of multiplication

6. $$\begin{aligned}30 \div (-6) + 3 - 2(-5) &= -5 + 3 - 2(-5)\\ &= -5 + 3 + 10\\ &= -2 + 10\\ &= 8\end{aligned}$$

7. $6\left(-\frac{2}{3}\right) + (-5)(-2) = -4 + 10 = 6$

8. $\frac{15-4(3)}{2-8} = \frac{15-12}{-6} = \frac{3}{-6} = -\frac{1}{2}$

9. $\frac{-5+(-5)}{-15} = \frac{-10}{-15} = \frac{2}{3}$

10. $-9 + 6(-2) - (-3) = -9 - 12 + 3 = -21 + 3 = -18$

11. $\sqrt{\frac{16}{49}} = \frac{\sqrt{16}}{\sqrt{49}} = \frac{4}{7}$

12. $\sqrt{0.81} = 0.9$

13. $4^4 = 4 \cdot 4 \cdot 4 \cdot 4 = 256$

14. $$\begin{aligned}12 - \sqrt{3^3 + 6(-3)} &= 12 - \sqrt{27 + (-18)}\\ &= 12 - \sqrt{9}\\ &= 12 - 3\\ &= 9\end{aligned}$$

15. $$\begin{aligned}(-4)^3 + 2(3^2 - 2^2) &= -64 + 2(9 - 4)\\ &= -64 + 2(5)\\ &= -64 + 10\\ &= -54\end{aligned}$$

16. $\frac{4-6^2}{3+\sqrt{16+9}} = \frac{4-36}{3+\sqrt{25}} = \frac{-32}{3+5} = \frac{-32}{8} = -4$

17. $\left|2^2 - 5 - 6\right| = |4 - 5 - 6| = |-7| = 7$

18. $\frac{\sqrt{(-2)^2+5}}{|12-15|} = \frac{\sqrt{4+5}}{|-3|} = \frac{\sqrt{9}}{3} = \frac{3}{3} = 1$

1.4 Exercises

1. $3^{-2} = \frac{1}{3^2} = \frac{1}{9}$

3. $x^{-5} = \frac{1}{x^5}$

5. $(-7)^{-2} = \dfrac{1}{(-7)^2} = \dfrac{1}{49}$

7. $\left(-\dfrac{1}{9}\right)^{-1} = \dfrac{1}{\left(-\frac{1}{9}\right)^1} = -9$

9. $x^4 \cdot x^8 = x^{4+8} = x^{12}$

11. $17^4 \cdot 17 = 17^{4+1} = 17^5$

13. $(3x)(-2x^5) = 3(-2)x^{1+5} = -6x^6$

15. $(-11x^2y^2)(-x^4y^7) = -11(-1)x^{2+4}y^{2+7}$
$= 11x^6y^9$

17. $4x^0y = 4 \cdot 1 \cdot y = 4y$

19. $(3xy)^0(7xy) = 1(7xy) = 7xy$

21. $(-6x^2yz^0)(-4x^0y^2z) = (-6x^2y)(-4y^2z)$
$= (-6)(-4)x^2y^{1+2}z$
$= 24x^2y^3z$

23. $\left(-\dfrac{3}{5}m^{-2}n^4\right)(5m^2n^{-5}) = -\dfrac{3(5)m^{-2+2}n^{4-5}}{5}$
$= -\dfrac{3m^0n^{-1}}{1}$
$= -\dfrac{3}{n}$

25. $\dfrac{x^{16}}{x^5} = x^{16-5} = x^{11}$

27. $\dfrac{a^{20}}{a^{25}} = a^{20-25} = a^{-5} = \dfrac{1}{a^5}$

29. $\dfrac{2^8}{2^5} = 2^{8-5} = 2^3 = 8$

31. $\dfrac{2x^3}{x^8} = 2x^{3-8} = 2x^{-5} = \dfrac{2}{x^5}$

33. $\dfrac{-15x^4yz}{3xy} = -5x^{4-1}y^{1-1}z = -5x^3y^0z = -5x^3z$

35. $\dfrac{-20a^{-3}b^{-8}}{14a^{-5}b^{-12}} = -\dfrac{10}{7}a^{-3-(-5)}b^{-8-(-12)}$
$= -\dfrac{10}{7}a^{-3+5}b^{-8+12}$
$= -\dfrac{10}{7}a^2b^4$

37. $(x^2)^8 = x^{2\cdot 8} = x^{16}$

39. $(3a^5b)^4 = 3^4a^{5\cdot 4}b^4 = 81a^{20}b^4$

41. $\left(\dfrac{x^2y^3}{z}\right)^6 = \dfrac{x^{2\cdot 6}y^{3\cdot 6}}{z^6} = \dfrac{x^{12}y^{18}}{z^6}$

43. $\left(\dfrac{3ab^{-2}}{4a^0b^4}\right)^2 = \left(\dfrac{3a}{4b^6}\right)^2 = \dfrac{3^2a^2}{4^2b^{6\cdot 2}} = \dfrac{9a^2}{16b^{12}}$

45. $\left(\dfrac{2xy^2}{x^{-3}y^{-4}}\right)^{-3} = (2x^4y^6)^{-3}$
$= \dfrac{1}{(2x^4y^6)^3}$
$= \dfrac{1}{2^3x^{4\cdot 3}y^{6\cdot 3}}$
$= \dfrac{1}{8x^{12}y^{18}}$

47. $(x^{-1}y^3)^{-2}(2x)^2 = x^2y^{-6}(4x^2) = \dfrac{4x^4}{y^6}$

49. $\dfrac{(-3m^5n^{-1})^3}{(mn)^2} = \dfrac{(-3)^3m^{5\cdot 3}n^{-1(3)}}{m^2n^2}$
$= \dfrac{-27m^{15}n^{-3}}{m^2n^2}$
$= \dfrac{-27m^{13}}{n^5}$
$= -\dfrac{27m^{13}}{n^5}$

51. $\dfrac{2^{-3}a^2}{2^{-4}a^{-2}} = 2^{-3-(-4)}a^{2-(-2)} = 2^{-3+4}a^{2+2} = 2a^4$

53. $\left(\frac{1}{3}y\right)^{-3} = \left(\frac{1}{3}\right)^{-3} y^{-3} = \frac{3^3}{y^3} = \frac{27}{y^3}$

55. $\left(\frac{y^{-3}}{x}\right)^{-3} = \frac{y^{-3(-3)}}{x^{-3}} = x^3y^9$

57. $\frac{a^0b^{-4}}{a^{-3}b} = \frac{1\cdot b^{-4}}{a^{-3}b^1} = \frac{a^3}{b^{1+4}} = \frac{a^3}{b^5}$

59. $\left(\frac{14x^{-3}y^{-3}}{7x^{-4}y^{-3}}\right)^{-2} = (2x)^{-2} = \frac{1}{(2x)^2} = \frac{1}{4x^2}$

61. $\frac{7^{-8}\cdot 5^{-6}}{7^{-9}\cdot 5^{-5}} = 7^{-8-(-9)}5^{-6-(-5)} = 7^1 5^{-1} = \frac{7}{5}$

63. $(9x^{-2}y)\left(-\frac{2}{3}x^3y^{-2}\right) = -\frac{9\cdot 2}{3}\cdot x^{-2+3}y^{1-2}$

$= -6xy^{-1}$

$= -\frac{6x}{y}$

65. $(-3.6982x^3y^4)^7 = (-3.6982)^7 x^{3\cdot 7}y^{4\cdot 7}$

$\approx -9460.906704x^{21}y^{28}$

67. $38 = 3.8\times 10^1$

69. $1{,}730{,}000 = 1.73\times 10^6$

71. $0.83 = 8.3\times 10^{-1}$

73. $0.0008125 = 8.125\times 10^{-4}$

75. $7.13\times 10^5 = 713{,}000$

77. $3.07\times 10^{-1} = 0.307$

79. $9.01\times 10^{-7} = 0.000000901$

81. $(3.1\times 10^{-4})(1.5\times 10^{-2}) = (3.1\times 1.5)\times 10^{-4-2}$

$= 4.65\times 10^{-6}$

83. $\frac{3.6\times 10^{-5}}{1.2\times 10^{-6}} = 3\times 10^{-5-(-6)} = 3\times 10^1$

85. $7{,}200{,}000\frac{\text{ft}^3}{\text{s}}\cdot\frac{0.305^3\ \text{m}^3}{\text{ft}^3} = 204{,}282.9\ \frac{\text{m}^3}{\text{s}}$

The mouth of the Amazon pours out $204{,}282.9\ \text{m}^3$/sec.

87. $\left(\frac{5.1\times 10^4\ \text{cycles}}{\text{sec}}\right)(1.5\times 10^2\ \text{sec})$

$= 7.65\times 10^6$ cycles

In 1.5×10^2 seconds, 7.65×10^6 cycles would occur.

89. $\left(\frac{5.3\times 10^{-23}\ \text{gram}}{\text{molecule}}\right)(2\times 10^4\ \text{molecule})$

$= 10.6\times 10^{-19}$ gram

$= 1.06\times 10^{-18}$ gram

It would weigh 1.06×10^{-18} gram.

Cumulative Review

91. $\frac{\text{mass of Jupiter}}{\text{mass of Mercury}} = \frac{2.09\times 10^{24}}{3.64\times 10^{20}}$

$= 0.5742\times 10^4$

$= 5742$ times greater

The mass of Jupiter is 5742 times greater than the mass of Mercury.

92. $\frac{1\ \text{cal}}{2.78\times 10^{-7}\ \text{kW-hr}}\cdot 5.56\times 10^3\ \text{kW-hr}$

$= 2\times 10^{10}$ calories

There are 2×10^{10} calories.

93. $-9+14\div(-2)+5^2 = -9+14\div(-2)+25$

$= -9+(-7)+25$

$= -16+25$

$= 9$

94. $-6^2+16\div 2 = -36+16\div 2 = -36+8 = -28$

Quick Quiz 1.4

1. $\frac{20x^4y^3}{25x^{-2}y^6} = \frac{20}{25}x^{4-(-2)}y^{3-6} = \frac{4}{5}x^6y^{-3} = \frac{4x^6}{5y^3}$

2. $\left(\dfrac{3a^{-4}b^2}{a^3}\right)^2 = (3a^{-4-3}b^2)^2$
$= (3a^{-7}b^2)^2$
$= 3^2a^{-7(2)}b^{2(2)}$
$= 9a^{-14}b^4$
$= \dfrac{9b^4}{a^{14}}$

3. $0.000578 = 5.78\times10^{-4}$

4. Answers may vary. Possible solution: Multiply the constants. Then simplify variables by adding exponents. Then rewrite using only positive exponents.
$(3x^2y^{-3})(2x^4y^2) = 3\cdot2x^{2+4}y^{-3+2}$
$= 6x^6y^{-1}$
$= \dfrac{6x^6}{y}$

1.5 Exercises

1. It is $-5x^2$ since y is multiplied by $-5x^2$.

3. $5x^3, -6x^2, 4x, 8$

5. The coefficient of x^5 is 1, of $-3x$ is 3, and of $-8y$ is -8.

7. The coefficient of $5x^3$ is 5, of $-3x^2$ is -3, and of x is 1.

9. The coefficient of $6.5x^3y^3$ is 6.5, of $-0.02x^2y$ is -0.02, and of $3.05y$ is 3.05.

11. $3ab + 8ab = (3 + 8)ab = 11ab$

13. $4y - 7x + 2x - 6y = -2y - 5x$

15. $-4x^2+3+x^2-7=-3x^2-4$

17. $4ab-3b^2-5ab+3b^2=-ab$

19. $0.1x^2+3x-0.5x^2=-0.4x^2+3x$

21. $\frac{2}{3}m+\frac{5}{6}n-\frac{1}{3}m+\frac{1}{3}n=\frac{1}{3}m+\frac{7}{6}n$

23. $\frac{2}{3}a^2+2b+\frac{1}{3}a^2-8b=a^2-6b$

25. $1.2x^2-5.6x-8.9x^2+2x=-7.7x^2-3.6x$

27. $6x(3x+y)=(6x)(3x)+(6x)(y)=18x^2+6xy$

29. $-y(y^2-3y+5)=(-y)(y^2)-y(-3y)-y(5)$
$=-y^3+3y^2-5y$

31. $-2a^2(a-3a^2+2ab)$
$=-2a^2(a)-2a^2(-3a^2)-2a^2(2ab)$
$=-2a^3+6a^4-4a^3b$

33. $2xy(x^2-3xy+4y^2)$
$=2xy(x^2)+2xy(-3xy)+2xy(4y^2)$
$=2x^3y-6x^2y^2+8xy^3$

35. $\frac{3}{4}(8x^2-4x+2)=6x^2-3x+\frac{3}{2}$

37. $\frac{x}{5}(5x^2-2x+1)=x^3-\frac{2}{5}x^2+\frac{x}{5}$

39. $3ab(a^4b-3a^2+a-b)$
$=3a^5b^2-9a^3b+3a^2b-3ab^2$

41. $0.5x^2(2x-4y+3y^2)=x^3-2x^2y+1.5x^2y^2$

43. $2(x-1)-x(x+1)+3(x^2+2)$
$=2x-2-x^2-x+3x^2+6$
$=2x^2+x+4$

45. $2\{3x-2[x-4(x+1)]\}=2\{3x-2[x-4x-4]\}$
$=2\{3x-2[-3x-4]\}$
$=2\{3x+6x+8\}$
$=2\{9x+8\}$
$=18x+16$

47. $a(a-4b)-5a(a+b)=a^2-4ab-5a^2-5ab$
$=-4a^2-9ab$

49. $6x^3-2x^2-9x^3-12x^2=-3x^3-14x^2$

51. $2[-3(2x+4)+8(2x-4)]$
$= 2[-6x-12+16x-32]$
$= 2[10x-44]$
$= 20x-88$

53. $3y[y-(x-5)] = 3y[y-x+5] = 3y^2-3xy+15y$

Cumulative Review

55. $3(-2)^3-5(-6) = 3(-8)-5(-6) = -24+30 = 6$

56. $\sqrt{81}-5(3-5+2) = 9-5(3-5+2)$
$= 9-5(0)$
$= 9-0$
$= 9$

57. $\dfrac{5(-2)-8}{3+4-(-3)} = \dfrac{-10-8}{7+3} = \dfrac{-18}{10} = \dfrac{-9}{5} = -1.8$

58. $(-3)^5+2(-3) = -243-6 = -249$

59. $\dfrac{1{,}893{,}500 \text{ organisms}}{\text{inch}} \cdot \dfrac{1 \text{ inch}}{0.0254 \text{ meters}}$
$= \dfrac{1{,}893{,}500 \text{ organisms}}{0.0254 \text{ meters}} \cdot \dfrac{1000 \text{ meter}}{\text{kilometer}}$
$\approx \dfrac{7.4547\times10^{10} \text{ organisms}}{\text{kilometer}}$
You would find 7.4547×10^{10} organisms in 1 kilometer.

60. $4167 \text{ meter}\left(\dfrac{\text{ft}}{0.305 \text{ meter}}\right)\left(\dfrac{2\%}{1000 \text{ ft}}\right)$
$\approx 27.3\%$ loss
There would be a 27.3% loss.

Quick Quiz 1.5

1. $2xy(-3x^2+4xy-5x)$
$= 2xy(-3x^2)+2xy(4xy)+2xy(-5x)$
$= -6x^3y+8x^2y^2-10x^2y$

2. $3x^2(x+4y)-2(5x^3-2x^2y)$
$= 3x^3+12x^2y-10x^3+4x^2y$
$= -7x^3+16x^2y$

3. $3[-4(x+2)+3(5x-1)] = 3(-4x-8+15x-3)$
$= 3(11x-11)$
$= 33x-33$

4. Answers may vary. Possible solution:
Combine like terms:
$2x^2-3x+4y-2x^2y-8x-5y$
$= 2x^2+(-3-8)x+(4-5)y-2x^2y$
$= 2x^2-11x-y-2x^2y$

1.6 Exercises

1. $14 + 6x = 14 + 6(-2) = 14 - 12 = 2$

3. $x^2+2x-9 = (-4)^2+2(-4)-9$
$= 16+(-8)-9$
$= -1$

5. $3+7x-x^2 = 3+7(1)-1^2$
$= 3+7-1$
$= 10-1$
$= 9$

7. $-2x^2+5x-3 = -2(-4)^2+5(-4)-3$
$= -2(16)+(-20)-3$
$= -32+(-20)-3$
$= -52-3$
$= -55$

9. $(-3y)^4 = (-3(-1))^4 = (3)^4 = 81$

11. $-3y^4 = -3(-1)^4 = -3(1) = -3$

13. $-ay+4bx-b = -(-1)(-4)+4(3)(2)-3$
$= -4+24-3$
$= 17$

15. $\sqrt{b^2-4ac} = \sqrt{5^2-4(1)(-14)}$
$= \sqrt{24+56}$
$= \sqrt{81}$
$= 9$

17. $2x^2-5x+6 = 2(-3.52176)^2-5(-3.52176)+6$
$= 48.41439$ to five decimal places

19. $F = \dfrac{9}{5}(-60)+32 = -76°\text{F}$

21. $C = \dfrac{5(122)-160}{9} = 50°\text{C}$

23. $T = 2\pi\sqrt{\frac{L}{g}}$

$T = 2(3.14)\sqrt{\frac{32}{32}}$

$T = 6.28$

It will take 6.28 seconds.

25. $A = p(1 + rt)$

$A = \$4800[1 + (0.12)(1.5)]$

$A = \$5664$

The amount is \$5664.

27. $A = p(1 + rt)$

$A = 2200[1 + (0.04)(4)] = \$2552$

The amount to be repaid is \$2552.

29. $S = \frac{1}{2}gt^2$

$S = \frac{1}{2}(32)(3)^2 = 144$

The distance is 144 feet.

31. $S = \frac{1}{2}gt^2$

$S = \frac{1}{2}(32)(7)^2 = 784$

The distance is 784 feet.

33. $z = \frac{(36)(4)}{36 + 4} = \frac{144}{40} = \frac{18}{5} = 3\frac{3}{5}$

z is $3\frac{3}{5}$.

35. $m = \frac{cx}{c + 12}$

$m = \frac{8 \cdot 400}{8 + 12} = \frac{3200}{20} = 160$

Give the child 160 milligrams.

37. $A = \pi r^2 = 3.14 \cdot 0.5^2 = 0.785$

The area is 0.785 in.^2.

39. $A = \frac{1}{2}ab = \frac{1}{2} \cdot 16 \cdot 7 = 56$

The area is 56 cm^2.

41. $A = ab = 4\left(\frac{7}{8}\right) = \frac{7}{2}$

The area is $\frac{7}{2} \text{ yd}^2$.

43. $A = lw = (6.1)(4.05)$

$A = 24.705$

The area is 24.705 cm^2.

45. $A = \frac{1}{2}a(b + c) = \frac{1}{2} \cdot 6(5 + 7)$

$A = \frac{1}{2} \cdot 6(12) = 3(12)$

$A = 36$

The area is 36 cm^2.

47. **a.** $V = \pi r^2 h = 3.14(4)^2(11)$

$V = 552.64$

The volume is 552.64 cm^3.

b. $S = 2\pi rh + 2\pi r^2$

$S = 2(3.14)(4)(11) + 2(3.14)(4)^2$

$S = 376.8$

The surface area is 376.8 cm^2.

49. $A = \pi r^2 = 3.14(8)^2 = 200.96$

$C = \pi(16) = 3.14(16) = 50.24$

The area is 200.96 cm^2 and the circumference is 50.24 cm.

Cumulative Review

51. $(6x^{-4}y^3z^0)^2 = \left(\frac{6y^3}{x^4}\right)^2 = \frac{6^2 y^{3\cdot 2}}{x^{4\cdot 2}} = \frac{36y^6}{x^8}$

52. $\left(\frac{2x^3}{3y}\right)^3 = \frac{2^3 x^{3\cdot 3}}{3^3 y^3} = \frac{8x^9}{27y^3}$

53. $$\begin{aligned} 2\{5 - 2[x - 3(2x + 1)]\} &= 2\{5 - 2[x - 6x - 3]\} \\ &= 2\{5 - 2[-5x - 3]\} \\ &= 2\{5 + 10x + 6\} \\ &= 2\{11 + 10x\} \\ &= 22 + 20x \\ &= 20x + 22 \end{aligned}$$

54. $2^3 - 4^2 + \sqrt{9 \cdot 2 - 2} = 8 - 16 + \sqrt{18 - 2}$
$= -8 + \sqrt{16}$
$= -8 + 4$
$= -4$

55. 12,000,000 – 0.30(19,000,000)
= 6,300,000
There were 6,300,000 people on the Internet at 9 P.M. but were not on the Internet at 2 P.M.

56. 180(0.95) = 171 living graduates
171(0.69) = 118 graduates attending
118 + 0.77(118) = 209 including spouses
209 + 22 faculty = 231 people attending
The total number of people at the reunion was 231.

Quick Quiz 1.6

1. $3x^2 - 5x - 2 = 3(-3)^2 - 5(-3) - 2$
$= 3(9) - 5(-3) - 2$
$= 27 + 15 - 2$
$= 40$

2. $-2x^2 + 4xy - y^2 = -2(-2)^2 + 4(-2)(4) - 4^2$
$= -2(4) + 4(-2)(4) - 16$
$= -8 - 32 - 16$
$= -56$

3. $A = \pi r^2$
$= 3.14(5)^2$
$= 3.14(25)$
$= 78.5$
The area is 78.5 square centimeters.

4. Answers may vary. Possible solution:
Substitute $p = 5000$, $r = 0.08$, and $t = 2$ into the formula and simplify.
$A = p(1 + rt)$
$= 5000(1 + 0.08 \cdot 2)$
$= 5000(1 + 0.16)$
$= 5000(1.16)$
$= \$5800$

Putting Your Skills to Work

1. 16" diameter pizza: $r = 8$
$A = \pi r^2 \approx 3.14(8)^2 = 200.96$
one slice $= \dfrac{200.96}{6} \approx 33.5$
Each slice has about 33.5 square inches.

2. 20" diameter pizza: $r = 10$
$A = \pi r^2 \approx 3.14(10)^2 = 314$
one slice $= \dfrac{314}{8} \approx 39.3$
Each slice has about 39.3 square inches.

3. Poppolo's: $\dfrac{\$1.75}{33.5} \approx \$0.052$ per square inch
Joe's: $\dfrac{\$2.10}{39.3} \approx \$0.053$ per square inch

4. Answers may vary.

5. One year = 52 weeks
$20 × 52 = $1040
He will save a total of $1040 per year.

Chapter 1 Review Problems

1. –5: Integer, Rational, Real

2. $\frac{7}{8}$: Rational, Real

3. 3: Natural, Whole, Integer, Rational, Real

4. $0.\overline{3}$: Rational, Real

5. 2.1652384...: Irrational, Real

6. Commutative property of addition

7. Associative property of multiplication

8. Yes, all rational numbers are real numbers.

9. $-15 - (-20) = -15 + 20 = 5$

10. $-7.3 + (-16.2) = -23.5$

11. $-8(-6) = 48$

12. $-12 \div 3 = -4$

13. $-\frac{4}{5} \div \left(-\frac{12}{5}\right) = -\frac{4}{5} \cdot \left(-\frac{5}{12}\right) = \frac{1}{3}$

14. $-\frac{5}{6}\left(\frac{7}{10}\right) = -\frac{7}{12}$

15. $4(-3)(-10) = -12(-10) = 120$

16. $5 + 6 - 2 - 5 = 11 - 2 - 5 = 9 - 5 = 4$

17. $-3.6(-1.5) = 5.4$

18. $0 \div (-14) = 0$

19. $7 \div 0$ undefined

20. $-17 + (+17) = 0$

21. $17 - 3(6) = 17 - 18 = -1$

22. $\dfrac{5-8}{2-7-(-2)} = \dfrac{-3}{-5+2} = \dfrac{-3}{-3} = 1$

23. $$\begin{aligned}2\sqrt{49} - 3^2 + 5 &= 2(7) - 3^2 + 5\\ &= 14 - 9 + 5\\ &= 5 + 5\\ &= 10\end{aligned}$$

24. $$\begin{aligned}4(6) - |-8| + (-1)^3 &= 24 - 8 + (-1)\\ &= 16 - 1\\ &= 15\end{aligned}$$

25. $4 - 2 + 6\left(-\dfrac{1}{3}\right) = 4 - 2 + (-2) = 2 + (-2) = 0$

26. $$\begin{aligned}\sqrt{(-1)^2 + 6(4)} + 8 \div (-2) &= \sqrt{1+24} + (-4)\\ &= \sqrt{25} + (-4)\\ &= 5 + (-4)\\ &= 1\end{aligned}$$

27. $\sqrt{\dfrac{25}{36}} - 2\left(\dfrac{1}{12}\right) = \dfrac{5}{6} - \left(\dfrac{1}{6}\right) = \dfrac{4}{6} = \dfrac{2}{3}$

28. $$\begin{aligned}6|-3-1| + 5(-3)(0) - 4^2 &= 6|-4| + 0 - 16\\ &= 6(4) - 16\\ &= 24 - 16\\ &= 8\end{aligned}$$

29. $(-0.4)^3 = (-0.4)(-0.4)(-0.4) = -0.064$

30. $$\begin{aligned}(3xy^2)(-2x^0y)(4x^3y^3) &= 3(-2)(4)x^{1+0+3}y^{2+1+3}\\ &= -24x^4y^6\end{aligned}$$

31. $$\begin{aligned}(5a^4bc^2)(-6ab^2) &= 5(-6)a^{4+1}b^{1+2}c^2\\ &= -30a^5b^3c^2\end{aligned}$$

32. $\dfrac{36x^5y^4}{60x^2y^5} = \left(\dfrac{36}{60}\right)x^{5-2}y^{4-5} = \dfrac{3}{5}x^3y^{-1} = \dfrac{3x^3}{5y}$

33. $$\begin{aligned}\frac{16abc^0}{48ab^4c^2} &= \left(\frac{16}{48}\right)a^{1-1}b^{1-4}c^{0-2}\\ &= \frac{1}{3}a^0b^{-3}c^{-2}\\ &= \frac{1}{3b^3c^2}\end{aligned}$$

34. $$\begin{aligned}\left(\frac{-3x^3y}{2x^4z^2}\right)^4 &= \frac{(-3)^4(x^3)^4y^4}{2^4(x^4)^4(z^2)^4}\\ &= \frac{81x^{12}y^4}{16x^{16}z^8}\\ &= \frac{81y^4}{16x^{16-12}z^8}\\ &= \frac{81y^4}{16x^4z^8}\end{aligned}$$

35. $(-2xy^6z^0)^3 = (-2)^3x^3y^{6(3)}(1)^3 = -8x^3y^{18}$

36. $$\begin{aligned}(2x^2y^{-4})(-5x^{-1}y) &= -10x^{2-1}y^{-4+1}\\ &= -10xy^{-3}\\ &= -\frac{10x}{y^3}\end{aligned}$$

37. $\dfrac{3x^5y^{-6}}{12x^{-2}y} = \dfrac{x^{5-(-2)}}{4y \cdot y^6} = \dfrac{x^7}{4y^{1+6}} = \dfrac{x^7}{4y^7}$

38. $\dfrac{(5^{-1}x^{-2})^{-1}}{(2^{-2}y)^{-3}} = \dfrac{5^{-1(-1)}x^{-2(-1)}}{2^{-2(-3)}y^{-3}} = \dfrac{5^1x^2}{2^6y^{-3}} = \dfrac{5x^2y^3}{64}$

39. $\dfrac{(3a)^{-2}}{(4b^{-3})^{-2}} = \dfrac{3^{-2}a^{-2}}{4^{-2}b^{-3(-2)}} = \dfrac{4^2}{3^2a^2b^6} = \dfrac{16}{9a^2b^6}$

40. $$\begin{aligned}\left(\frac{a^5b^2}{3^{-1}a^{-5}b^{-4}}\right)^3 &= \left(\frac{3a^{5-(-5)}b^{2-(-4)}}{1}\right)^3\\ &= (3a^{10}b^6)^3\\ &= 3^3a^{10\cdot 3}b^{6\cdot 3}\\ &= 27a^{30}b^{18}\end{aligned}$$

41. $\left(\dfrac{x^3y^4}{5x^6y^8}\right)^3 = \dfrac{x^9y^{12}}{5^3x^{18}y^{24}}$
$= \dfrac{1}{125x^{18-9}y^{24-12}}$
$= \dfrac{1}{125x^9y^{12}}$

42. $0.00721 = 7.21\times10^{-3}$

43. $(5,300,000)(2,000,000,000)$
$= (5.3\times10^6)(2.0\times10^9)$
$= 10.6\times10^{15}$
$= 1.06\times10^{16}$

44. $3.48\times10^{-7} = 0.000000348$

45. $5.82\times10^{13} = 58,200,000,000,000$

46. $-x+8+6x^2+7x-4 = 6x^2+(7-1)x+8-4$
$= 6x^2+6x+4$

47. $-5ab^2(a^3+2a^2b-3b-4)$
$= -5a^4b^2-10a^3b^3+15ab^3+20ab^2$

48. $3x(x-7)-(x^2+1) = 3x^2-21x-x^2-1$
$= 2x^2-21x-1$

49. $2x^2-\{2+x[3-2(x-1)]\}$
$= 2x^2-\{2+x[3-2x+2]\}$
$= 2x^2-\{2+x[5-2x]\}$
$= 2x^2-\{2+5x-2x^2\}$
$= 2x^2-2-5x+2x^2$
$= 4x^2-5x-2$

50. $5x^2-3xy-2y^3 = 5(2)^2-3(2)(-1)-2(-1)^3$
$= 5(4)-6(-1)-2(-1)$
$= 20+6+2$
$= 28$

51. $V = \pi r^2h$
$V = 3.14(3)^2(8)$
$V = 226.08$
The volume is 226.08 cubic inches.

52. $A = \frac{1}{2}bh$
$A = \frac{1}{2}(52)(88)$
$A = 2288$
The area is 2288 square yards.

53. $A = \frac{1}{2}(b_1+b_2)h$
$A = \frac{1}{2}(26+34)14$
$A = 420$
The area is 420 square inches.

54. $9(-2)+(-28\div7)^3-5 = 9(-2)+(-4)^3-5$
$= -18+(-64)-5$
$= -82-5$
$= -87$

55. $\sqrt{20+5}-12(2)\div8 = \sqrt{25}-12(2)\div8$
$= 5-24\div8$
$= 5-3$
$= 2$

56. $(-7a^2b)(-2a^0b^3c^2) = -7(-2)a^{2+0}b^{1+3}c^2$
$= 14a^2b^4c^2$

57. $\dfrac{(3x^{-1}y^2)^3}{(4x^2y^{-2})^2} = \dfrac{3^3x^{-1(3)}y^{2(3)}}{4^2x^{2(2)}y^{-2(2)}}$
$= \dfrac{27x^{-3}y^6}{16x^4y^{-4}}$
$= \dfrac{27y^{6-(-4)}}{16x^{4-(-3)}}$
$= \dfrac{27y^{10}}{16x^7}$

58. $\dfrac{4x^3y^2}{-16x^2y^{-3}} = -\dfrac{x^{3-2}y^{2-(-3)}}{4} = -\dfrac{xy^5}{4}$

59. $\left(\dfrac{3a^{-5}b^0}{2a^{-2}b^3}\right)^2 = \dfrac{3^2a^{-5(2)}(1)^2}{2^2a^{-2(2)}b^{3(2)}}$
$= \dfrac{9a^{-10-(-4)}}{4b^6}$
$= \dfrac{9a^{-6}}{4b^6}$
$= \dfrac{9}{4a^6b^6}$

60. $0.000058 = 5.8\times10^{-5}$

61. $8.95\times10^7 = 89,500,000$

62. $4x^2 - x^3 + 7x - 5x^2 + 6x^3 - 2x$
$= (-1+6)x^3 + (4-5)x^2 + (7-2)x$
$= 5x^3 - x^2 + 5x$

63. $2a^3b(5a - ab - 3) = 10a^4b - 2a^4b^2 - 6a^3b$

64. $-2\{x + 3[y - 5(x+y)]\} = -2\{x + 3[y - 5x - 5y]\}$
$= -2\{x + 3[-4y - 5x]\}$
$= -2\{x - 12y - 15x\}$
$= -2\{-14x - 12y\}$
$= 28x + 24y$

65. $5a^2 - 3ab + 4b = 5(-3)^2 - 3(-3)(-2) + 4(-2)$
$= 5(9) + 9(-2) + (-8)$
$= 45 - 18 - 8$
$= 19$

66. $A = \pi r^2$
$A = 3.14(4)^2$
$A = 50.24$
The area is 50.24 m^2.

67. $T = 2\pi\sqrt{\dfrac{L}{g}} = 2(3.14)\sqrt{\dfrac{512}{32}}$
$T = 25.12$
The period of the pendulum is 25.12 seconds.

68. $\dfrac{2}{3}(6x - 9y) - (x - 2y) = 4x - 6y - x + 2y$
$= 3x - 4y$

69. $A = p(1 + rt)$
$= 3200(1 + 0.09\times2)$
$= 3200(1 + 0.18)$
$= 3200(1.18)$
$= 3776$
The amount to be repaid is \$3776.

How Am I Doing? Chapter 1 Test

1. $\pi,\ 2\sqrt{5}$

2. $-2,\ 12,\ \dfrac{9}{3},\ \dfrac{25}{25},\ 0,\ \sqrt{4}$

3. $(8\cdot x)3 = 3(8\cdot x)$
Commutative property of multiplication

4. $(7-5)^2 - 18\div(-3) + \sqrt{10+6}$
$= 2^3 + (-18)\div(-3) + \sqrt{16}$
$= 8 + (-18)\div(-3) + 4$
$= 8 + 6 + 4$
$= 14 + 4$
$= 18$

5. $(4-5)^2 - 3(-2)\div3 = (-1)^2 - (-6)\div3$
$= 1 - (-2)$
$= 1 + 2$
$= 3$

6. $\dfrac{16x^3y}{20x^{-1}y^5} = \dfrac{4x^{3-(-1)}}{5y^{5-1}} = \dfrac{4x^4}{5y^4}$

7. $(5x^{-3}y^{-5})(-2x^3y^0) = 5(-2)x^{-3+3}y^{-5+0}$
$= -10x^0y^{-5}$
$= -\dfrac{10}{y^5}$

8. $\left(\dfrac{5a^{-2}b}{a}\right)^2 = \dfrac{5^2a^{-2(2)}b^2}{a^2}$
$= \dfrac{25a^{-4}b^2}{a^2}$
$= \dfrac{25b^2}{a^{2-(-4)}}$
$= \dfrac{25b^2}{a^6}$

9. $7x-9x^2-12x-8x^2+5x$
$=(7-12+5)x+(-9-8)x^2$
$=0x-17x^2$
$=-17x^2$

10. $5a+4b-6a^2+b-7a-2a^2$
$=(-6-2)a^2+(5-7)a+(4+1)b$
$=-8a^2-2a+5b$

11. $3xy^2(4x-3y+2x^2)=12x^2y^2-9xy^3+6x^3y^2$

12. $0.000002186=2.186\times10^{-6}$

13. $2.158\times10^9=2,158,000,000$

14. $(3.8\times10^{-5})(4\times10^{-2})=(3.8\times4)\times10^{-5-2}$
$=15.2\times10^{-7}$
$=1.52\times10^{-6}$

15. $2x^2(x-3y)-x(4-8x^2)$
$=2x^3-6x^2y-4x+8x^3$
$=10x^3-6x^2y-4x$

16. $2[-3(2x+4)+8(3x-2)]$
$=2[-6x-12+24x-16]$
$=2[18x-28]$
$=36x-56$

17. $2x^2-3x-6=2(-4)^2-3(-4)-6$
$=2(16)+(-3)(-4)-6$
$=32+12-6$
$=44-6$
$=38$

18. $5x^2+3xy-y^2=5(3)^2+3(3)(-3)-(-3)^2$
$=5(9)+(-27)-9$
$=45-27-9$
$=18-9$
$=9$

19. $A=\frac{1}{2}(b_1+b_2)$
$A=\frac{1}{2}(6+7)(12)=78$

The area is 78 m^2.

20. $A=\pi r^2=3.14(6)^2=113.04$

The area is 113.04 m^2.

21. $A=p(1+rt)$
$A=\$8000(1+0.05(3))$
$A=\$9200$
The amount to be repaid is \$9200.

Chapter 2

2.1 Exercises

1. $3x - 15 = 3(-20) - 15 = -75 \neq 45$
No; when you replace x by -20 in the equation, you do not get a true statement.

3. $7x - 8 = 7 \cdot \frac{2}{7} - 8 = 2 - 8 = -6$

Yes; when you replace x with $\frac{2}{7}$ in the equation, you get a true statement.

5. Multiply each term of the equation by the LCD, 12, to clear the fractions.

7. No; it would be easier to subtract 3.6 from both sides of the equation since the coefficient of x is 1.

9.
$$\begin{aligned} -11 + x &= -3 \\ -11 + 11 + x &= -3 + 11 \\ x &= 8 \end{aligned}$$
Check: $-11 + 8 \stackrel{?}{=} -3$
$-3 = -3$

11.
$$\begin{aligned} -8x &= 56 \\ \frac{-8x}{-8} &= \frac{56}{-8} \\ x &= -7 \end{aligned}$$
Check: $-8(-7) \stackrel{?}{=} 56$
$56 = 56$

13.
$$\begin{aligned} -14x &= -70 \\ \frac{-14x}{-14} &= \frac{-70}{-14} \\ x &= 5 \end{aligned}$$
Check: $-14(5) \stackrel{?}{=} -70$
$-70 = -70$

15.
$$\begin{aligned} 8x - 1 &= 11 \\ 8x - 1 + 1 &= 11 + 1 \\ 8x &= 12 \\ \frac{8x}{8} &= \frac{12}{8} \\ x &= \frac{12}{8} = \frac{3}{2} \\ x &= 1\frac{1}{2} \\ x &= 1.5 \end{aligned}$$
Check: $8\left(\frac{3}{2}\right) - 1 \stackrel{?}{=} 11$
$11 = 11$

17.
$$\begin{aligned} 8x + 5 &= 2x - 13 \\ 8x - 2x + 5 &= 2x - 2x - 13 \\ 6x + 5 &= -13 \\ 6x + 5 - 5 &= -13 - 5 \\ 6x &= -18 \\ \frac{6x}{6} &= \frac{-18}{6} \\ x &= -3 \end{aligned}$$
Check: $8(-3) + 5 \stackrel{?}{=} 2(-3) - 13$
$-24 + 5 \stackrel{?}{=} -6 - 13$
$-19 = -19$

19.
$$\begin{aligned} 16 - 2x &= 5x - 5 \\ 16 - 2x - 5x &= 5x - 5x - 5 \\ 16 - 7x &= -5 \\ 16 - 7x - 16 &= -5 - 16 \\ -7x &= -21 \\ \frac{-7x}{-7} &= \frac{-21}{-7} \\ x &= 3 \end{aligned}$$
Check: $16 - 2(3) \stackrel{?}{=} 5(3) - 5$
$16 - 6 \stackrel{?}{=} 15 - 5$
$10 = 10$

21.
$$\begin{aligned} 3a - 5 - 2a &= 2a - 3 \\ a - 5 &= 2a - 3 \\ a - 2a - 5 &= 2a - 2a - 3 \\ -a - 5 &= -3 \\ -a - 5 + 5 &= -3 + 5 \\ -a &= 2 \\ a &= -2 \end{aligned}$$
Check: $3(-2) - 5 - 2(-2) \stackrel{?}{=} 2(-2) - 3$
$-6 - 5 + 4 \stackrel{?}{=} -4 - 3$
$-11 + 4 \stackrel{?}{=} -7$
$-7 = -7$

23.
$$\begin{aligned} 4(y-1) &= -2(3+y) \\ 4y-4 &= -6-2y \\ 4y-4+2y &= -6-2y+2y \\ 6y-4 &= -6 \\ 6y-4+4 &= -6+4 \\ 6y &= -2 \\ \frac{6y}{6} &= \frac{-2}{6} \\ y &= -\frac{1}{3} \end{aligned}$$

Check: $4\left(-\frac{1}{3}-1\right) \stackrel{?}{=} -2\left(3+\left(-\frac{1}{3}\right)\right)$
$$4\left(\frac{-4}{3}\right) \stackrel{?}{=} -2\left(\frac{8}{3}\right)$$
$$-\frac{16}{3} = -\frac{16}{3}$$

25.
$$\begin{aligned} 3+y &= 10-3(y+1) \\ 3+y &= 10-3y-3 \\ 3+y &= 7-3y \\ 3+y+3y &= 7-3y+3y \\ 3+4y &= 7 \\ 3-3+4y &= 7-3 \\ 4y &= 4 \\ \frac{4y}{4} &= \frac{4}{4} \\ y &= 1 \end{aligned}$$

Check: $3+1 \stackrel{?}{=} 10-3(1+1)$
$$4 \stackrel{?}{=} 10-3(2)$$
$$4 \stackrel{?}{=} 10-6$$
$$4 = 4$$

27.
$$\begin{aligned} \frac{2}{3}x &= 8 \\ \frac{3}{2}\cdot\frac{2}{3}x &= \frac{3}{2}\cdot 8 \\ x &= 12 \end{aligned}$$

Check: $\frac{2}{3}(12) \stackrel{?}{=} 8$
$$8 = 8$$

29.
$$\begin{aligned} \frac{y}{2}+4 &= \frac{1}{6} \\ 6\left(\frac{y}{2}+4\right) &= 6\left(\frac{1}{6}\right) \\ 3y+24 &= 1 \\ 3y+24-24 &= 1-24 \\ 3y &= -23 \\ \frac{3y}{3} &= \frac{-23}{3} \\ y &= -\frac{23}{3} \text{ or } -7\frac{2}{3} \end{aligned}$$

Check: $\frac{-\frac{23}{3}}{2}+4 \stackrel{?}{=} \frac{1}{6}$
$$-\frac{23}{6}+\frac{24}{6} \stackrel{?}{=} \frac{1}{6}$$
$$\frac{1}{6} = \frac{1}{6}$$

31.
$$\begin{aligned} \frac{2}{3}-\frac{x}{6} &= 1 \\ 6\cdot\left(\frac{2}{3}-\frac{x}{6}\right) &= 1\cdot 6 \\ 6\cdot\frac{2}{3}-6\cdot\frac{x}{6} &= 6 \\ 4-x &= 6 \\ (-1)(4-x) &= 6(-1) \\ -4+x &= -6 \\ -4+x+4 &= -6+4 \\ x &= -2 \end{aligned}$$

Check: $\frac{2}{3}-\frac{-2}{6} \stackrel{?}{=} 1$
$$\frac{4}{6}+\frac{2}{6} \stackrel{?}{=} 1$$
$$1 = 1$$

33.
$$\begin{aligned} \frac{1}{2}(x+3)-2 &= 1 \\ 2\cdot\left(\frac{1}{2}(x+3)-2\right) &= 2\cdot 1 \\ x+3-4 &= 2 \\ x-1 &= 2 \\ x-1+1 &= 2+1 \\ x &= 3 \end{aligned}$$

Check: $\frac{1}{2}(3+3)-2 \stackrel{?}{=} 1$
$$3-2 \stackrel{?}{=} 1$$
$$1 = 1$$

35.
$$5-\frac{2x}{7}=1-(x-4)$$
$$7\left(5-\frac{2x}{7}\right)=7[1-(x-4)]$$
$$35-2x=7-7x+28$$
$$35-2x=35-7x$$
$$35-35-2x=35-35-7x$$
$$-2x=-7x$$
$$-2x+7x=-7x+7x$$
$$5x=0$$
$$\frac{5x}{5}=\frac{0}{5}$$
$$x=0$$
Check: $5-\frac{2(0)}{7}\stackrel{?}{=}1-(0-4)$
$$5-0\stackrel{?}{=}1+4$$
$$5=5$$

37.
$$0.3x+0.4=0.5x-0.8$$
$$10(0.3x+0.4)=10(0.5x-0.8)$$
$$3x+4=5x-8$$
$$3x+4-3x=5x-8-3x$$
$$4=2x-8$$
$$4+8=2x-8+8$$
$$12=2x$$
$$\frac{12}{2}=\frac{2x}{2}$$
$$6=x$$
Check: $0.3(6)+0.4\stackrel{?}{=}0.5(6)-0.8$
$$1.8+0.4\stackrel{?}{=}3.0-0.8$$
$$2.2=2.2$$

39.
$$0.3-0.05x=0.2x-0.1$$
$$100(0.3-0.05x)=100(0.2x-0.1)$$
$$30-5x=20x-10$$
$$30-5x+5x=20+5x-10$$
$$30=25x-10$$
$$30+10=25x-10+10$$
$$40=25x$$
$$\frac{40}{25}=\frac{25x}{25}$$
$$1.6=x \text{ or } x=\frac{8}{5} \text{ or } 1\frac{3}{5}$$
Check: $0.3-0.05(1.6)\stackrel{?}{=}0.2(1.6)-0.1$
$$0.3-0.08\stackrel{?}{=}0.32-0.1$$
$$0.22=0.22$$

41.
$$0.2(x-4)=3$$
$$10[0.2(x-4)]=10(3)$$
$$2(x-4)=30$$
$$2x-8=30$$
$$2x-8+8=30+8$$
$$2x=38$$
$$\frac{2x}{2}=\frac{38}{2}$$
$$x=19$$
Check: $0.2(19-4)\stackrel{?}{=}3$
$$0.2(15)\stackrel{?}{=}3$$
$$3=3$$

43.
$$0.05x-2=0.3(x-5)$$
$$100(0.05x-2)=100[0.3(x-5)]$$
$$5x-200=30(x-5)$$
$$5x-200=30x-150$$
$$5x-30x-200=30x-30x-150$$
$$-25x-200=-150$$
$$-25x-200+200=-150+200$$
$$-25x=500$$
$$\frac{-25x}{-25}=\frac{50}{-25}$$
$$x=-2$$
Check: $0.05(-2)-2\stackrel{?}{=}0.3(-2-5)$
$$-0.1-2\stackrel{?}{=}0.3(-7)$$
$$-2.1=-2.1$$

45.
$$2x+12=3+4x-7$$
$$2x+12=4x-4$$
$$2x-4x+12=4x-4x-4$$
$$-2x+12=-4$$
$$-2x+12-12=-4-12$$
$$-2x=-16$$
$$\frac{-2x}{-2}=\frac{-16}{-2}$$
$$x=8$$

47. $\frac{1}{2}-\frac{x}{3}=\frac{2x-3}{3}$

$$6\left(\frac{1}{2}-\frac{x}{3}\right)=6\left(\frac{2x-3}{3}\right)$$
$$3-2x=2(2x-3)$$
$$3-2x=4x-6$$
$$3-2x-4x=4x-4x-6$$
$$3-6x=-6$$
$$3-3-6x=-6-3$$
$$-6x=-9$$
$$\frac{-6x}{-6}=\frac{-9}{-6}$$
$$x=\frac{3}{2} \text{ or } 1\frac{1}{2} \text{ or } 1.5$$

49. $\frac{1}{3}-\frac{x+1}{5}=\frac{x}{3}$

$$15\cdot\left(\frac{1}{3}-\frac{x+1}{5}\right)=15\cdot\frac{x}{3}$$
$$15\cdot\frac{1}{3}-15\cdot\frac{x+1}{5}=5x$$
$$5-3(x+1)=5x$$
$$5-3x-3=5x$$
$$2-3x=5x$$
$$2-3x+3x=5x+3x$$
$$2=8x$$
$$x=\frac{2}{8}=\frac{1}{4} \text{ or } 0.25$$

51. $2+0.1(5-x)=1.3x-(0.4x-2.5)$

$$2+0.5-0.1x=1.3x-0.4x+2.5$$
$$2.5-0.1x=0.9x+2.5$$
$$2.5-2.5=0.9x+0.1x$$
$$0=x$$
$$x=0$$

53. $6x-8=9-2x+5-3x$

$$6x-8=14-5x$$
$$6x+5x-8=14-5x+5x$$
$$11x-8=14$$
$$11x-8+8=14+8$$
$$11x=22$$
$$\frac{11x}{11}=\frac{22}{11}$$
$$x=2$$

55. $2x-4(x+1)=-2x+14$

$$2x-4x-4=-2x+14$$
$$-2x-4=-2x+14$$
$$-2x+2x-4=-2x+2x+14$$
$$-4=14, \text{ since } -4\neq 14, \text{ no solution}$$

57. $-(x-2)+1=2x+3(1-x)$

$$-x+2+1=2x+3-3x$$
$$-x+3=-x+3$$
$$-x+x+3=-x+x+3$$
$$3=3$$

Any real number is a solution.

59. $6+8(x-2)=10x-2(x+4)$

$$6+8x-16=10x-2x-8$$
$$8x-10=8x-8$$
$$8x-8x-10=8x-8x-8$$
$$-10=-8, \text{ since } -10\neq -8, \text{ no solution}$$

61. $x-2+\frac{2x}{5}=-2+\frac{7x}{5}$

$$5\left(x-2+\frac{2x}{5}\right)=5\left(-2+\frac{7x}{5}\right)$$
$$5x-10+2x=-10+7x$$
$$7x-10=7x-10$$
$$7x-7x-10=7x-7x-10$$
$$-10=-10$$

Any real number is a solution.

Cumulative Review

63. $5-(4-2)^2+3(-2)=5-(2)^2+(-6)$

$$=5-4+(-6)$$
$$=1+(-6)$$
$$=-5$$

64. $(-2)^4-12-6(-2)=16-12+(-6)(-2)$

$$=16-12+12$$
$$=4+12$$
$$=16$$

65. $\left(\dfrac{3xy^2}{2x^2y}\right)^3 = \dfrac{3^3x^3y^{2\cdot3}}{2^3x^{2\cdot3}y^3}$

$= \dfrac{27x^3y^6}{8x^6y^3}$

$= \dfrac{27y^{6-3}}{8x^{6-3}}$

$= \dfrac{27y^3}{8x^3}$

66. $(2x^{-2}y^{-3})^2(4xy^{-2})^{-2}$

$= 2^2x^{-2\cdot2}y^{-3\cdot2}\cdot4^{-2}x^{-2}y^{-2(-2)}$

$= 4x^{-4}y^{-6}\cdot\dfrac{1}{16}\cdot x^{-2}y^4$

$= \dfrac{4}{16}x^{-4-2}y^{-6+4}$

$= \dfrac{1}{4}x^{-6}y^{-2}$

$= \dfrac{1}{4x^6y^2}$

Quick Quiz 2.1

1.
$$\begin{aligned}
4(7-3x) &= 16-3(3x+1)\\
28-12x &= 16-9x-3\\
28-12x &= 13-9x\\
28-12x+9x &= 13-9x+9x\\
28-3x &= 13\\
28-28-3x &= 13-28\\
-3x &= -15\\
\frac{-3x}{-3} &= \frac{-15}{-3}\\
x &= 5
\end{aligned}$$

2.
$$\begin{aligned}
\frac{2}{3}(2x-1)+3 &= 4(2x-4)\\
3\left[\frac{2}{3}(2x-1)+3\right] &= 3[4(2x-4)]\\
2(2x-1)+3\cdot3 &= 12(2x-4)\\
4x-2+9 &= 24x-48\\
4x+7 &= 24x-48\\
4x-24x+7 &= 24x-24x-48\\
-20x+7 &= -48\\
-20x+7-7 &= -48-7\\
-20x &= -55\\
\frac{-20x}{-20} &= \frac{-55}{-20}\\
x &= \frac{11}{4} \text{ or } 2\frac{3}{4} \text{ or } 2.75
\end{aligned}$$

3.
$$\begin{aligned}
0.7x+1.3 &= 5x-2.14\\
100(0.7x)+100(1.3) &= 100(5x)-100(2.14)\\
70x+130 &= 500x-214\\
70x-500x+130 &= 500x-500x-214\\
-430x+130 &= -214\\
-430x+130-130 &= -214-130\\
-430x &= -344\\
\frac{-430x}{-430} &= \frac{-344}{-430}\\
x &= \frac{4}{5} \text{ or } 0.8
\end{aligned}$$

4. Answers may vary. Possible solution: Multiply both sides by the LCD, 2 · 3 · 5 = 30. Then simplify by obtaining all numerical values on one side and all variable terms on the other side. Divide both sides by –21.

$$\begin{aligned}
\frac{3x+1}{2}+\frac{2}{3} &= \frac{4x}{5}\\
30\left(\frac{3x+1}{2}+\frac{2}{3}\right) &= 30\left(\frac{4x}{5}\right)\\
15(3x+1)+10(2) &= 6(4x)\\
45x+15+20 &= 24x\\
45x+35 &= 24x\\
45x-45x+35 &= 24x-45x\\
35 &= -21x\\
\frac{35}{-21} &= \frac{-21x}{-21}\\
-\frac{5}{3} &= x
\end{aligned}$$

2.2 Exercises

1. $6x+5y=3$
$6x=3-5y$
$x=\dfrac{3-5y}{6}$

3. $4x+y=18-3x$
$4x+3x=18-y$
$7x=18-y$
$x=\dfrac{18-y}{7}$

5. $y=\dfrac{2}{3}x-4$
$3y=3\left(\dfrac{2}{3}x-4\right)$
$3y=2x-12$
$2x-12=3y$
$2x=3y+12$
$x=\dfrac{3y+12}{2}$

7. $x=-\dfrac{3}{4}y+\dfrac{2}{3}$
$12(x)=12\left(-\dfrac{3}{4}y+\dfrac{2}{3}\right)$
$12x=-9y+8$
$9y=-12x+8$
$y=\dfrac{-12x+8}{9}$

9. $A=lw$
$\dfrac{A}{w}=\dfrac{lw}{w}$
$\dfrac{A}{w}=l$ or $l=\dfrac{A}{w}$

11. $A=\dfrac{h}{2}(B+b)$
$2A=2\left[\dfrac{h}{2}(B+b)\right]$
$2A=hB+hb$
$hB+hb=2A$
$hB=2A-hb$
$B=\dfrac{2A-hb}{h}$

13. $A=2\pi rh$
$\dfrac{A}{2\pi h}=\dfrac{2\pi rh}{2\pi h}$
$r=\dfrac{A}{2\pi h}$

15. $H=\dfrac{2}{3}(a+2b)$
$3H=3\left[\dfrac{2}{3}(a+2b)\right]$
$3H=2a+4b$
$4b=3H-2a$
$b=\dfrac{3H-2a}{4}$

17. $2(2ax+y)=3ax-4y$
$4ax+2y=3ax-4y$
$4ax-3ax=-4y-2y$
$ax=-6y$
$x=-\dfrac{6y}{a}$

19. a. $A=\dfrac{1}{2}ab$
$2A=2\left(\dfrac{1}{2}ab\right)$
$2A=ab$
$\dfrac{2A}{a}=\dfrac{ab}{a}$
$\dfrac{2A}{a}=b$

b. $b=\dfrac{2A}{a}$
$b=\dfrac{2(18)}{\frac{3}{2}}$
$b=24$

21. a. $A=a+d(n-1)$
$A=a+dn-d$
$A-a+d=dn$
$dn=A-a+d$
$n=\dfrac{A-a+d}{d}$

b. $n=\dfrac{A-a+d}{d}$
$n=\dfrac{28-3+15}{15}=\dfrac{8}{3}$

23. $y = -7.4x + 322$
$7.4x = 322 - y$
$x = \frac{322 - y}{7.4}$ or $\frac{3220 - 10y}{74}$
$x = \frac{322 - 137}{7.4} = 25$
1990 + 25 = 2015
In the year 2015, the approximate heart rate will be 137.

25. a. $\frac{m}{1.15} = k$
$m = 1.15k$

b. $m = 1.15(29)$
$m = 33.35$
The ship is traveling 33.35 miles per hour.

27. a. $C = 0.6547D + 5.8263$
$C - 5.8263 = 0.6547D$
$0.6547D = C - 5.8263$
$D = \frac{C - 5.8263}{0.6547}$

b. $D = \frac{9.56 - 5.8263}{0.6547}$
$D \approx 5.7$
The disposable income is about \$5.7 billion.

Cumulative Review

29. $(2x^{-3}y)^{-2} = 2^{-2}x^{-3(-2)}y^{-2}$
$= 2^{-2}x^6y^{-2}$
$= \frac{x^6}{2^2y^2}$
$= \frac{x^6}{4y^2}$

30. $\left(\frac{5x^2y^{-3}}{x^{-4}y^2}\right)^{-3} = \frac{5^{-3}x^{2(-3)}y^{-3(-3)}}{x^{-4(-3)}y^{2(-3)}}$
$= \frac{5^{-3}x^{-6}y^9}{x^{12}y^{-6}}$
$= \frac{y^{9+6}}{5^3x^{12+6}}$
$= \frac{y^{15}}{125x^{18}}$

31. $1 + 16 \div (2-4)^3 - 3 = 1 + 16 \div (-2)^3 - 3$
$= 1 + 16 \div (-8) - 3$
$= 1 + (-2) - 3$
$= -1 - 3$
$= -4$

32. $2[a - (3 - 2b)] + 5a = 2(a - 3 + 2b) + 5a$
$= 2a - 6 + 4b + 5a$
$= 7a + 4b - 6$

33. \$5000(1.05) + \$4000(1.09) = \$9610
They will have \$9610.

34. $\frac{46,622.1 - 45,711.3}{9.9 + 11.7 + 10.6 + 5.8 + 8} = \frac{910.8}{46} = 19.8$
The car got 19.8 miles per gallon.

Quick Quiz 2.2

1. $A = d + g(b+3)$
$A = d + gb + 3g$
$A - d - 3g = gb$
$\frac{A - d - 3g}{g} = \frac{gb}{g}$
$b = \frac{A - d - 3g}{g}$

2. $V = \frac{1}{3}xy$
$3V = 3\left(\frac{1}{3}xy\right)$
$3V = xy$
$\frac{3V}{y} = \frac{xy}{y}$
$\frac{3V}{y} = x$ or $x = \frac{3V}{y}$

3. $G = 4x + \frac{1}{2}w + \frac{3}{4}$
$4G = 4\left(4x + \frac{1}{2}w + \frac{3}{4}\right)$
$4G = 16x + 2w + 3$
$4G - 16x - 3 = 2w$
$\frac{4G - 16x - 3}{2} = \frac{2w}{2}$
$w = \frac{4G - 16x - 3}{2}$

4. Answers may vary. Possible solution: Isolate all terms containing x on one side of the equals sign. Then divide both sides by the coefficient of x.

$$3(3ax+y)=2ax-5y$$
$$9ax+3y=2ax-5y$$
$$9ax-2ax=-5y-3y$$
$$7ax=-8y$$
$$\frac{7ax}{7a}=\frac{-8y}{7a}$$
$$x=-\frac{8y}{7a}$$

2.3 Exercises

1. It will always have two solutions. One solution is when $x = b$ and one when $x = -b$. Since $b > 0$ the values of b and $-b$ are always different numbers.

3. You must first isolate the absolute value expression. To do this you add 2 to each side of the equation. The result will be $|x + 7| = 10$. Then you solve the two equations $x + 7 = 10$ and $x + 7 = -10$. The final answer is $x = -17$ and $x = 3$.

5. $|x| = 30$

$x = 30$ or $x = -30$

Check: $|30| \stackrel{?}{=} 30$, $30 = 30$ $\qquad$ $|-30| \stackrel{?}{=} 30$, $30 = 30$

7. $|x + 4| = 10$

$x+4=10$ or $x+4=-10$

$x=6$ $\qquad$ $x=-14$

Check: $|6+4| \stackrel{?}{=} 10$, $|10| \stackrel{?}{=} 10$, $10=10$ $\qquad$ $|-14+4| \stackrel{?}{=} 10$, $|-10| \stackrel{?}{=} 10$, $10=10$

9. $|2x - 5| = 13$

$2x-5=13$ or $2x-5=-13$

$2x=18$ $\qquad$ $2x=-8$

$x=9$ $\qquad$ $x=-4$

Check: $|2\cdot 9-5| \stackrel{?}{=} 13$, $|18-5| \stackrel{?}{=} 13$, $|13| \stackrel{?}{=} 13$, $13=13$ $\qquad$ $|2(-4)-5| \stackrel{?}{=} 13$, $|-8-5| \stackrel{?}{=} 13$, $|-13| \stackrel{?}{=} 13$, $13=13$

11. $|5 - 4x| = 11$

$5-4x=11$ or $5-4x=-11$

$-4x=6$ $\qquad$ $-4x=-16$

$x=-\frac{3}{2}$ $\qquad$ $x=4$

Check: $\left|5-4\left(-\frac{3}{2}\right)\right| \stackrel{?}{=} 11$, $|5+6| \stackrel{?}{=} 11$, $|11| \stackrel{?}{=} 11$, $11=11$ $\qquad$ $|5-4(4)| \stackrel{?}{=} 11$, $|5-16| \stackrel{?}{=} 11$, $|-11| \stackrel{?}{=} 11$, $11=11$

13. $\left|\frac{1}{2}x-3\right|=2$

$\frac{1}{2}x-3=2$ or $\frac{1}{2}x-3=-2$

$x-6=4$ $\qquad$ $x-6=-4$

$x=10$ $\qquad$ $x=2$

Check: $\left|\frac{1}{2}\cdot 10-3\right| \stackrel{?}{=} 2$, $|5-3| \stackrel{?}{=} 2$, $|2| \stackrel{?}{=} 2$, $2=2$ $\qquad$ $\left|\frac{1}{2}\cdot 2-3\right| \stackrel{?}{=} 2$, $|1-3| \stackrel{?}{=} 2$, $|-2| \stackrel{?}{=} 2$, $2=2$

15. $|1.8 - 0.4x| = 1$

$1.8-0.4x=1$ or $1.8-0.4x=-1$

$-0.4x=-0.8$ $\qquad$ $-0.4x=-2.8$

$x=2$ $\qquad$ $x=7$

Check: $|1.8-0.4(2)| \stackrel{?}{=} 1$, $|1.8-0.8| \stackrel{?}{=} 1$, $|1| \stackrel{?}{=} 1$, $1=1$ $\qquad$ $|1.8-0.4(7)| \stackrel{?}{=} 1$, $|1.8-2.8| \stackrel{?}{=} 1$, $|-1| \stackrel{?}{=} 1$, $1=1$

17. $|x+2|-1=7$

$|x+2|=8$

$x+2=8$ or $x+2=-8$

$x=6$ $\qquad$ $x=-10$

Check: $|6+2|-1 \stackrel{?}{=} 7$, $|8|-1 \stackrel{?}{=} 7$, $8-1 \stackrel{?}{=} 7$, $7=7$ $\qquad$ $|-10+2|-1 \stackrel{?}{=} 7$, $|-8|-1 \stackrel{?}{=} 7$, $8-1 \stackrel{?}{=} 7$, $7=7$

19. $\left|\frac{1}{2}-\frac{3}{4}x\right|+1=3$

$\left|\frac{1}{2}-\frac{3}{4}x\right|=2$

$\frac{1}{2}-\frac{3}{4}x=2$ or $\frac{1}{2}-\frac{3}{4}x=-2$

$-\frac{3}{4}x=\frac{3}{2}$ $\qquad$ $-\frac{3}{4}x=-\frac{5}{2}$

$x=-2$ $\qquad$ $x=\frac{10}{3}$

Check: $\left|\frac{1}{2}-\frac{3}{4}(-2)\right|+1 \stackrel{?}{=} 3$ $\left|\frac{1}{2}-\frac{3}{4}\cdot\frac{10}{3}\right|+1 \stackrel{?}{=} 3$

$|2|+1 \stackrel{?}{=} 3$ $|-2|+1 \stackrel{?}{=} 3$

$2+1 \stackrel{?}{=} 3$ $2+1 \stackrel{?}{=} 3$

$3=3$ $3=3$

21. $\left|2-\frac{2}{3}x\right|-3=5$

$\left|2-\frac{2}{3}x\right|=8$

$2-\frac{2}{3}x=8$ or $2-\frac{2}{3}x=-8$

$-\frac{2}{3}x=6$ $-\frac{2}{3}x=-10$

$x=-9$ $x=15$

Check: $\left|2-\frac{2}{3}(-9)\right|-3 \stackrel{?}{=} 5$

$|2+6|-3 \stackrel{?}{=} 5$

$|8|-3 \stackrel{?}{=} 5$

$8-3 \stackrel{?}{=} 5$

$5=5$

$\left|2-\frac{2}{3}(15)\right|-3 \stackrel{?}{=} 5$

$|2-10|-3 \stackrel{?}{=} 5$

$|-8|-3 \stackrel{?}{=} 5$

$8-3 \stackrel{?}{=} 5$

$5=5$

23. $\left|\frac{1-3x}{2}\right|=\frac{4}{5}$

$\frac{1-3x}{2}=\frac{4}{5}$ or $\frac{1-3x}{2}=-\frac{4}{5}$

$5-15x=8$ $5-15x=-8$

$-15x=3$ $-15x=-13$

$x=-\frac{1}{5}$ $x=\frac{13}{15}$

Check: $\left|\frac{1-3\left(-\frac{1}{5}\right)}{2}\right| \stackrel{?}{=} \frac{4}{5}$ $\left|\frac{1-3\left(\frac{13}{15}\right)}{2}\right| \stackrel{?}{=} \frac{4}{5}$

$\left|\frac{4}{5}\right| \stackrel{?}{=} \frac{4}{5}$ $\left|-\frac{4}{5}\right| \stackrel{?}{=} \frac{4}{5}$

$\frac{4}{5}=\frac{4}{5}$ $\frac{4}{5}=\frac{4}{5}$

25. $|x+4|=|2x-1|$

$x+4=2x-1$ or $x+4=-(2x-1)$

$4=x-1$ $x+4=-2x+1$

$5=x$ $3x+4=1$

$x=5$ $3x=-3$

$x=-1$

27. $\left|\frac{x-1}{2}\right|=|2x+3|$

$\frac{x-1}{2}=2x+3$ or $\frac{x-1}{2}=-(2x+3)=-2x-3$

$x-1=4x+6$ $x-1=-4x-6$

$-3x=7$ $5x=-5$

$x=-\frac{7}{3}$ $x=-1$

29. $|1.5x-2|=|x-0.5|$

$1.5x-2=x-0.5$ or $1.5x-2=-(x-0.5)$

$15x-20=10x-5$ $15x-20=-10x+5$

$5x=15$ $25x=25$

$x=3$ $x=1$

31. $|3-x|=\left|\frac{x}{2}+3\right|$

$3-x=\frac{x}{2}+3$ or $3-x=-\left(\frac{x}{2}+3\right)=-\frac{x}{2}-3$

$6-2x=x+6$ $6-2x=-x-6$

$-3x=0$ $-x=-12$

$x=0$ $x=12$

33. $|1.62x+3.14|=2.19$

$1.62x+3.14=2.19$ or $1.62x+3.14=-2.19$

$1.62x=-0.95$ $1.62x=-5.33$

$x\approx-0.59$ $x\approx-3.29$

35. $|3(x+4)|+2=14$

$|3x+12|=12$

$3x+12=12$ or $3x+12=-12$

$3x=0$ $3x=-24$

$x=0$ $x=-8$

Check: $|3(0+4)|+2 \stackrel{?}{=} 14$ $|3(-8+4)|+2 \stackrel{?}{=} 14$

$|12|+2 \stackrel{?}{=} 14$ $|-12|+2 \stackrel{?}{=} 14$

$12+2 \stackrel{?}{=} 14$ $12+2 \stackrel{?}{=} 14$

$14=14$ $14=14$

37. $\left|\frac{5x}{3}-1\right|=0$

$$\frac{5x}{3}-1=0$$
$$5x=3$$
$$x=\frac{3}{5}$$

Check: $\left|\frac{5\cdot\frac{3}{5}}{3}-1\right| \stackrel{?}{=} 0$

$$|1-1| \stackrel{?}{=} 0$$
$$|0| \stackrel{?}{=} 0$$
$$0=0$$

39. $\left|\frac{4}{3}x-\frac{1}{8}\right|=-5$

No solution, since absolute value is nonnegative.

41. $\left|\frac{3x-1}{3}\right|=\frac{2}{5}$

$$\frac{3x-1}{3}=\frac{2}{5} \quad \text{or} \quad \frac{3x-1}{3}=-\frac{2}{5}$$
$$5(3x-1)=3(2) \qquad 5(3x-1)=3(-2)$$
$$15x-5=6 \qquad 15x-5=-6$$
$$15x=11 \qquad 15x=-1$$
$$x=\frac{11}{15} \qquad x=-\frac{1}{15}$$

Check: $\left|\frac{3\cdot\frac{11}{15}-1}{3}\right| \stackrel{?}{=} \frac{2}{5}$ $\qquad \left|\frac{3\left(-\frac{1}{15}\right)-1}{3}\right| \stackrel{?}{=} \frac{2}{5}$

$$\left|\frac{\frac{11}{5}-\frac{5}{5}}{3}\right| \stackrel{?}{=} \frac{2}{5} \qquad \left|\frac{-\frac{1}{5}-\frac{5}{5}}{3}\right| \stackrel{?}{=} \frac{2}{5}$$
$$\left|\frac{6}{5}\cdot\frac{1}{3}\right| \stackrel{?}{=} \frac{2}{5} \qquad \left|-\frac{6}{5}\cdot\frac{1}{3}\right| \stackrel{?}{=} \frac{2}{5}$$
$$\left|\frac{2}{5}\right| \stackrel{?}{=} \frac{2}{5} \qquad \left|-\frac{2}{5}\right| \stackrel{?}{=} \frac{2}{5}$$
$$\frac{2}{5}=\frac{2}{5} \qquad \frac{2}{5}=\frac{2}{5}$$

Cumulative Review

43. $(3x^{-3}yz^0)\left(\frac{5}{3}x^4y^2\right)=5x^{-3+4}y^{1+2}\cdot 1=5xy^3$

44. $\frac{\sqrt{3-2\cdot 1^2}+5}{4^2-2\cdot 3}=\frac{\sqrt{3-2}+5}{16-6}$

$$=\frac{\sqrt{1}+5}{10}$$
$$=\frac{1+5}{10}$$
$$=\frac{6}{10}$$
$$=\frac{3}{5}$$

Quick Quiz 2.3

1. $|3x-4|=49$

$$3x-4=49 \quad \text{or} \quad 3x-4=-49$$
$$3x=53 \qquad 3x=-45$$
$$x=\frac{53}{3} \qquad x=-15$$

2. $\left|\frac{2}{3}x+1\right|-3=5$

$$\left|\frac{2}{3}x+1\right|=8$$
$$\frac{2}{3}x+1=8 \quad \text{or} \quad \frac{2}{3}x+1=-8$$
$$\frac{2}{3}x=7 \qquad \frac{2}{3}x=-9$$
$$x=\frac{21}{2} \qquad x=-\frac{27}{2}$$

3. $|2x+5|=|x-4|$

$$2x+5=x-4 \quad \text{or} \quad 2x+5=-(x-4)$$
$$x+5=-4 \qquad 2x+5=-x+4$$
$$x=-9 \qquad 3x+5=4$$
$$3x=-1$$
$$x=-\frac{1}{3}$$

4. Answers may vary. Possible solution: Since $|x|=x$ if $x\geq 0$ and $|x|=-x$ if $x<0$, write two separate equations by removing the absolute value bars and letting the right side of the equation equal $\frac{1}{2}$ and $-\frac{1}{2}$. Then simplify each.

$$|2x+4| = \frac{1}{2}$$

$$2x+4 = \frac{1}{2} \quad \text{or} \quad 2x+4 = -\frac{1}{2}$$

$$2x = -4+\frac{1}{2} \qquad 2x = -4-\frac{1}{2}$$

$$2x = -\frac{7}{2} \qquad 2x = -\frac{9}{2}$$

$$x = -\frac{7}{4} \qquad x = -\frac{9}{4}$$

2.4 Exercises

1. Let x = the number.
$$\frac{3}{5}x = -54$$
$$3x = -270$$
$$x = -90$$
The number is –90.

3. Let x = the monthly membership fee.
$$12x - 50 = 526$$
$$12x = 576$$
$$x = 48$$
The monthly membership is \$48.

5. Let x = the weight of the package.
$$3 + 0.8x = 17.40$$
$$0.8x = 14.4$$
$$x = 18$$
The package weighed 18 pounds.

7. Let x = the number of weeks for the laundromat cost to equal the cost of a new washer and dryer.
$$11.75x = 846$$
$$x = 72$$
It would take 72 weeks.

9. Profit = Revenue – Cost.
For one year the profit must be \$129,000.
The revenue for one week is
$(2000 \cdot 8 \cdot 15) = 240{,}000$.
The cost for one week is
$14{,}000 \cdot 8 + 85{,}000 = 197{,}000$.
The profit for one week is
$240{,}000 - 197{,}000 = 43{,}000$.

11. Let x = the width of the field.
$$2W + 2L = P$$
$$2x + 2(3x-6) = 340$$
$$2x + 6x - 12 = 340$$
$$8x = 352$$
$$x = 44$$
$3x - 6 = 3(44) - 6 = 126$
The width of the field is 44 yards and the length of the field is 126 yards.

13. Let x = the length of the shortest side.
$2x - 7$ = the length of the longest side.
$x + 6$ = the length of the third side.
$$2x - 7 + x + 6 + x = 59$$
$$4x - 1 = 59$$
$$4x = 60$$
$$x = 15 \text{ ft, short side}$$
$2x - 7 = 23$ ft, longest side
$x + 6 = 21$ ft, third side
The longest side is 23 feet, the shortest side is 15 feet, and the third side is 21 feet.

Cumulative Review

15. $57 + 0 = 57$
Identity property of addition

16. $(2 \cdot 3) \cdot 9 = 2 \cdot (3 \cdot 9)$
Associative property of multiplication

17.
$$\begin{aligned} 7(-2) \div 7(-3) - 3 &= -14 \div 7(-3) - 3 \\ &= (-2)(-3) - 3 \\ &= 6 - 3 \\ &= 3 \end{aligned}$$

18.
$$\begin{aligned} (7-12)^3 - (-4) + 3^3 &= (-5)^3 + (4) + 27 \\ &= -125 + 4 + 27 \\ &= -94 \end{aligned}$$

Quick Quiz 2.4

1. Let x = the number.
$$\frac{3}{4}x = -69$$
$$\frac{4}{3} \cdot \frac{3}{4}x = \frac{4}{3} \cdot (-69)$$
$$x = -92$$
The number is –92.

2. Let x = length of second side.
$2x$ = length of first side.
$x + 8$ = length of third side.
$$x + 2x + x + 8 = 100$$
$$4x + 8 = 100$$
$$4x = 92$$
$$x = 23$$
$2x = 2(23) = 46$
$x + 8 = 23 + 8 = 31$
The first side is 46 yards, the second side is 23 yards, and the third side is 31 yards.

3. Let x = number of hours he parked at the airport.

$$\begin{aligned}6+3.50(x-1)&=65.50\\6+3.5x-3.5&=65.5\\3.5x+2.5&=65.5\\3.5x&=63\\x&=18\end{aligned}$$

He parked at the airport for 18 hours.

4. Answers may vary. Possible solution:
Let the width be represented by the variable x. Then define the length in terms of x. Substitute the variable expressions in the perimeter formula, P = 2(width) + 2(length), and simplify to solve for x. Then substitute x back in the expressions for the width and length.
Let x = width, then $4x + 7$ = length.

$$\begin{aligned}212&=2x+2(4x+7)\\212&=2x+8x+14\\212&=10x+14\\198&=10x\\19.8&=x\end{aligned}$$

$4x + 7 = 4(19.8) + 7 = 86.2$
The length is 86.2 feet and the width is 19.8 feet.

How Am I Doing? Sections 2.1–2.4

1.

$$\begin{aligned}2x-1&=12x+36\\2x-12x-1&=12x-12x+36\\-10x-1&=36\\-10x-1+1&=36+1\\-10x&=37\\\frac{-10x}{-10}&=\frac{37}{-10}\\x&=-3.7\text{ or }-\frac{37}{10}\text{ or }-3\frac{7}{10}\end{aligned}$$

2.

$$\begin{aligned}\frac{x-2}{4}&=\frac{1}{2}x+4\\4\left(\frac{x-2}{4}\right)&=4\left(\frac{1}{2}x+4\right)\\x-2&=2x+16\\x-2x-2&=2x-2x+16\\-x-2&=16\\-x-2+2&=16+2\\-x&=18\\x&=-18\end{aligned}$$

3.

$$\begin{aligned}4(x-3)&=x+2(5x-1)\\4x-12&=x+10x-2\\4x-12&=11x-2\\4x-11x-12&=11x-11x-2\\-7x-12&=-2\\-7x-12+12&=-2+12\\-7x&=10\\x&=-\frac{10}{7}=-1\frac{3}{7}\end{aligned}$$

4.

$$\begin{aligned}0.6x+3&=0.5x-7\\10(0.6x+3)&=10(0.5x-7)\\6x+30&=5x-70\\6x-5x+30&=5x-5x-70\\x+30&=-70\\x+30-30&=-70-30\\x&=-100\end{aligned}$$

5.

$$\begin{aligned}3x-7y&=14\\3x-3x-7y&=14-3x\\-7y&=14-3x\\\frac{-7y}{-7}&=\frac{14-3x}{-7}\\y&=\frac{3x-14}{7}\text{ or }y=\frac{3x}{7}-2\end{aligned}$$

6.

$$\begin{aligned}5ab-2b&=16ab-3(8+b)\\5ab-2b&=16ab-24-3b\\-11ab&=-b-24\\11ab&=b+24\\a&=\frac{b+24}{11b}\end{aligned}$$

7.

$$\begin{aligned}A&=P+Prt\\Prt&=A-P\\\frac{Prt}{Pt}&=\frac{A-P}{Pt}\\r&=\frac{A-P}{Pt}\end{aligned}$$

8.

$$\begin{aligned}r&=\frac{A-P}{Pt}\\r&=\frac{118-100}{(100)3}\\r&=\frac{18}{300}\\r&=\frac{3}{50}\\r&=0.06\end{aligned}$$

9. $|3x-2|=7$

$$3x-2=7 \quad \text{or} \quad 3x-2=-7$$
$$3x=9 \qquad 3x=-5$$
$$x=3 \qquad x=-\frac{5}{3}$$

10. $$|9-x|+2=5$$
$$|9-x|+2-2=5-2$$
$$|9-x|=3$$
$$9-x=3 \quad \text{or} \quad 9-x=-3$$
$$-x=-6 \qquad -x=-12$$
$$x=6 \qquad x=12$$

11. $$\left|\frac{2x+3}{4}\right|=2$$
$$\frac{2x+3}{4}=2 \quad \text{or} \quad \frac{2x+3}{4}=-2$$
$$2x+3=8 \qquad 2x+3=-8$$
$$2x=5 \qquad x=-11$$
$$x=\frac{5}{2}=2.5 \qquad x=-\frac{11}{2}=-5.5$$

12. $|5x-8|=|3x+2|$

$$5x-8=3x+2 \quad \text{or} \quad 5x-8=-3x-2$$
$$2x=10 \qquad 8x=6$$
$$x=5 \qquad x=\frac{6}{8}=0.75$$

13. Let W = width, then $W+20$ = length.

$$P=2L+2W$$
$$280=2(W+20)+2W$$
$$280=2W+40+2W$$
$$280=4W+40$$
$$240=4W$$
$$60=W$$
$$80=W+20$$

The dimensions are 60 in. × 80 in.

14. Let n = the number of checks.

$$6+0.12n=9.12$$
$$0.12n=3.12$$
$$n=26$$

He used 26 checks.

15. Let x = number of lb Cindi picked up.

$$x+\frac{x}{2}+80=455$$
$$2x+x+160=910$$
$$3x=750$$
$$x=250$$

$\frac{x}{2}+80=205$ pounds for Alan

Cindi picked up 250 pounds and Alan picked up 205 pounds.

16. Let x = length of short side.
Then $2x-5$ = length of long side and
$x+9$ = length of third side.

$$2x-5+x+9+x=62$$
$$4x+4=62$$
$$4x=58$$
$$x=14.5 \text{ ft, short side}$$

$x+9=14.5+9=23.5$ ft, third side
$2x-5=2(14.5)-5=24$ ft, long side

2.5 Exercises

1. Let x = population in 1990.

$$x+0.21x=301.1$$
$$1.21x=301.1$$
$$x\approx 248.8$$

The population in 1990 was approximately 248.8 million people.

3. Let x = the original price.
sale price = 80% of original price

$$0.80x=340$$
$$x=425$$

The original price was $425.

5. Let x = number of unemployed.

$$x-0.15x=969$$
$$0.85x=969$$
$$x=1140$$

1140 residents were unemployed in the previous month.

7. Let x = the number of hemlocks.
$2x$ = the number of spruce, and
$3x+20$ = the number of balsams.

$$x+2x+3x+20=1400$$
$$6x=1380$$
$$x=230$$

$2x=460$
$3x+20=710$
230 hemlocks, 460 spruces, and 710 balsams were planted.

9. Let x = Walker's salary 25 years ago.
Then $500-x$ = Angela's salary 25 years ago.

$$3x+4(500-x)=1740$$
$$3x+2000-4x=1740$$
$$-x=-260$$
$$x=260$$

$500 - x = 240$
Walker earned \$260 a week 25 years ago.
Angela earned \$240 a week 25 years ago.

11. Let x = number of mg in packet A.
Then $8 - x$ = number of mg in packet B.
$$17x + 14(8 - x) = 127$$
$$17x + 112 - 14x = 127$$
$$3x = 15$$
$$x = 5$$
$8 - x = 3$
5 mg are contained in packet A. 3 mg are contained in packet B.

13. $I = prt$
$I = 1500(0.09)(2)$
$I = 270$
The interest is \$270.

15. $I = prt$
$I = 5000(0.031)(1.5)$
$I = 232.50$
The interest is \$232.50.

17. Let x = amount invested at 5%.
Then $6400 - x$ = amount invested at 8%.
$$0.05x + 0.08(6400 - x) = 395$$
$$0.05x + 512 - 0.08x = 395$$
$$-0.03x = -117$$
$$x = 3900$$
$6400 - x = 2500$
She invested \$3900 at 5% and \$2500 at 8%.

19. Let x = amount invested at 3.5%.
Then $18{,}000 - x$ = amount invested at 2.2%.
$$0.035x + 0.022(18{,}000 - x) = 552$$
$$0.035x + 396 - 0.022x = 552$$
$$0.013x = 156$$
$$x = 12{,}000$$
$18{,}000 - x = 6000$
He invested \$12,000 in the certificate of deposit and \$6000 in the fixed interest account.

21. Let x = number of grams with 45% fat.
Then $30 - x$ = number of grams with 20% fat.
$$0.45x + 0.20(30 - x) = 0.30(30)$$
$$0.45x + 6 - 0.2x = 9$$
$$0.25x = 3$$
$$x = 12$$
$30 - x = 18$
She should use 12 grams of the 45% fat cheese and 18 grams of the 20% fat cheese.

23. Let x = number of pounds with 30% fat.
Then $100 - x$ = number of pounds with 10% fat.
$$0.30x + 0.10(100 - x) = 0.25(100)$$
$$0.3x + 10 - 0.1x = 25$$
$$0.2x = 15$$
$$x = 75$$
$100 - 75 = 25$
She should use 75 pounds of the 30% fat hamburger and 25 pounds of the 10% fat hamburger.

25. Let x = number of gal of 25% fertilizer.
Then $150 - x$ = number of gal of 15% fertilizer.
$$0.25x + 0.15(150 - x) = 0.18(150)$$
$$0.25x + 22.5 - 0.15x = 27$$
$$0.1x = 4.5$$
$$x = 45$$
$150 - x = 105$
They should mix 45 gal of 25% fertilizer with 105 gal of 15% fertilizer.

27. Let x = speed on secondary roads.
Then $x + 20$ = speed on the highway.
$$4x + 2(x + 20) = 250$$
$$4x + 2x + 40 = 250$$
$$6x = 210$$
$$x = 35$$
Speed on the secondary roads was 35 miles per hour.

29. Let x = time they walked on treadmill.
$$5x = 4.2x + 0.6$$
$$0.8x = 0.6$$
$$x = \frac{3}{4}$$
They each walked $\frac{3}{4}$ of an hour or 0.75 hour.

Cumulative Review

31. $5a - 2b + c = 5(1) - 2(-3) + (-4)$
$= 5 + 6 - 4$
$= 11 - 4$
$= 7$

32. $2x^2 - 3x + 1 = 2(-2)^2 - 3(-2) + 1$
$= 2 \cdot 4 + 6 + 1$
$= 8 + 6 + 1$
$= 14 + 1$
$= 15$

33. $\frac{5+8(-2)+2^4}{|2-7|}=\frac{5+(-16)+16}{|-5|}=\frac{5}{5}=1$

34. $\frac{\sqrt{7^2-24}}{2^3(-1)+7(4)}=\frac{\sqrt{49-24}}{8(-1)+7(4)}$
$=\frac{\sqrt{25}}{-8+28}$
$=\frac{5}{20}$
$=\frac{1}{4}$

Quick Quiz 2.5

1. Let x = price of television last year.
$x-0.08x=2208$
$0.92x=2208$
$x=2400$
The television was \$2400 last year.

2. Let x = amount of 50% fertilizer.
Then $200-x$ = amount of 30% fertilizer.
$0.50x+0.30(200-x)=0.45(200)$
$0.50x+60-0.30x=90$
$0.20x=30$
$x=150$
$200-x=50$
They should mix 150 gallons of the 50% fertilizer and 50 gallons of the 30% fertilizer.

3. Let x = amount invested at 5% and $4000-x$ = amount invested at 7%.
$0.05x+0.07(4000-x)=250$
$0.05x+280-0.07x=250$
$280-0.02x=250$
$-0.02x=-30$
$x=1500$
$4000-x=2500$
She invested \$1500 at 5% and \$2500 at 7%.

4. Answers may vary. Possible solution: Let a variable represent one quantity, such as x = price the previous year. Then write an equation in terms of x from the problem. Solve the equation and state the answer to the question asked in the problem.
$x+0.12x=39,200$
$1.12x=39,200$
$x=35,000$
The price was \$35,000 the previous year.

2.6 Exercises

1. True, $6<8$ and $8>6$ convey the same information.

3. True, dividing both sides of an inequality by -4 reverses the direction of the inequality.

5. False, the graph of $x\le 6$ does include the point at 6 on the number line.

7. $6>-3$, because 6 is to the right of -3 on a number line.

9. $-7<-2$ because -7 is to the left of -2 on a number line.

11. $\frac{3}{4}>\frac{2}{3}$ because $\frac{3}{4}$ is to the right of $\frac{2}{3}$ on a number line.

13. $-\frac{2}{9}=-0.\overline{2}<-0.2\overline{142857}=-\frac{3}{14}$

15. $-3.4>-3.41$ because -3.4 is to the right of -3.41 on a number line.

17. $|3-7|=|-4|=4$
$|9-2|=|7|=7$
$|3-7|<|9-2|$ since $4<7$.

19. $x\ge -2$
−4 −2 0

21. $x<15$
13 15 17

23. $2x-7\le -5$
$2x-7+7\le -5+7$
$2x\le 2$
$\frac{2x}{2}\le\frac{2}{2}$
$x\le 1$
−1 1 3

25.
$$3x-7>9x+5$$
$$3x-7-9x>9x-9x+5$$
$$-6x-7>5$$
$$-6x-7+7>5+7$$
$$-6x>12$$
$$\frac{-6x}{-6}<\frac{12}{-6}$$
$$x<-2$$

−4 −2 0

27.
$$0.5x+0.1<1.1x+0.7$$
$$10(0.5x+0.1)<10(1.1x+0.7)$$
$$5x+1<11x+7$$
$$5x-11x+1<11x-11x+7$$
$$-6x+1<7$$
$$-6x+1-1<7-1$$
$$-6x<6$$
$$\frac{-6x}{-6}>\frac{6}{-6}$$
$$x>-1$$

−3 −1 1

29.
$$4x-1>15$$
$$4x-1+1>15+1$$
$$4x>16$$
$$\frac{4x}{4}>\frac{16}{4}$$
$$x>4$$

31.
$$5x+3\le 2x-9$$
$$5x-2x+3\le 2x-2x-9$$
$$3x+3\le -9$$
$$3x+3-3\le -9-3$$
$$3x\le -12$$
$$\frac{3x}{3}\le\frac{-12}{3}$$
$$x\le -4$$

33.
$$2x+\frac{5}{3}>\frac{2}{5}x-1$$
$$15\left(2x+\frac{5}{3}\right)>15\left(\frac{2}{5}x-1\right)$$
$$30x+25>6x-15$$
$$30x-6x+25>6x-6x-15$$
$$24x+25>-15$$
$$24x+25-25>-15-25$$
$$24x>-40$$
$$\frac{24x}{24}>\frac{-40}{24}$$
$$x>-\frac{5}{3}$$

35.
$$3x-11+4(x+8)<0$$
$$3x-11+4x+32<0$$
$$7x+21<0$$
$$7x<-21$$
$$\frac{7x}{7}<\frac{-21}{7}$$
$$x<-3$$

37.
$$\frac{3}{5}x-(x+2)\ge -2$$
$$\frac{3}{5}x-x-2\ge -2$$
$$5\left(\frac{3}{5}x-x-2\right)\ge 5(-2)$$
$$3x-5x-10\ge -10$$
$$-2x-10\ge -10$$
$$-2x\ge 0$$
$$\frac{-2x}{-2}\le\frac{0}{-2}$$
$$x\le 0$$

39.
$$0.4x+1\le 2.6$$
$$10(0.4x+1)\le 10(2.6)$$
$$4x+10\le 26$$
$$4x\le 16$$
$$\frac{4x}{4}\le\frac{16}{4}$$
$$x\le 4$$

41.
$$0.1(x-2)\ge 0.5x-0.2$$
$$0.1x-0.2\ge 0.5x-0.2$$
$$0.1x-0.5x\ge -0.2-0.2$$
$$-0.4x\ge 0$$
$$\frac{-0.4x}{-0.4}\le\frac{0}{-0.4}$$
$$x\le 0$$

43. $$2-\frac{1}{5}(x-1)\ge\frac{2}{3}(2x+1)$$
$$15\left[2-\frac{1}{5}(x-1)\right]\ge 15\left[\frac{2}{3}(2x+1)\right]$$
$$30-3(x-1)\ge 10(2x+1)$$
$$30-3x+3\ge 20x+10$$
$$-3x+33\ge 20x+10$$
$$-23x\ge -23$$
$$\frac{-23x}{-23}\le\frac{-23}{-23}$$
$$x\le 1$$

45. $$\frac{2x-3}{5}+1\ge\frac{1}{2}x+3$$
$$10\left(\frac{2x-3}{5}+1\right)\ge 10\left(\frac{1}{2}x+3\right)$$
$$2(2x-3)+10\ge 5x+30$$
$$4x-6+10\ge 5x+30$$
$$4x+4\ge 5x+30$$
$$4x-5x\ge -4+30$$
$$-x\ge 26$$
$$x\le -26$$

47. Let x = number of tables.
$$4(4.50)+6x>72$$
$$18+6x>72$$
$$6x>54$$
$$\frac{6x}{6}>\frac{54}{6}$$
$$x>9$$
She would have to serve more than 9 tables.

49. Let x = the number of minutes he talks after the first minute.
$$3.95+0.55x\le 13.30$$
$$0.55x\le 9.35$$
$$x\le 17$$
He can talk for a maximum of 17 + 1 = 18 minutes.

51. Let x = number of computers.
$$130+155+59x\le 1100$$
$$59x\le 815$$
$$x\le 13.81355932$$
A maximum of 13 computers can be taken up.

Cumulative Review

53. $3xy(x+2)-4x^2(y-1)$
$$=3x^2y+6xy-4x^2y+4x^2$$
$$=6xy-x^2y+4x^2$$

54. $\frac{2}{3}ab(6a-2b+9)$
$$=\frac{2}{3}ab(6a)-\frac{2}{3}ab(2b)+\frac{2}{3}ab(9)$$
$$=4a^2b-\frac{4}{3}ab^2+6ab$$

55. $$\left(\frac{4x^2}{3yw^{-1}}\right)^3=\frac{4^3x^{2\cdot 3}}{3^3y^3w^{-1(3)}}=\frac{64x^6}{27y^3w^{-3}}=\frac{64x^6w^3}{27y^3}$$

56. $(-3a^0b^{-3}c^5)^{-2}=(-3b^{-3}c^5)^{-2}$
$$=(-3)^{-2}b^{-3(-2)}c^{5(-2)}$$
$$=\frac{1}{9}b^6c^{-10}$$
$$=\frac{b^6}{9c^{10}}$$

Quick Quiz 2.6

1. $$5x+7>3x-9$$
$$5x-3x+7>3x-3x-9$$
$$2x+7>-9$$
$$2x+7-7>-9-7$$
$$2x>-16$$
$$\frac{2x}{2}>\frac{-16}{2}$$
$$x>-8$$

2. $$-3(x+4)<6x-8$$
$$-3x-12<6x-8$$
$$-3x-6x-12<6x-6x-8$$
$$-9x-12<-8$$
$$-9x-12+12<-8+12$$
$$-9x<4$$
$$\frac{-9x}{-9}>\frac{4}{-9}$$
$$x>-\frac{4}{9}$$

3. $\frac{1}{5}(x-3) \geq \frac{1}{4}(x-5)+1$

$$20\left[\frac{1}{5}(x-3)\right] \geq 20\left[\frac{1}{4}(x-5)+1\right]$$
$$4(x-3) \geq 5(x-5)+20$$
$$4x-12 \geq 5x-25+20$$
$$4x-12 \geq 5x-5$$
$$4x-5x \geq 12-5$$
$$-x \geq 7$$
$$\frac{-1x}{-1} \leq \frac{7}{-1}$$
$$x \leq -7$$

4. Answers may vary. Possible solution: If both sides of an inequality are multiplied or divided by the same negative number, the inequality symbol is reversed. Therefore, when solving $-3x < 9$, the inequality symbol is reversed:

$$-3x < 9$$
$$\frac{-3x}{-3} > \frac{9}{-3}$$
$$x > -3$$

2.7 Exercises

1. $3 < x$ and $x < 8$

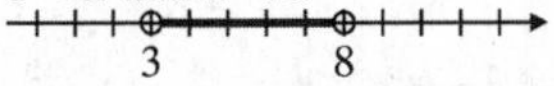

3. $-4 < x$ and $x < 2$

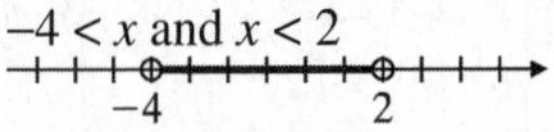

5. $7 < x < 9$

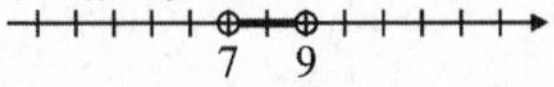

7. $-2 < x \leq \frac{1}{2}$

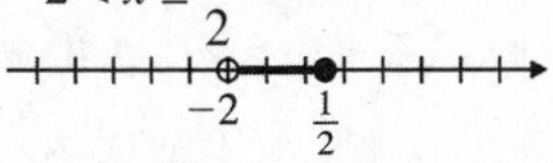

9. $x > 8$ or $x < 2$

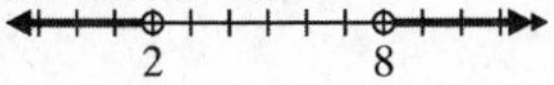

11. $x \leq -\frac{5}{2}$ or $x > 4$

13. $x \leq -10$ or $x \geq 40$

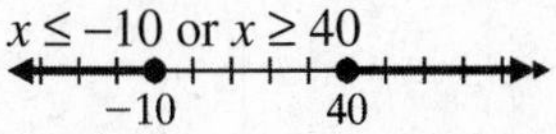

15. $2x+3 \leq 5$ and $x+1 \geq -2$

$2x \leq 2$ $x \geq -3$

$x \leq 1$

$-3 \leq x \leq 1$

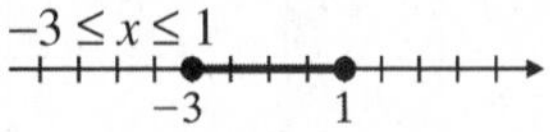

17. $2x-3 > 0$ or $x-2 < -7$

$2x > 3$ $x < -5$

$x > \frac{3}{2}$

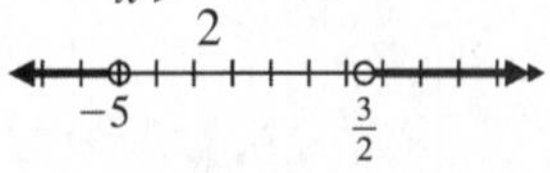

19. $x < 8$ and $x > 10$ do not overlap. No solution

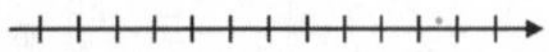

21. $t < 10.9$ or $t > 11.2$

23. $5000 \leq c \leq 12{,}000$

25.
$$-20 \leq C \leq 11$$
$$-20 \leq \frac{5}{9}(F-32) \leq 11$$
$$9(-20) \leq 5(F-32) \leq 9(11)$$
$$-180 \leq 5F-160 \leq 99$$
$$-20 \leq 5F \leq 259$$
$$-4° \leq F \leq 51.8°$$

27.
$$20{,}000 \leq Y \leq 34{,}000$$
$$20{,}000 \leq 107(d-5) \leq 34{,}000$$
$$186.92 \leq d-5 \leq 317.76$$
$$\$191.92 \leq d \leq \$322.76$$

29. $x-3 > -5$ and $2x+4 < 8$

$x > -2$ $2x < 4$

$-2 < x$ $x < 2$

$-2 < x < 2$ is the solution.

31. $-6x+5 \geq -1$ and $2-x \leq 5$

$-6x \geq -6$ $-x \leq 3$

$x \leq 1$ $x \geq -3$

$-3 \leq x \leq 1$ is the solution.

33. $4x-3 < -11$ or $7x+2 \geq 23$

$4x < -8$ $7x \geq 21$

$x < -2$ $x \geq 3$

$x < -2$ or $x \geq 3$ is the solution.

35. $-0.3x + 1 \geq 0.2x$ or $-0.2x + 0.5 > 0.7$
Multiply by 10 on both sides of both inequalities to clear decimals.

$$\begin{array}{rcl} -3x+10 \geq 2x & \text{or} & -2x+5>7 \\ -5x \geq -10 & & -2x > 2 \\ x \leq 2 & & x < -1 \end{array}$$

$x \leq 2$ contains $x < -1$. $x \leq 2$ is the solution.

37. $$\begin{array}{rcl} \frac{5x}{2}+1 \geq 3 & \text{and} & x-\frac{2}{3} \geq \frac{4}{3} \\ 5x+2 \geq 6 & & 3x-2 \geq 4 \\ 5x \geq 4 & & 3x \geq 6 \\ x \geq \frac{4}{5} & & x \geq 2 \end{array}$$

$x \geq 2$ is the solution.

39. $$\begin{array}{rcl} 2x+5<3 & \text{and} & 3x-1>-1 \\ 2x<-2 & & 3x>0 \\ x<-1 & & x>0 \end{array}$$

$x < -1$ and $x > 0$ do not overlap.
No solution

41. $$\begin{array}{rcl} 2x-3 \geq 7 & \text{and} & 5x-8 \leq 2x+7 \\ 2x \geq 10 & & 3x \leq 15 \\ x \geq 5 & & x \leq 5 \end{array}$$

$x \geq 5$ and $x \leq 5$ overlap at $x = 5$.
$x = 5$ is the solution.

43. $$\begin{array}{rcl} \frac{1}{4}(x+2)+\frac{1}{8}(x-3) \leq 1 & \text{and} & \frac{3}{4}(x-1) > -\frac{1}{4} \\ 2x+4+x-3 \leq 8 & & 3x-3>-1 \\ 3x+1 \leq 8 & & 3x>2 \\ 3x \leq 7 & & x > \frac{2}{3} \\ x \leq \frac{7}{3} & & \end{array}$$

$\frac{2}{3} < x \leq \frac{7}{3}$ is the solution.

Cumulative Review

45. $$\begin{aligned} 3y-5x &= 8 \\ -5x &= 8-3y \\ (-1)(-5x) &= (-1)(8-3y) = -8+3y \\ 5x &= 3y-8 \\ x &= \frac{3y-8}{5} \end{aligned}$$

46. $$\begin{aligned} 7x+6y &= -12 \\ 6y &= -12-7x \\ y &= \frac{-12-7x}{6} \end{aligned}$$

47. $$\begin{aligned} x^2+5x-|x+3| &= (-2)^2+5(-2)-|-2+3| \\ &= (-2)^2+5(-2)-|1| \\ &= 4-10-1 \\ &= -6-1 \\ &= -7 \end{aligned}$$

48. $$\begin{aligned} 3x^3-x^2-\sqrt{8x+9} &= 3(2)^3-2^2-\sqrt{8\cdot 2+9} \\ &= 3(2)^3-2^2-\sqrt{16+9} \\ &= 3(8)-4-\sqrt{25} \\ &= 24-4-5 \\ &= 20-5 \\ &= 15 \end{aligned}$$

Quick Quiz 2.7

1. $$\begin{array}{rcl} 3x+2<8 & \text{and} & 3x>-16 \\ 3x<6 & & x > -\frac{16}{3} \\ x<2 & & \end{array}$$

$-\frac{16}{3} < x < 2$ is the solution.

2. $$\begin{array}{rcl} x>5 & \text{and} & 2x-1<23 \\ & & 2x<24 \\ & & x<12 \end{array}$$

$5 < x < 12$ is the solution.

3. $$\begin{array}{rcl} x-7 \leq -15 & \text{or} & 2x+3 \geq 5 \\ x \leq -8 & & 2x \geq 2 \\ & & x \geq 1 \end{array}$$

$x \leq -8$ or $x \geq 1$ is the solution.

4. Answers may vary. Possible solution:

$$\begin{array}{rcl} x+8<3 & \text{and} & 2x-1>5 \\ x<-5 & & 2x>6 \\ & & x>3 \end{array}$$

Since $x < -5$ and $x > 3$ do not overlap, there is no solution.

2.8 Exercises

1. $|x| \leq 8$
$-8 \leq x \leq 8$

−8 −4 0 4 8

3. $|x + 4.5| < 5$
$-5 < x + 4.5 < 5$
$-9.5 < x < 0.5$

−16 −12 −8 −4 0 4 8

5. $|x-3|\le 5$
$-5\le x-3\le 5$
$-2\le x\le 8$

7. $|2x-1|\le 5$
$-5\le 2x-1\le 5$
$-4\le 2x\le 6$
$-2\le x\le 3$

9. $|5x-2|\le 4$
$-4\le 5x-2\le 4$
$-2\le 5x\le 6$
$-\frac{2}{5}\le x\le \frac{6}{5}$
$-\frac{2}{5}\le x\le 1\frac{1}{5}$

11. $|0.5-0.1x|<1$
$-1<0.5-0.1x<1$
$-1.5<-0.1x<0.5$
$15>x>-5$
$-5<x<15$

13. $\left|\frac{1}{4}x+2\right|<6$
$-6<\frac{1}{4}x+2<6$
$-24<x+8<24$
$-32<x<16$

15. $\left|\frac{2}{3}(x-2)\right|<4$
$-4<\frac{2}{3}(x-2)<4$
$-6<x-2<6$
$-4<x<8$

17. $\left|\frac{3x-2}{4}\right|<3$
$-3<\frac{3x-2}{4}<3$
$-12<3x-2<12$
$-10<3x<14$
$-\frac{10}{3}<x<\frac{14}{3}$
$-3\frac{1}{3}<x<4\frac{2}{3}$

19. $|x|>5$
$x<-5$ or $x>5$

21. $|x+2|>5$
$x+2<-5$ or $x+2>5$
$x<-7$ $\quad x>3$

23. $|x-1|\ge 2$
$x-1\le -2$ or $x-1\ge 2$
$x\le -1$ $\quad x\ge 3$

25. $|4x-7|\ge 9$
$4x-7\le -9$ or $4x-7\ge 9$
$4x\le -2$ $\quad 4x\ge 16$
$x\le -\frac{1}{2}$ $\quad x\ge 4$

27. $|6-0.1x|>5$
$6-0.1x>5$ or $6-0.1x<-5$
$-0.1x>-1$ $\quad -0.1x<-11$
$x<10$ $\quad x>110$

29. $\left|\frac{1}{5}x-\frac{1}{10}\right|>2$
$\frac{1}{5}x-\frac{1}{10}<-2$ or $\frac{1}{5}x-\frac{1}{10}>2$
$2x-1<-20$ $\quad 2x-1>20$
$2x<-19$ $\quad 2x>21$
$x<-\frac{19}{2}$ $\quad x>\frac{21}{2}$
$x<-9\frac{1}{2}$ $\quad x>10\frac{1}{2}$

31. $\left|\frac{1}{3}(x-2)\right|<5$
$-5<\frac{1}{3}(x-2)<5$
$-15<x-2<15$
$-13<x<17$

33. $|3x+5|<17$
$-17<3x+5<17$
$-22<3x<12$
$-\frac{22}{3}<x<4$
$-7\frac{1}{3}<x<4$

35. $|3 - 8x| > 19$

$3-8x<-19$ or $3-8x>19$

$-8x<-22$ $\quad -8x>16$

$x<\frac{11}{4}$ $\quad x<-2$

$x<-2$ or $x>2\frac{3}{4}$

37.

$$|m-s|\le 0.12$$
$$|m-18.65|\le 0.12$$
$$-0.12\le m-18.65\le 0.12$$
$$18.53\le m\le 18.77$$

39.

$$|n-p|\le 0.05$$
$$|n-9.68|\le 0.05$$
$$-0.05\le n-9.68\le 0.05$$
$$9.63\le n\le 9.73$$

Cumulative Review

41.
$$\begin{aligned}4^2+(5-2)^3\div(-9)&=4^2+(3)^3\div(-9)\\&=16+27\div(-9)\\&=16-3\\&=13\end{aligned}$$

42.
$$\begin{aligned}(-4)(7)\div 2+(-8)-12&=-28\div 2+(-8)-12\\&=-14+(-8)-12\\&=-22-12\\&=-34\end{aligned}$$

43.
$$\begin{aligned}\text{distance}&=2\left[\frac{1}{8}\cdot\text{circumference}\right]\\&=2\left[\frac{1}{8}(2\pi\cdot\text{radius})\right]\\&=2\left[\frac{1}{3}(2\cdot 3.14\cdot 19)\right]\\&=29.83\end{aligned}$$

The end of the rope travels 29.83 meters.

44.
$$\begin{aligned}\text{distance}&=2\cdot\frac{1}{6}(2\pi\cdot 30)\\&=2\cdot\frac{1}{6}(2\cdot 3.14\cdot 30)\\&=62.8\end{aligned}$$

The end of the wire travels 62.8 feet.

Quick Quiz 2.8

1. $\left|\frac{1}{2}x+\frac{1}{4}\right|<6$

$$-6<\frac{1}{2}x+\frac{1}{4}<6$$
$$4(-6)<4\left(\frac{1}{2}x+\frac{1}{4}\right)<4(6)$$
$$-24<2x+1<24$$
$$-25<2x<23$$
$$-\frac{25}{2}<x<\frac{23}{2}$$
$$-12\frac{1}{2}<x<11\frac{1}{2}$$

2. $|8x - 4| \le 20$

$$-20\le 8x-4\le 20$$
$$-16\le 8x\le 24$$
$$-2\le x\le 3$$

3. $|5x + 2| > 7$

$5x+2<-7$ or $5x+2>7$

$5x<-9$ $\quad 5x>5$

$x<-\frac{9}{5}$ $\quad x>1$

$x<-1\frac{4}{5}$

4. Answers may vary. Possible solution: Since the absolute value is always nonnegative, there is no solution. $|7x + 3|$ cannot be <-4.

Putting Your Skills To Work

1. 950 + 200 + 90 + 60 + 300 = 1600
The monthly costs would total $1600.

2. $\frac{1600}{2}=800$
Her new expected costs would be $800.
800 – 475 = 325
It is $325 more than she expected to pay.

3. 15 × 130 = 1950
She will earn $1950.
Yes, she can pay for these expenses since $1950 > $800.

4. Answers may vary.

5. Answers may vary.

Chapter 2 Review Problems

1.
$$\begin{aligned}7x-3&=-5x-18\\7x+5x-3&=-5x+5x-18\\12x-3&=-18\\12x-3+3&=-18+3\\12x&=-15\\\frac{12x}{12}&=\frac{-15}{12}\\x&=-\frac{5}{4}\\x&=-1.25\text{ or }-1\frac{1}{4}\end{aligned}$$

2.
$$\begin{aligned}8-2(x+3)&=24-(x-6)\\8-2x-6&=24-x+6\\2-2x&=30-x\\-2x+x&=30-2\\-x&=28\\x&=-28\end{aligned}$$

3.
$$\begin{aligned}5(x-2)+4&=x+9-2x\\5x-10+4&=-x+9\\5x-6&=-x+9\\5x+x-6&=-x+x+9\\6x-6&=9\\6x-6+6&=9+6\\6x&=15\\\frac{6x}{6}&=\frac{15}{6}\\x&=\frac{5}{2}\text{ or }2\frac{1}{2}\text{ or }2.5\end{aligned}$$

4.
$$\begin{aligned}x-\frac{4}{3}&=\frac{11}{12}+\frac{3}{4}x\\12\left(x-\frac{4}{3}\right)&=12\left(\frac{11}{12}+\frac{3}{4}x\right)\\12x-16&=11+9x\\12x-9x&=11+16\\3x&=27\\x&=9\end{aligned}$$

5.
$$\begin{aligned}\frac{1}{9}x-1&=\frac{1}{2}\left(x+\frac{1}{3}\right)\\18\left(\frac{1}{9}x-1\right)&=18\left[\frac{1}{2}\left(x+\frac{1}{3}\right)\right]\\2x-18&=9x+3\\2x-9x&=3+18\\-7x&=21\\x&=-3\end{aligned}$$

6.
$$\begin{aligned}\frac{x-4}{2}-\frac{1}{5}&=\frac{7x+1}{20}\\20\left(\frac{x-4}{2}-\frac{1}{5}\right)&=20\left(\frac{7x+1}{20}\right)\\10(x-4)-4&=7x+1\\10x-40-4&=7x+1\\10x-44&=7x+1\\10x-7x&=1+44\\3x&=45\\x&=15\end{aligned}$$

7.
$$\begin{aligned}5x&=3(1.6x-4.2)\\5x&=4.8x-12.6\\0.2x&=-12.6\\x&=-63\end{aligned}$$

8.
$$\begin{aligned}1.2x-1&=2(1.6x+1.5)\\1.2x-1&=3.2x+3\\1.2x-3.2x&=3+1\\-2x&=4\\x&=-2\end{aligned}$$

9.
$$\begin{aligned}6x-11y&=8\\-11y&=-6x+8\\y&=\frac{-6x+8}{-11}\\y&=\frac{6x-8}{11}\end{aligned}$$

10.
$$\begin{aligned}P&=\frac{1}{2}ab\\2P&=ab\\\frac{2P}{b}&=\frac{ab}{b}\\\frac{2P}{b}&=a\text{ or }a=\frac{2P}{b}\end{aligned}$$

11.
$$\begin{aligned}2(3ax-2y)-6ax&=-3(ax+2y)\\6ax-4y-6ax&=-3ax-6y\\-4y&=-3ax-6y\\2y&=-3ax\\3ax&=-2y\\a&=-\frac{2y}{3x}\end{aligned}$$

12.
$$\begin{aligned}\frac{1}{2}a+3b&=\frac{2}{3}(2b-1)\\3a+18b&=8b-4\\10b&=-3a-4\\b&=\frac{-3a-4}{10}\end{aligned}$$

13. a.
$$\begin{aligned} C &= \frac{5F-160}{9} \\ 9C &= 5F-160 \\ 5F-160 &= 9C \\ 5F &= 9C+160 \\ F &= \frac{9C+160}{5} \end{aligned}$$

b. $F = \frac{9(10)+160}{5} = \frac{250}{5} = 50$
$F = 50°$ when $C = 10°$.

14. a.
$$\begin{aligned} P &= 2W+2L \\ P-2L &= 2W \\ 2W &= P-2L \\ W &= \frac{P-2L}{2} \end{aligned}$$

b.
$$\begin{aligned} W &= \frac{100-2(20.5)}{2} \\ &= \frac{100-41}{2} \\ &= \frac{59}{2} \\ &= 29.5 \end{aligned}$$
$W = 29.5$ meters

15. $|2x-7| = 9$
$2x-7=9$ or $2x-7=-9$
$2x=16$ $\quad$ $2x=-2$
$x=8$ $\quad$ $x=-1$

16. $|5x+2| = 7$
$5x+2=7$ or $5x+2=-7$
$5x=5$ $\quad$ $5x=-9$
$x=1$ $\quad$ $x=-\frac{9}{5}$

17. $|3-x| = |5-2x|$
$3-x=5-2x$ or $3-x=-(5-2x)$
$x=2$ $\quad$ $3-x=-5+2x$
$-3x=-8$
$x=\frac{8}{3}$

18. $|x+8| = |2x-4|$
$x+8=2x-4$ or $x+8=-2x+4$
$-x=-12$ $\quad$ $3x=-4$
$x=12$ $\quad$ $x=-\frac{4}{3}$

19. $\left|\frac{1}{4}x-3\right| = 8$
$\frac{1}{4}x-3=8$ or $\frac{1}{4}x-3=-8$
$x-12=32$ $\quad$ $x-12=-32$
$x=44$ $\quad$ $x=-20$

20. $|4-7x| = 25$
$4-7x=25$ or $4-7x=-25$
$-7x=21$ $\quad$ $-7x=-29$
$x=-3$ $\quad$ $x=\frac{29}{7}$

21. $|2x-8|+7=12$
$|2x-8|=5$
$2x-8=5$ or $2x-8=-5$
$2x=13$ $\quad$ $2x=3$
$x=\frac{13}{2}$ $\quad$ $x=\frac{3}{2}$

22. $|0.2x-1|+1.2=2.3$
$|0.2x-1|=1.1$
$0.2x-1=1.1$ or $0.2x-1=-1.1$
$0.2x=2.1$ $\quad$ $0.2x=-0.1$
$x=\frac{21}{2}$ $\quad$ $x=-\frac{1}{2}$

23.
$$\begin{aligned} P &= 2L+2W \\ 42 &= 2(2W+3)+2W \\ 21 &= 2W+3+W \\ 3W &= 18 \\ W &= 6 \end{aligned}$$
$2W+3=15$
The width is 6 feet and the length is 15 feet.

24. Let x = the number of women.
Then $2x-200$ = the number of men.
$$\begin{aligned} 2x-200+x &= 280 \\ 3x-200 &= 280 \\ 3x &= 480 \\ x &= 160 \end{aligned}$$
$2x-200=120$
There are 160 women and 120 men attending Western Tech.

25. Let x = miles she drove.

$$\begin{aligned} 3(38)+0.15x &= 150 \\ 114+0.15x &= 150 \\ 0.15x &= 36 \\ x &= 240 \end{aligned}$$

She drove 240 miles.

26. Let x = number of miles from airport to hotel.

$$\begin{aligned} \$2.50+\frac{\$0.35}{\frac{1}{5}\text{ mile}}\left(x-\frac{1}{5}\right)\text{ mile} &= \$14.75 \\ 0.5+0.35\left(x-\frac{1}{5}\right) &= 2.95 \\ 0.35x-0.07 &= 2.45 \\ 0.35x &= 2.52 \\ x &= 7.2 \end{aligned}$$

It is 7.2 miles or $7\frac{1}{5}$ miles from the airport to the hotel.

27. Let x = the amount withheld for retirement. Then $x + 13$ = the amount withheld for state tax, and $3(x + 13)$ = the amount withheld for federal tax.

$$\begin{aligned} x+x+13+3(x+13) &= 102 \\ 2x+13+3x+39 &= 102 \\ 5x+52 &= 102 \\ 5x &= 50 \\ x &= 10 \end{aligned}$$

$x + 13 = 23$

$3(x + 13) = 69$

\$10 is withheld for retirement, \$23 for state tax, and \$69 for federal tax.

28. Let x = the number of tickets Nicholas sold. Then $2x - 5$ = the number of tickets Emma sold, and $2x + 10$ = the number of tickets Jackson sold.

$$\begin{aligned} x+2x-5+2x+10 &= 180 \\ 5x &= 175 \\ x &= 35 \end{aligned}$$

$2x - 5 = 65$

$2x + 10 = 80$

Nicholas sold 35 tickets, Emma sold 65 tickets, and Jackson sold 80 tickets.

29. Let x = the number of students enrolled five years ago.

$$\begin{aligned} x+0.15x &= 2415 \\ 1.15x &= 2415 \\ x &= 2100 \end{aligned}$$

2100 students were enrolled five years ago.

30. Let x = the number of two-door sedans. Then $3x$ = the number of four-door sedans.

$$\begin{aligned} 3x+x &= 260,000 \\ 4x &= 260,000 \\ x &= 65,000 \end{aligned}$$

$3x = 195,000$

They should manufacture 65,000 two-door sedans and 195,000 four-door sedans.

31. Let x = amount invested at 11%. Then $9000 - x$ = the amount invested at 6%.

$$\begin{aligned} 0.11x+0.06(9000-x) &= 815 \\ 0.11x+540-0.06x &= 815 \\ 540+0.05x &= 815 \\ 0.05x &= 275 \\ x &= 5500 \end{aligned}$$

$9000 - x = 3500$

He invested \$5500 at 11% and \$3500 at 6%.

32. Let x = the number of liters of 2% acid. Then $24 - x$ = the number of liters of 5% acid.

$$\begin{aligned} 0.02x+0.05(24-x) &= 0.04(24) \\ 0.02x+1.2-0.05x &= 0.96 \\ -0.03x &= -0.24 \\ x &= 8 \end{aligned}$$

$24 - x = 16$

He should use 8 liters of the 2% acid and 16 liters of the 5% acid.

33. Let x = the number of pounds of the \$4.25 a pound coffee. Then $30 - x$ = the number of pounds of the \$4.50 a pound coffee.

$$\begin{aligned} 4.25x+4.50(30-x) &= 4.40(30) \\ 4.25x+135-4.5x &= 132 \\ -0.25x &= -3 \\ x &= 12 \end{aligned}$$

$30 - x = 18$

12 pounds of \$4.25 and 18 pounds of \$4.50 should be used.

34. Let x = current full-time students.

$$\begin{aligned} \frac{1}{2}x+\frac{1}{3}(890-x) &= 380 \\ 3x+1780-2x &= 2280 \\ x &= 500 \end{aligned}$$

$890 - 500 = 390$

The present number of students is 500 full-time and 390 part-time.

35.
$$\begin{aligned} 7x+8 &< 5x \\ 2x &< -8 \\ \frac{2x}{2} &< \frac{-8}{2} \\ x &< -4 \end{aligned}$$

36.
$$\begin{aligned} 9x+3 &< 12x \\ -3x &< -3 \\ \frac{-3x}{-3} &> \frac{-3}{-3} \\ x &> 1 \end{aligned}$$

37.
$$\begin{aligned} 4x-1 &< 3(x+2) \\ 4x-1 &< 3x+6 \\ 4x-3x &< 6+1 \\ x &< 7 \end{aligned}$$

38.
$$\begin{aligned} 3(3x-2) &< 4x-16 \\ 9x-6 &< 4x-16 \\ 9x-4x &< -16+6 \\ 5x &< -10 \\ \frac{5x}{5} &< \frac{-10}{5} \\ x &< -2 \end{aligned}$$

39.
$$\begin{aligned} \frac{7}{8}x-\frac{1}{4} &> \frac{1}{2} \\ 8\left(\frac{7}{8}x-\frac{1}{4}\right) &> 8\left(\frac{1}{2}\right) \\ 7x-2 &> 4 \\ 7x &> 6 \\ x &> \frac{6}{7} \end{aligned}$$

40.
$$\begin{aligned} \frac{5}{3}-x &\ge -\frac{1}{6}x+\frac{5}{6} \\ 6\left(\frac{5}{3}-x\right) &\ge 6\left(-\frac{1}{6}x+\frac{5}{6}\right) \\ 10-6x &\ge -x+5 \\ -6x+x &\ge 5-10 \\ -5x &\ge -5 \\ \frac{-5x}{-5} &\le \frac{-5}{-5} \\ x &\le 1 \end{aligned}$$

41.
$$\begin{aligned} \frac{1}{3}(x-2) &< \frac{1}{4}(x+5)-\frac{5}{3} \\ 12\left[\frac{1}{3}(x-2)\right] &< 12\left[\frac{1}{4}(x+5)-\frac{5}{3}\right] \\ 4(x-2) &< 3(x+5)-20 \\ 4x-8 &< 3x+15-20 \\ 4x-8 &< 3x-5 \\ 4x-3x &< -5+8 \\ x &< 3 \end{aligned}$$

42.
$$\begin{aligned} \frac{1}{3}(x+2) &> 3x-5(x-2) \\ 3\left[\frac{1}{3}(x+2)\right] &> 3[3x-5(x-2)] \\ x+2 &> 9x-15(x-2) \\ x+2 &> 9x-15x+30 \\ x+2 &> -6x+30 \\ x+6x &> 30-2 \\ 7x &> 28 \\ x &> 4 \end{aligned}$$

43.
$$\begin{aligned} 7x-6 &\le \frac{1}{3}(-2x+5) \\ 3(7x-6) &\le 3\left[\frac{1}{3}(-2x+5)\right] \\ 21x-18 &\le -2x+5 \\ 21x+2x &\le 5+18 \\ 23x &\le 23 \\ x &\le 1 \end{aligned}$$

44. $-3 \le x < 2$

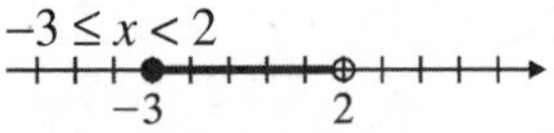

45. $-4 < x \le 5$

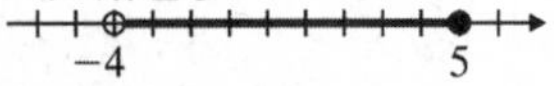

46. $-8 \le x \le -4$

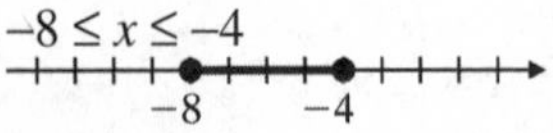

47. $-9 \le x \le -6$

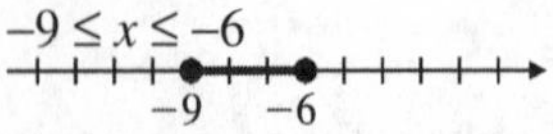

48. $x < -2$ or $x \ge 5$

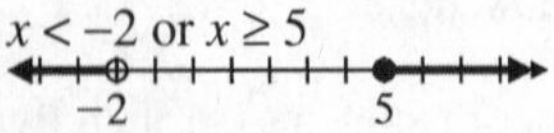

49. $x < -3$ or $x \ge 6$

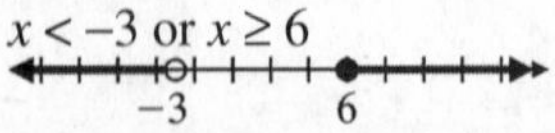

50. $x > -5$ and $x < -1$

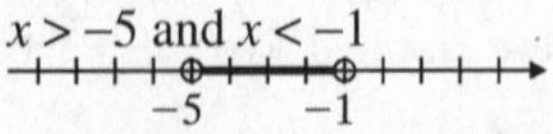

51. $x > -8$ and $x < -3$

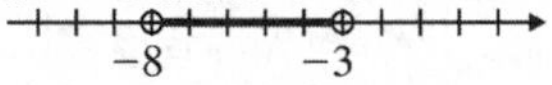

52. $x+3>8$ or $x+2<6$

$x>5 \qquad x<4$

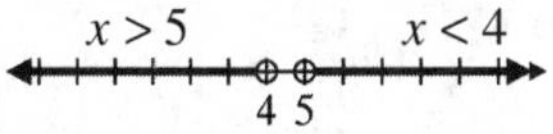

53. $x-2>7$ or $x+3<2$

$x>9 \qquad x<-1$

54. $x+3>8$ and $x-4<-2$

$x>5 \qquad x<2$

Since x cannot be both > 5 and < 2, there is no solution.

55. $-1<x+5<8$

$-6<x<3$

56. $0 \le 5-3x \le 17$

$-5 \le -3x \le 12$

$\frac{5}{3} \ge x \ge -4$

$-4 \le x \le \frac{5}{3}$

$-4 \le x \le 1\frac{2}{3}$

57. $2x-7<3$ and $5x-1 \ge 8$

$2x<10 \qquad 5x \ge 9$

$x<5 \qquad x \ge \frac{9}{5}$

$\frac{9}{5} \le x < 5$

$1\frac{4}{5} \le x < 5$

58. $4x-2<8$ or $3x+1>4$

$4x<10 \qquad 3x>3$

$x<\frac{5}{2} \qquad x>1$

The solution is all real numbers.

59. $|x+7|<15$

$-15<x+7<15$

$-22<x<8$

60. $|x+9|<18$

$-18<x+9<18$

$-27<x<9$

61. $\left|\frac{1}{2}x+2\right|<\frac{7}{4}$

$-\frac{7}{4}<\frac{1}{2}x+2<\frac{7}{4}$

$-7<2x+8<7$

$-15<2x<-1$

$-\frac{15}{2}<x<-\frac{1}{2}$

$-7\frac{1}{2}<x<-\frac{1}{2}$

62. $\left|\frac{1}{5}x+3\right|<\frac{11}{5}$

$-\frac{11}{5}<\frac{1}{5}x+3<\frac{11}{5}$

$-11<x+15<11$

$-26<x<-4$

63. $|2x-1| \ge 9$

$2x-1 \le -9$ or $2x-1 \ge 9$

$2x \le -8 \qquad 2x \ge 10$

$x \le -4 \qquad x \ge 5$

64. $|3x-1| \ge 2$

$3x-1 \le -2$ or $3x-1 \ge 2$

$3x \le -1 \qquad 3x \ge 3$

$x \le -\frac{1}{3} \qquad x \ge 1$

65. $|4(x+1)| \ge 3$

$4(x+1) \le -3$ or $4(x+1) \ge 3$

$4x+4 \le -3 \qquad 4x+4 \ge 3$

$4x \le -7 \qquad 4x \ge -1$

$x \le -\frac{7}{4} \qquad x \ge -\frac{1}{4}$

$x \le -1\frac{3}{4}$

66. $|2(x-5)| \ge 2$

$2(x-5) \le -2$ or $2(x-5) \ge 2$

$2x-10 \le -2 \qquad 2x-10 \ge 2$

$2x \le 8 \qquad 2x \ge 12$

$x \le 4 \qquad x \ge 6$

67. Let x = the number of minutes he talks.

$3.95+0.64(x-1) \le 13.05$

$3.95+0.64x-0.64 \le 13.05$

$0.64x \le 9.74$

$x \le 15.21875$

He can talk for a maximum of 15 minutes.

68. Let x = the number of packages.
$$170+200+77.5x \le 1765$$
$$77.5x \le 1395$$
$$x \le 18$$
A maximum of eighteen packages can be carried.

69. Let x = number of cubic yards.
$$40+28x \le 250$$
$$28x \le 210$$
$$x \le 7.5$$
He can order a maximum of 7 cubic yards.

70. Let x = the weight of the envelope.
$$0.41+0.28(x-1) \le 4.6$$
$$0.41+0.28x-0.28 \le 4.6$$
$$0.13+0.28x \le 4.6$$
$$0.28x \le 4.47$$
$$x \le 15.96$$
The envelope could weigh a maximum of 15 ounces.

71. Let n = number of bolts per box.
$$1.5+2.5n \le 14$$
$$2.5n \le 12.5$$
$$n \le 5$$
5 is the maximum number of bolts per box.

72. $$1.04(2,312,000) \le x \le 1.06(2,854,000)$$
$$2,404,480 \le x \le 3,025,240$$

73. $$4-7x = 3(x+3)$$
$$4-7x = 3x+9$$
$$-7x-3x = 9-4$$
$$-10x = 5$$
$$\frac{-10x}{-10} = \frac{5}{-10}$$
$$x = -\frac{1}{2} \text{ or } -0.5$$

74. $$H = \frac{3}{4}B-16$$
$$\frac{3}{4}B = H+16$$
$$B = \frac{4}{3}(H+16)$$
$$B = \frac{4H+64}{3}$$

75. Let x = number of grams of 77% copper.
Then $100-x$ = number of grams of 92% copper.
$$0.77x+0.92(100-x) = 0.80(100)$$
$$0.77x+92-0.92x = 80$$
$$-0.15x = -12$$
$$x = 80$$
$$100-x = 20$$
She should use 80 grams of 77% copper and 20 grams of 92% copper.

76. $$7x+12 < 9x$$
$$-2x < -12$$
$$x > 6$$
6
$x > 6$

77. $$\frac{2}{3}x-\frac{5}{6}x-3 \le \frac{1}{2}x-5$$
$$4x-5x-18 \le 3x-30$$
$$-x-18 \le 3x-30$$
$$-4x \le -12$$
$$x \ge 3$$
3
$x \ge 3$

78. $$-2 \le x+1 \le 4$$
$$-3 \le x \le 3$$
−3 0 3

79. $$2x+3 < -5 \quad \text{or} \quad x-2 > 1$$
$$2x < -8 \qquad x > 3$$
$$x < -4$$
−4 0 3

80. $$|2x-7|+4 = 5$$
$$|2x-7| = 1$$
$$2x-7 = -1 \quad \text{or} \quad 2x-7 = 1$$
$$2x = 6 \qquad 2x = 8$$
$$x = 3 \qquad x = 4$$

81. $$\left|\frac{2}{3}x-\frac{1}{2}\right| \le 3$$
$$-3 \le \frac{2}{3}x-\frac{1}{2} \le 3$$
$$-18 \le 4x-3 \le 18$$
$$-15 \le 4x \le 21$$
$$-\frac{15}{4} \le x \le \frac{21}{4}$$

82. $|2-5x-4|>13$

$$2-5x-4>13 \quad\text{or}\quad 2-5x-4<-13$$
$$-5x>15 \qquad\qquad -5x<-11$$
$$x<-3 \qquad\qquad x>\frac{11}{5}$$

How Am I Doing? Chapter 2 Test

1.
$$\begin{aligned}5x-8&=-6x-10\\5x+6x-8&=-6x+6x-10\\11x-8&=-10\\11x-8+8&=-10+8\\11x&=-2\\\frac{11x}{11}&=\frac{-2}{11}\\x&=-\frac{2}{11}\end{aligned}$$

2.
$$\begin{aligned}3(7-2x)&=14-8(x-1)\\21-6x&=14-8x+8\\21-6x&=22-8x\\21-6x+8x&=22-8x+8x\\21+2x&=22\\21-21+2x&=22-21\\2x&=1\\x&=\frac{1}{2}\text{ or }0.5\end{aligned}$$

3.
$$\begin{aligned}\frac{1}{3}(-x+1)+4&=4(3x-2)\\3\left[\frac{1}{3}(-x+1)+4\right]&=3[4(3x-2)]\\1(-x+1)+12&=12(3x-2)\\-x+1+12&=36x-24\\-x+13&=36x-24\\-x-36x&=-24-13\\-37x&=-37\\x&=1\end{aligned}$$

4.
$$\begin{aligned}0.5x+1.2&=4x-3.05\\100(0.5x+1.2)&=100(4x-3.05)\\50x+120&=400x-305\\120+305&=400x-50x\\425=350x&\Rightarrow 350x=425\\x&=\frac{425}{350}=\frac{17(25)}{14(25)}=\frac{17}{14}\\x&=1\frac{3}{14}\end{aligned}$$

5.
$$\begin{aligned}L&=a+d(n-1)\\L&=a+dn-d\\L-a+d&=dn\\n&=\frac{L-a+d}{d}\end{aligned}$$

6.
$$\begin{aligned}A&=\frac{1}{2}bh\\2A&=bh\\bh&=2A\\b&=\frac{2A}{h}\end{aligned}$$

7.
$$\begin{aligned}b&=\frac{2A}{h}\\b&=\frac{2(15)\text{ cm}^2}{10\text{ cm}}\\b&=3\text{ cm}\end{aligned}$$

8.
$$\begin{aligned}H&=\frac{1}{2}r+3b-\frac{1}{4}\\4H&=2r+12b-1\\2r&=4H-12b+1\\r&=\frac{4H-12b+1}{2}\end{aligned}$$

9. $|5x-2|=37$

$$5x-2=37 \quad\text{or}\quad 5x-2=-37$$
$$5x=39 \qquad\qquad 5x=-35$$
$$x=\frac{39}{5} \qquad\qquad x=-7$$

10.
$$\left|\frac{1}{2}x+3\right|-2=4$$
$$\left|\frac{1}{2}x+3\right|=6$$
$$\frac{1}{2}x+3=6 \quad\text{or}\quad \frac{1}{2}x+3=-6$$
$$x+6=12 \qquad\qquad x+6=-12$$
$$x=6 \qquad\qquad x=-18$$

11. Let x = the length of first side.
Then $2x$ = the length of the second side,
and $x + 5$ = the length of the third side.

$$\begin{aligned}x+2x+x+5&=69\\4x&=64\\x&=16\end{aligned}$$
$2x = 32$
$x + 5 = 21$
The first side is 16 meters, the second side is 32 meters, and the third side is 21 meters.

12. Let x = electric bill for August.

$$x - 0.05x = 2489$$
$$0.95x = 2489$$
$$x = 2620$$

The electric bill for August was \$2620.

13. Let x = gallons of 50% antifreeze.
Then $10 - x$ = gallons of 90% antifreeze.

$$0.50x + 0.90(10 - x) = 0.60(10)$$
$$0.5x + 9 - 0.9x = 6$$
$$-0.4x = -3$$
$$x = 7.5$$

$10 - 7.5 = 2.5$
She should use 2.5 gallons of 90% and 7.5 gallons of 50%.

14. Let x = amount invested at 6%.
Then $5000 - x$ = amount invested at 10%.

$$0.06x + 0.10(5000 - x) = 428$$
$$0.06x + 500 - 0.1x = 428$$
$$-0.04x = -72$$
$$x = 1800$$

$5000 - x = 3200$
\$1800 was invested at 6% and \$3200 was invested at 10%.

15.
$$5 - 6x < 2x + 21$$
$$-8x < 16$$
$$\frac{-8x}{-8} > \frac{16}{-8}$$
$$x > -2$$

−5 −4 −3 −2 −1 0 1 2 3

16.
$$-\frac{1}{2} + \frac{1}{3}(2 - 3x) \geq \frac{1}{2}x + \frac{5}{3}$$
$$6\left[-\frac{1}{2} + \frac{1}{3}(2 - 3x)\right] \geq 6\left(\frac{1}{2}x + \frac{5}{3}\right)$$
$$-3 + 4 - 6x \geq 3x + 10$$
$$1 - 6x \geq 3x + 10$$
$$-6x - 3x \geq 10 - 1$$
$$-9x \geq 9$$
$$\frac{-9x}{-9} \leq \frac{9}{-9}$$
$$x \leq -1$$

−6 −5 −4 −3 −2 −1 0 1 2

17.
$$-11 < 2x - 1 \leq -3$$
$$-10 < 2x \leq -2$$
$$-5 < x \leq -1$$

18. $x - 4 \leq -6$ or $2x + 1 \geq 3$
$x \leq -2$ $\qquad 2x \geq 2$
$\qquad x \geq 1$

19.
$$|7x - 3| \leq 18$$
$$-18 \leq 7x - 3 \leq 18$$
$$-15 \leq 7x \leq 21$$
$$-\frac{15}{7} \leq x \leq 3$$

20. $|3x + 1| > 7$
$3x + 1 < -7$ or $3x + 1 > 7$
$3x < -8$ $\qquad 3x > 6$
$x < -\frac{8}{3}$ $\qquad x > 2$

Cumulative Test for Chapters 1–2

1. $-12, -3, 0, \frac{1}{4}, 2.16, 2.333..., -\frac{5}{8}, 3$

2. $7 + (6 + 3) = (7 + 6) + 3$
Associative property of addition

3.
$$\begin{aligned}\sqrt{100} + 4(3-5)^3 - (-20) &= 10 + 4(-2)^3 - (-20)\\ &= 10 + 4(-8) - (-20)\\ &= 10 - 32 + 20\\ &= -22 + 20\\ &= -2\end{aligned}$$

4.
$$\begin{aligned}\left(-\frac{2}{3}x^4y^{-2}z^0\right)(6x^{-1}y^6z^2) &= -4x^{4-1}y^{2+6}z^{0+2}\\ &= -4x^3y^4z^2\end{aligned}$$

5.
$$\frac{6a^{-1}b^3}{-18a^5b} = -\frac{6}{18}a^{-1-5}b^{3-1} = -\frac{1}{3}a^{-6}b^2 = -\frac{b^2}{3a^6}$$

6.
$$\begin{aligned}2x^2 + 3xy - y^2 &= 2(-2)^2 + 3(-2)(1) - 1^2\\ &= 2(4) - 6 - 1\\ &= 8 - 6 - 1\\ &= 2 - 1\\ &= 1\end{aligned}$$

7. $A = \pi r^2 = 3.14(7)^2 = 153.86$ sq in.

8.
$$\begin{aligned}2x - [6x - 3(x + 5y)] &= 2x - [6x - 3x - 15y]\\ &= 2x - [3x - 15y]\\ &= 2x - 3x + 15y\\ &= -x + 15y\end{aligned}$$

9. $\frac{1}{4}(x+5)-\frac{5}{3}=\frac{1}{3}(x-2)$

$$12\left[\frac{1}{4}(x+5)-\frac{5}{3}\right]=12\left[\frac{1}{3}(x-2)\right]$$
$$3(x+5)-4(5)=4(x-2)$$
$$3x+15-20=4x-8$$
$$3x-5=4x-8$$
$$3x-4x=-8+5$$
$$-x=-3$$
$$x=3$$

10. $h=\frac{2}{3}(b+d)$

$$3h=3\left[\frac{2}{3}(b+d)\right]$$
$$3h=2(b+d)$$
$$3h=2b+2d$$
$$2b=3h-2d$$
$$b=\frac{3h-2d}{2}$$

11. Let x = length of first side.
Then $x + 15$ = length of second side and
$2x - 7$ = length of third side.

$$x+x+15+2x-7=112$$
$$4x+8=112$$
$$4x=104$$
$$x=26$$

$x + 15 = 41$
$2x - 7 = 45$
1st side = 26 inches
2nd side = 41 inches
3rd side = 45 inches

12. Let x = original price.

$$0.85x=\$68$$
$$\frac{0.85x}{0.85}=\frac{\$68}{0.85}$$
$x = \$80$, original price of saw

13. Let x = the number of gallons at 50%.
Then $9 - x$ = the number of gallons at 80%.

$$0.50x+0.80(9-x)=0.70(9)$$
$$0.5x+7.2-0.8x=6.3$$
$$-0.3x=-0.9$$
$$x=3$$

$9-x=6$
He should use 3 gallons of 50% and 6 gallons of 80%.

14. Let x = amount invested at 12%.
Then $6500 - x$ = amount invested at 10%.

$$0.12x+0.10(6500-x)=690$$
$$0.12x+650-0.1x=690$$
$$0.02x=40$$
$$x=2000$$

$6500 - x = 4500$
She invested \$2000 at 12% and \$4500 at 10%.

15. $-4-3x<-2x+6$

$$-x<10$$
$$x>-10$$

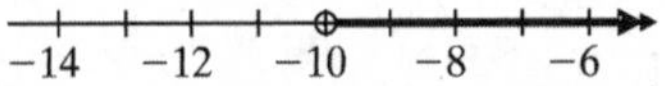

16. $\frac{1}{3}(x+2)\le\frac{1}{5}(x+6)$

$$15\left[\frac{1}{3}(x+2)\right]\le 15\left[\frac{1}{5}(x+6)\right]$$
$$5(x+2)\le 3(x+6)$$
$$5x+10\le 3x+18$$
$$5x-3x\le 18-10$$
$$2x\le 8$$
$$\frac{2x}{2}\le\frac{8}{2}$$
$$x\le 4$$

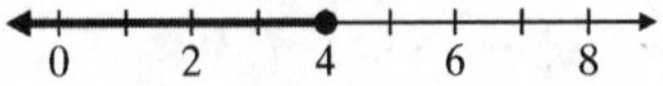

17. $-13<4x-5<3$

$$-8<4x<8$$
$$-2<x<2$$

18. $x+5\le -4$ or $2-7x\le 16$

$x\le -9$ $-7x\le 14$

$x\ge -2$

19. $\left|\frac{1}{2}x+2\right|\le 8$

$$-8\le\frac{1}{2}x+2\le 8$$
$$-16\le x+4\le 16$$
$$-20\le x\le 12$$

20. $|3x-4|>11$

$3x-4<-11$ or $3x-4>11$

$3x<-7$ $3x>15$

$x<-\frac{7}{3}$ $x>5$

$x<-2\frac{1}{3}$

Chapter 3

3.1 Exercises

1. Graphs are used to show the relationships among the variables in an equation.

3. To locate the point (a, b), assuming that $a, b > 0$, we move a units to the right and b units up. To locate the point (b, a) we move b units right and a units up. If $a \neq b$ the graphs of the points will be different, Thus, the order of the numbers in (a, b) matters. (1, 3) is not the same as (3, 1).

5. $y = 3x - 7$
$y = 3(-2) - 7 = -6 - 7 = -13$
$(-2, -13)$ is a solution of $y = 3x - 7$.

7.
$$\begin{aligned} 5x + 12y &= -17 \\ 5x + 12\left(\frac{1}{4}\right) &= -17 \\ 5x + 3 &= -17 \\ 5x &= -17 - 3 \\ 5x &= -20 \\ x &= -4 \end{aligned}$$
$\left(-4, \frac{1}{4}\right)$ is a solution of $5x + 12y = -17$.

9. $y = 2x - 3$

x	$y = 2x - 3$	y
0	$y = 2(0) - 3$	-3
2	$y = 2(2) - 3$	1
4	$y = 2(4) - 3$	5

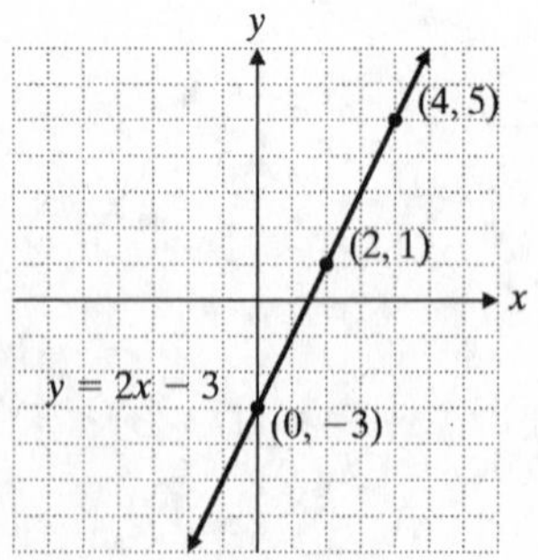

11. $y = 4 - 2x$

x	$y = 4 - 2x$	y
-1	$y = 4 - 2(-1)$	6
0	$y = 4 - 2(0)$	4
2	$y = 4 - 2(2)$	0

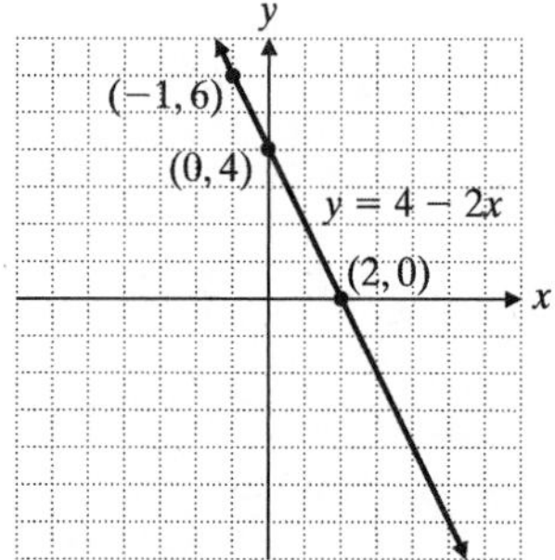

13. $y = \frac{2}{3}x - 4$

x	$y = \frac{2}{3}x - 4$	y
-3	$y = \frac{2}{3}(-3) - 4$	-6
0	$y = \frac{2}{3}(0) - 4$	-4
3	$y = \frac{2}{3}(3) - 4$	-2

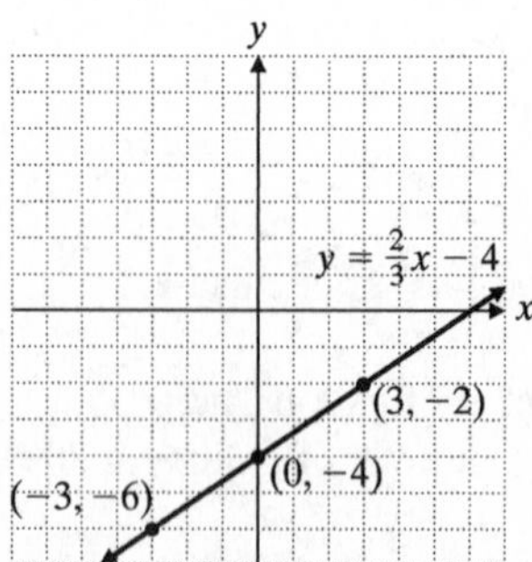

15. $2y - 3x = 6$
Let $y = 0$: $2(0) - 3x = 6$
$x = -2$
Let $x = 0$: $2y - 3(0) = 6$
$y = 3$
Let $x = 2$: $2y - 3(2) = 6$
$y = 6$

x	y
−2	0
0	3
2	6

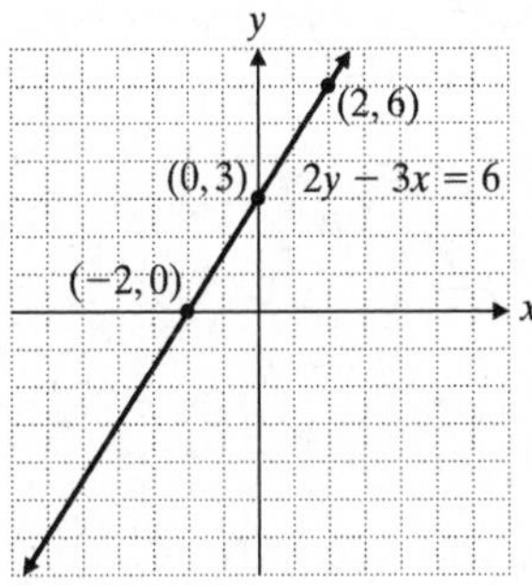

17. $2x - y = 6$

Let $x = 0$: $2(0) - y = 6$
$y = -6$

Let $y = 0$: $2x - 0 = 6$
$x = 3$

Let $x = 2$: $2(2) - y = 6$
$y = -2$

x	y
0	−6
2	−2
3	0

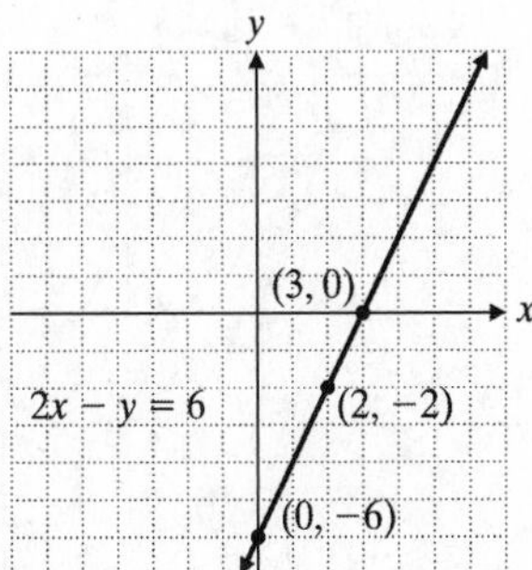

19. $-4x - 3y = 6$

Let $x = 0$: $-4(0) - 3y = 6$
$y = -2$

Let $y = 0$: $-4x - 3(0) = 6$
$x = -\frac{3}{2}$

Let $x = 3$: $-4(3) - 3y = 6$
$y = -6$

x	y
0	−2
$-1\frac{1}{2}$	0
3	−6

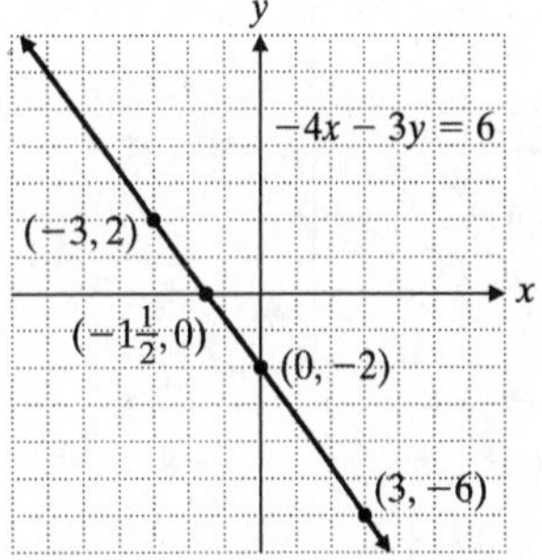

21. $5y - 4 = 3x - 4$
$3x - 5y = 0$

Let $x = 0$: $3(0) - 5y = 0$
$y = 0$

Let $x = -5$: $3(-5) - 5y = 0$
$y = -3$

Let $x = 5$: $3(5) - 5y = 0$
$y = 3$

x	y
0	0
−5	−3
5	3

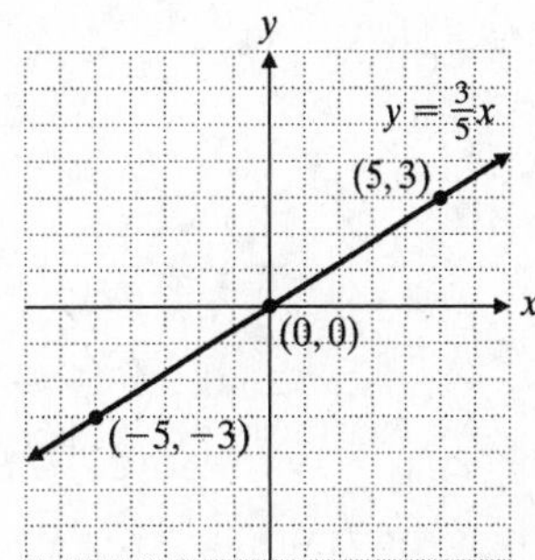

23. $x = -5$, vertical line

x	y
−5	3
−5	0
−5	−2

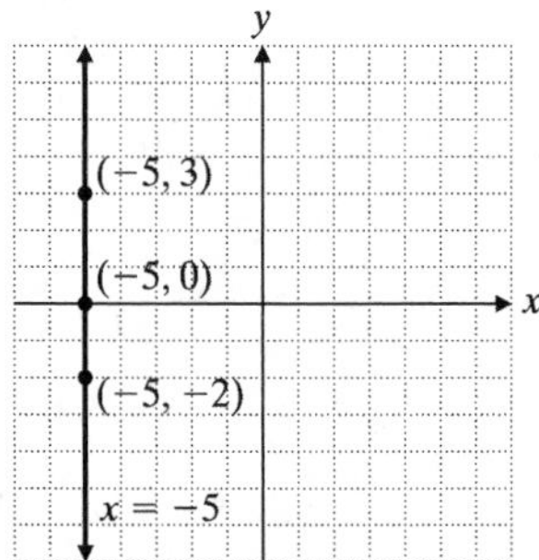

25. $4x - 16 = 0$

$x = 4$; vertical line

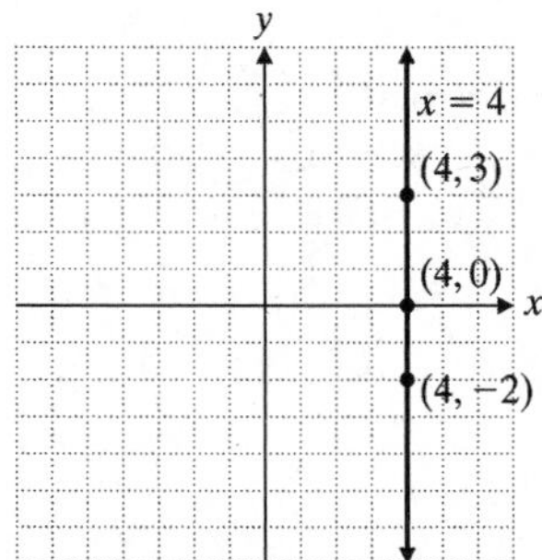

27. $2y + 8 = 0$, $2y = -8$, $y = -4$

The graph is a horizontal line.

x	y
0	−4
1	−4
2	−4

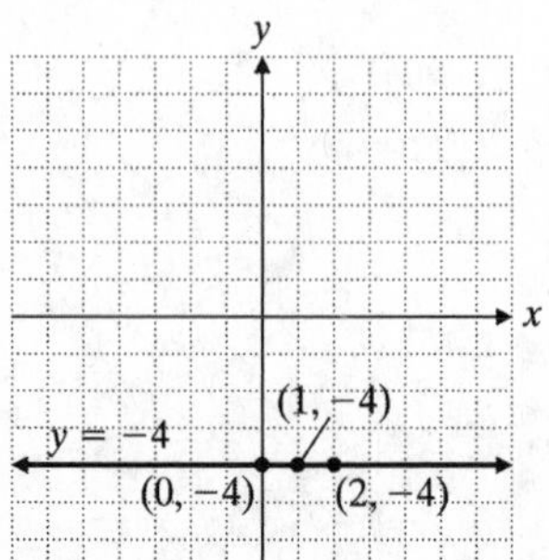

29. $y = -1.5x + 2$

Let $x = -2$: $y = -1.5(-2) + 2$
$y = 5$

Let $x = 0$: $y = -1.5(0) + 2$
$y = 2$

Let $x = 2$: $y = -1.5(2) + 2$
$y = -1$

x	y
−2	5
0	2
2	−1

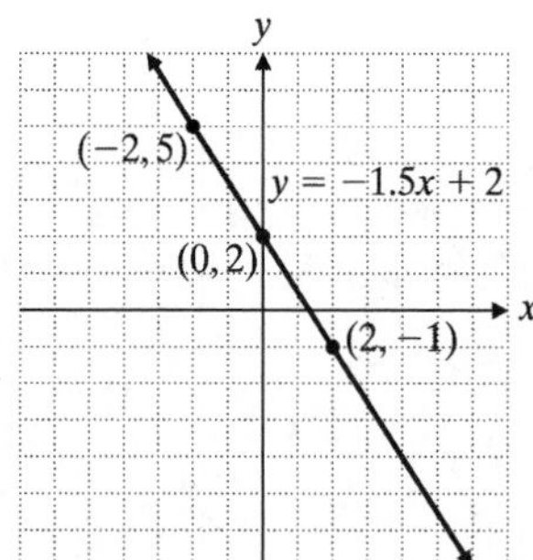

31. $2x + 5y = -5$

Let $x = -5$: $2(-5) + 5y = -5$
$y = 1$

Let $x = 0$: $2(0) + 5y = -5$
$y = -1$

Let $x = 5$: $2(5) + 5y = -5$
$y = -3$

x	y
−5	1
0	−1
5	−3

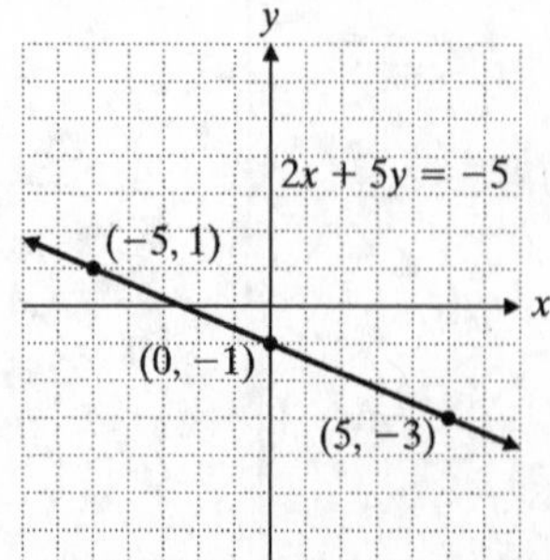

33. $5x + y + 4 = 8x$
$y = 3x - 4$

x	$y = 3x - 4$	y
0	$y = 3(0) - 4$	-4
1	$y = 3(1) - 4$	-1
2	$y = 3(2) - 4$	2

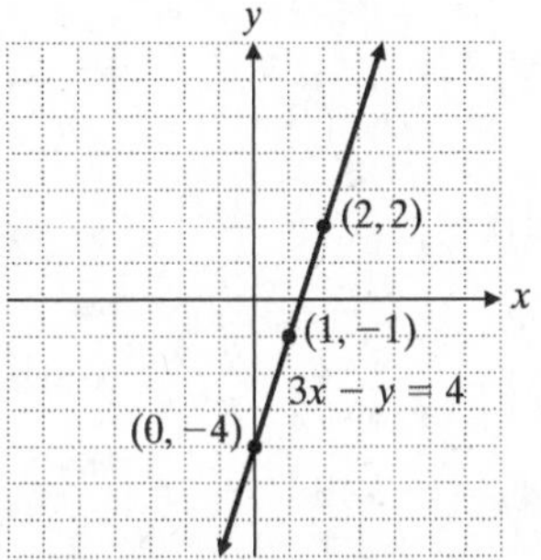

35. $y = 82x + 150$

x	$y = 82x + 150$	y
-1	$y = 82(-1) + 150$	68
0	$y = 82(0) + 150$	150
1	$y = 82(1) + 150$	232

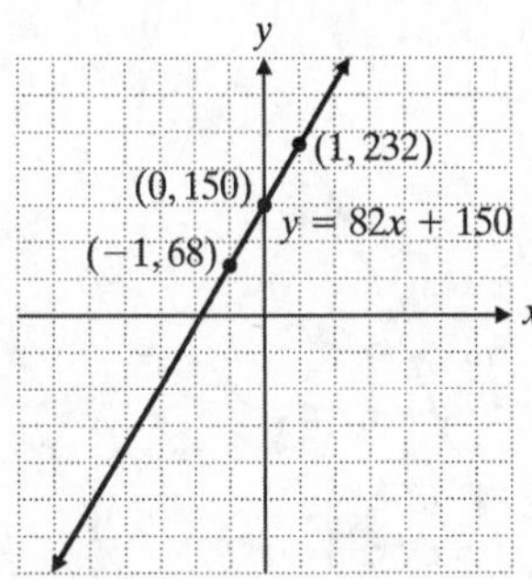

Vertical scale: 1 square = 50 units

37. From the table the greatest increase in median weekly earnings of men occurred between 1999 and 2001.

2 year period	increase
1997–1999	$619 - 579 = 40$
1999 – 2001	$670 - 619 = 51$
2001–2003	$695–670 = 25$
2003–2005	$722 - 695 = 27$
2005–2007	$766 - 722 = 44$

39. From the table the median weekly earnings of men and women had the largest difference in 2001.

year	difference
1997	$579 - 431 = 148$
1999	$619 - 473 = 146$
2001	$670 - 512 = 158$
2003	$695 - 552 = 143$
2005	$722 - 585 = 137$
2007	$766 - 614 = 152$

41. $(614 - 431) \div 431 \cdot 100\% = 42.5\%$
The percent increase was 42.5%.

43. a. $V(T) = 120 - 32T$

T	$V(T) = 120 - 32T$
0	$120 - 32(0) = 120$
1	$120 - 32(1) = 88$
2	$120 - 32(2) = 56$
3	$120 - 32(3) = 24$
4	$120 - 32(4) = -8$

b.

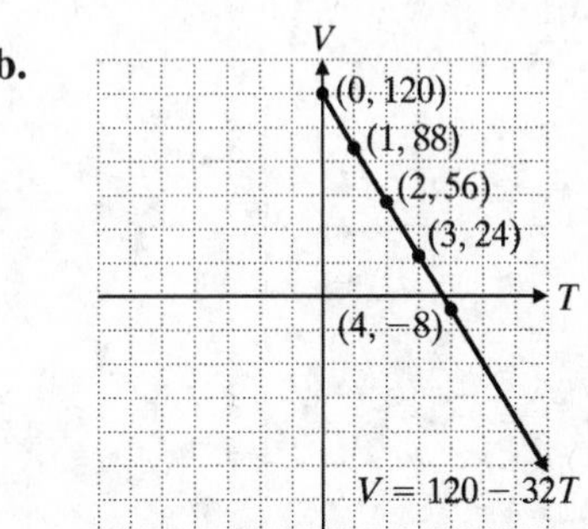

Vertical scale: 1 square = 20 units

c. The baseball is moving downward instead of upward at $T = 4$ seconds.

Cumulative Review

45. $36 \div (8-6)^2 + 3(-4) = 36 \div 2^2 + (-12)$
$= 36 \div 4 + (-12)$
$= 9 + (-12)$
$= -3$

46. $3(x-6)+2 \le 4(x+2)-21$
$3x-18+2 \le 4x+8-21$
$3x-16 \le 4x-13$
$3x-4x \le 16-13$
$-x \le 3$
$x \ge -3$

47. $R = 2G$
$B = 3R$
$2W = Y$
$2Y = R$
$W = 130$
$Y = 2W = 260$
$R = 2Y = 520$
$G = \frac{1}{2}R = 260$
$B = 3R = 1560$
There are 520 red, 1560 blue, 260 yellow, 260 green, and 130 white.

48. Let x = sales price of house.
$9100 = 0.07(100{,}000) + 0.03(x-100{,}000)$
$9100 = 7000 + 0.03x - 3000$
$0.03x = 5100$
$x = 170{,}000$
The selling price was $170,000.

Quick Quiz 3.1

1. $y = -\frac{2}{3}x + 4$

x	$y = -\frac{2}{3}x+4$	y
0	$y = -\frac{2}{3}(0)+4$	4
6	$y = -\frac{2}{3}(6)+4$	0
−3	$y = -\frac{2}{3}(-3)+4$	6

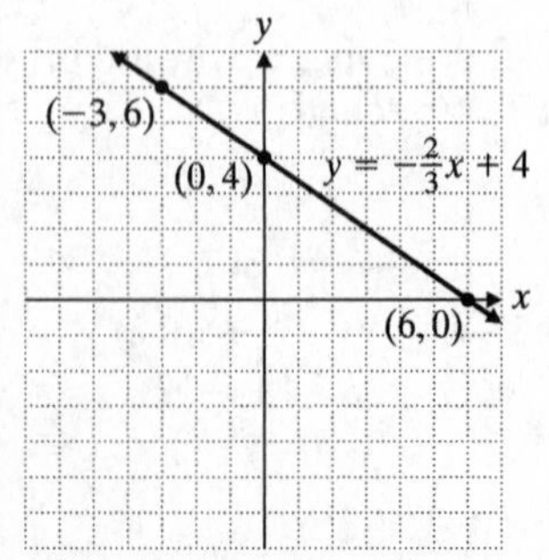

2. $7y - 5 = 16$
$7y = 21$
$y = 3$
This is a horizontal line.

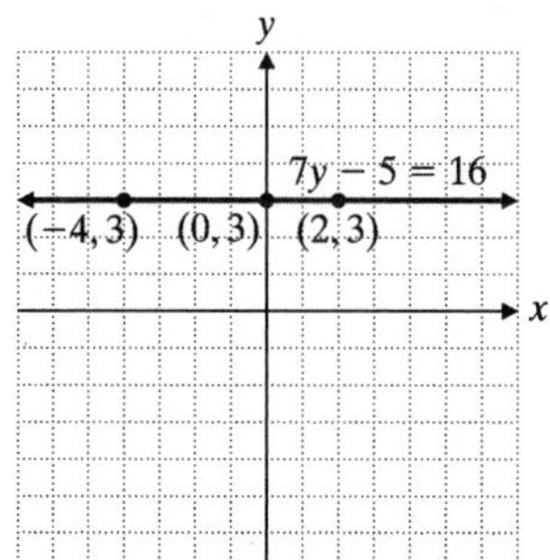

3. $-4x + 2y = -12$
Let $x = 0$: $-4(0) + 2y = -12$
$y = -6$
Let $y = 0$: $-4x + 2(0) = -12$
$x = 3$
Let $x = 2$: $-4(2) + 2y = -12$
$y = -2$

x	y
0	−6
3	0
2	−2

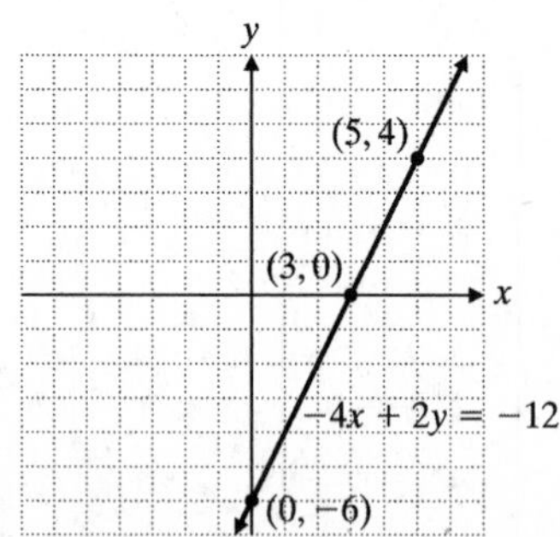

4. Answers may vary. Possible solution:
To find the x-intercept of $7x + 3y = -14$, set $y = 0$ and solve for x.
$7x + 3(0) = -14$
$7x = -14$
$x = -2$
To find the y-intercept of $7x + 3y = -14$, set

$x = 0$ and solve for y.

$$7(0)+3y=-14$$
$$3y=-14$$
$$y=-4\frac{2}{3}$$

3.2 Exercises

1. Slope measures vertical change (rise) versus horizontal change (run).

3. The slope of a horizontal line is 0.

5. $m=\frac{y_2-y_1}{x_2-x_1}=\frac{-7-5}{-3-(-3)}=\frac{-12}{0}$

The line passing through (–3, –7) and (–3, 5) does not have a slope because division by 0 is undefined.

7. $m=\frac{y_2-y_1}{x_2-x_1}$

$m=\frac{2-(-6)}{2-6}=\frac{8}{-4}=-2$

9. $m=\frac{y_2-y_1}{x_2-x_1}=\frac{4-0}{\frac{3}{2}-(-2)}=\frac{4}{\frac{7}{2}}=\frac{8}{7}$

11. $m=\frac{y_2-y_1}{x_2-x_1}=\frac{-1.5-(-2.3)}{6.8-5.6}=\frac{0.8}{1.2}=\frac{2}{3}$

13. $m=\frac{y_2-y_1}{x_2-x_1}=\frac{-2-\frac{1}{4}}{\frac{3}{2}-\frac{3}{2}}=\frac{-\frac{9}{4}}{0}$ undefined slope

15. $m=\frac{y_2-y_1}{x_2-x_1}=\frac{-3-(-3)}{-7-10}$

$m=\frac{0}{-17}=0$

17. $m=\frac{y_2-y_1}{x_2-x_1}=\frac{-2-(-4)}{-\frac{1}{3}-1}=\frac{2}{-\frac{4}{3}}$

$m=-\frac{3}{2}$

19. $m=\frac{\text{rise}}{\text{run}}=\frac{48\text{ ft}}{80\text{ ft}}=\frac{3}{5}=0.6$

The grade of the hill is $\frac{3}{5}$ or 0.6.

21. $m=\frac{\text{rise}}{\text{run}}=\frac{35.7}{142.8}=0.25$ or $\frac{1}{4}$

The pitch of the rock is 0.25 or $\frac{1}{4}$.

23. $m = 0.35$

$0.35=\frac{\text{rise}}{120}$

rise = 42

It rises vertically 42 feet.

25. $m=\frac{y_2-y_1}{x_2-x_1}=\frac{7-3}{6-24}=\frac{4}{-18}=-\frac{2}{9}$

$m_{||}=m=-\frac{2}{9}$

27. $m=\frac{y_2-y_1}{x_2-x_1}=\frac{4-2}{5-5.5}=\frac{2}{-0.5}$

$m=-4$

$m_{||}=m=-4$

29. $m=\frac{y_2-y_1}{x_2-x_1}$

$m=\frac{\frac{1}{2}-5}{-9-(-6)}=\frac{-\frac{9}{2}}{-3}=\frac{3}{2}$

$m_{||}=m=\frac{3}{2}$

31. $m=\frac{y_2-y_1}{x_2-x_1}$

$m=\frac{12-9}{8-3}=\frac{3}{5}$

$m_\perp$ is the negative reciprocal of m.

$m_\perp=-\frac{5}{3}$

33. $m=\frac{y_2-y_1}{x_2-x_1}=\frac{\frac{1}{2}-\left(-\frac{3}{2}\right)}{3-2}$

$m=\frac{\frac{1}{2}+\frac{3}{2}}{1}$

$m=2$

$m_\perp$ is the negative reciprocal of m.

$m_\perp=-\frac{1}{2}$

35. $m = \dfrac{y_2 - y_1}{x_2 - x_1}$

$m = \dfrac{0 - 4.2}{-8.4 - 0} = \dfrac{1}{2}$

$m_\perp$ is the negative reciprocal of m.

$m_\perp = -2$

37. $m_k = \dfrac{-9 - 11}{-3 - 1} = 5$

$m_h = \dfrac{-13 - 7}{-2 - 2} = 5$

Lines k and h are parallel because they have the same slope.

39. To be a parallelogram, AD must be parallel to BC and AB must be parallel to DC.

$m_{AD} = \dfrac{1 - 2}{2 - (-4)} = -\dfrac{1}{6}$

$m_{BC} = \dfrac{-2 - (-1)}{-1 - (-7)} = -\dfrac{1}{6}$

$m_{AB} = \dfrac{1 - (-2)}{2 - (-1)} = 1$

$m_{CD} = \dfrac{-1 - 2}{-7 - (-4)} = 1$

Since $m_{AD} = m_{BC}$ and $m_{AB} = m_{CD}$ the opposite sides are parallel and $ABCD$ is a parallelogram.

41. a. $m = \dfrac{\text{rise}}{\text{run}} = \dfrac{5}{60} = \dfrac{1}{12}$

The slope of the ramp is $\dfrac{1}{12}$.

b. $\dfrac{1}{12} = \dfrac{\text{rise}}{24}$, $24 = 12 \cdot \text{rise}$

$12 \cdot \text{rise} = 24 \Rightarrow \text{rise} = \dfrac{24}{2} = 2$

The maximum height is 2 feet.

c. $\dfrac{1}{12} = \dfrac{1.7}{\text{run}}$, run = 20.4

The ramp will have a horizontal distance of 20.4 feet.

Cumulative Review

43. $\dfrac{5 + 3\sqrt{9}}{|2 - 9|} = \dfrac{5 + 3(3)}{|-7|} = \dfrac{5 + 9}{7} = \dfrac{14}{7} = 2$

44. $$\begin{aligned} 2(3-6)^3 + 20 \div (-10) &= 2(-3)^3 + 20 \div (-10) \\ &= 2(-27) + 20 \div (-10) \\ &= -54 + 20 \div (-10) \\ &= -54 + (-2) \\ &= -56 \end{aligned}$$

45. $\dfrac{-15x^6y^3}{-3x^{-4}y^6} = \dfrac{5x^{6+4}}{y^{6-3}} = \dfrac{5x^{10}}{y^3}$

46. $$\begin{aligned} 8x(x-1) - 2(x+y) &= 8x^2 - 8x - 2x - 2y \\ &= 8x^2 - 10x - 2y \end{aligned}$$

Quick Quiz 3.2

1. $m = \dfrac{y_2 - y_1}{x_2 - x_1}$

$m = \dfrac{-2 - \frac{1}{2}}{-7 - 3}$

$m = \dfrac{-\frac{5}{2}}{-10}$

$m = \dfrac{1}{4}$

2. $m = \dfrac{y_2 - y_1}{x_2 - x_1}$

$m = \dfrac{-3 - (-3)}{-8 - 5}$

$m = \dfrac{0}{-13}$

$m = 0$

3. $m = \dfrac{y_2 - y_1}{x_2 - x_1} = \dfrac{4 - (-7)}{19 - (-3)}$

$m = \dfrac{11}{22}$

$m = \dfrac{1}{2}$

The slope of a line perpendicular is the negative reciprocal.

$m = -2$

4. Answers may vary. Possible solution:

$m = \dfrac{\text{rise}}{\text{run}}$

$m = \dfrac{8.5}{120}$

$m \approx 0.07$ or 7%

3.3 Exercises

1. First determine the slope from the coordinates of the points. Then substitute the slope and the coordinates of one of the points into the point-slope form of the equation of a line. This may be rewritten in standard form or slope-intercept form.

3. $y = mx + b,\ y = \frac{3}{4}x - 9$

5. $y = mx + b \Rightarrow y = \frac{3}{4}x + \frac{1}{2}$
   $4y = 3x + 2 \Rightarrow 3x + 2 = 4y \Rightarrow 3x = 4y - 2$
   $3x - 4y = -2$

7. From the graph, $m = \frac{1}{2},\ b = 3.$
   $y = \frac{1}{2}x + 3$

9. From the graph, $m = -1, b = 3.$
   $y = -x + 3$

11. From the graph, $m = 3, b = 0.$
    $y = 3x$

13. From the graph, $m = -\frac{3}{2},\ b = 0.$
    $y = -\frac{3}{2}x$

15. From the graph, $m = \frac{3}{4},\ b = -4.$
    $y = \frac{3}{4}x - 4$

17. $x - y = 5$
    $y = x - 5,\ m = 1,\ b = -5,$ y-intercept $(0, -5)$

19. $5x - 4y = -20$
    $4y = 5x + 20$
    $y = \frac{5}{4}x + 5,\ m = \frac{5}{4},\ b = 5,$
    y-intercept $(0, 5)$

21. $2x + \frac{3}{4}y = -3$
    $\frac{3}{4}y = -2x - 3$
    $y = \frac{4}{3}(-2x) - 3\left(\frac{4}{3}\right)$
    $y = -\frac{8}{3}x - 4$
    $m = -\frac{8}{3}$
    $b = -4,$ y-intercept $(0, -4)$

23. $y = \frac{1}{2}x - 3,\ m = \frac{1}{2},\ b = -3,$ y-intercept $(0, -3)$

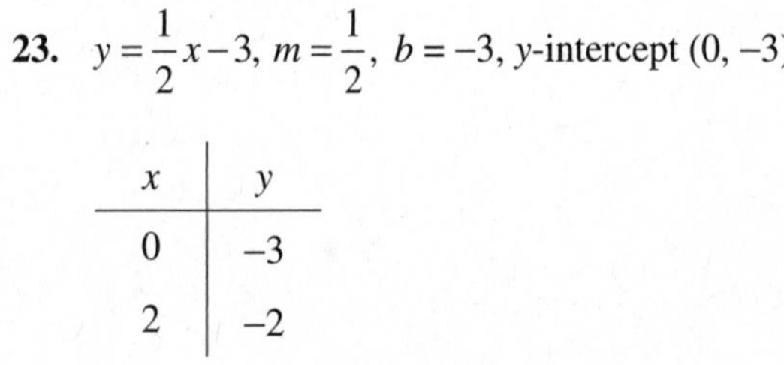

x	y
0	−3
2	−2

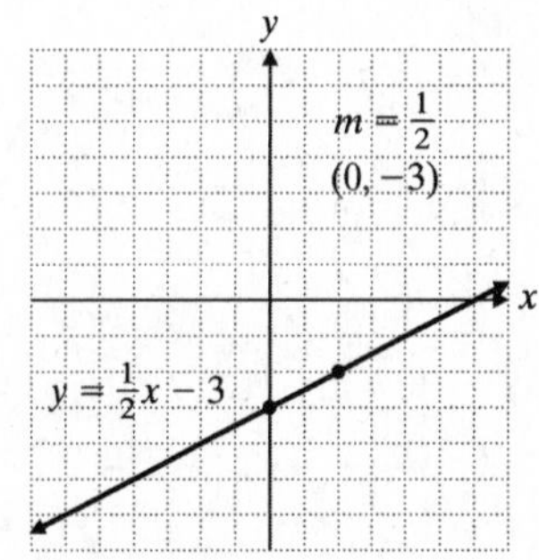

25. $5x + 3y = 18$
    $3y = -5x + 18$
    $y = -\frac{5}{3}x + 6$
    $m = -\frac{5}{3},\ b = 6,$ y-intercept $(0, 6)$

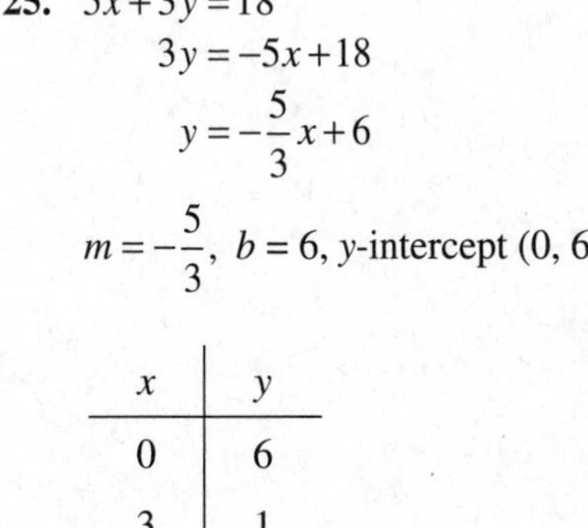

x	y
0	6
3	1

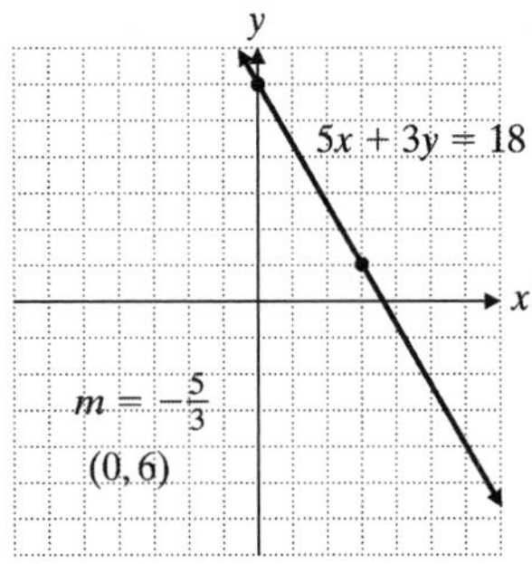

27. $y - y_1 = m(x - x_1)$
$y - 4 = -\frac{1}{2}(x - 4)$
$y - 4 = -\frac{1}{2}x + 2$
$y = -\frac{1}{2}x + 6$

29. $y - y_1 = m(x - x_1)$
$y - (-2) = 5(x - (-7))$
$y + 2 = 5x + 35$
$y = 5x + 33$

31. $y - y_1 = m(x - x_1)$
$y - 0 = -\frac{1}{5}(x - 6)$
$y = -\frac{1}{5}x + \frac{6}{5}$

33. $m = \frac{y_2 - y_1}{x_2 - x_1}$
$m = \frac{4+1}{3+4} = \frac{5}{7}$
$y - y_1 = m(x - x_1)$
$y - 4 = \frac{5}{7}(x - 3)$
$y - 4 = \frac{5}{7}x - \frac{15}{7}$
$y = \frac{5}{7}x + \frac{13}{7}$

35. $m = \frac{y_2 - y_1}{x_2 - x_1}$
$m = \frac{-3-(-5)}{\frac{1}{2} - \frac{7}{2}} = -\frac{2}{3}$
$y - y_1 = m(x - x_1)$
$y + 3 = -\frac{2}{3}\left(x - \frac{1}{2}\right)$
$y + 3 = -\frac{2}{3}x + \frac{1}{3}$
$y = -\frac{2}{3}x - \frac{8}{3}$

37. $m = \frac{y_2 - y_1}{x_2 - x_1}$
$m = \frac{-3-(-3)}{12-7} = 0$
$y - y_1 = m(x - x_1)$
$y - (-3) = 0(x - 12)$
$y + 3 = 0$
$y = -3$

39. $5x - y = 4$
$y = 5x - 4,\ m = 5$
$m_{\parallel} = 5$
$y - y_1 = m(x - x_1)$
$y - 0 = 5(x - (-2))$
$y = 5x + 10$
$5x - y = -10$

41. $x = 3y - 8$
$3y = x + 8,\ y = \frac{1}{3}x + \frac{8}{3},\ m = \frac{1}{3}$
$m_{\parallel} = \frac{1}{3}$
$y - (-1) = \frac{1}{3}(x - 5)$
$3y + 3 = x - 5$
$x - 3y = 8$

43. $2y = -3x,\ y = -\frac{3}{2}x,\ m = -\frac{3}{2}$
$m_{\parallel} = \frac{2}{3}$
$y - (-1) = \frac{2}{3}(x - 6)$
$3y + 3 = 2x - 12$
$2x - 3y = 15$

45. $x+7y=-12,\ y=-\frac{1}{7}x-\frac{12}{7},\ m=-\frac{1}{7}$

$m_{\perp}=7$

$y-(-1)=7(x+4)$

$y+1=7x+28$

$7x-y=-27$

47. $-3x+5y=40,\ y=\frac{3}{5}x+8,\ m_1=\frac{3}{5}$

$5y+3x=17,\ y=-\frac{3}{5}x+\frac{17}{3},\ m_2=-\frac{3}{5}$

$m_1m_2=\frac{3}{5}\left(-\frac{3}{5}\right)=-\frac{9}{25}$

The lines are neither parallel nor perpendicular.

49. $y=-\frac{3}{4}x-2,\ m_1=-\frac{3}{4}$

$6x+8y=-5$

$8y=-6x-5$

$y=-\frac{3}{4}x-\frac{5}{8}$

$m_2=-\frac{3}{4}$

$m_1=m_2 \Rightarrow$ lines are parallel

51. $y=\frac{5}{6}x-\frac{1}{3},\ m_1=\frac{5}{6}$

$6x+5y=-12$

$5y=-6x-12$

$y=-\frac{6}{5}x-\frac{12}{5}$

$m_2=-\frac{6}{5}$

$m_1\neq m_2$

$m_1m_2=\frac{5}{6}\left(-\frac{6}{5}\right)=-1$

The lines are perpendicular.

53. $y = 1.43x - 2.17,\ y = 1.43x + 0.39$

3
−3 3
−3

Lines appear parallel.

55. $y = 10.7x + 133.9$

$y = 10.7(20) + 153.9$

$y = 347{,}900$

The expected sale price is $347,900.

57. Answers may vary; from the graph, the estimate is approximately $294,000.

Cumulative Review

58. $11-(x+2)=7(3x+6)$

$11-x-2=21x+42$

$22x=-33$

$x=-\frac{3}{2}$

59. $0.3x+0.1=0.27x-0.02$

$30x+10=27x-2$

$3x=-12$

$x=-4$

60. $70+70(0.01x)+3=82.10$

$70+0.7x+3=82.1$

$0.7x=9.1$

$x=13$

61. $\frac{5}{4}-\frac{3}{4}(2x+1)=x-2$

$5-3(2x+1)=4x-8$

$5-6x-3=4x-8$

$10x=10$

$x=1$

Quick Quiz 3.3

1. $-3x+7y=-9$

$7y=3x-9$

$y=\frac{3}{7}x-\frac{9}{7}$

$m=\frac{3}{7},\ b=-\frac{9}{7}$

y-intercept $\left(0,-\frac{9}{7}\right)$

2. $m=\frac{y_2-y_1}{x_2-x_1}=\frac{2-(-7)}{-3-(-4)}=\frac{9}{1}=9$

$y-y_1=m(x-x_1)$

$y-2=9(x+3)$

$y-2=9x+27$

$y=9x+29$

$9x-y=-29$

3. $m = -\frac{3}{4}$

y-intercept $(0, -5)$; $b = -5$

$y = -\frac{3}{4}x - 5$

4. Answers may vary. Possible solution:
Because the line is horizontal the slope is 0, the y-intercept is $(0, -8)$.

$y = 0(x) - 8$

$y = -8$

How Am I Doing? Sections 3.1–3.3

1.
$$5x + 2y = -12$$
$$5a + 2(6) = -12$$
$$5a + 12 = -12$$
$$5a = -24$$
$$a = -\frac{24}{5} = -4.8$$

2. $y = -\frac{1}{2}x + 5$

x	$y = -\frac{1}{2}x + 5$	y
0	$y = -\frac{1}{2}(0) + 5$	5
2	$y = -\frac{1}{2}(2) + 5$	4
4	$y = -\frac{1}{2}(4) + 5$	3

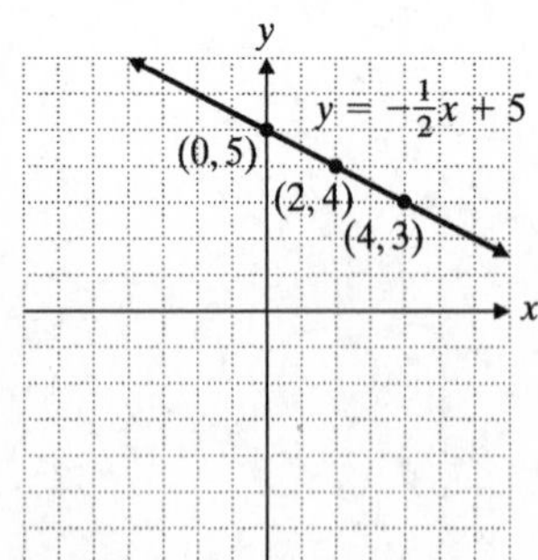

3. $5x + 3y = -15$

Let $x = -6$: $5(-6) + 3y = -15$
$$y = 5$$
Let $y = 0$: $5x + 3(0) = -15$
$$x = -3$$
Let $x = 0$: $5(0) + 3y = -15$
$$y = -5$$

x	y
−6	5
−3	0
0	−5

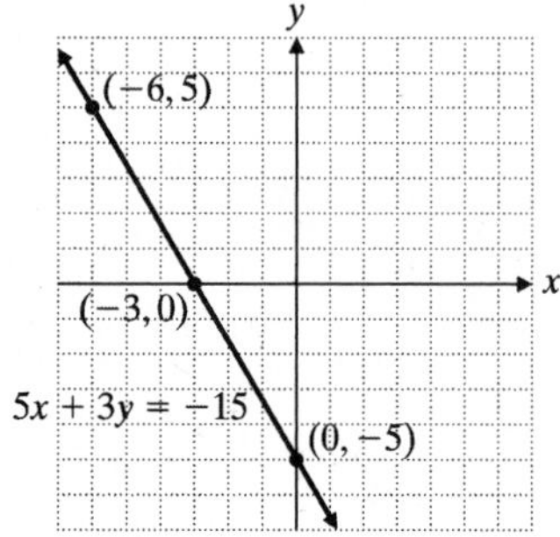

4.
$$4y + 6x = -8 + 9x$$
$$4y = 3x - 8$$
$$y = \frac{3}{4}x - 2$$

x	$y = \frac{3}{4}x - 2$	y
−4	$y = \frac{3}{4}(-4) - 2$	−5
0	$y = \frac{3}{4}(0) - 2$	−2
4	$y = \frac{3}{4}(4) - 2$	1

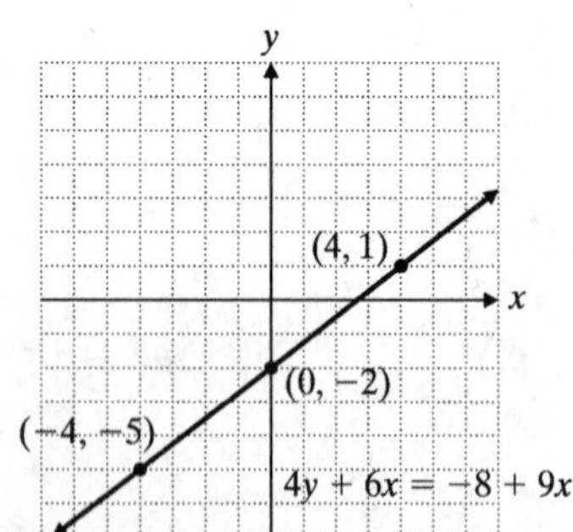

5. $m = \frac{y_2 - y_1}{x_2 - x_1}$

$m = \frac{3 - (-6)}{-2 - (-1)} = \frac{9}{-1}$

$m = -9$

6. $m = \frac{y_2 - y_1}{x_2 - x_1}$

$m = \frac{4-(-2)}{\frac{2}{3}-\frac{5}{6}} = \frac{6}{-\frac{1}{6}}$

$m = -36$

$m_{\parallel} = m = -36$

7. $m = \frac{y_2 - y_1}{x_2 - x_1}$

$m = \frac{0-(-4)}{5-(-13)} = \frac{4}{18}$

$m = \frac{2}{9}$

$m_{\perp}$ is the negative reciprocal of m.

$m_{\perp} = -\frac{9}{2}$

8. $m = 0.13 = \frac{\text{rise}}{\text{run}}$

$0.13 = \frac{\text{rise}}{500 \text{ ft}}$

$\text{rise} = 0.13(500)$

$\text{rise} = 65$

The road rises vertically 65 feet.

9. $5x + 3y = 9$

$3y = -5x + 9$

$y = -\frac{5}{3}x + 3$

$m = -\frac{5}{3},\ b = 3$

y-intercept (0, 3)

10. $y - y_1 = m(x - x_1)$

$y - (-3) = -2(x - 7)$

$y + 3 = -2x + 14$

$y = -2x + 11$

11. $3x - 5y = 10$

$5y = 3x - 10$

$y = \frac{3}{5}x - 2,\ m = \frac{3}{5},\ m_{\perp} = -\frac{5}{3}$

$y - y_1 = m(x - x_1)$

$y - (-2) = -\frac{5}{3}(x - (-1))$

$3y + 6 = -5x - 5$

$3y = -5x - 11$

$y = -\frac{5}{3}x - \frac{11}{3}$

12. $m = \frac{y_2 - y_1}{x_2 - x_1} = \frac{-10-2}{-3-1}$

$m = \frac{-12}{-4}$

$m = 3$

$y - y_1 = m(x - x_1)$

$y + 10 = 3(x + 3)$

$y = 3x + 9 - 10$

$y = 3x - 1$

3.4 Exercises

1. You need to use a dashed line when graphing a linear inequality that contains the > or < symbols.

3. To graph the region $x > 5$ you should shade the region to the right of the line $x = 5$.

5. When graphing $3x - 2y \geq 0$ the point (0, 0) cannot be used as a test point because it is on the line. Using (–4, 2) as a test point gives $3x - 2y = 3(-4) - 2(2) = -16 \geq 0$ which is a false statement.

7. $y > -2x + 4$, graph $y = -2x + 4$ using a dashed line. Test point: (0, 0)
$0 > -2(0) + 4,\ 0 > 4$ False
Shade the region not containing (0, 0).

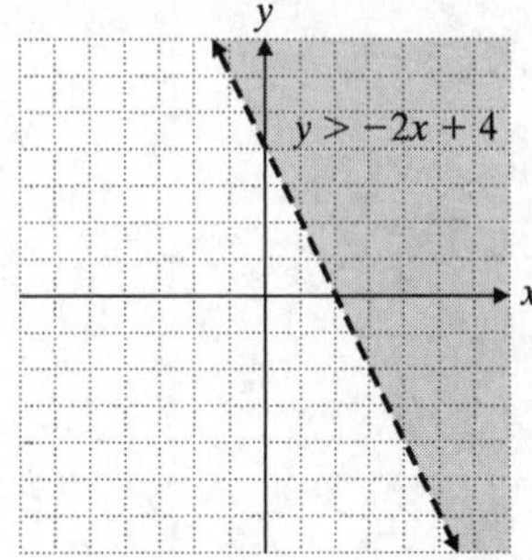

9. $y < \frac{2}{3}x - 2$, graph $y = \frac{2}{3}x - 2$ using a dashed line. Test point: (0, 0)

$0 < \frac{2}{3}(0) - 2,\ 0 < -2$ False

Shade the region not containing (0, 0).

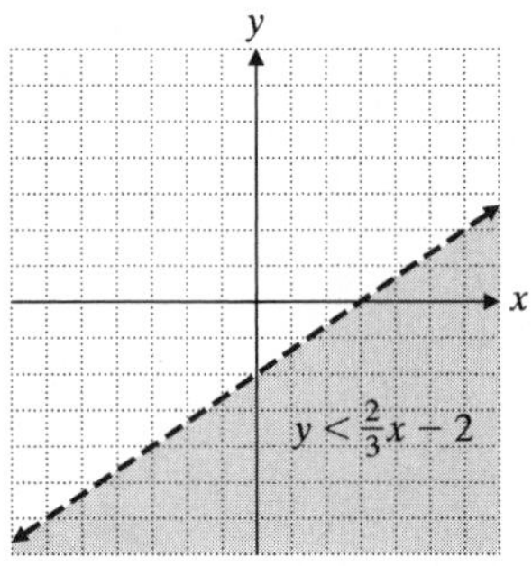

11. $y \geq \frac{3}{4}x + 4$

Graph $y = \frac{3}{4}x + 4$ using a solid line.

x	y
0	4
$-\frac{16}{3}$	0

Test point: (–4, 2)

$2 \geq \frac{3}{4}(-4) + 4$

$2 \geq -3 + 4$

$2 \geq 1$ True

Shade the region containing (–4, 2).

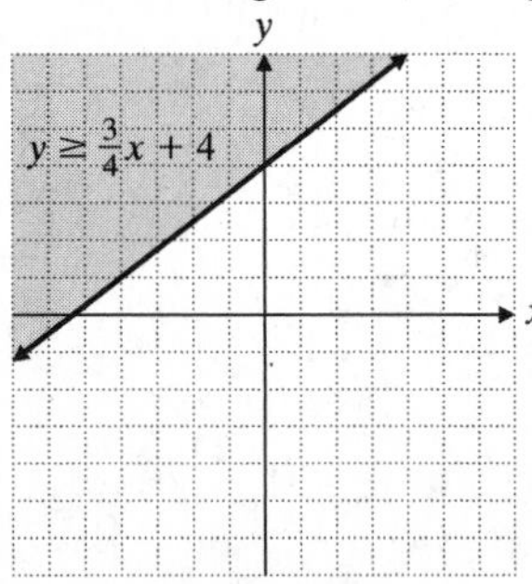

13. $5y - x \leq 15$

Graph $5y - x = 15$ using a solid line.

Test point: (0, 0)

$5(0) - 0 \leq 15$

$0 \leq 15$ True

Shade the region containing (0, 0).

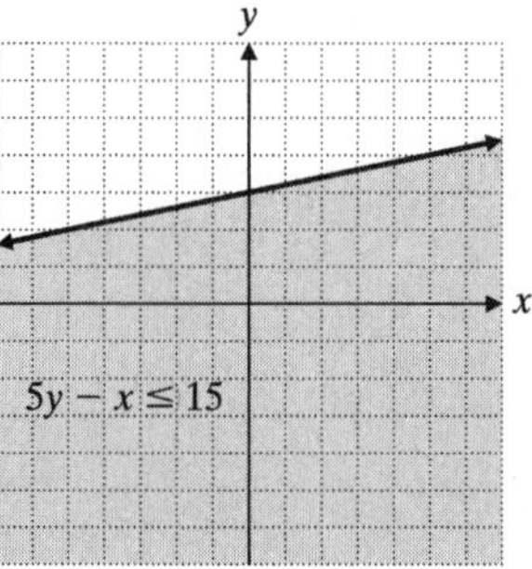

15. $-3x + y > 0$

Graph $-3x + y = 0$ using a dashed line.

x	y
0	0
–1	–3

Test point: (–2, 2)

$-3(-2) + 2 > 0$

$6 + 2 > 0$ True

Shade the region containing (–2, 2).

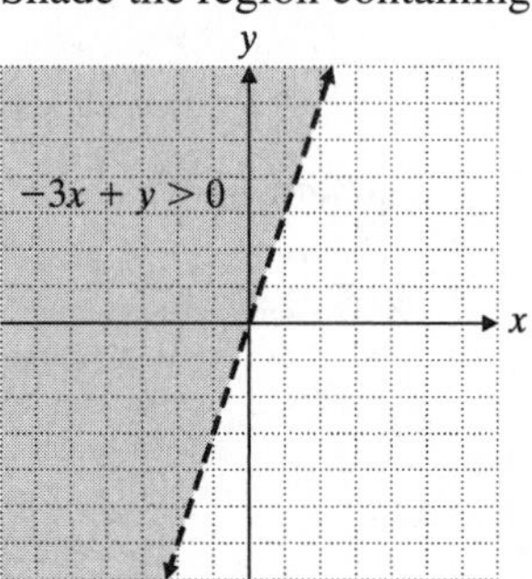

17. $5x - 2y \geq 0$

Graph $5x - 2y = 0$ using a solid line.

Test point: (1, 1)

$5(1) - 2(1) \geq 0$

$3 \geq 0$ True

Shade the region containing (1, 1).

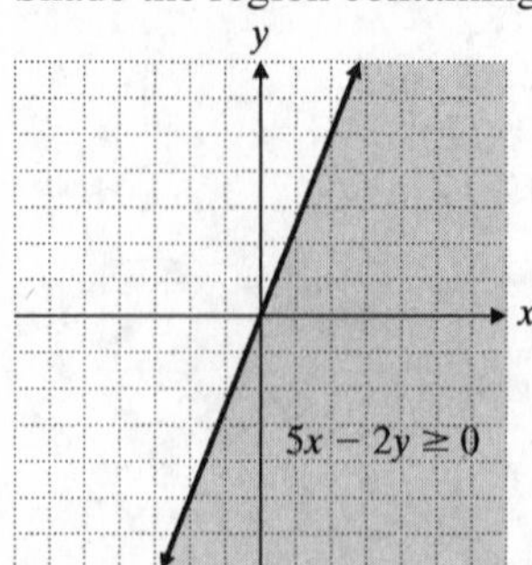

19. $x > -4$
Graph $x = 4$ using a dashed line. Shade to the right of the line.

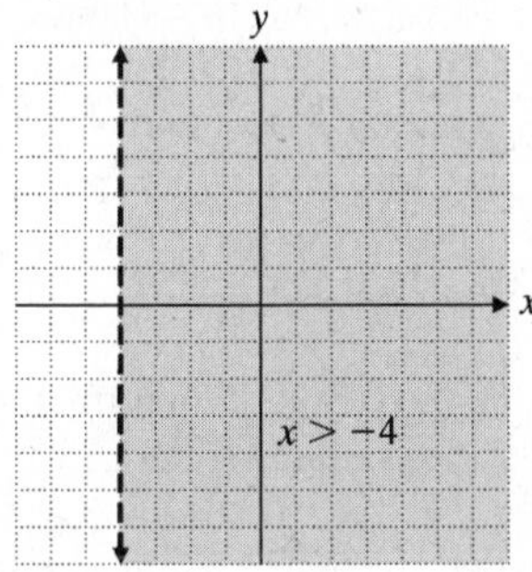

21. $y \le -1$
Graph $y = -1$ using a solid line. Shade below the line.

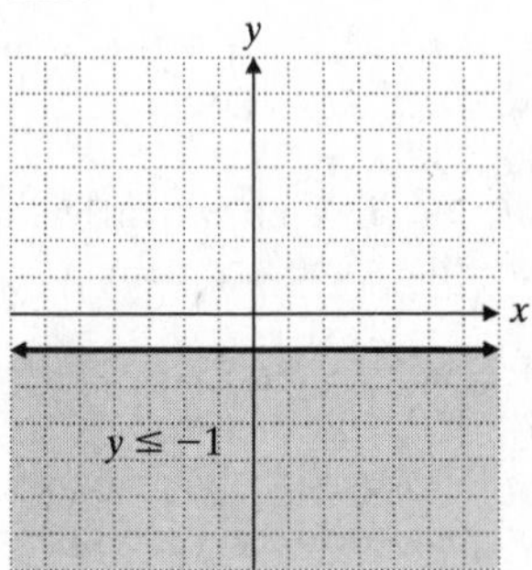

23. $-8x \le -12,\ x \ge \frac{3}{2}$

Graph $x = \frac{3}{2}$ using a solid line. Shade to the right of the line.

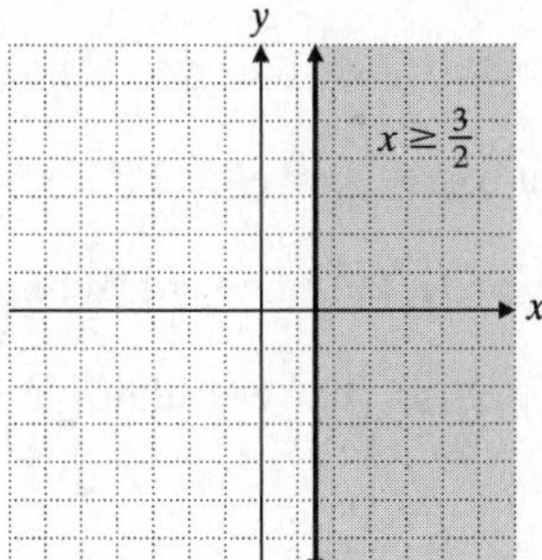

25. $4y \ge 2,\ y \ge \frac{1}{2}$

Graph $y = \frac{1}{2}$ using a solid line. Shade above the line.

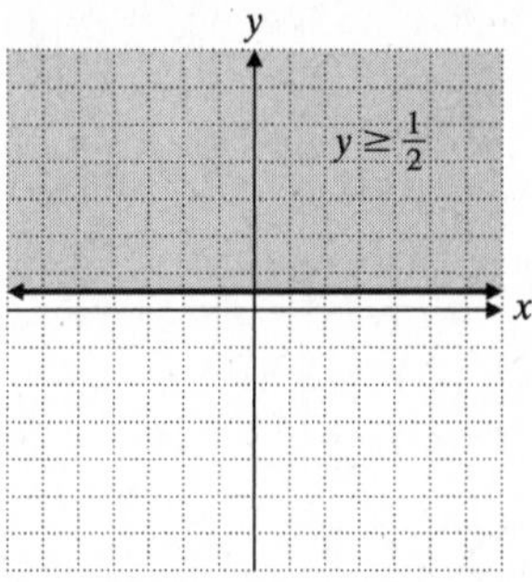

Cumulative Review

27. Let $x = -3$

$$\begin{aligned} 2x^2 - 5x + 4 &= 2(-3)^2 - 5(-3) + 4 \\ &= 18 + 15 + 4 \\ &= 37 \end{aligned}$$

28. $A = \frac{1}{2}a(b+c)$

$A = \frac{1}{2}(4)(1.5+7.5)$

$A = \frac{1}{2}(4)(9)$

$A = 18$

29. $\frac{20(18)}{32} = 11.25$ which requires 12 packages
They will need to purchase 12 packages.

Quick Quiz 3.4

1. $y > 3x + 4$
Test point (3, –4)

$-4 \overset{?}{>} 3(3) + 4$

$-4 \overset{?}{>} 13$

false
The graph of the region lies above the line.

2. $y < -\frac{3}{5}x + 3$

Graph $y = -\frac{3}{5}x + 3$ using a dashed line.

x	y
0	3
5	0

Test point: (0, 0)

$0 < -\frac{3}{5}0 + 3$

$0 < 3$ True

Shade the region containing (0, 0).

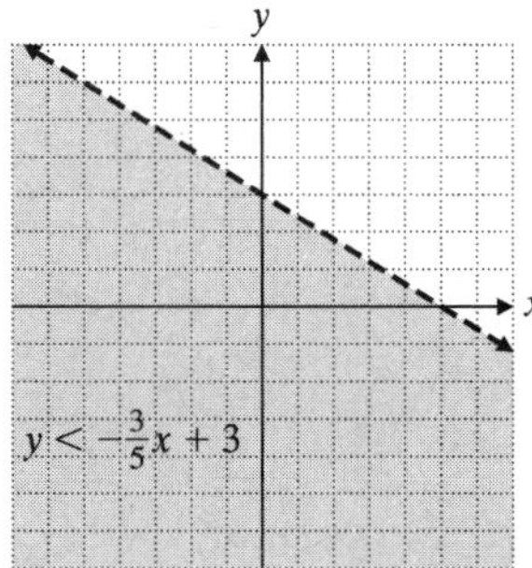

3. $3x - 5y \le -10$

Graph $3x - 5y = -10$ using a solid line.

x	y
0	2
$-\frac{10}{3}$	0

Test point: (0, 0)

$3(0) - 5(0) \le -10$

$0 \le -10$ False

Shade the region not containing (0, 0).

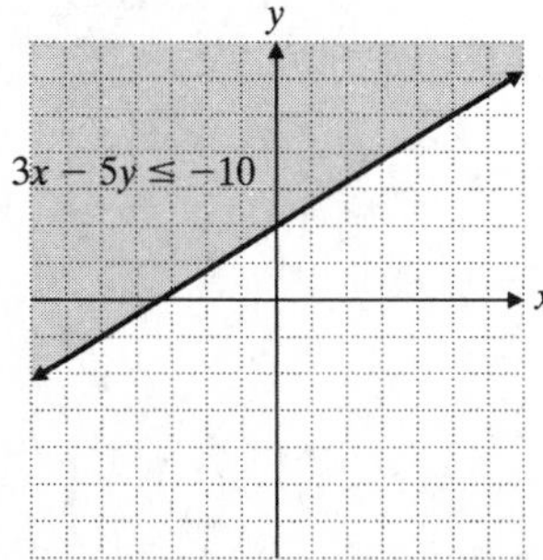

4. Answers may vary. Possible solution:
First simplify the inequality.

$4y - 8 < 0$

$4y < 8$

$y < 2$

The graph lies below a horizontal line of $y = 2$.

3.5 Exercises

1. A relation is any set of ordered pairs. A function is a set of ordered pairs in which no two different ordered pairs have the same first coordinate.

3. A function may be described as a set of ordered pairs, as an equation, and as a graph.

5. Domain = {0, 5, 7}
Range = {0, 11, 13}
Relation is *not* a function since (5, 13) and (5, 0) have the same first coordinate.

7. Domain = {85, 16, −102, 62}
Range = {−12, 4, 48}
Relation is a function since no two different ordered pairs have the same first item.

9. Domain = {6, 8, 10, 12, 14}
Range = {38, 40, 42, 44, 46}
Relation is a function since no two different ordered pairs have the same first item.

11. Domain = {Jan., Feb., Mar., April, May, June, July, Aug., Sept., Oct., Nov., Dec.}
Range = {81, 80, 79}
Relation is a function since no two different ordered pairs have the same first item.

13. Domain = {Chicago, New York}
Range = {1454, 1350, 1250, 1136, 1127, 1046}
Relation is *not* a function since there exists two different ordered pairs with the same first item.

15. Domain = {10, 20, 30, 40, 50}
Range = {11.51, 23.02, 34.53, 46.04, 57.55}
Relation is a function since no two different ordered pairs have the same first item.

17. Function since it passes the vertical line test.

19. Not a function since it fails the vertical line test.

21. Function since it passes the vertical line test.

23. Not a function since it fails the vertical line test.

25. Function since it passes the vertical line test.

27. $g(x) = 2x - 5$
$g(-2.4) = 2(-2.4) - 5 = -4.8 - 5 = -9.8$

29. $g(x) = 2x - 5$
$g\left(\frac{1}{2}\right) = 2\left(\frac{1}{2}\right) - 5 = 1 - 5 = -4$

31. $y(x) = \frac{3}{4}x + 1$

$h(-8) = \frac{3}{4}(-8) + 1$

$h(-8) = -6 + 1$

$h(-8) = -5$

33. $h(x) = \frac{3}{4}x + 1$

$h(3) = \frac{3}{4}(3) + 1$

$h(3) = \frac{9}{4} + \frac{4}{4}$

$h(3) = \frac{13}{4}$

35. $r(x) = 2x^2 - 4x + 1$

$r(-1) = 2(-1)^2 - 4(-1) + 1 = 2 + 4 + 1 = 7$

37. $r(x) = 2x^2 - 4x + 1$

$r(-0.1) = 2(-0.1)^2 - 4(-0.1) + 1$

$r(-0.1) = 0.02 + 0.4 + 1$

$r(-0.1) = 1.42$

39. $t(x) = x^3 - 3x^2 + 2x - 3$

$t(4) = 4^3 - 3(4)^2 + 2(4) - 3$

$t(4) = 64 - 48 + 8 - 3$

$t(4) = 21$

41. $t(x) = x^3 - 3x^2 + 2x - 3$

$t(-2) = (-2)^3 - 3(-2)^2 + 2(-2) - 3$

$t(-2) = -8 - 3(4) - 4 - 3$

$t(-2) = -8 - 12 - 4 - 3$

$t(-2) = -27$

43. $f(x) = \sqrt{2 - x}$

$f(-2) = \sqrt{2 - (-2)}$

$f(-2) = \sqrt{4}$

$f(-2) = 2$

45. $g(x) = |6 - x^2|$

$g(2) = |6 - 2^2|$

$g(2) = |6 - 4|$

$g(2) = |2| = 2$

47. $g(x) = x^2 + 3$

Substitute given domain values for x to find $f(x)$.

x	$g(x) = x^2 + 3$	$g(x) = x^2 + 3$
-2	$g(-2) = (-2)^2 + 3$	7
-1	$g(-1) = (-1)^2 + 3$	4
0	$g(0) = 0^2 + 3$	3
1	$g(1) = 1^2 + 3$	4
2	$g(2) = 2^2 + 3$	7

Range = {3, 4, 7}

49. $d(x) = 4 - \frac{1}{3}x$

$y = 4 - \frac{1}{3}x \Rightarrow x = -3y + 12$

Substitute given range values for y to find x.

$x = -3y + 12$	$x = -3y + 12$	y
12	$x = -3(0) + 12$	0
8	$x = -3\left(\frac{4}{3}\right) + 12$	$\frac{4}{3}$
0	$x = -3(4) + 12$	4
-3	$x = -3(5) + 12$	5

Domain = {12, 8, 0, −3}

Cumulative Review

50. $|3x - 2| = 1$

$3x - 2 = 1$ and $3x - 2 = -1$

$3x = 1 + 2$ $\quad$ $3x = -1 + 2$

$3x = 3$ $\quad$ $3x = 1$

$x = 1$ $\quad$ $x = \frac{1}{3}$

51. $|x - 5| \le 3$

$-3 \le x - 5 \le 3$

$-3 + 5 \le x - 5 + 5 \le 3 + 5$

$2 \le x \le 8$

Quick Quiz 3.5

1. Domain = {9, −2, 4, −4}
   Range = {3, 5}

2. $p(x)=\left|-\frac{3}{4}x+2\right|$

   $p(-8)=\left|-\frac{3}{4}(-8)+2\right|=|6+2|=|8|=8$

3. $f(x)=-2x^3+4x^2-x+4$
   $f(-2)=-2(-2)^3+4(-2)^2-(-2)+4$
   $f(-2)=16+16+2+4$
   $f(-2)=38$

4. Answers may vary. Possible solution:
   If a vertical line can intersect the graph of a relation more than once, the relation is not a function. If no such line can be drawn, the relation is a function.

3.6 Exercises

1. $f(x)=\frac{3}{4}x+2$

x	$f(x)$
−4	−1
0	2
4	5

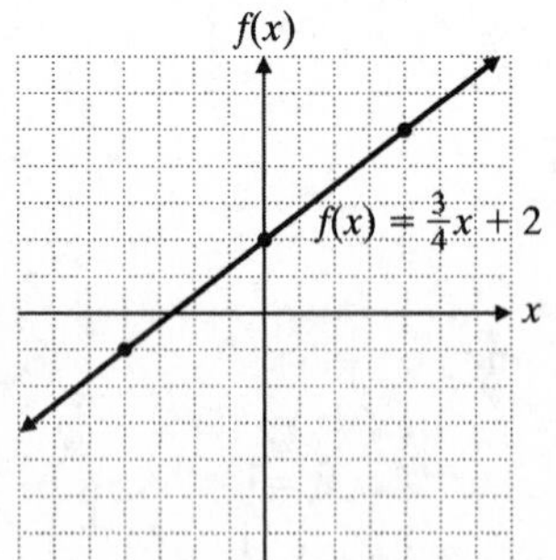

3. $f(x) = -3x - 1$

x	$f(x)$
−1	2
0	−1
1	−4

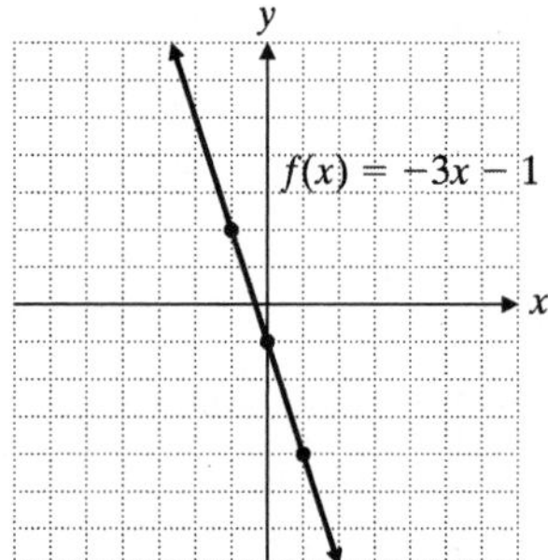

5. $C(x) = 0.15x + 25$

x	$C(x)$
0	25
100	40
200	55
300	70

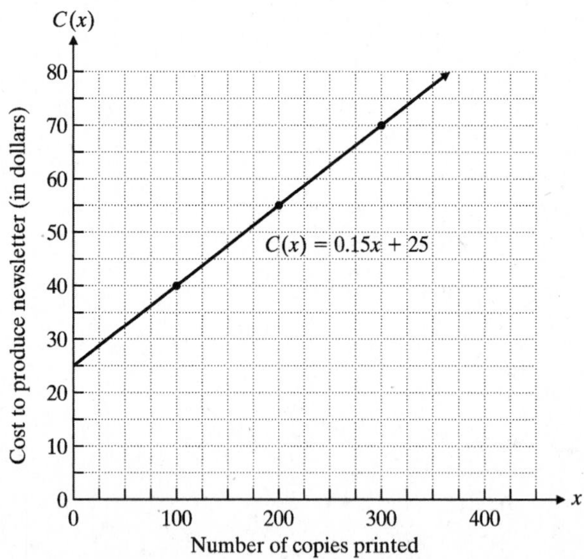

7. $P(x) = -1500x + 45{,}000$

x	$P(x)$
0	45,000
10	30,000
30	0

With 30 tons of pollutants, there will be no fish.

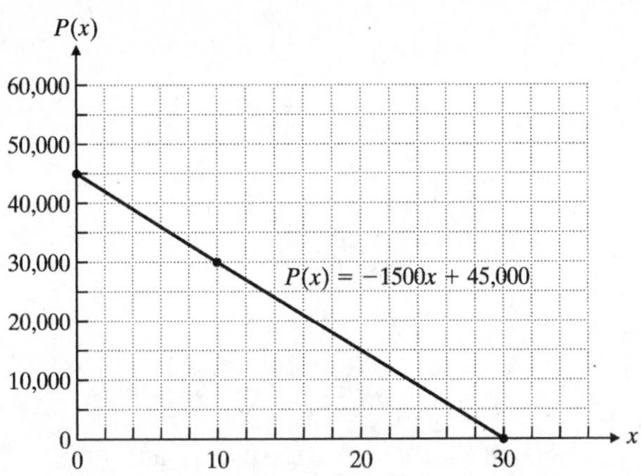

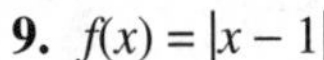

9. $f(x) = |x - 1|$

x	−1	0	1	2	3	4
$f(x)$	2	1	0	1	2	3

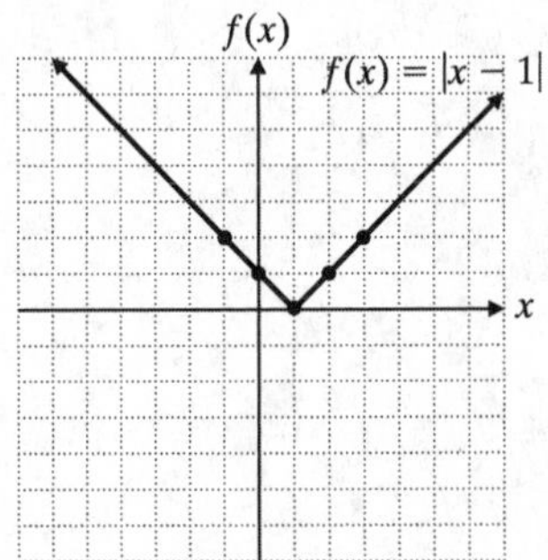

11. $g(x) = |x| - 5$

x	−2	−1	0	1	2
$g(x)$	−3	−4	−5	−4	−3

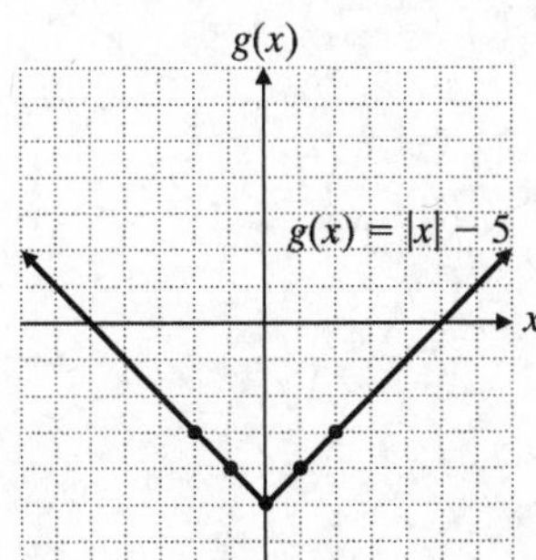

13. $g(x) = x^2 - 4$

x	−3	−2	−1	0	1	2	3
$g(x)$	5	0	−3	−4	−3	0	5

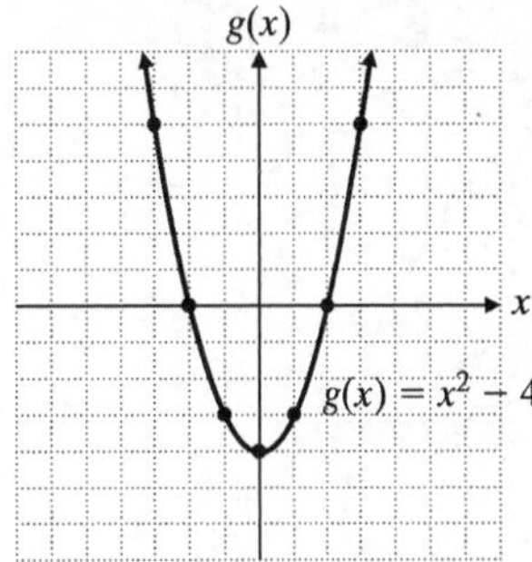

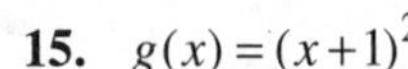

15. $g(x) = (x+1)^2$

x	−3	−2	−1	0	1
$g(x)$	4	1	0	1	4

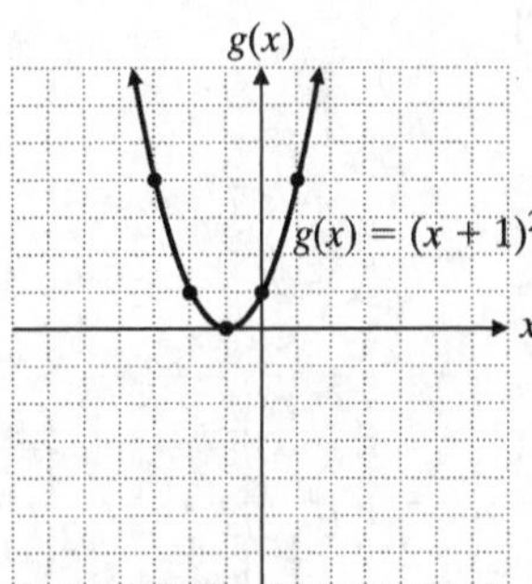

17. $g(x) = x^3 - 3$

x	$g(x)$
−1	−4
0	−3
1	−2

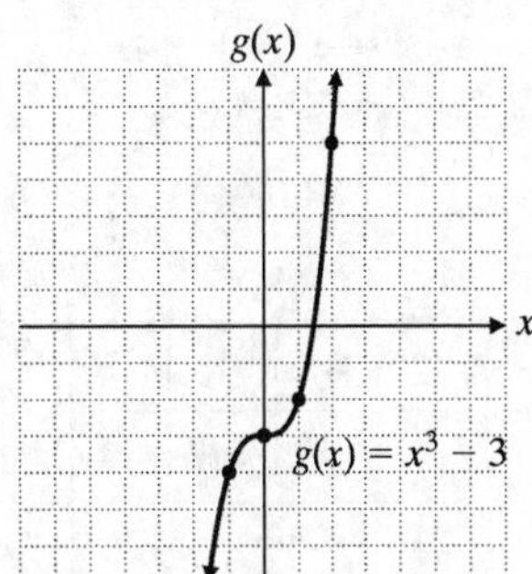

19. $p(x) = -x^3$

x	$p(x)$
−1	1
0	0
1	−1

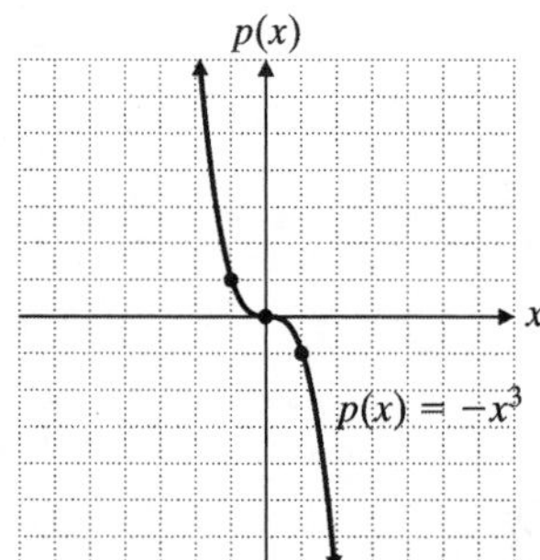

21. $f(x) = \frac{2}{x}$

x	−4	−2	−1	$-\frac{1}{2}$	$\frac{1}{2}$	1	2	4
$f(x)$	$-\frac{1}{2}$	−1	−2	−4	4	2	1	$\frac{1}{2}$

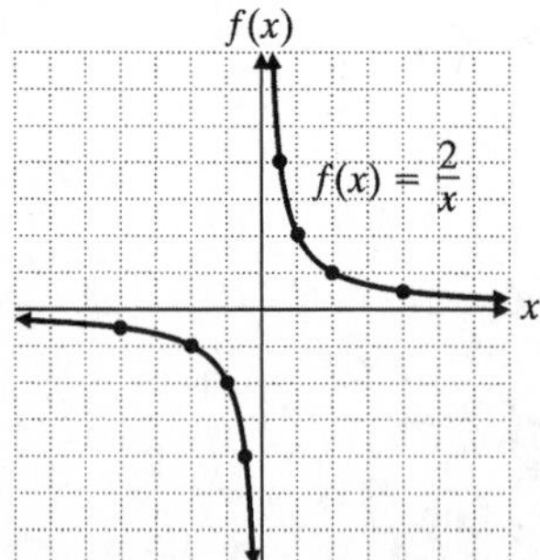

23. $h(x) = -\frac{6}{x}$

x	−6	−3	−2	−1	1	2	3	6
$h(x)$	1	2	3	6	−6	−3	−2	−1

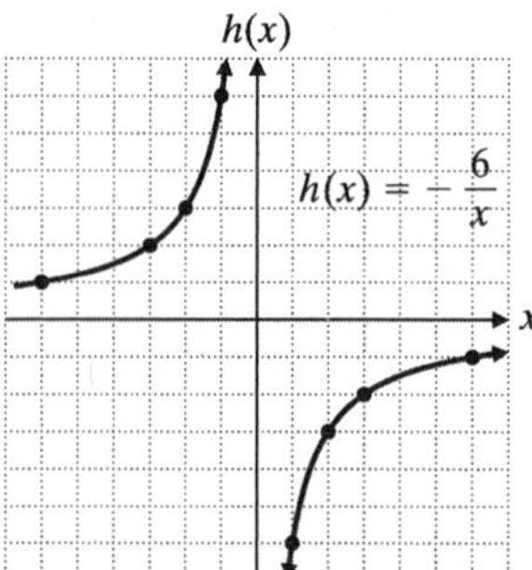

25. $y_1 = x^2$, $y_2 = 0.4x^2$, $y_3 = 2.6x^2$

From the graphs in we see that the larger the coefficient of x^2 the faster the graph rises.

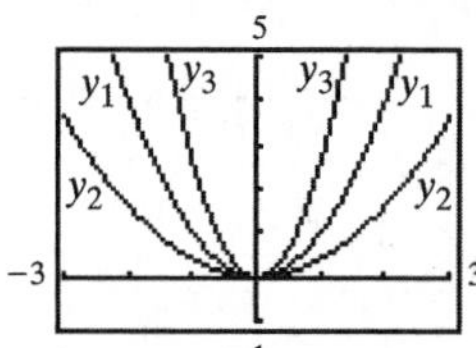

27. From the graph $f(2) = 1\frac{1}{2}$

x	$f(x)$
−5	−2
3	2
5	3

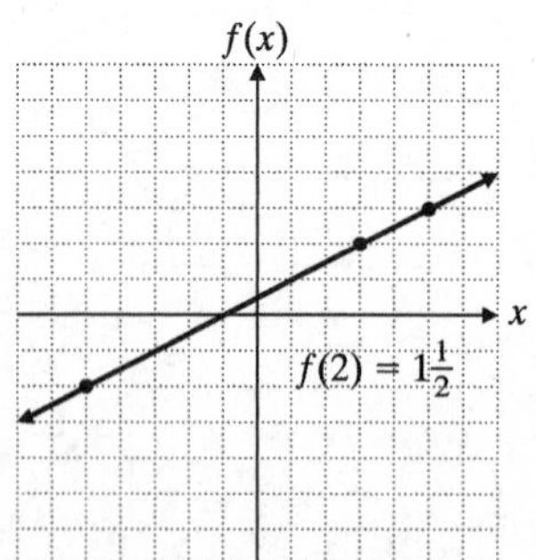

29. From the graph $f(2) = 2.4$

x	$f(x)$
0	3
1	2.5
3	2.2
−1	4
−2	6

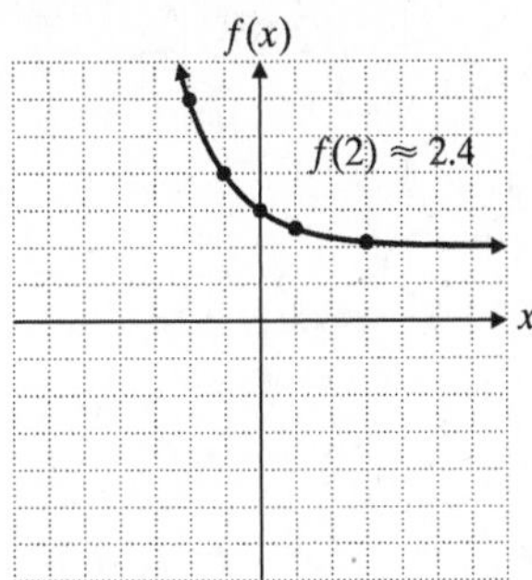

31. From the graph $f(2) = 0$

x	$f(x)$
–1	–3
0	0
1	1
3	–3

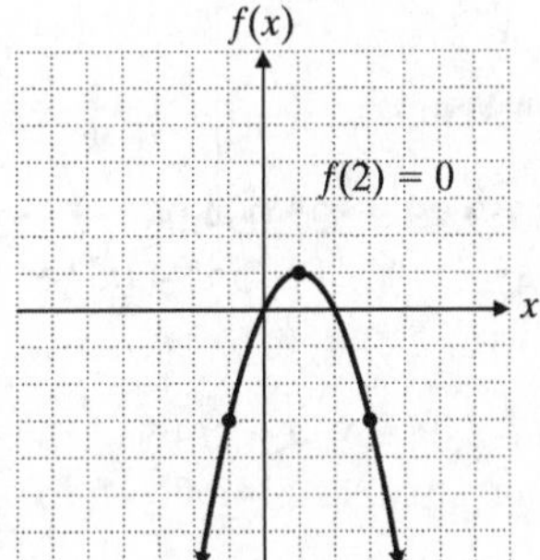

33. a.

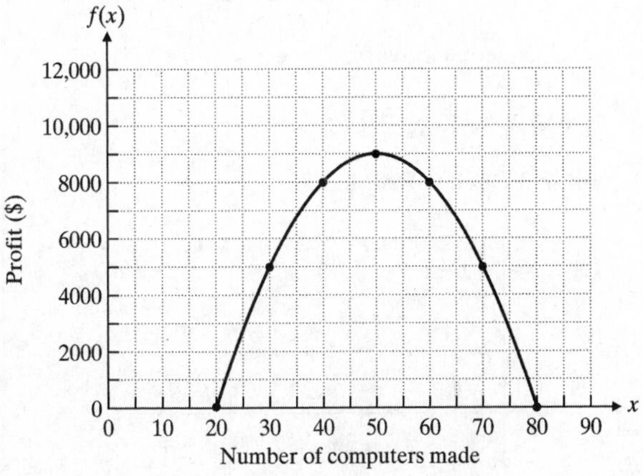

b. From the graph, 50 computers should be made each day for maximum profit.

c. To earn a profit of \$8000 or more each day the company should manufacture between 40 and 60 computers.

d. If the company manufactures 82 computers per day they will operate at a loss.

e. If the company manufactures forty-five computers per day the profit will be approximately \$8700.

Cumulative Review

35.
$$\begin{aligned}2(3ax-4y) &= 5(ax+3)\\ 6ax-8y &= 5ax+15\\ ax &= 8y+15\\ x &= \frac{8y+15}{a}\end{aligned}$$

36.
$$\begin{aligned}0.12(x-4) &= 1.16x-8.02\\ 0.12x-0.48 &= 1.16x-8.02\\ 1.04x &= 7.54\\ x &= 7.25\end{aligned}$$

37.
$$\begin{aligned}\frac{1}{2}(x+2)-5 &= x-\frac{3}{4}(x+4)\\ \frac{1}{2}x+1-5 &= x-\frac{3}{4}x-3\\ \frac{1}{2}x-4 &= \frac{1}{4}x-3\\ \frac{1}{4}x &= 1\\ x &= 4\end{aligned}$$

38.
$$\frac{10\text{ times}}{\text{hour}}\cdot\frac{24\text{ hours}}{\text{day}}(31+31+29)\text{ days}$$
$= 21{,}840$ times
It will breathe 21,840 times.

Quick Quiz 3.6

1. $f(x) = |x-4|$

x	$y = \|x-4\|$	
0	4	
–2	6	
2	2	
4	0	(vertex)

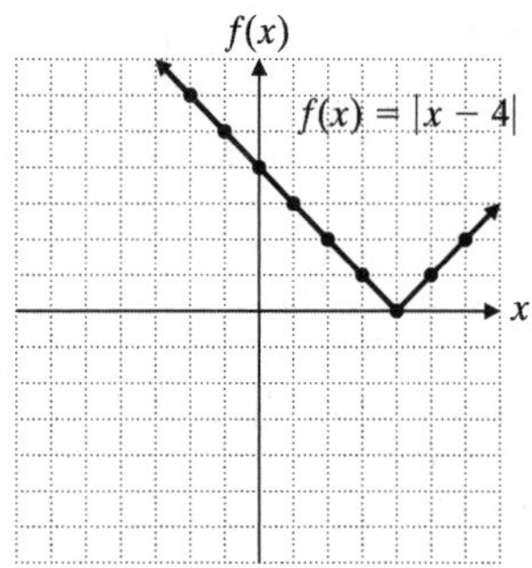

2. $g(x) = 6 - x^2$

x	$y = 6 - x^2$	
0	6	(vertex)
1	5	
−1	5	
3	−3	
−3	−3	

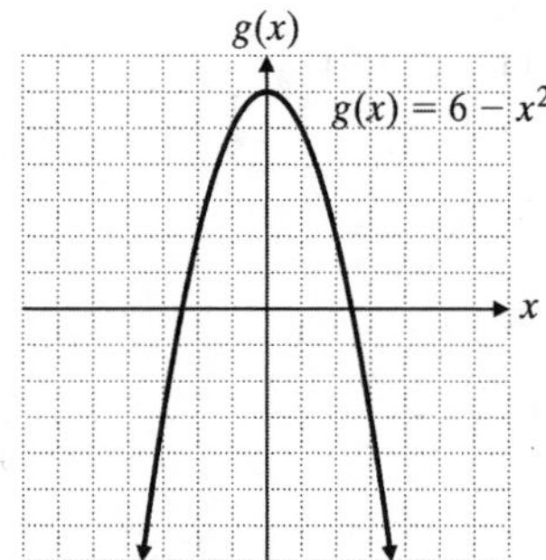

3. $h(x) = x^3 - 2$

x	$y = x^3 - 2$
0	−2
1	−1
−1	−3
2	6
−2	−10

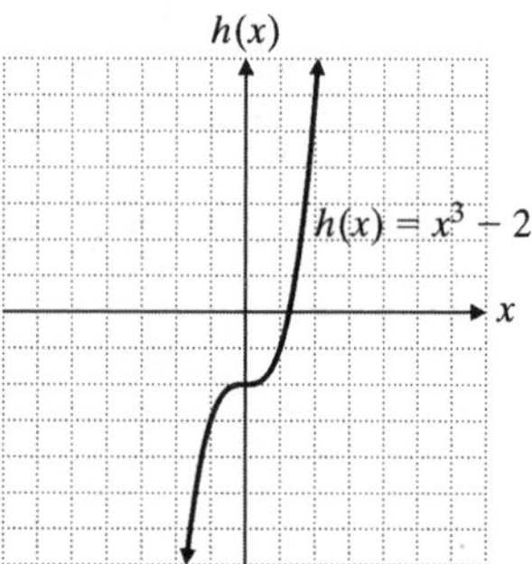

4. Answers may vary. Possible solution:

To evaluate for $f\left(\frac{1}{4}\right)$, substitute all values of x with $\frac{1}{4}$.

$$f(x) = \frac{5}{8x - 5}$$

$$f\left(\frac{1}{4}\right) = \frac{5}{8\left(\frac{1}{4}\right) - 5} = -\frac{5}{3}$$

Putting Your Skills to Work

1. $1000 + 36 × $669.28 = $25,094.08
Louvy would pay a total of $25,094.08 for the loan.

2. $1000 + 36 × $388.06 = $14,970.16
Louvy would pay a total of $14,970.16 for the lease.

3. $14,970.16 + $11,000 = $25,970.16
Louvy would pay a total of $25,970.16.

4. Answers may vary.

5. Answers may vary.

Chapter 3 Review Problems

1. $y = -\frac{1}{4}x - 1$

x	$x = -\frac{1}{4}x - 1$	y
−4	$x = -\frac{1}{4}(-4) - 1$	0
0	$y = -\frac{1}{4}(0) - 1$	−1
4	$y = -\frac{1}{4}(4) - 1$	−2

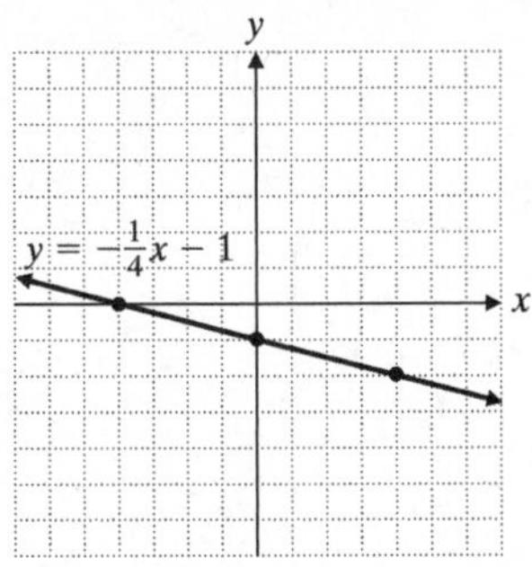

2. $y = -\dfrac{3}{2}x + 5$

x	$x = -\frac{3}{2}x + 5$	y
0	$y = -\frac{3}{2}(0) + 5$	5
2	$y = -\frac{3}{2}(2) + 5$	2
4	$y = -\frac{3}{2}(4) + 5$	−1

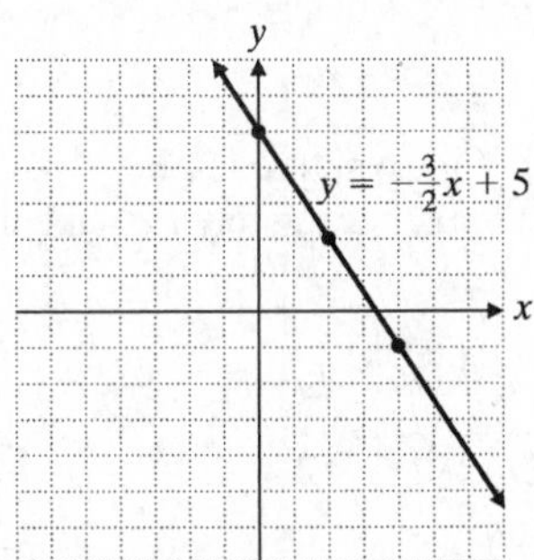

3. $y - 2x + 4 = 0$
$y = 2x - 4$

x	$y = 2x - 4$	y
0	$y = 2(0) - 4$	−4
2	$y = 2(2) - 4$	0
4	$y = 2(4) - 4$	4

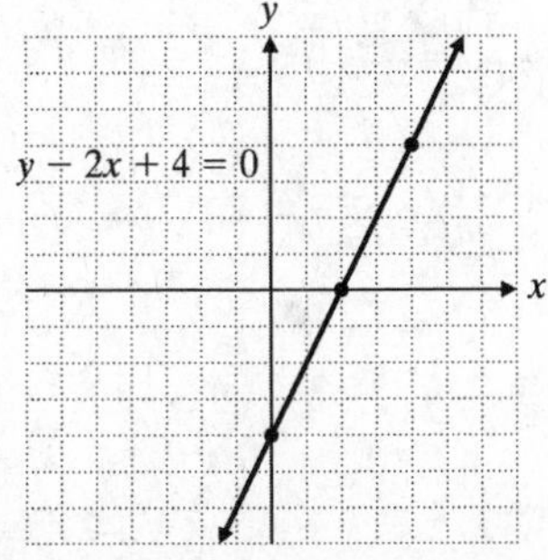

4. $7x = x + 2y$
$6x = 2y$
$y = 3x$

x	$y = 3x$	y
−2	$y = 3(-2)$	−6
0	$y = 3(0)$	0
2	$y = 3(2)$	6

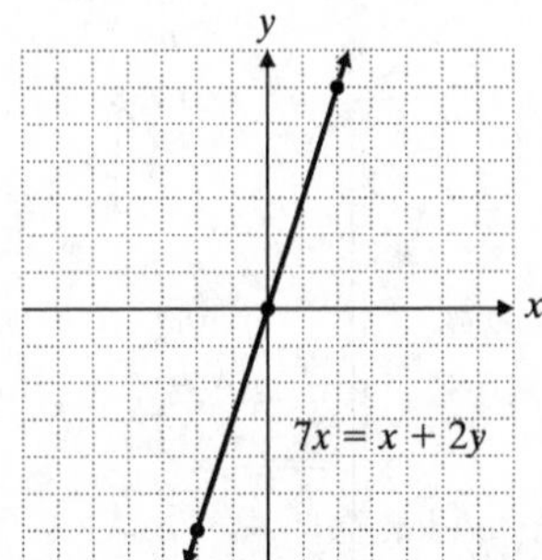

5. $5x - 3 = 9x + 13$
$x = -4$
The graph of $x = -4$ is a vertical line.

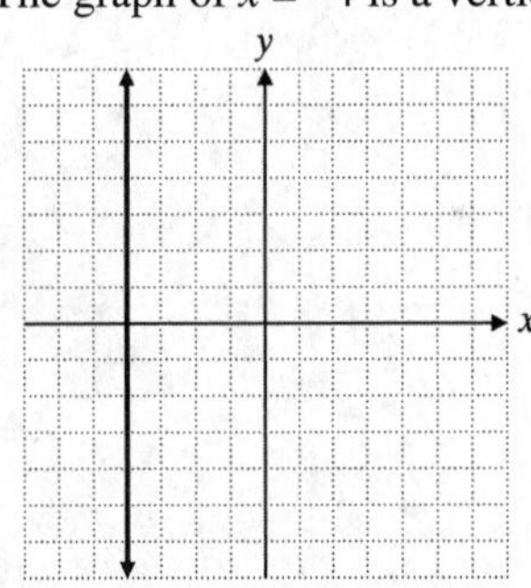

6. $8y + 5 = 10y + 1$
$y = 2$
The graph of $y = 2$ is a horizontal line.

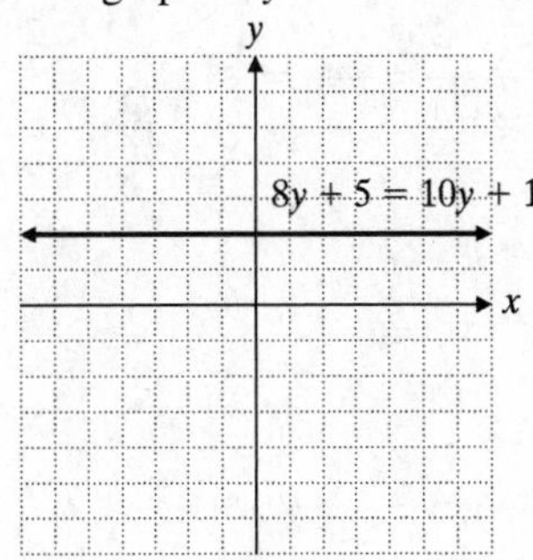

7. $m = \dfrac{y_2 - y_1}{x_2 - x_1}$

$m = \dfrac{3-1}{\frac{1}{2} - \frac{3}{2}} = \dfrac{2}{-1}$

$m = -2$

8. $m = \dfrac{y_2 - y_1}{x_2 - x_1}$

$m = \dfrac{-3-(-4)}{2-(-8)} = \dfrac{-3+4}{2+8}$

$m = \dfrac{1}{10}$

9. $m = \dfrac{y_2 - y_1}{x_2 - x_1}$

$m = \dfrac{6-1.8}{-3-(-3)}$

$m = \dfrac{4.2}{0}$ slope is undefined

10. $m = \dfrac{y_2 - y_1}{x_2 - x_1}$

$m = \dfrac{-1-(-1)}{7.5-0.3}$

$m = \dfrac{0}{7.2} = 0$

11. $m = \dfrac{y_2 - y_1}{x_2 - x_1}$

$m = \dfrac{-2-(-4)}{9-1}$

$m = \dfrac{2}{8}$

$m = \dfrac{1}{4}$

12. $m = \dfrac{y_2 - y_1}{x_2 - x_1} = \dfrac{\frac{1}{3} - 2}{\frac{2}{3} - 4} = \dfrac{-\frac{5}{3}}{-\frac{10}{3}} = \dfrac{1}{2}$

$m_\perp = -\dfrac{1}{m}$

$m_\perp = -\dfrac{1}{\frac{1}{2}} = -2$

13. $m = \dfrac{1}{3},\ b = 5$

$y = \dfrac{1}{3}x + 5$

$x - 3y = -15$

14. $y - y_1 = m(x - x_1)$

$y - (-2) = -4\left(x - \dfrac{1}{2}\right)$

$y + 2 = -4x + 2$

$4x + y = 0$

15. $y - y_1 = m(x - x_1)$

$y - 1 = 0(x - (-3))$

$y - 1 = 0$

$y = 1$

16. a. $P = 140x - 2000$ has $m = 140$. The slope is 140.

b. $P = 140x - 2000 = 0$

$140x = 2000$

$x = 14\dfrac{2}{7}$

The company must sell at least 15 microcomputers each day to make a profit.

17. $m = \dfrac{y_2 - y_1}{x_2 - x_1} = \dfrac{6 - \frac{-1}{2}}{5-(-1)} = \dfrac{13}{12}$

$y - y_1 = m(x - x_1)$

$y - 6 = \dfrac{13}{12}(x - 5)$

$12y - 72 = 13x - 65$

$13x - 12y = -7$

18. A line with undefined slope is a vertical line, $x = -6$.

19. $7x + 8y - 12 = 0$

$8y = -7x + 12$

$y = -\dfrac{7}{8}x + \dfrac{3}{2}$

$m = -\dfrac{7}{8},\ m_\perp = -\dfrac{1}{m} = -\dfrac{1}{-\frac{7}{8}} = \dfrac{8}{7}$

$y - y_1 = m_{\perp}(x - x_1)$

$y - 5 = \frac{8}{7}(x - (-2))$

$7y - 35 = 8x + 16$

$8x - 7y = -51$

20. $3x - 2y = 8$

$2y = 3x - 8$

$y = \frac{3}{2}x - 4$

$m = \frac{3}{2} \Rightarrow m_{\parallel} = \frac{3}{2}$

$y - y_1 = m_{\parallel}(x - x_1)$

$y - 1 = \frac{3}{2}(x - 5)$

$2y - 2 = 3x - 15$

$3x - 2y = 13$

21. From the graph $m = -2$ and y-intercept (0, 6) or $b = 6$.
$y = mx + b$
$y = -2x + 6$

22. From the graph $m = \frac{2}{3}$ and y-intercept (0, −5) or

$b = -5$.
$y = mx + b$
$y = \frac{2}{3}x - 5$

23. From the graph $m = 0$ and $b = 2$.
$y = mx + b$
$y = 0x + 2$
$y = 2$

24. $y < 2x + 4$
Graph $y = 2x + 4$ with a dashed line.
Test point: (0, 0)
$0 < 2(0) + 4$, $0 < 4$ True
Shade the region containing (0, 0).

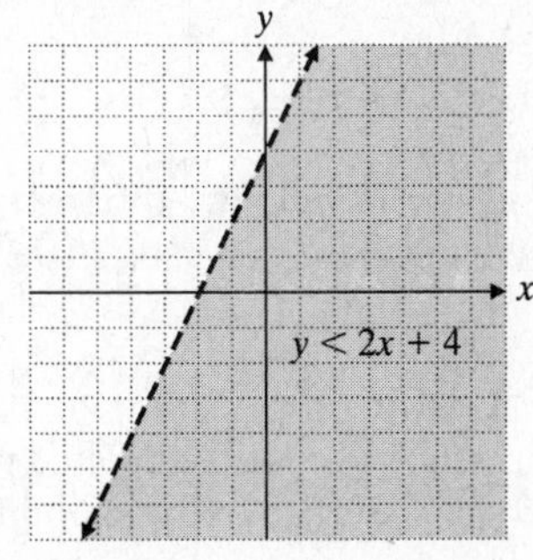

25. $y < 3x + 1$
Graph $y = 3x + 1$ as a dashed line.
Test point: (0, 0), $0 < 3(0) + 1$, $0 < 1$, True
Shade the region containing (0, 0).

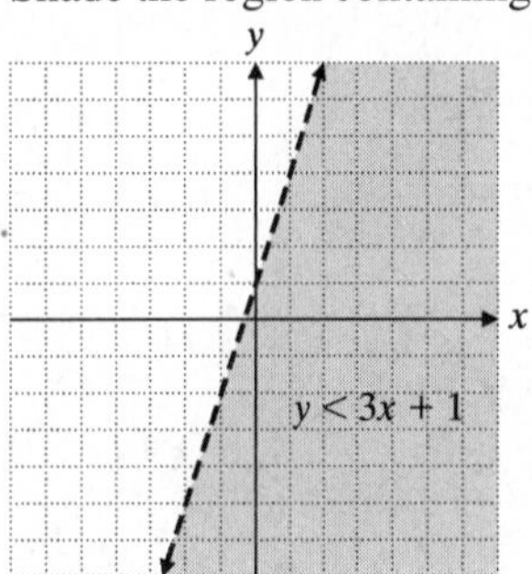

26. $y > -\frac{1}{2}x + 3$

Graph $y = -\frac{1}{2}x + 3$ with a dashed line.

Test point: (0, 0)

$0 > -\frac{1}{2}(0, 0) + 3$

$0 > 3$ False

Shade the region not containing (0, 0).

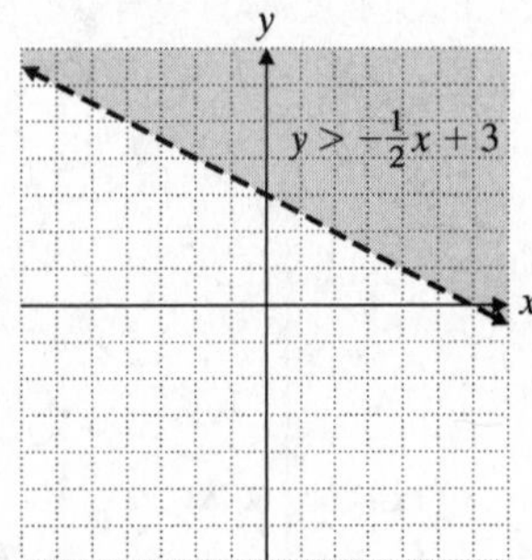

27. $y > -\frac{2}{3}x + 1$

Graph $y = -\frac{2}{3}x + 1$ as a dashed line.

Test point: (0, 0)

$0 > -\frac{2}{3}(0) + 1$

$0 > 1$ False

Shade the region not containing (0, 0).

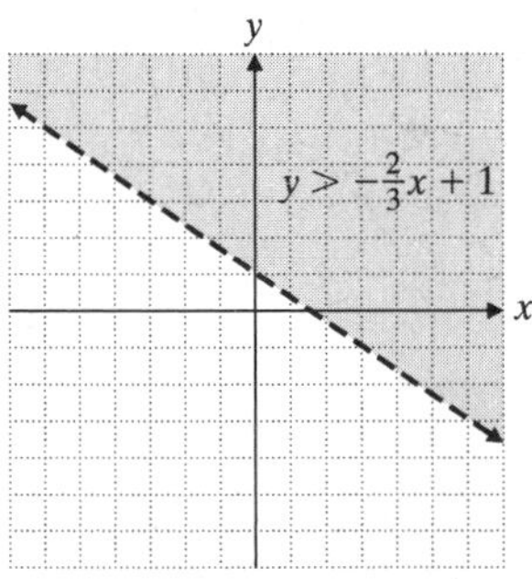

28. $3x + 4y \le -12$
Graph $3x + 4y = -12$ with a solid line.
Test point: (0, 0)
$3(0) + 4(0) \le -12$
$0 \le -12$ False
Shade the region not containing (0, 0).

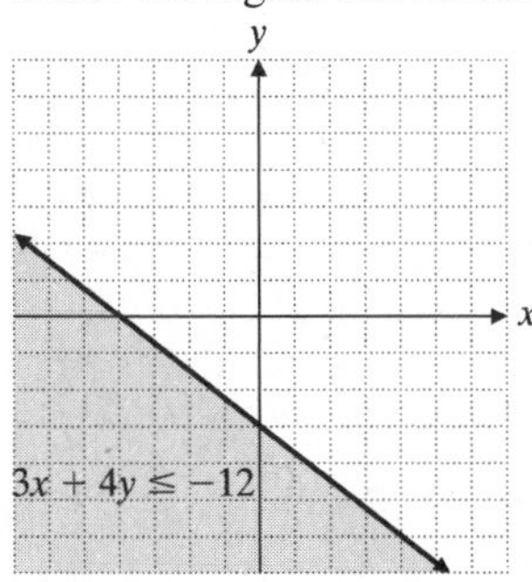

29. $x \le 3y$
Graph $x = 3y$ with a solid line.
Test point: (0, 3)
$0 \le 3(3)$
$0 \le 9$ True
Shade the region containing (0, 3).

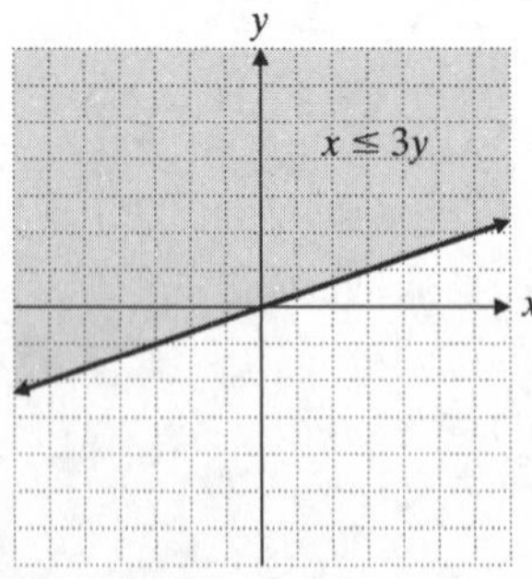

30. $5x + 3y \le -15$
Graph $5x + 3y = -15$ with a solid line.
Test point: (0, 0)
$5(0) + 3(0) \le -15$
$0 \le -15$ False
Shade the region not containing (0, 0).

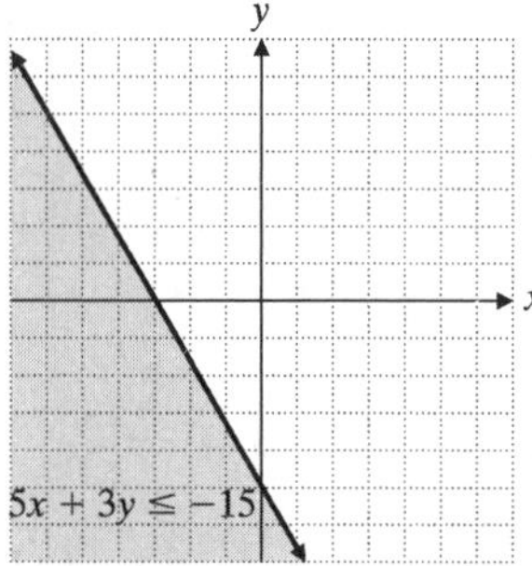

31. $3x - 5 < 7$
$3x < 12$
$x < 4$
Graph $x = 4$ with a dashed line. Shade the region to the left of the line.

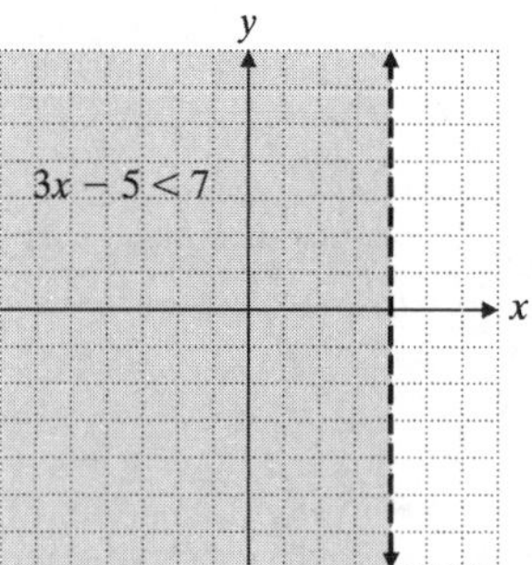

32. $5y - 2 > 3y - 10$
$2y > -8$
$y > -4$
Graph $y = -4$ with a dashed line. Shade the region above the line.

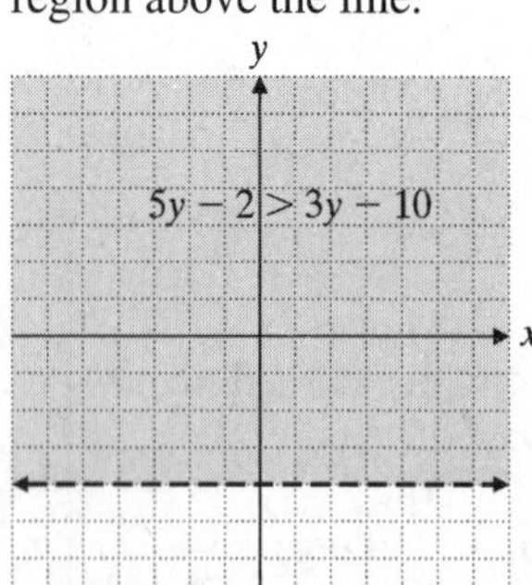

33. Domain = {−20, −18, −16, −12}
Range: {14, 16, 18}
Relation is a function since no two different ordered pairs have the same first coordinate.

34. Domain: {0, 1, 2, 3}
Range: {0, 1, 4, 9, 16}
Relation is not a function since the ordered pairs (1, 1) and (1, 16) have the same first coordinates.

35. Function, no two ordered pairs have the same first coordinate by the vertical line test.

36. Function, no two ordered pairs have the same first coordinate by the vertical line test.

37. Not a function, from the vertical line test at least two ordered pairs have the same first coordinate.

38. $f(x) = -2x + 10$

$f(-1) = -2(-1) + 10$
$f(-1) = 12$

$f(-5) = -2(-5) + 10$
$f(-5) = 20$

39. $g(x) = 2x^2 + x - 4$

$g(3) = 2(3)^2 + 3 - 4$
$g(3) = 18 + 3 - 4$
$g(3) = 17$

$g(-1) = 2(-1)^2 + (-1) - 4$
$g(-1) = 2 - 1 - 4$
$g(-1) = -3$

40. $h(x) = x^3 + 2x^2 - 5x + 8$
$h(-1) = (-1)^3 + 2(-1)^2 - 5(-1) + 8$
$h(-1) = -1 + 2 + 5 + 8$
$h(-1) = 14$

41. $p(x) = |-6x - 3|$
$p(3) = |-6(3) - 3|$
$p(3) = |-21|$
$p(3) = 21$

42. $f(x) = 2|x - 1|$

x	$f(x) = 2\|x - 1\|$
−2	6
0	2
1	0
2	2
3	4

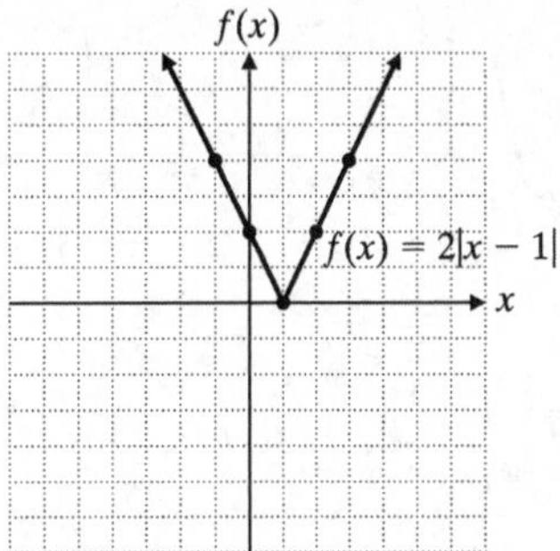

43. $g(x) = x^2 - 5$

x	$y = g(x)$
−3	4
−2	−1
−1	−4
0	−5
1	−4
2	−1
3	4

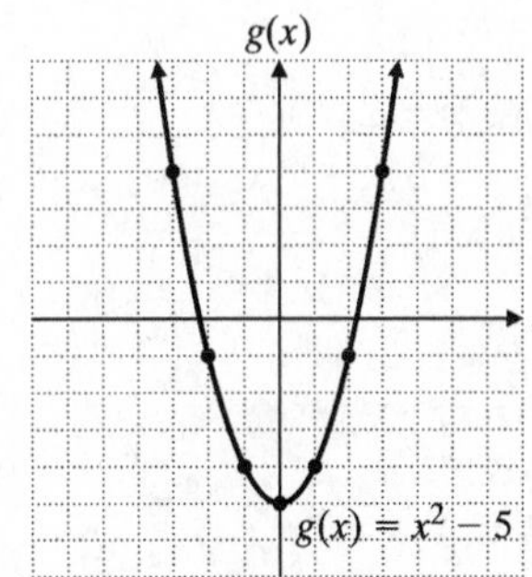

44. $h(x) = x^3 + 3$

x	$y = h(x)$
−2	−5
−1	2
0	3
1	4

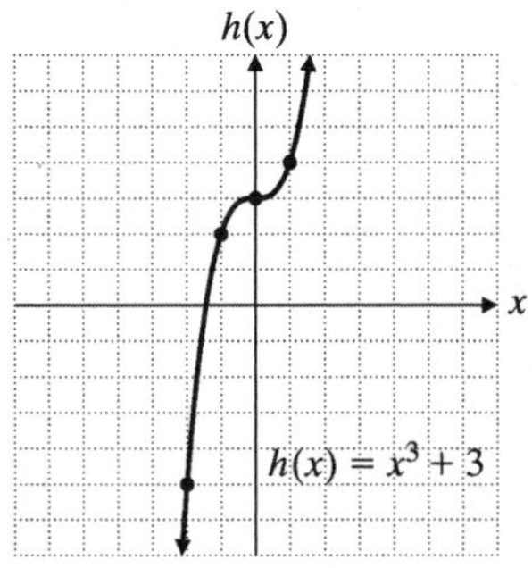

45. From the graph $f(-2) = 0$

x	−1	−3	−4	−5	−6	−7
$f(x)$	5	−3	−4	−3	0	5

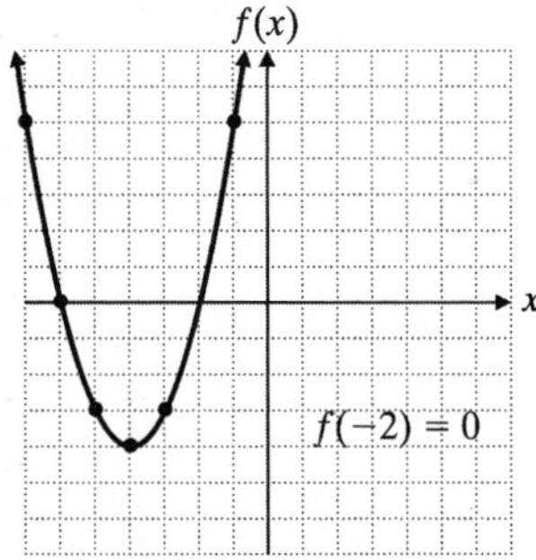

46. From the graph $f(-2) = 3$

x	$f(x)$
−3	4
−1	2
0	1
1	0
2	−1
3	0
4	1

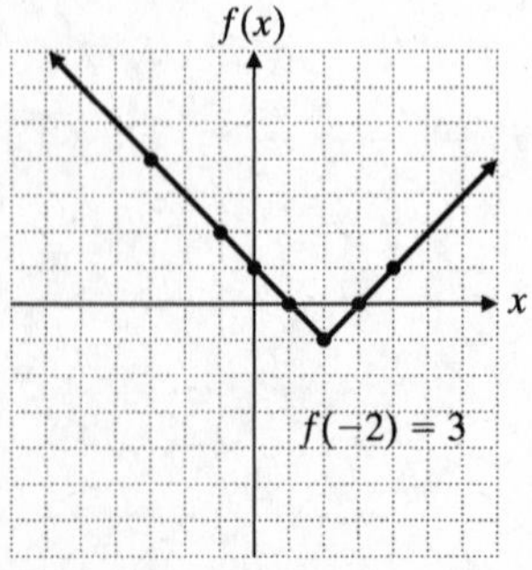

47. $f(x) = -\frac{4}{5}x + 3$

x	$f(x) = -\frac{4}{5}x + 3$
−5	7
0	3
10	−5

48. $f(x) = 2x^2 - 3x + 4$

x	$f(x) = 2x^2 - 3x + 4$
−3	31
0	4
4	24

49. $f(x) = 3x^3 - 4$

x	$f(x) = 3x^3 - 4$
−1	−7
0	−4
2	20

50. $f(x) = \dfrac{7}{2x+3}$

x	$f(x) = \frac{7}{2x+3}$
−2	−7
0	$\frac{7}{3}$
2	1

51. $m = \dfrac{y_2 - y_1}{x_2 - x_1}$

$$m = \frac{0-2}{5-(-4)}$$

$$m = -\frac{2}{9}$$

52. $m = \frac{y_2 - y_1}{x_2 - x_1} = \frac{6-9}{-3-1} = \frac{3}{4}$

m_n is the negative reciprocal of m.

$m_n = -\frac{4}{3}$

53. $m = \frac{y_2 - y_1}{x_2 - x_1} = \frac{8-(-6)}{4.5-2.5} = \frac{14}{2} = 7$

$m_p = m = 7$

54.
$$\begin{aligned} y &= mx + b \\ y &= \frac{5}{6}x + (-5) \\ 6y &= 5x - 30 \\ 5x - 6y &= 30 \end{aligned}$$

55. $y = 5x - 2,\ m = 5,\ m_{||} = 5$
$$\begin{aligned} y - y_1 &= m_{||}(x - x_1) \\ y - 10 &= 5(x - 4) \\ y - 10 &= 5x - 20 \\ y &= 5x - 10 \end{aligned}$$

56. $m = \frac{y_2 - y_1}{x_2 - x_1} = \frac{6-3}{5-(-7)} = \frac{3}{12} = \frac{1}{4}$
$$\begin{aligned} y - y_1 &= m(x - x_1) \\ y - 6 &= \frac{1}{4}(x - 5) \\ y - 6 &= \frac{1}{4}x - \frac{5}{4} \\ y &= \frac{1}{4}x + \frac{19}{4} \end{aligned}$$

57.
$$\begin{aligned} 3x - 6y &= 9 \\ 6y &= 3x - 9 \\ y &= \frac{1}{2}x - \frac{3}{2},\ m = \frac{1}{2},\ m_{\perp} = -2 \end{aligned}$$
$$\begin{aligned} y - y_1 &= m(x - x_1) \\ y - (-1) &= -2(x - (-2)) \\ y + 1 &= -2x - 4 \\ 2x + y &= -5 \end{aligned}$$

58. Vertical lines have the form $x = c$.
$(5, 6) \rightarrow x = 5$

59. $m = 0.20$
$b = 40$
$f(x) = 0.20x + 40$

60. $m = 200$
$b = 24{,}000$
$f(x) = 200x + 24{,}000$

61. $f(x) = 18{,}000 - 65x$

How Am I Doing? Chapter 3 Test

1. $y = \frac{1}{3}x - 2$

x	$y = \frac{1}{3}x - 2$	y
−3	$y = \frac{1}{3}(-3) - 2$	−3
0	$y = \frac{1}{3}(0) - 2$	−2
3	$y = \frac{1}{3}(3) - 2$	−1

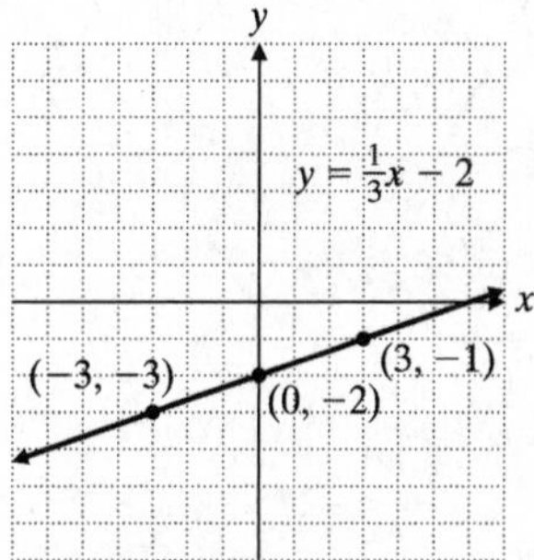

2.
$$\begin{aligned} 2x - 3 &= 1 \\ 2x &= 4 \\ x &= 2 \end{aligned}$$

x	y
2	−2
2	0
2	2

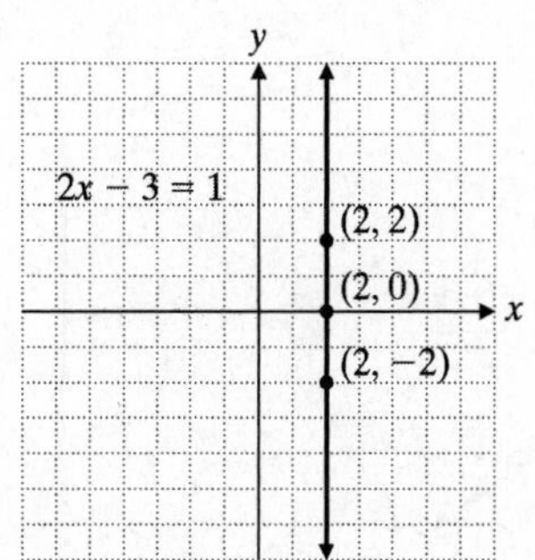

3. $5x + 3y = 9$

Let $x = 0$: $5(0)+3y=9$
$y=3$

Let $y=-2$: $5x+3(-2)=9$
$x=3$

Let $x=6$: $5(6)+3y=9$
$y=-7$

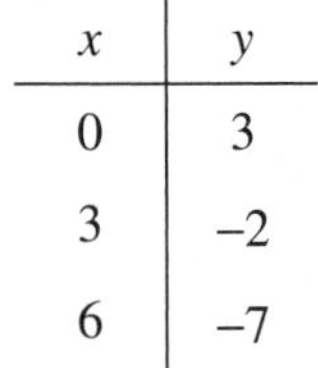

x	y
0	3
3	−2
6	−7

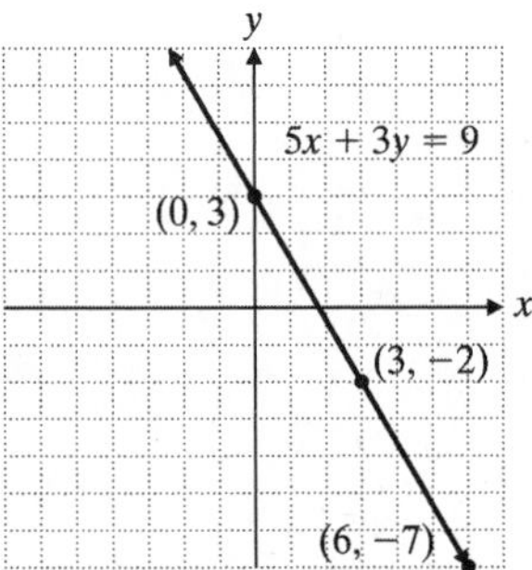

4. $2x + 3y = -10$

Let $x=-2$: $2(-2)+3y=-10$
$y=-2$

Let $x=1$: $2(1)+3y=-10$
$y=-4$

Let $x=4$: $2(4)+3y=-10$
$y=-6$

x	y
−2	−2
1	−4
4	−6

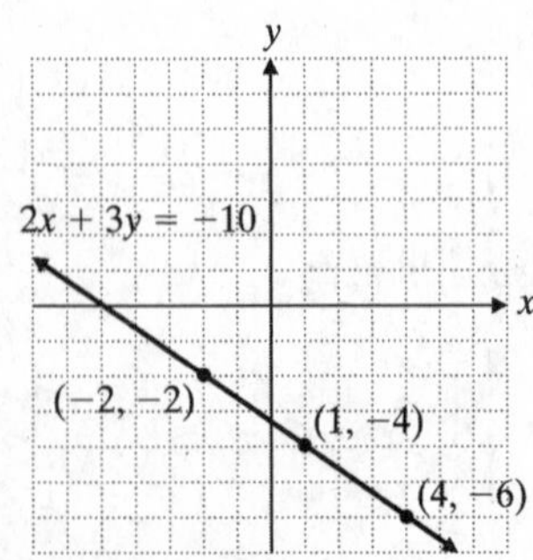

5. $m=\dfrac{y_2-y_1}{x_2-x_1}$

$m=\dfrac{-3-(-6)}{2-\frac{1}{2}}$

$m=\dfrac{3}{\frac{3}{2}}$

$m=2$

6. $m=\dfrac{y_2-y_1}{x_2-x_1}$

$m=\dfrac{5-5}{-7-6}$

$m=0$

7. $9x+7y=13$

$7y=-9x+13$

$y=-\dfrac{9}{7}x+\dfrac{13}{7}$

$m=-\dfrac{9}{7}$

8. $6x-7y-1=0$

$7y=6x-1$

$y=\dfrac{6}{7}x-\dfrac{1}{7}$

$m=\dfrac{6}{7},\ m_{\perp}=-\dfrac{7}{6}$

$y-y_1=m_{\perp}(x-x_1)$

$y-(-2)=-\dfrac{7}{6}(x-0)$

$y+2=-\dfrac{7}{6}x$

$6y+12=-7x$

$7x+6y=-12$

9. $m=\dfrac{y_2-y_1}{x_2-x_1}$

$m=\dfrac{-2-(-1)}{5-(-3)}=-\dfrac{1}{8}$

$y-y_1=m(x-x_1)$

$y-(-2)=-\dfrac{1}{8}(x-5)$

$8y+16=-x+5$

$x+8y=-11$

10. $y=2$

11. $y = mx + b$
$y = -5x + (-8)$
$y = -5x - 8$

12. $y \ge -4x$
Graph $y = -4x$ with a solid line.
Test point: (2, 2)
$2 \ge -4(2)$
$2 \ge -8$ True
Shade the region containing (2, 2).

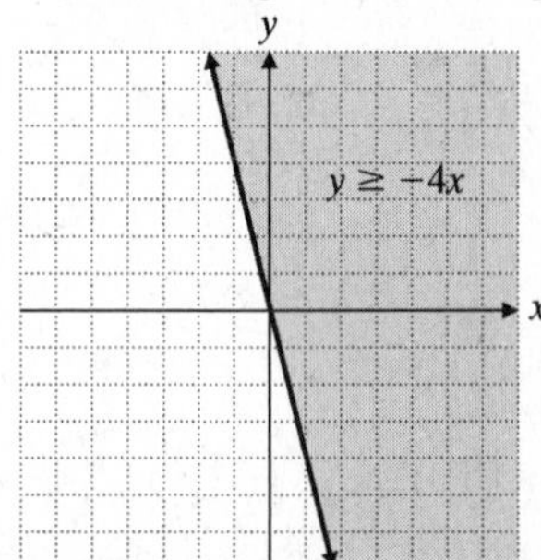

13. $4x - 2y < -6$
Graph $4x - 2y = -6$ with a dashed line.
Test point: (0, 0)
$4(0) - 2(0) < -6$
$0 < -6$ False
Shade the region not containing (0, 0).

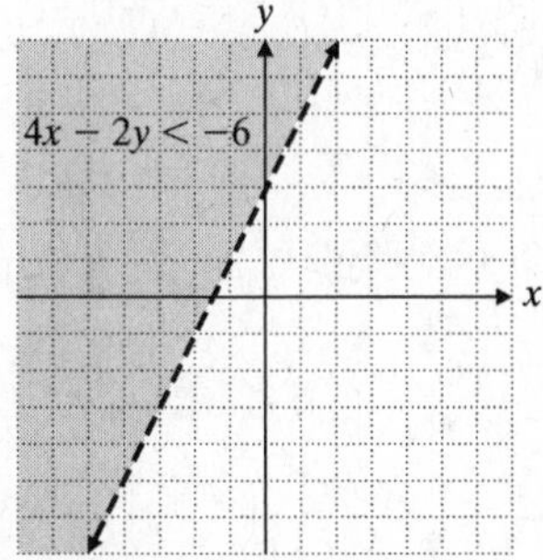

14. Domain = {0, 1, 2}
Range = {−4, −1, 0, 1, 4}

15. $f(x) = 2x - 3$
$f\left(\frac{3}{4}\right) = 2\left(\frac{3}{4}\right) - 3$
$f\left(\frac{3}{4}\right) = \frac{3}{2} - \frac{6}{2}$
$f\left(\frac{3}{4}\right) = -\frac{3}{2}$

16. $g(x) = \frac{1}{2}x^2 + 3$
$g(-4) = \frac{1}{2}(-4)^2 + 3$
$g(-4) = 8 + 3$
$g(-4) = 11$

17. $h(x) = \left|-\frac{2}{3}x + 4\right|$
$h(-9) = \left|-\frac{2}{3}(-9) + 4\right|$
$h(-9) = |10|$
$h(-9) = 10$

18. $p(x) = -2x^3 + 3x^2 + x - 4$
$p(-2) = -2(-2)^3 + 3(-2)^2 + (-2) - 4$
$p(-2) = 16 + 12 - 2 - 4$
$p(-2) = 22$

19. $g(x) = 5 - x^2$

x	g(x)
−2	1
−1	4
0	5
1	4
2	1

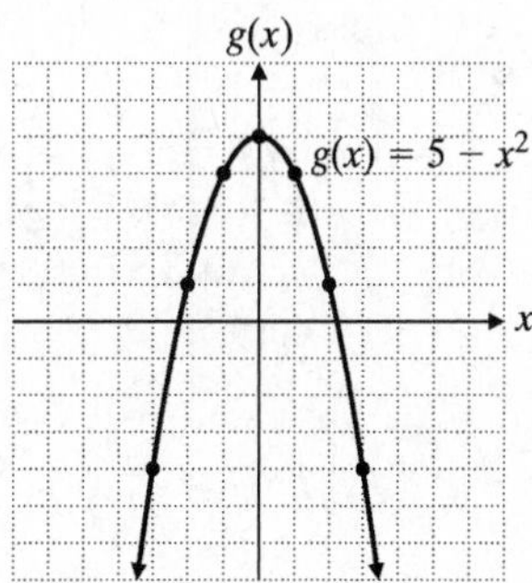

20. $h(x) = x^3 - 4$

x	y
0	−4
1	−3
2	4

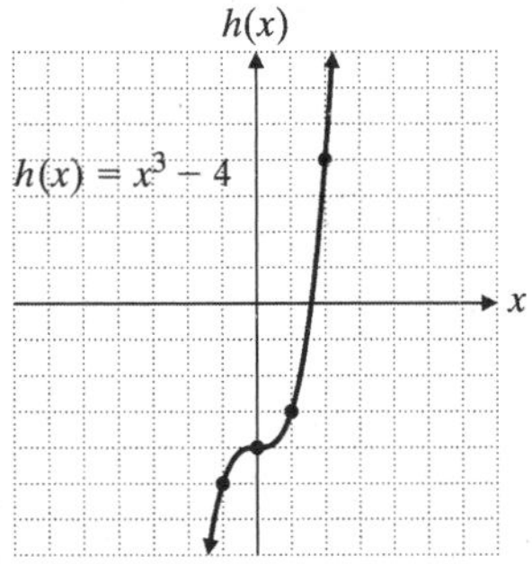

21.

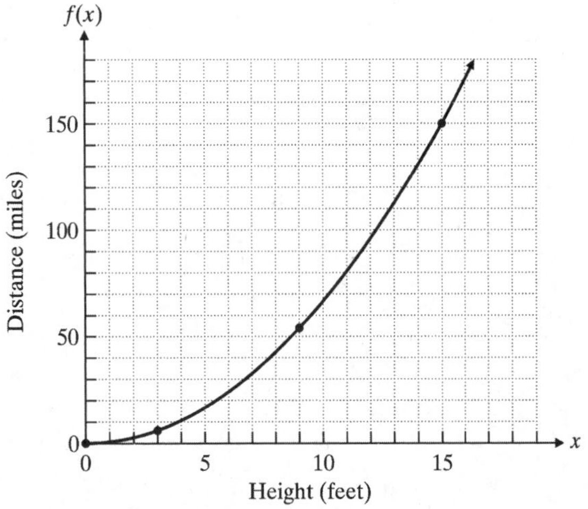

From the graph, $f(4) = 10$.
You can see 10 miles if you are 4 feet above the water.

Cumulative Test for Chapters 1–3

1. Inverse property of addition

2. $3(4-6)^2+\sqrt{16}+12\div(-3) = 3(-2)^2+4+(-4)$
$= 3(4)+4+(-4)$
$= 12+4+(-4)$
$= 12$

3. $(2a^{-3}b^4)^{-2} = 2^{-2}a^{-3(-2)}b^{4(-2)} = \dfrac{a^6}{2^2b^8} = \dfrac{a^6}{4b^8}$

4. $4(2x^2-1)-2x(3x-5y) = 8x^2-4-6x^2+10xy$
$= 2x^2+10xy-4$

5. $0.000437 = 4.37\times10^{-4}$

6. $3(x-2)>6$ or $5-3(x+1)>8$
$3x-6>6$ $\quad 5-3x-3>8$
$3x>12$ $\quad -3x>6$
$x>4$ $\quad x<-2$

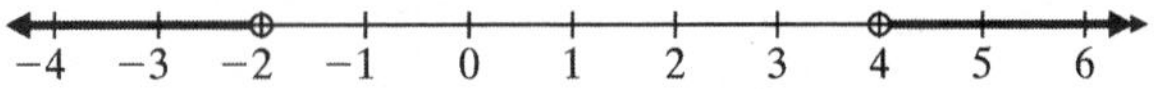

7. $2a-x=\dfrac{1}{3}(6x-y)$
$6a-3x=6x-y$
$-9x=-6a-y$
$9x=6a+y$
$x=\dfrac{6a+y}{9}$

8. Let w = the width.
$P=2l+2w$
$92=2(2w+1)+2w$
$46=(2w+1)+w=2w+1+w$
$46=3w+1 \Rightarrow 3w+1=46$
$3w=45$
$w=15$
$2w+1=31$
The width is 15 cm and the length is 31 cm.

9. $0.04x+0.07(7000-x)=391$
$0.04x+490-0.07x=391$
$-0.03x=391-490$
$x=3300$
$7000-x=3700$
Marissa invested \$3300 at 4% and \$3700 at 7%.

10. $A=\dfrac{1}{2}\pi r^2$
$A=\dfrac{1}{2}(3.14)(3)^2$
$A=14.13$
The area of the semicircle is 14.13 in^2.

11. $4x-6y=10$
Let $x=-2$: $4(-2)-6y=10$
$y=-3$
Let $x=1$: $4(1)-6y=10$
$y=-1$
Let $x=4$: $4(4)-6y=10$
$y=1$

x	y
−2	−3
1	−1
4	1

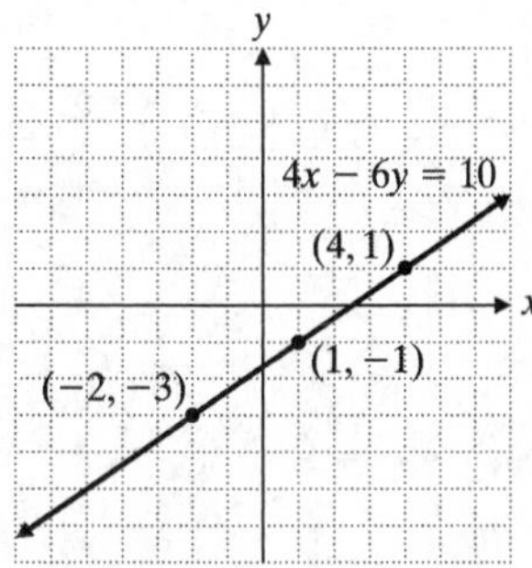

12. $m = \frac{y_2 - y_1}{x_2 - x_1} = \frac{5-1}{6-(-2)} = \frac{4}{8} = \frac{1}{2}$

13. $m = \frac{y_2 - y_1}{x_2 - x_1} = \frac{3-1}{4-5} = \frac{2}{-1} = -2$

$y - y_1 = m(x - x_1)$
$y - 1 = -2(x - 5)$
$y - 1 = -2x + 10$
$2x + y = 11$

14. $y = -\frac{1}{3}x + 6$

$m = -\frac{1}{3},\ m_\perp = 3$
$y - y_1 = m_\perp(x - x_1)$
$y - 4 = 3(x + 1)$
$y = 3x + 3 + 4$
$y - 3x = 7$
$3x - y = -7$

15. $D = \left\{\frac{1}{2}, 2, 3, 5\right\}$

$R = \{-1, 2, 7, 8\}$
Yes, the relation is a function since no two different ordered pairs have the same first coordinate.

16. $f(x) = -2x^2 - 4x + 1$

$f(-3) = -2(-3)^2 - 4(-3) + 1$
$f(-3) = -18 + 12 + 1$
$f(-3) = -5$

17. $p(x) = -\frac{1}{3}x + 2$

x	$p(x) = -\frac{1}{3}x + 2$	$p(x) = -\frac{1}{3}x + 2$
−3	$-\frac{1}{3}(-3) + 2$	3
0	$-\frac{1}{3}(0) + 2$	2
3	$-\frac{1}{3}(3) + 2$	1

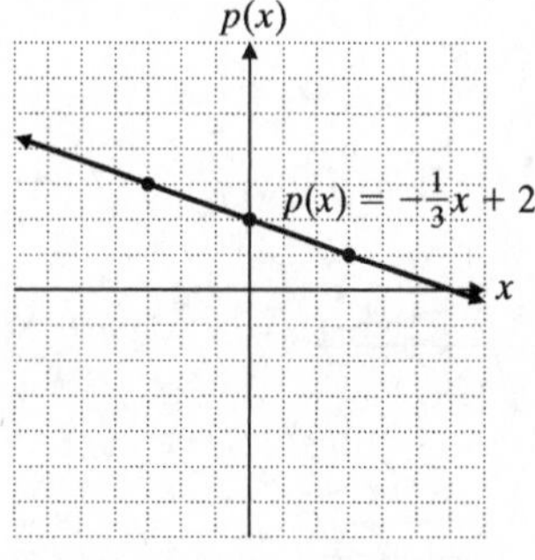

18. $h(x) = |x - 2|$

x	$h(x) = \|x - 2\|$
0	2
1	1
2	0
3	1
4	2

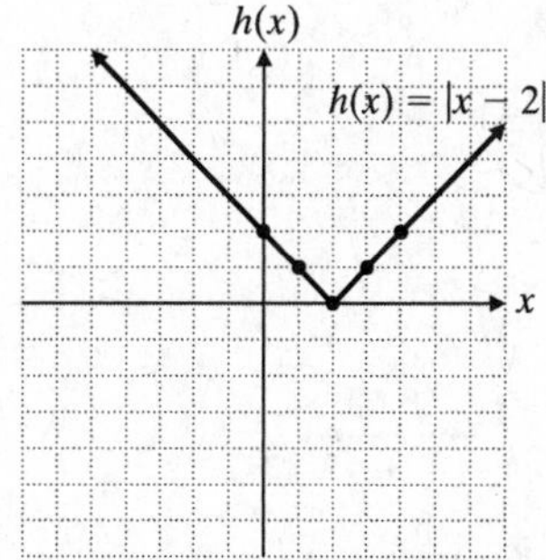

19. $r(x) = \frac{3}{x}$

x	$r(x) = \frac{3}{x}$		x	$r(x) = \frac{3}{x}$
-3	-1		1	3
-2	$-\frac{3}{2}$		$\frac{3}{2}$	2
$-\frac{3}{2}$	-2		2	$\frac{3}{2}$
-1	-3		3	1

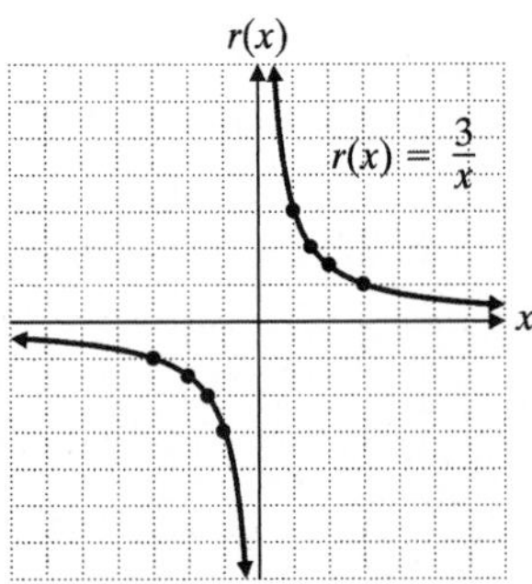

20. $f(x) = x^2 - 3$

x	$f(x) = x^2 - 3$
-2	1
-1	-2
0	-3
1	-2
2	1

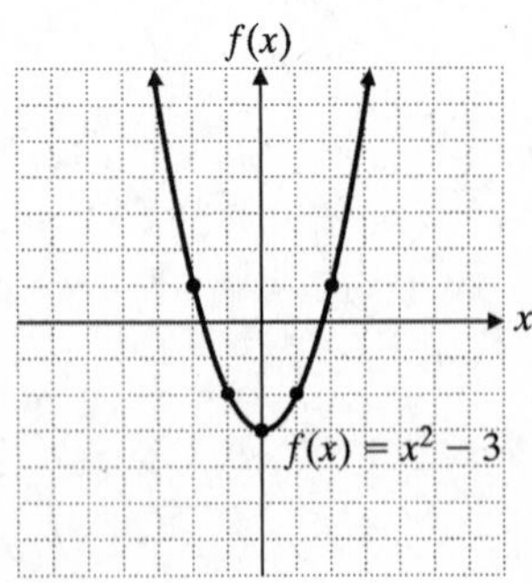

21. $y \le -\frac{3}{2}x + 3$

Graph $y = -\frac{3}{2}x + 3$ with a solid line.
Test point: (0, 0)
$0 \le -\frac{3}{2}(0) + 3$
$0 \le 3$ True
Shade the region containing (0, 0).

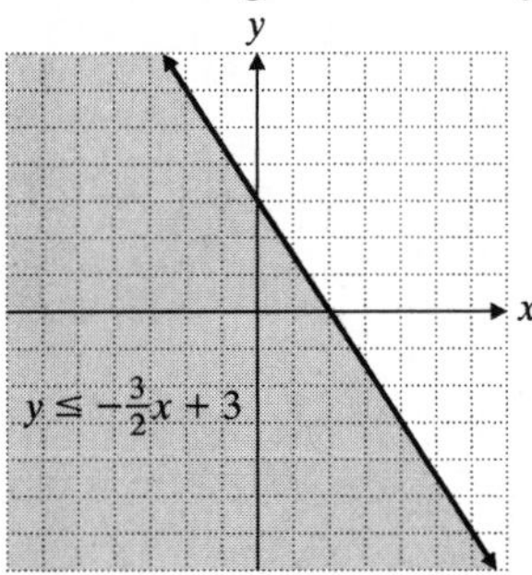

22. $f(x) = -2x^3 + 4$

x	$f(x) = -2x^3 + 4$	$f(x) = -2x^3 + 4$
-2	$-2(-2)^3 + 4$	20
0	$-2(0)^3 + 4$	4
3	$-2(3)^3 + 4$	-50

23. $m = 1.5$
$b = 31.1$
$f(x) = 1.5x + 31.1$

24. $x = 2012 - 2000 = 12$
$f(x) = 1.5x + 31.1$
$f(12) = 1.5(12) + 31.1 = 491.1$
There is a predicted 491.1 million people that will be living below poverty level in 2012.

Chapter 4

4.1 Exercises

1. There is no solution. There is no point (x, y) that satisfies both equations. The graph of such a system yields two parallel lines.

3. A system of two linear equations in two unknowns can have an infinite number of solutions, one solution, or no solutions.

5.
$$4x+1=6-y$$
$$4\left(\frac{3}{2}\right)+1 \stackrel{?}{=} 6-(-1)$$
$$7=7$$
$$2x-5y=8$$
$$2\left(\frac{3}{2}\right)-5(-1) \stackrel{?}{=} 8$$
$$8=8$$
$\left(\frac{3}{2}, -1\right)$ is a solution.

7. 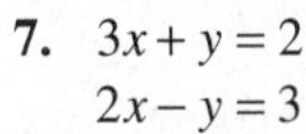
$$3x+y=2$$
$$2x-y=3$$

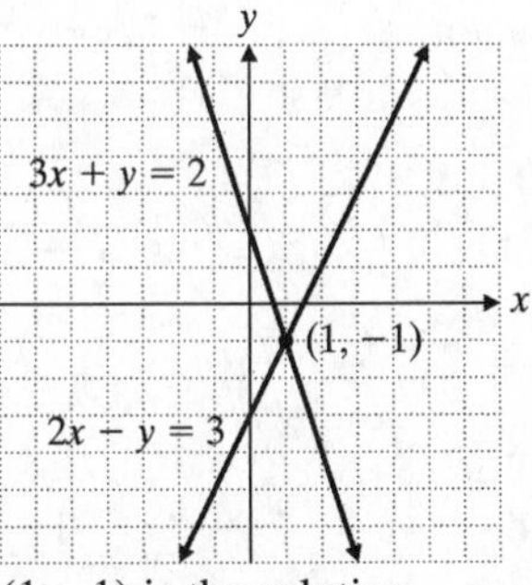

(1, −1) is the solution.

9.
$$3x-2y=6$$
$$4x+y=-3$$

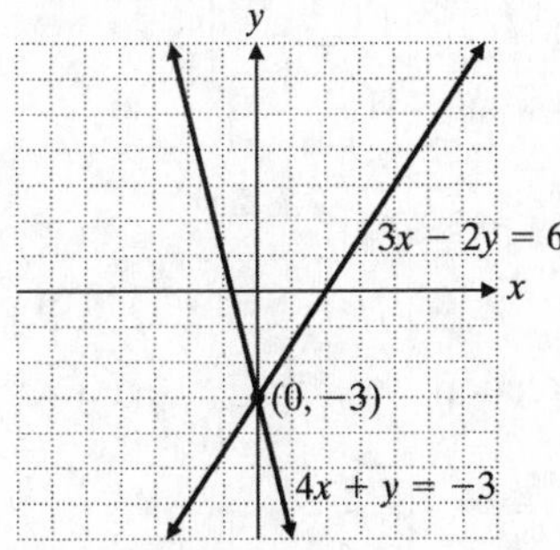

(0, −3) is the solution.

11.
$$y=-x+3$$
$$x+y=-\frac{2}{3}$$

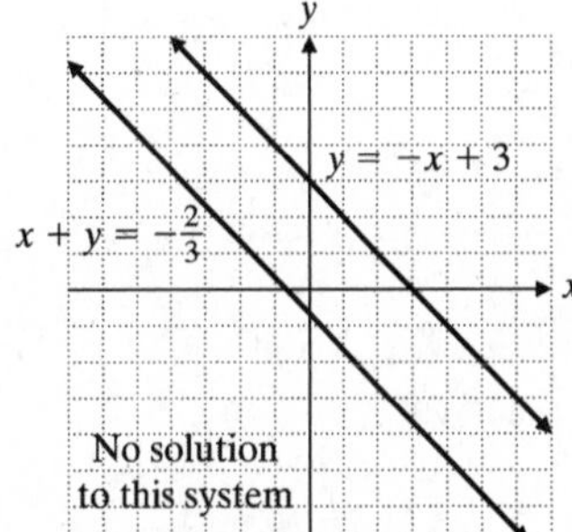

The system has no solution.

13.
$$y=-2x+5$$
$$3y+6x=15$$

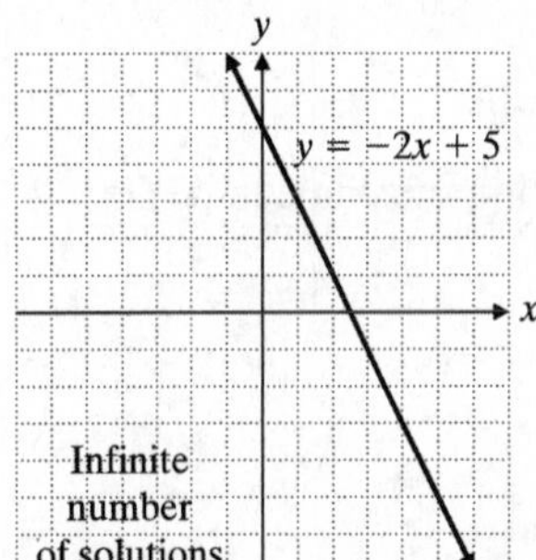

There are an infinite number of solutions.

15. $3x+4y=14$ **(1)**
$x+2y=-2$ **(2)**

Solve **(2)** for x and substitute into **(1)**.
$$x=-2-2y$$
$$3(-2-2y)+4y=14$$
$$-6-6y+4y=14$$
$$-2y=20$$
$$y=-10$$

Substitute $y = -10$ into **(2)** and solve for x.
$$x+2(-10)=-2$$
$$x-20=-2$$
$$x=18$$

(18, −10) is the solution.

Check:

(1) $3(18) + 4(-10) \stackrel{?}{=} 14, \quad 14 = 14$

(2) $18 + 2(-10) \stackrel{?}{=} -2, \quad -2 = -2$

17. $-x+3y=-8$ **(1)**
$2x-y=6$ **(2)**
Solve **(2)** for y and substitute into **(1)**.

$$\begin{aligned} y &= 2x-6 \\ -x+3(2x-6) &= -8 \\ -x+6x-18 &= -8 \\ 5x &= 10 \\ x &= 2 \end{aligned}$$

Substitute $x = 2$ into **(1)**.

$$\begin{aligned} -2+3y &= -8 \\ 3y &= -6 \\ y &= -2 \end{aligned}$$

(2, –2) is the solution.
Check:
(1) $-2+3(-2) \stackrel{?}{=} -8,\ -8=-8$
(2) $2(2)-(-2) \stackrel{?}{=} 6,\ 6=6$

19. $2x-\frac{1}{2}y=-3$ **(1)**
$\frac{x}{5}+2y=\frac{19}{5}$ **(2)**
Solve **(2)** for y and substitute into **(1)**.

$$\begin{aligned} \frac{x}{5}+2y &= \frac{19}{5} \\ x+10y &= 19 \\ y &= \frac{19-x}{10} \\ 2x-\frac{1}{2}\left(\frac{19-x}{10}\right) &= -3 \\ 40x-19+x &= -60 \\ 41x &= -41 \\ x &= -1 \end{aligned}$$

Substitute $x = -1$ into **(1)** and solve for y.

$$\begin{aligned} 2(-1)-\frac{1}{2}y &= -3 \\ -4-y &= -6 \\ -y &= -2 \\ y &= 2 \end{aligned}$$

(–1, 2) is the solution.
Check:
(1) $2(-1)-\frac{1}{2}(2) \stackrel{?}{=} -3,\ -3=-3$
(2) $\frac{-1}{5}+2(2) \stackrel{?}{=} \frac{19}{5},\ \frac{19}{5}=\frac{19}{5}$

21. $\frac{1}{2}x-\frac{1}{8}y=3$ **(1)**
$\frac{2}{3}x+\frac{3}{4}y=4$ **(2)**
Solve **(2)** for x and substitute into **(1)**.

$$\begin{aligned} 12\left(\frac{2}{3}x+\frac{3}{4}y\right) &= 12(4) \\ 8x+9y &= 48 \\ 8x &= 48-9y \\ x &= 6-\frac{9}{8}y \end{aligned}$$

$$\begin{aligned} \frac{1}{2}\left(6-\frac{9}{8}y\right)-\frac{1}{8}y &= 3 \\ 3-\frac{9}{16}y-\frac{1}{8}y &= 3 \\ -\frac{11}{16}y &= 0 \\ y &= 0 \end{aligned}$$

Substitute $y = 0$ into **(2)** and solve for x.

$$\begin{aligned} \frac{2}{3}x+\frac{3}{4}(0) &= 4 \\ x &= 4\left(\frac{3}{2}\right) \\ x &= 6 \end{aligned}$$

(6, 0) is the solution.
Check:
(1) $\frac{1}{2}(6)-\frac{1}{8}(0) \stackrel{?}{=} 3,\ 3=3$
(2) $\frac{2}{3}(6)-\frac{3}{4}(0) \stackrel{?}{=} 4,\ 4=4$

23. $9x+2y=2$ **(1)**
$3x+5y=5$ **(2)**
Multiply **(2)** by –3 and add to **(1)**.

$$\begin{aligned} 9x+2y &= 2 \\ -9x-15y &= -15 \\ \hline -13y &= -13 \\ y &= 1 \end{aligned}$$

Substitute $y = 1$ into **(1)**.

$$\begin{aligned} 9x+2(1) &= 2 \\ 9x+2 &= 2 \\ 9x &= 0 \\ x &= 0 \end{aligned}$$

(0, 1) is the solution.
Check:
(1) $9(0)+2(1) \stackrel{?}{=} 2,\ 2=2$
(2) $3(0)+5(1) \stackrel{?}{=} 5,\ 5=5$

25. $3s+3t=10$ **(1)**
$4s-9t=-4$ **(2)**
Multiply **(1)** by 3 and add to **(2)**.

$$\begin{aligned}9s+9t&=30\\+\ 4s-9t&=-4\\\hline 13s&=26\\s&=2\end{aligned}$$

Substitute $s=2$ into **(1)** and solve for t.

$$\begin{aligned}3(2)+3t&=10\\6+3t&=10\\3t&=4\\t&=\frac{4}{3}\end{aligned}$$

$\left(2,\frac{4}{3}\right)$ is the solution.

Check:

(1) $3(2)+3\left(\frac{4}{3}\right)\stackrel{?}{=}10,\ 10=10$

(2) $4(2)-9\left(\frac{4}{3}\right)\stackrel{?}{=}-4,\ -4=-4$

27. $\frac{7}{2}x+\frac{5}{2}y=-4$ **(1)**
$3x+\frac{2}{3}y=1$ **(2)**
Clear fractions.
$7x+5y=-8$ **(1)**
$9x+2y=3$ **(2)**
Multiply **(1)** by 2 and **(2)** by −5 and add.

$$\begin{aligned}14x+10y&=-16\\-45x-10y&=-15\\\hline -31x&=-31\\x&=1\end{aligned}$$

Substitute $x=1$ into **(2)**.

$$\begin{aligned}9(1)+2y&=3\\2y&=-6\\y&=-3\end{aligned}$$

(1, −3) is the solution.

29. $1.6x+1.5y=1.8\rightarrow 16x+15y=18$ **(1)**
$0.4x+0.3y=0.6\rightarrow 4x+3y=6$ **(2)**
Multiply **(2)** by −5 and add to **(1)**.

$$\begin{aligned}16x+15y&=18\\-20x-15y&=-30\\\hline -4x&=-12\\x&=3\end{aligned}$$

Substitute $x=3$ into **(1)**.

$$\begin{aligned}16(3)+15y&=18\\48+15y&=18\\15y&=-30\\y&=-2\end{aligned}$$

(3, −2) is the solution.

31. $7x-y=6$ **(1)**
$3x+2y=22$ **(2)**
Solve **(1)** for y and substitute into **(2)**.

$$\begin{aligned}7x-y&=6\\y&=7x-6\end{aligned}$$

$$\begin{aligned}3x+2(7x-6)&=22\\3x+14x-12&=22\\17x&=34\\x&=2\end{aligned}$$

Substitute $x=2$ into **(1)**.

$$\begin{aligned}7(2)-y&=6\\y&=8\end{aligned}$$

(2, 8) is the solution.

33. $3x+4y=8$ **(1)**
$5x+6y=10$ **(2)**
Multiply **(1)** 5 and **(2)** by −3 and add.

$$\begin{aligned}15x+20y&=40\\-15x-18y&=-30\\\hline 2y&=10\\y&=5\end{aligned}$$

Substitute $y=5$ into **(1)**.

$$\begin{aligned}3x+4(5)&=8\\3x+20&=8\\3x&=-12\\x&=-4\end{aligned}$$

(−4, 5) is the solution.

35. $2x+y=4$ **(1)**
$\frac{2}{3}x+\frac{1}{4}y=2\xrightarrow{\times 12}8x+3y=24$ **(2)**
Solve **(1)** for y and substitute into **(2)**.
$y=4-2x$

$$\begin{aligned}8x+3(4-2x)&=24\\8x+12-6x&=24\\2x&=12\\x&=6\end{aligned}$$

Substitute $x=6$ into **(1)**.

$$\begin{aligned}2(6)+y&=4\\12+y&=4\\y&=-8\end{aligned}$$

(6, −8) is the solution.

37. $0.2x = 0.1y - 1.2$ **(1)**
$2x - y = 6$ **(2)**
Solve **(2)** for y and substitute into **(1)**.
$y = 2x - 6$
$0.2x = 0.1(2x - 6) - 1.2$
$0.2x = 0.2x - 0.6 - 1.2$
$0 = -1.8$
This is an inconsistent system of equations and has no solution.

39. $5x - 7y = 12$ **(1)**
$-10x + 14y = -24$ **(2)**
Multiply **(1)** by 2 and add to **(2)**.
$$\begin{array}{r} 10x - 14y = 24 \\ \underline{-10x + 14y = -24} \\ 0 = 0 \end{array}$$
This is a dependent system of equations and has an infinite number of solutions.

41. $0.8x + 0.9y = 1.3$
$0.6x - 0.5y = 4.5$
Multiply both equations by 10 to clear decimals.
$8x + 9y = 13$ **(1)**
$6x - 5y = 45$ **(2)**
Multiply **(1)** by 5 and **(2)** by 9 and add.
$$\begin{array}{r} 40x + 45y = 65 \\ \underline{54x - 45y = 405} \\ 94x = 470 \\ x = 5 \end{array}$$
Substitute $x = 5$ into **(1)**.
$8(5) + 9y = 13$
$40 + 9y = 13$
$9y = -27$
$y = -3$
$(5, -3)$ is the solution.

43. $-9a + 15b = 3$ **(1)**
$9a - 15b = 2$ **(2)**
$\frac{5}{3}b = \frac{1}{3} + a$
$9a - 15b = 2$
Multiply the first equation by –9, rearrange, and add to the second equation.
$0 = 5$
No solution; inconsistent system of equations

45. $\frac{2}{3}x - y = 4$ **(1)**
$2x - \frac{3}{4}y = 21$ **(2)**
Solve **(1)** for y and substituting into **(2)**,
$y = \frac{2}{3}x - 4$
$2x - \frac{3}{4}\left(\frac{2}{3}x - 4\right) = 21$
$2x - \frac{6}{12}x + 3 = 21$
$2x - \frac{1}{2}x = 18$
$\frac{3}{2}x = 18$
$x = 12$
Substituting $x = 12$ in **(1)** and solving for y yields:
$\frac{2}{3}(12) - y = 4$
$8 - y = 4$
$-y = -4$
$y = 4$
$(12, 4)$ is the solution.

47. $3.2x - 1.5y = -3 \Rightarrow 32x - 15y = -30$ **(1)**
$0.7x + y = 2 \Rightarrow 7x + 10y = 20$ **(2)**
Multiply **(1)** by 2 and **(2)** by 3 and add.
$$\begin{array}{r} 64x - 30y = -60 \\ \underline{21x + 30y = 60} \\ 85x = 0 \\ x = 0 \end{array}$$
Substitute $x = 0$ into **(2)**.
$7(0) + 10y = 20$
$10y = 20$
$y = 2$
$(0, 2)$ is the solution.

49. $3 - (2x + 1) = y + 6$ **(1)**
$x + y + 5 = 1 - x$ **(2)**
Solve **(1)** for y and substitute in **(2)**.
$3 - (2x + 1) = y + 6$
$y = 3 - 2x - 1 - 6$
$y = -2x - 4$
$x - 2x - 4 + 5 = 1 - x$
$1 = 1$
This is a dependent system and has an infinite number of solutions.

51. a. $y = 200 + 50x$ for Old World Tile
$y = 300 + 30x$ for Modern Bath

b.

x	$y = 300 + 30x$
0	300
4	420
8	540

x	$y = 200 + 50x$
0	200
4	400
8	600

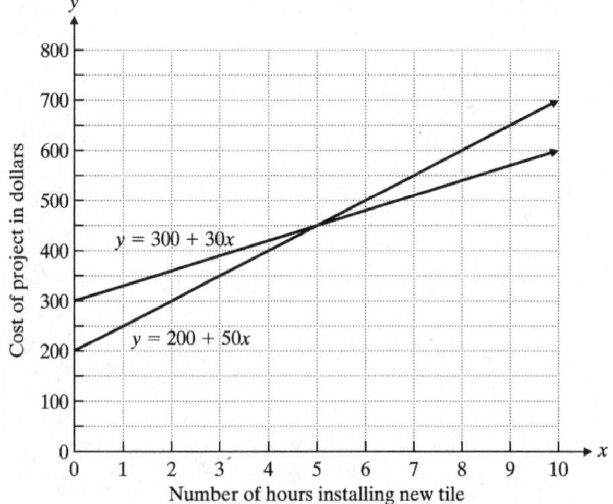

c. From the graph, the cost will be the same for 5 hours of installing new tile.

d. From the graph, Modern Bathroom Headquarters will cost less to remove old tile and install new tile if the time needed to install the new tile is 6 hours.

53. $y_1 = -1.7x + 3.8$
$y_2 = 0.7x - 2.1$

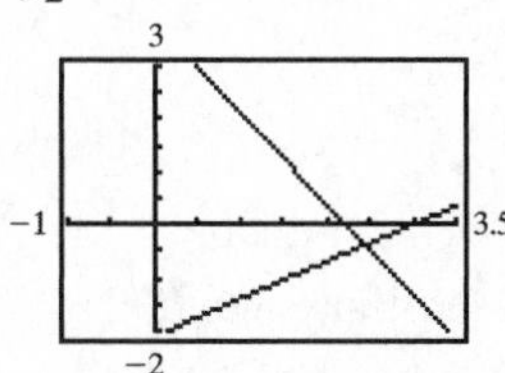

To the nearest hundredth, the point of intersection is (2.46, −0.38).

55. $0.5x + 1.1y = 5.5 \Rightarrow y_1 = \dfrac{(5.5 - 0.5x)}{1.1}$
$-3.1x + 0.9y = 13.1 \Rightarrow y_2 = \dfrac{13.1 + 3.1x}{0.9}$

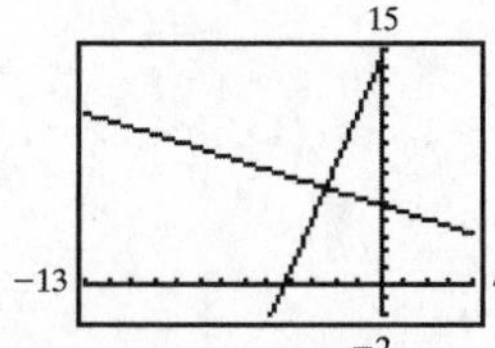

The point of intersection is (−2.45, 6.11).

Cumulative Review

56. $\dfrac{200{,}000{,}000}{9{,}000{,}000 \text{ tons}}\left(\dfrac{1 \text{ ton}}{2000 \text{ pounds}}\right)$
≈ 0.01
This is approximately \$0.01 per pound.

57. $\dfrac{4}{5}x = 273{,}511$
$x = 341{,}888.75$
During rush hour about 341,889 cars enter the city.

Quick Quiz 4.1

1. $5x - 3y = 14$ **(1)**
$2x - y = 6$ **(2)**
Solve (2) for y and substitute into **(1)**.
$y = 2x - 6$
$5x - 3(2x - 6) = 14$
$5x - 6x + 18 = 14$
$-x = -4$
$x = 4$
Substitute $x = 4$ into **(2)** and solve for y.
$2(4) - y = 6$
$8 - y = 6$
$-y = -2$
$y = 2$
(4, 2) is the solution.
Check:
(1) $5(4) - 3(2) \stackrel{?}{=} 14$, $14 = 14$
(2) $2(4) - 2 \stackrel{?}{=} 6$, $6 = 6$

2. $6x + 7y = 26$ **(1)**
$5x - 2y = 6$ **(2)**
Multiply **(1)** by 2 and **(2)** by 7 then add equations.
$12x + 14y = 52$
$35x - 14y = 42$
$47x = 94$
$x = 2$
Substitute $x = 2$ into **(2)** and solve for y.
$5(2) - 2y = 6$
$10 - 2y = 6$
$-2y = 6 - 10$
$y = 2$
(2, 2) is the solution.
Check:
(1) $6(2) + 7(2) \stackrel{?}{=} 26$, $26 = 26$
(2) $5(2) - 2(2) \stackrel{?}{=} 6$, $6 = 6$

3. $\frac{2}{3}x+\frac{3}{5}y=-17$ **(1)**

$\frac{1}{2}x-\frac{1}{3}y=-1$ **(2)**

Solve **(2)** for x then substitute into **(1)**.

$\frac{1}{2}x-\frac{1}{3}y=-1$

$\frac{1}{2}x=\frac{1}{3}y-1$

$x=\frac{2}{3}y-2$

$\frac{2}{3}\left(\frac{2}{3}y-2\right)+\frac{3}{5}y=-17$

$\frac{4}{9}y-\frac{4}{3}+\frac{3}{5}y=-17$

$\frac{20}{45}y+\frac{27}{45}y=-17+\frac{4}{3}$

$\frac{47}{45}y=-\frac{47}{3}$

$y=-15$

Substitute $y=-15$ into **(2)** then solve for x.

$\frac{1}{2}x-\frac{1}{3}(-15)=-1$

$\frac{1}{2}x+5=-1$

$\frac{1}{2}x=-6$

$x=-12$

$(-12, -15)$ is the solution.

Check:

(1) $\frac{2}{3}(-12)+\frac{3}{5}(-15)\stackrel{?}{=}-17,\ -17=-17$

(2) $\frac{1}{2}(-12)-\frac{1}{3}(-15)\stackrel{?}{=}-1,\ -1=-1$

4. Answers may vary. Possible solution: The equation results in an identity $(0 = 0)$. Therefore the system is dependent and there is an infinite number of solutions.

4.2 Exercises

1. $2x-3y+2z=-7$ **(1)**

$x+4y-z=10$ **(2)**

$3x+2y+z=4$ **(3)**

Check:

(1) $2(2)-3(1)+2(-4)\stackrel{?}{=}-7,\ -7=-7$

(2) $2+4(1)-(-4)\stackrel{?}{=}10,\ 10=10$

(3) $3(2)+2(1)+(-4)\stackrel{?}{=}4,\ 4=4$

$(2, 1, -4)$ is a solution.

3. $3x+2y-z=6$ **(1)**

$x-y-2z=-8$ **(2)**

$4x+y+2z=5$ **(3)**

Check:

(1) $3(-1)+2(5)-1\stackrel{?}{=}6,\ 6=6$

(2) $-1-5-2(1)\stackrel{?}{=}-8,\ -8=-8$

(3) $4(-1)+5+2(1)\stackrel{?}{=}5,\ 3\neq 5$

$(-1, 5, 1)$ is not a solution to the system.

5. $x+y+2z=0$ **(1)**

$2x-y-z=1$ **(2)**

$x+2y+3z=1$ **(3)**

Add **(1)** and **(2)**.

$3x+z=1$ **(4)**

Add $2\cdot$ **(2)** and **(3)**.

$5x+z=3$ **(5)**

Subtract **(5)** from **(4)**.

$-2x=-2$

$x=1$

Substitute $x = 1$ into **(5)**.

$5(1)+z=3$

$z=-2$

Substitute $x = 1$, $z = -2$ into **(1)**.

$1+y+2(-2)=0$

$1+y-4=0$

$y=3$

$(1, 3, -2)$ is the solution.

7. $-x+2y-z=-1$ **(1)**

$2x+y+z=2$ **(2)**

$x-y-2z=-13$ **(3)**

Add **(1)** and **(2)**.

$x+3y=1$ **(4)**

Add $2\cdot$ **(2)** and **(3)**.

$5x+y=-9$ **(5)**

Solve **(4)** for x and substitute into **(5)**.

$x=1-3y$

$5(1-3y)+y=-9$

$5-15y+y=-9$

$-14y=-14$

$y=1$

Substitute $y = 1$ into **(4)** then solve for x.

$x+3(1)=1$

$x=1-3$

$x=-2$

Substitute $x = -2$, $y = 1$ into **(1)**, then solve for z.

$$\begin{aligned} -(-2)+2(1)-z &= -1 \\ 4-z &= -1 \\ -z &= -5 \\ z &= 5 \end{aligned}$$

(−2, 1, 5) is the solution.

9. $8x-5y+z=15$ **(1)**
$3x+y-z=-7$ **(2)**
$x+4y+z=-3$ **(3)**

Add **(1)** and **(2)**.

$$\begin{array}{r} 8x-5y+z=15 \\ 3x+y-z=-7 \\ \hline 11x-4y=8 \end{array} \quad \textbf{(4)}$$

Add **(2)** and **(3)**.

$$\begin{array}{r} 3x+y-z=-7 \\ x+4y+z=-3 \\ \hline 4x+5y \quad =-10 \end{array} \quad \textbf{(5)}$$

Add 5 times **(4)** and 4 times **(5)**.

$$\begin{array}{r} 55x-20y=40 \\ 16x+20y=-40 \\ \hline 76x \quad =0 \\ x=0 \end{array}$$

Substitute $x = 0$ into **(5)**.

$$\begin{aligned} 4(0)+5y &= -10 \\ 5y &= -10 \\ y &= -2 \end{aligned}$$

Substitute $x = 0$, $y = -2$ into **(1)**.

$$\begin{aligned} 8(0)-5(-2)+z &= 15 \\ 10+z &= 15 \\ z &= 5 \end{aligned}$$

(0, −2, 5) is the solution.

11. $x+4y-z=-5$ **(1)**
$-2x-3y+2z=5$ **(2)**
$x-\frac{2}{3}y+z=\frac{11}{3}$ **(3)**

Add **(1)** and **(3)**.

$2x+\frac{10}{3}y=-\frac{4}{3}$ **(4)**

Add 2 · **(1)** and **(2)**.

$$\begin{aligned} 5y &= -5 \\ y &= -1 \end{aligned}$$

Substitute $y = -1$ into **(4)**.

$$\begin{aligned} 2x+\frac{10}{3}(-1) &= -\frac{4}{3} \\ 6x-10 &= -4 \\ 6x &= 6 \\ x &= 1 \end{aligned}$$

Substitute $x = 1$ and $y = -1$ into **(3)**.

$$1-\frac{2}{3}(-1)+z=\frac{11}{3}$$
$$3+2+3z=11$$
$$3z=6$$
$$z=2$$

(1, –1, 2) is the solution.

13. $2x+2z=-7+3y \Rightarrow 2x-3y+2z=-7$ **(1)**

$2\cdot\left(\frac{3}{2}x+y+\frac{1}{2}z\right)=2\cdot 2 \Rightarrow 3x+2y+z=4$ **(2)**

$x+4y=10+z \Rightarrow x+4y-z=10$ **(3)**

Add **(1)** and 2 times **(3)**.

$4x + 5y = 13$ **(4)**

Add **(3)** and **(2)**.

$4x + 6y = 14$ **(5)**

Subtract **(5)** from **(4)**.

$$-y=-1$$
$$y=1$$

Substitute $y = 1$ into **(5)**.

$$4x+6(1)=14$$
$$4x=8$$
$$x=2$$

Substitute $x = 2$, $y = 1$ into **(3)**.

$$2+4(1)-z=10$$
$$6-z=10$$
$$-z=4$$
$$z=-4$$

(2, 1, –4) is the solution.

15. $a=8+3b-2c \Rightarrow a-3b+2c=8$ **(1)**

$4a+2b-3c=10$ **(2)**

$c=10+b-2a \Rightarrow 2a-b+c=10$ **(3)**

Add $3\cdot$ **(1)** and $2\cdot$ **(2)**.

$$\begin{array}{l} 3a-9b+6c=24 \\ \underline{8a+4b-6c=20} \\ 11a-5b \quad\;\; =44 \end{array}$$ **(4)**

Add **(2)** and $3\cdot$ **(3)**.

$$\begin{array}{l} 4a+2b-3c=10 \\ \underline{6a-3b+3c=30} \\ 10a\;-b \quad\;\; =40 \end{array}$$ **(5)**

Add **(4)** and –5 times **(5)**.

$$\begin{array}{l} 11a-5b=44 \\ \underline{-50a+5b=-200} \\ -39a \quad\; =-156 \end{array}$$
$$a=4$$

Substitute $a = 4$ into **(5)**.

$$10(4)-b=40$$
$$40-b=40$$
$$-b=0$$
$$b=0$$

Substitute $a = 4$, $b = 0$ into **(3)**.
$c = 10 + 0 - 2(4) = 2$
(4, 0, 2) is the solution.

17. Multiply all three equations by 10 to clear decimals.

$$2a + b + 2c = 1 \quad \textbf{(1)}$$
$$3a + 2b + 4c = -1 \quad \textbf{(2)}$$
$$6a + 11b + 2c = 3 \quad \textbf{(3)}$$

Subtract **(3)** from **(1)**.

$$-4a - 10b = -2$$
$$2a + 5b = 1 \quad \textbf{(4)}$$

Add $-2 \cdot$ **(1)** and **(2)**.

$$-a = -3$$
$$a = 3$$

Substitute $a = 3$ into **(4)**.

$$2(3) + 5b = 1$$
$$6 + 5b = 1$$
$$5b = -5$$
$$b = -1$$

Substitute $a = 3$, $b = -1$ into **(1)**.

$$2(3) + (-1) + 2c = 1$$
$$6 - 1 + 2c = 1$$
$$2c = -4$$
$$c = -2$$

(3, −1, −2) is the solution.

19. When a calculator is used it is convenient to keep all three equations together as the operations are performed.

$$x - 4y + 4z = -3.72186$$
$$-x + 3y - z = 5.98115$$
$$2x - y + 5z = 7.93645$$

Now perform two operations on the system; first, add the first equation to the second and add −2 times the first to the third. Note that this *does not* change the first equation but the second and third.

$$x - 4y + 4z = -3.72186$$
$$-y + 3z = 2.25929$$
$$7y - 3z = 15.38017$$

Add the second equation to the third.

$$x - 4y + 4z = -3.72186$$
$$-y + 3z = 2.25929$$
$$6y = 17.63946$$

From the third equation $y = 2.93991$ which may be substituted into the second equation to give $z = 1.73307$. Substituting these values for x and y into the first equation gives $x = 1.10551$. (1.10551, 2.93991, 1.73307) is the solution.

21.

$$x - y = 5 \quad \textbf{(1)}$$
$$2y - z = 1 \quad \textbf{(2)}$$
$$3x + 3y + z = 6 \quad \textbf{(3)}$$

Add **(2)** and **(3)**.
$3x + 5y = 7$ **(4)**
Solve **(1)** for x and substitute into **(4)**.

$$x = 5 + y$$
$$3(5 + y) + 5y = 7$$
$$15 + 3y + 5y = 7$$
$$8y = -8$$
$$y = -1$$

Substitute $y = -1$ into **(1)**.

$$x + 1 = 5$$
$$x = 4$$

Substitute $y = -1$ into **(2)**.

$$2(-1) - z = 1$$
$$-2 - z = 1$$
$$-z = 3$$
$$z = -3$$

(4, −1, −3) is the solution.

23.

$$-y + 2z = 1 \quad \textbf{(1)}$$
$$x + y + z = 2 \quad \textbf{(2)}$$
$$-x + 3z = 2 \quad \textbf{(3)}$$

Add **(2)** and **(3)**.
$y + 4z = 4$ **(4)**
Add **(1)** and **(4)**.
$6z = 5$

Substitute $z = \frac{5}{6}$ into **(4)**.

$$y + 4\left(\frac{5}{6}\right) = 4$$
$$6y + 20 = 24$$
$$6y = 4$$
$$y = \frac{2}{3}$$

Substitute $z = \frac{5}{6}$, $y = \frac{2}{3}$ into **(2)**.

$$x + \frac{2}{3} + \frac{5}{6} = 2$$
$$x = \frac{1}{2}$$

$\left(\frac{1}{2}, \frac{2}{3}, \frac{5}{6}\right)$ is the solution.

25.

$$x - 2y + z = 0 \quad \textbf{(1)}$$
$$-3x - y = -6 \quad \textbf{(2)}$$
$$y - 2z = -7 \quad \textbf{(3)}$$

Multiply **(1)** by 2 and add to **(3)**.

$2x - 3y = -7$ **(4)**
Multiply **(2)** by −3 and add to **(4)**.
$11x = 11$
$x = 1$
Substitute $x = 1$ into **(2)**.
$-3(1) - y = -6$
$y = 3$
Substitute $x = 1$, $y = 3$ into **(1)**.
$1 - 6 + z = 0$
$z = 5$
(1, 3, 5) is the solution.

27. $\frac{3}{2}a - b + 2c = 2$ **(1)**
$\frac{a}{2} + 2b - 2c = 4$ **(2)**
$-a + b = -6$ **(3)**
Add **(1)** and **(2)** then solve for *b*.
$2a + b = 6$
$b = 6 - 2a$ **(4)**
Substitute **(4)** into **(3)**.
$-a + (6 - 2a) = -6$
$-a + 6 - 2a = -6$
$-3a = -12$
$a = 4$
Substitute $a = 4$ into **(3)**.
$-4 + b = -6$
$b = -2$
Substitute $a = 4$, $b = -2$ into **(2)**.
$\frac{4}{2} + 2(-2) - 2c = 4$
$2 - 4 - 2c = 4$
$-2c = 6$
$c = -3$
(4, −2, −3) is the solution.

29. $2x + y = -3$ **(1)**
$2y + 16z = -18$ **(2)**
$-7x - 3y + 4z = 6$ **(3)**
Add **(2)** and $-4 \cdot$ **(3)**.
$2y + 16z = -18$
$\underline{28x + 12y - 16z = -24}$
$28x + 14y = -42$ and dividing by 14
$2x + y = -3$ which is **(1)**.
The system of equations is a dependent system and has an infinite number of solutions.

31. $3x + 3y - 3z = -1$ **(1)**
$4x + y - 2z = 1$ **(2)**
$-2x + 4y - 2z = -8$ **(3)**
Subtract **(3)** form **(2)**.
$6x - 3y = 9$ **(4)**
Multiply **(1)** by −2 and add to 3 times **(2)**.
$6x - 3y = 5$ **(5)**
Comparing **(4)** and **(5)** gives 5 = 9 which is false. This is an inconsistent system of equations and has no solution.

Cumulative Review

33. $|2x - 1| = 7$
$2x - 1 = 7$ or $2x - 1 = -7$
$2x = 8$ $\quad 2x = -6$
$x = 4$ $\quad x = -3$
$x = -3, 4$

34. $76,300,000 = 7.63 \times 10^7$

35. $m = \frac{y_2 - y_1}{x_2 - x_1} = \frac{4-3}{1-(-2)} = \frac{1}{3}$
$y - y_1 = m(x - x_1)$
$y - 4 = \frac{1}{3}(x - 1)$
$3y - 12 = x - 1$
$x - 3y = -11$

36. $y = -\frac{2}{3}x + 4 \Rightarrow m = -\frac{2}{3},\ m_{\perp} = \frac{3}{2}$
$y - y_1 = m(x - x_1)$
$y - 2 = \frac{3}{2}(x - (-4))$
$2y - 4 = 3x + 12$
$3x - 2y = -16$

Quick Quiz 4.2

1. $4x - y + 2z = 0$ **(1)**
$2x + y + z = 3$ **(2)**
$3x - y + z = -2$ **(3)**
Add **(1)** and **(2)**.
$6x + 3z = 3$ **(4)**
Add **(2)** and **(3)**.
$5x + 2z = 1$ **(5)**
Solve **(4)** for *x* then substitute in **(5)**.
$6x + 3z = 3$
$6x = 3 - 3z$
$x = \frac{3}{6} - \frac{3}{6}z$
$x = \frac{1}{2} - \frac{1}{2}z$

$5\left(\frac{1}{2}-\frac{1}{2}z\right)+2z=1$
$\frac{5}{2}-\frac{5}{2}z+2z=1$
$-\frac{5}{2}z+2z=1-\frac{5}{2}$
$-\frac{1}{2}z=-\frac{3}{2}$
$z=\frac{3}{2}\cdot\frac{2}{1}$
$z=3$
Substitute $z = 3$ into **(4)**.
$5x+2(3)=1$
$5x+6=1$
$5x=-5$
$x=-1$
Substitute $x = -1$, $z = 3$ into **(3)**.
$3(-1)-y+3=-2$
$-3-y+3=-2$
$-y=-2$
$y=2$
$(-1, 2, 3)$ is the solution.

2. $x+2y+2z=-1$ **(1)**
$2x-y+z=1$ **(2)**
$x+3y-6z=7$ **(3)**
Add **(1)** and –**(3)** then solve for y.
$x+2y+2z=-1$
$-x-3y+6z=-7$
$-y+8z=-8$
$-y=-8-8z$
$y=8+8z$ **(4)**
Add -2**(1)** and **(2)**, substitute **(4)** into **(5)**.
$-2x-4y-4z=2$
$2x-y+z=1$
$-5y-3z=3$ **(5)**
$-5(8+8z)-3z=3$
$-40-40z-3z=3$
$-43z=43$
$z=-1$
Substitute $z = -1$ into **(5)**.
$-5y-3(-1)=3$
$-5y+3=3$
$-5y=0$
$y=0$
Substitute $y = 0$, $z = -1$ into **(1)**.
$x+2(0)+2(-1)=-1$
$x+0-2=-1$
$x=1$
$(1, 0, -1)$ is the solution.

3. $4x-2y+6z=0$ **(1)**
$6y+3z=3$ **(2)**
$x+2y-z=5$ **(3)**
Add **(1)** and -4**(3)** then solve for y.
$4x-2y+6z=0$
$-4x-8y+4z=-20$
$-10y+10z=-20$
$-10y=-20-10z$
$y=2+z$ **(4)**
Substitute **(4)** into **(2)**.
$6(2+z)+3z=3$
$12+6z+3z=3$
$9z=-9$
$z=-1$
Substitute $z = -1$ into **(2)**.
$6y+3(-1)=3$
$6y-3=3$
$6y=6$
$y=1$
Substitute $y = 1$, $z = -1$ into **(3)**.
$x+2(1)-(-1)=5$
$x+2+1=5$
$x+3=5$
$x=2$
$(2, 1, -1)$ is the solution.

4. Answers may vary. Possible solution:
To eliminate z and obtain two equations with only variables x and y one could do the following.
Add $5 \cdot$ **(1)** to $2 \cdot$ **(2)** and add -3**(2)** to $5 \cdot$ **(3)**

How Am I Doing? Sections 4.1–4.2

1. $4x-y=-1$ **(1)**
$3x+2y=13$ **(2)**
Solve **(1)** for y and substitute into **(2)**.
$4x-y=-1$
$-y=-1-4x$
$y=1+4x$
$3x+2(1+4x)=13$
$3x+2+8x=13$
$11x=11$
$x=1$

Substitute $x = 1$ into **(2)**.

$$\begin{aligned} 3(1)+2y &= 13 \\ 2y &= 10 \\ y &= 5 \end{aligned}$$

(1, 5) is the solution.

2. $5x+2y=0$ **(1)**
$-3x-4y=14$ **(2)**

Add 2(**1**) to (**2**). Then add.

$$\begin{aligned} 10x+4y &= 0 \\ -3x-4y &= 14 \\ \hline 7x \quad &= 14 \\ x &= 2 \end{aligned}$$

Substitute $x = 2$ into (**1**).

$$\begin{aligned} 5(2)+2y &= 0 \\ 2y &= -10 \\ y &= -5 \end{aligned}$$

(2, –5) is the solution.

3. $5x-2y=27$ **(1)**
$3x-5y=-18$ **(2)**

Multiply (**1**) by 5, and (**2**) by –2 and add.

$$\begin{aligned} 25x-10y &= 135 \\ -6x+10y &= 36 \\ \hline 19x \quad &= 171 \\ x &= 9 \end{aligned}$$

Substitute $x = 9$ into (**1**).

$$\begin{aligned} 5(9)-2y &= 27 \\ 45-2y &= 27 \\ 2y &= 18 \\ y &= 9 \end{aligned}$$

The solution is (9, 9).

4. $7x+3y=15$ **(1)**
$\frac{1}{3}x-\frac{1}{2}y=2$ **(2)**

Multiply (**2**) by 6 to clear fractions.

$2x - 3y = 12$ **(3)**

Add (**1**) and (**3**).

$$\begin{aligned} 7x+3y &= 15 \\ 2x-3y &= 12 \\ \hline 9x \quad &= 27 \\ x &= 3 \end{aligned}$$

Substitute $x = 3$ into (**1**).

$$\begin{aligned} 7(3)+3y &= 15 \\ 3y &= -6 \\ y &= -2 \end{aligned}$$

The solution is (3, –2).

5. $2x=3+y$ **(1)**
$3y=6x-9$ **(2)**

Solve (**1**) for y.

$y = 2x - 3$ and substitute into (**2**).

$$\begin{aligned} 3(2x-3) &= 6x-9 \\ 6x-9 &= 6x-9 \\ 0 &= 0 \end{aligned}$$

The equations are dependent and the system has an infinite number of solutions.

6. Multiply both equations by 10 to clear decimals.

$2x+7y=-10$ **(1)**
$5x+6y=-2$ **(2)**

Multiply (**1**) by 5 and (**2**) by –2 and add.

$$\begin{aligned} 10x+35y &= -50 \\ -10x-12y &= 4 \\ \hline 23y &= -46 \\ y &= -2 \end{aligned}$$

Substitute $y = -2$ into (**1**).

$$\begin{aligned} 2x+7(-2) &= -10 \\ 2x-14 &= -10 \\ 2x &= 4 \\ x &= 2 \end{aligned}$$

The solution is (2, –2).

7. $6x-9y=15$ **(1)**
$-4x+6y=8$ **(2)**

Add 2 times (**1**) to 3 times (**2**).

$$\begin{aligned} 12x-18y &= 30 \\ -12x+18y &= 54 \\ \hline 0 &= 84 \end{aligned}$$

This is an inconsistent system with no solution. The lines are parallel.

8. Substituting (2, 0, –3) into (**1**)

$$\begin{aligned} 4(2)-0+3(-3) &\overset{?}{=} -1 \\ -1 &= -1 \text{ yes} \end{aligned}$$

Substituting (2, 0, –3) into (**2**)

$$\begin{aligned} -2(2)+3(0)+5(-3) &\overset{?}{=} -19 \\ -19 &= -19 \text{ yes} \end{aligned}$$

Substituting (2, 0, –3) into (**3**)

$$\begin{aligned} 2-2(0)+4(-3) &\overset{?}{=} 10 \\ -10 &= 10 \text{ no} \end{aligned}$$

(2, 0, –3) does not satisfy all equations, it is not the solution.

9. $5x-2y+z=-1$ **(1)**
$3x+y-2z=6$ **(2)**
$-2x+3y-5z=7$ **(3)**

Multiply (**1**) by 2 and add to (**2**).

$$\begin{array}{r} 10x-4y+2z=-2 \\ 3x+y-2z=6 \\ \hline 13x-3y=4 \end{array} \quad \textbf{(4)}$$

Multiply **(1)** by 5 and add to **(3)**.

$$\begin{array}{r} 25x-10y+5z=-5 \\ -2x+3y-5z=7 \\ \hline 23x-7y=2 \end{array} \quad \textbf{(5)}$$

Multiply **(4)** by 7 and **(5)** by −3 and add.

$$\begin{array}{r} 91x-21y=28 \\ -69x+21y=-6 \\ \hline 22x \qquad =22 \\ x=1 \end{array}$$

Substitute $x = 1$ into **(5)**.

$$\begin{aligned} 23(1)-7y&=2 \\ 7y&=21 \\ y&=3 \end{aligned}$$

Substitute $x = 1$, $y = 3$ into **(1)**.

$$\begin{aligned} 5(1)-2(3)+z&=-1 \\ 5-6+z&=-1 \\ z&=0 \end{aligned}$$

The solution is (1, 3, 0).

10. $2x-y+3z=-1$ **(1)**
$5x+y+6z=0$ **(2)**
$2x-2y+3z=-2$ **(3)**

Add **(1)** and **(2)**.

$$\begin{array}{r} 2x-y+3z=-1 \\ 5x+y+6z=0 \\ \hline 7x \quad +9z=-1 \end{array} \quad \textbf{(4)}$$

Add 2 times **(2)** to **(3)**.

$$\begin{array}{r} 10x+2y+12z=0 \\ 2x-2y+3z=-2 \\ \hline 12x \quad +15z=-2 \end{array} \quad \textbf{(5)}$$

Add 5 times **(4)** to −3 times **(5)**.

$$\begin{array}{r} 35x+45z=-5 \\ -36x-45z=6 \\ \hline -x \qquad =1 \\ x=-1 \end{array}$$

Substitute $x = -1$ into **(4)**.

$$\begin{aligned} 7(-1)+9z&=-1 \\ 9z&=6 \\ z&=\frac{2}{3} \end{aligned}$$

Substitute $x=-1,\ z=\frac{2}{3}$ into **(2)**.

$$\begin{aligned} 5(-1)+y+6\left(\frac{2}{3}\right)&=0 \\ -5+y+4&=0 \\ y-1&=0 \\ y&=1 \end{aligned}$$

$\left(-1, 1, \frac{2}{3}\right)$ is the solution.

11. $x+y+2z=9$ **(1)**
$3x+2y+4z=16$ **(2)**
$2y+z=10$ **(3)**

Solve **(3)** for z and substitute into **(1)** and **(2)**.

$z = 10 - 2y$

$$\begin{aligned} x+y+2(10-2y)&=9 \\ x+y+20-4y&=9 \\ x-3y&=-11 \quad \textbf{(4)} \end{aligned}$$

$$\begin{aligned} 3x+2y+4(10-2y)&=16 \\ 3x+2y+40-8y&=16 \\ 3x-6y&=-24 \quad \textbf{(5)} \end{aligned}$$

Multiply **(4)** by −2 and add to **(5)**.

$$\begin{array}{r} -2x+6y=22 \\ 3x-6y=-24 \\ \hline x=-2 \end{array}$$

Substitute $x = -2$ into **(5)**.

$$\begin{aligned} 3(-2)-6y&=-24 \\ -6y&=-18 \\ y&=3 \end{aligned}$$

Substitute $x = -2$, $y = 3$ into **(1)**.

$$\begin{aligned} -2+3+2z&=9 \\ 2z&=8 \\ z&=4 \end{aligned}$$

The solution is (−2, 3, 4).

12. $x-2z=-5$ **(1)**
$y-3z=-3$ **(2)**
$2x-z=-4$ **(3)**

Solve **(3)** for z and substitute into **(1)**.

$z = 2x + 4$

$$\begin{aligned} x-2(2x+4)&=-5 \\ x-4x-8&=-5 \\ -3x&=3 \\ x&=-1 \end{aligned}$$

Substitute $x = -1$ into **(3)**.

$$\begin{aligned} 2(-1)-z&=-4 \\ -z&=-2 \\ z&=2 \end{aligned}$$

Substitute $z = 2$ into **(2)**.

$y-3(2)=-3$
$y-6=-3$
$y=3$
The solution is (−1, 3, 2).

4.3 Exercises

1. Let x = the smaller number and y = the larger number.
$x+y=87 \Rightarrow y=87-x$
$y-2x=12 \Rightarrow 87-x-2x=12$
$87-3x=12$
$-3x=-75$
$x=25$, the smaller number
$y=87-25=62$, the larger number

3. Let x = number of heavy equipment operators and y = number of laborers.
$x+y=35 \Rightarrow y=35-x$
$140x+90y=3950$
$140x+90(35-x)=3950$
$140x+3150-90x=3950$
$50x=800$
$x=16$
$y=35-x=35-16=19$
16 heavy equipment operators and 19 laborers were employed.

5. x = number of tickets for regular coach seats
y = number of tickets for sleeper car seats
$x+y=98$ **(1)**
$120x+290y=19{,}750$ **(2)**
Solve **(1)** for y and substitute into **(2)**.
$y=98-x$
$120x+290(98-x)=19{,}750$
$120x+28{,}420-290x=19{,}750$
$-170x=-8670$
$x=51$
Substitute 51 for x in **(1)**.
$51+y=98$
$y=47$
Number of coach tickets = 51
Number of sleeper tickets = 47

7. Let x = no. of experienced employees
y = no. of new employees
$3x+4y=115$ **(1)**
$4x+7y=170$ **(2)**
Solve **(1)** for x the substitute into **(2)**.
$3x=115-4y$
$x=\frac{115}{3}-\frac{4}{3}y$
$4\left(\frac{115}{3}-\frac{4}{3}y\right)+7y=170$
$\frac{460}{3}-\frac{16}{3}y+7y=170$
$\frac{5}{3}y=\frac{50}{3}$
$y=10$
Substitute $y=10$ into **(1)**.
$3x+4(10)=115$
$3x+40=115$
$3x=115-40$
$3x=75$
$x=25$
(25, 10) is the solution.
There were 25 experienced and 10 new employees.

9. x = number of packages of old fertilizer
y = number of packages of new fertilizer
$50x+65y=3125 \xrightarrow{\times 6} 300x+390y=18{,}750$
$60x+45y=2925 \xrightarrow{\times -5} -300x-225y=-14{,}625$
$165y=4125$
$y=25$ new packages
$50x+65(25)=3125$
$50x+1625=3125$
$50x=1500$
$x=30$ old packages
He should use 30 old packages and 25 new packages.

11. x = cost of one doughnut
y = cost of one large coffee
$3x+4y=4.91$ **(1)**
$5x+6y=7.59$ **(2)**
Multiply **(1)** by 5 and add to −3 times **(2)**.
$15x+20y=24.55$
$-15x-18y=-22.77$
$2y=1.78$
$y=0.89$
Substitute 0.89 for y in **(1)** and solve for x.
$3x+4(0.89)=4.91$
$3x=1.35$
$x=0.45$
The cost of one doughnut is \$0.45. The cost of one large coffee is \$0.89.

13. x = speed of plane in still air
y = speed of wind

$$(x-y)\left(\frac{7}{6}\right)=210$$
$$(x+y)\left(\frac{5}{6}\right)=210$$

Clear fractions.

$$(x-y)\left(\frac{7}{6}\right)\cdot\frac{6}{7}=210\cdot\frac{6}{7}$$
$$(x+y)\left(\frac{5}{6}\right)\cdot\frac{6}{5}=210\cdot\frac{6}{5}$$

$$\begin{aligned} x-y&=180\\ x+y&=252\\ \hline 2x\quad&=432\\ x&=216 \end{aligned}$$

$y = 252 - x = 36$
plane: 216 mph; wind: 36 mph

15. x = speed in still water
y = speed of current

$$(x-y)\cdot\frac{2}{3}=8 \Rightarrow x=y+12$$
$$(x+y)\cdot\frac{1}{2}=8 \Rightarrow (y+12+y)\cdot\frac{1}{2}=8$$

$2y + 12 = 16 \Rightarrow 2y = 4$
$y = 2$ mph, speed of current
$x = y + 12 = 14$ mph, speed of boat

17. Let x = no. of free throws
y = no. of 2 point shots

$x+2y=100$ **(1)**
$x+y=64$ **(2)**

Solve **(2)** for x then substitute into **(1)**.

$x = 64 - y$
$64-y+2y=100$
$y=36$

Substitute $y = 36$ into **(1)**.

$x = 64 - 36$
$x = 28$

(28, 36) is the solution.
There were 28 free throws and 36 regular shots.

19. x = weekend minutes
y = weekday minutes

$$x + y = 625 \Rightarrow y = 625 - x$$
$$0.05x+0.08y=43.40$$
$$0.05x+0.08(625-x)=43.4$$
$$-0.03x=-6.6$$
$$x=220$$

$y = 625 - x = 405$
Nick talked 405 minutes on the weekdays and 220 minutes on weekends.

21. x = cost of truck, y = cost of car
$256x + 183y = 5,791,948$

$$64x+107y=2,507,612 \xrightarrow{\times(-4)}$$
$$\begin{aligned} -256x-428y&=-10,030,448\\ 256x+183y&=5,791,948\\ \hline -245y&=-4,238,500\\ y&=17,300 \end{aligned}$$
$$\begin{aligned} 64x+107(17,300)&=2,507,612\\ 64x+1,851,100&=2,507,612\\ 64x&=656,512\\ x&=10,258 \end{aligned}$$

The department paid \$10,258 for a car and \$17,300 for a truck.

23. Let x = no. of pens
y = no. of notebooks
z = no. of highlighters

$x+y+z=15$ **(1)**
$(0.50)x+3y+(1.50)z=23$ **(2)**
$y=z+2$ **(3)**

Add **(1)** and -2**(2)** then solve for z.

$$\begin{aligned} x+y+z&=15\\ -x-6y-3z&=-46\\ \hline -5y-2z&=-31\\ -2z&=-31+5y\\ z&=\frac{31}{2}-\frac{5}{2}y \end{aligned}$$

Substitute $z=\frac{31}{2}-\frac{5}{2}y$ into **(3)**.

$$y=\frac{31}{2}-\frac{5}{2}y+2$$
$$\frac{7}{2}y=\frac{35}{2}$$
$$y=5$$

Substitute $y = 5$ into **(3)**.

$5 = z + 2$
$3 = z$

Substitute $z = 3$, $y = 5$ into **(1)**.

$x+5+3=15$
$x+8=15$
$x=7$

(7, 5, 3) is the solution.
She bought 7 pens, 5 notebooks, and 3 highlighters.

25. x = number of adults
y = number of high school students
z = number of children

$5x+3y+2z=1010$ **(1)**
$7x+4y+3z=1390$ **(2)**
$x+y+z=300$ **(3)**

Multiply –7 times **(3)** and add to **(2)**.
$-3y-4z=-710$, $3y+4z=710$ **(4)**
Multiply –5 times **(3)** and add to **(1)**.
$-2y-3z=-490$, $2y+3z=490$ **(5)**
Add $2\cdot$ **(4)** and $-3\cdot$ **(5)**.

$$-z=-50$$
$$z=50$$

Substitute $z=50$ into **(5)**.

$$2y+3(50)=490$$
$$2y=340$$
$$y=170$$

Substitute $y=170$, $z=50$ into **(3)**.

$$x+170+50=300$$
$$x=80$$

The number of adults attending was 80, the number of high school students was 170, and the number of children was 50.

27. x = number of children under 12
y = number of adults
z = number of senior citizens

$x+y+z=12{,}000$ **(1)**
$0.25x+y+0.5z=10{,}700$ **(2)**
$0.35x+1.5y+0.5z=15{,}820$ **(3)**

Add –0.25 times **(1)** to **(2)**.

$$-0.25x-0.25y-0.25z=-3000$$
$$0.25x+y+0.5z=10{,}700$$
$$0.75y+0.25z=7700 \quad \textbf{(4)}$$

Add –0.35 times **(1)** to **(3)**.

$$-0.35x-0.35y-0.35z=-4200$$
$$0.35x+1.5y+0.5z=15{,}820$$
$$1.15y+0.15z=11{,}620 \quad \textbf{(5)}$$

Add –1.15 times **(4)** to 0.75 times **(5)**.

$$-0.08626y-0.2875z=-8855$$
$$0.8625y+0.1125z=8715$$
$$-0.175z=-140$$
$$z=800$$

Substitute 800 for z in **(1)**.
$x+y=11{,}200$ **(6)**
Add –1 times **(3)** to **(2)**.

$$0.25x+y+0.5z=10{,}700$$
$$-0.35x-1.5y-0.5z=-15{,}820$$
$$-0.1x-0.5y=-5120 \quad \textbf{(7)}$$

Add 0.5 times **(6)** to **(7)**.

$$0.5x+0.5y=5600$$
$$-0.1x-0.5y=-5120$$
$$0.4x \quad =480$$
$$x=1200$$

Substitute 1200 for x in **(6)**.

$$1200+y=11{,}200$$
$$y=10{,}000$$

1200 children, 10,000 adults, and 800 senior citizens normally ride.

29. x = number of small pizzas
y = number of medium pizzas
z = number of large pizzas

$$x+y+z=20$$
$$3x+4y+5z=82$$
$$5x+9y+12z=181$$

Add –3 times first equation and –5 times first equation to third equation.

$$x+y+z=20$$
$$y+2z=22$$
$$4y+7z=81$$

Add –4 times second equation to third equation.

$$z=7$$
$$y+2z=22$$
$$y+2(7)=22$$
$$y=8 \text{ medium pizzas}$$
$$x+8+7=20$$
$$x=5 \text{ small pizzas}$$

There were 5 small, 8 medium, and 7 large pizzas delivered.

31. Let x = no. of A boxes
y = no. of B boxes
z = no. of C boxes

$12x+10y+5z=91$ **(1)**
$5x+6y+8z=63$ **(2)**
$3x+4y+5z=40$ **(3)**

Add **(1)** and –**(3)**.

$$12x+10y+5z=91$$
$$-3x-4y-5z=-40$$
$$9x+6y=51 \quad \textbf{(4)}$$

Add $-5\cdot$ **(2)** and $8\cdot$ **(3)** then solve for x.

$$-25x-30y-40z=-315$$
$$24x+32y+40z=320$$
$$-x+2y=5$$
$$-x=5-2y$$
$$x=2y-5 \quad \textbf{(5)}$$

Substitute **(5)** into **(4)**.

$$9(2y-5)+6y=51$$
$$18y-45+6y=51$$
$$24y=6$$
$$y=4$$

Substitute $y = 4$ into **(5)**.

$x = 2(4) - 5$

$x = 8 - 5$

$x = 3$

Substitute $x = 3$, $y = 4$ into **(3)**.

$$3(3)+4(4)+5z=40$$
$$9+16+5z=40$$
$$5z=15$$
$$z=3$$

(3, 4, 3) is the solution.

She can prepare 3 *A* boxes, 4 *B* boxes, and 3 *C* boxes.

Cumulative Review

32. $\frac{1}{3}(4-2x)=\frac{1}{2}x-3$

$$2(4-2x)=3x-18$$
$$8-4x=3x-18$$
$$7x=26$$
$$x=\frac{26}{7}=3\frac{5}{7}$$

33. $0.06x+0.15(0.5-x)=0.04$

$$0.06x+0.075-0.15x=0.04$$
$$-0.09x=-0.035$$
$$x=\frac{-0.035}{-0.09}=\frac{35}{90}$$
$$x=\frac{7(5)}{18(5)}=\frac{7}{18}$$

34. $2(y-3)-(2y+4)=-6y$

$$2y-6-2y-4=-6y$$
$$6y=10$$
$$y=\frac{5}{3}=1\frac{2}{3}$$

35. $6a(2x-3y)=7ax-3$

$$12ax-18ay=7ax-3$$
$$5ax=18ay-3$$
$$x=\frac{18ay-3}{5a}$$

Quick Quiz 4.3

1. Let x = speed of plane in still air
y = speed of wind

$2.5x+2.5y=1200$ **(1)**

$3x-3y=1200$ **(2)**

Solve **(2)** for x and substitute into **(1)**.

$$3x=1200+3y$$
$$x=400+y$$

$$2.5(x)+2.5y=1200$$
$$2.5(400+y)+2.5y=1200$$
$$1000+2.5y+2.5y=1200$$
$$5y=200$$
$$y=40$$

Substitute $y = 40$ into **(2)**.

$$3x-3(40)=1200$$
$$3x-120=1200$$
$$3x=1320$$
$$x=440$$

speed of plane in still air = 440 mph
speed of wind = 40 mph

2. Let x = mileage fee
y = daily fee

$300x+8y=355$ **(1)**

$260x+9y=380$ **(2)**

Add $9 \cdot$ **(1)** and $-8 \cdot$ **(2)**.

$$2700x+72y=3195$$
$$-2080x-72y=-3040$$
$$620x \quad =155$$
$$x=\frac{1}{4}$$

Substitute $x=\frac{1}{4}$ into **(1)**.

$$300\left(\frac{1}{4}\right)+8y=355$$
$$75+8y=355$$
$$8y=280$$
$$y=35$$

The mileage fee is \$0.25 per mile.
The daily fee is \$35 per day.

3. Let x = price of Drawing
y price of Carved Elephant
z = price of Drum Set

$3x+2y+z=55$ **(1)**

$2x+3y+z=65$ **(2)**

$4x+3y+2z=85$ **(3)**

Add **(1)** and –**(2)**.

$$\begin{array}{r} 3x+2y+z=55 \\ -2x-3y-z=-65 \\ \hline x-y \quad =-10 \end{array}$$
$$x=y-10 \quad \textbf{(4)}$$

Add −2(**2**) and (**3**).

$$\begin{array}{r} -4x-6y-2z=-130 \\ 4x+3y+2z=85 \\ \hline -3y \quad =-45 \end{array}$$
$$y=15$$

Substitute $y = 15$ into (**4**).

$x = 15 - 10$

$x = 5$

Substitute $x = 5$, $y = 15$ into (**1**).

$$3(5)+2(15)+z=55$$
$$15+30+z=55$$
$$z=10$$

Drawing price is $5

Carved Elephant price is $15

Drum Set price is $10

4. Answers may vary. Possible solution: The equations would be set up the same except the right sides would be set to 1500 rather than 1200.

4.4 Exercises

1. In the graph of the system $y > 3x + 1$, $y < -2x + 5$ the boundary lines should be dashed because they are not included in the solution. The system contains only < or > symbols.

3. Test the point in both regions.

$y < -2x + 3$, $-4 < -2(3) + 3 = -3$, True

$y > 5x - 3$, $-4 > 5(3) - 3 = 12$, False

The point (3, −4) does not lie in the solution region.

5. $y \geq 2x - 1$

Test point: (0, 0)

$0 \geq 2(0) - 1$

$0 \geq -1$ True

$x + y \leq 6$

Test point: (0, 0)

$0+0 \leq 6$

$0 \leq 6$ True

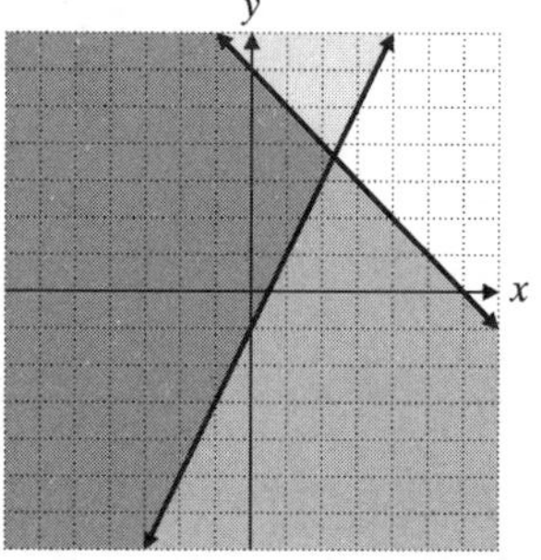

7. $y \geq -2x$

$y \geq 3x + 5$

Test point: (−1, 4)

$y \geq -2x$

$4 \geq -2(-1)$

$4 \geq -3$ True

$y \geq 3x+5$

$4 \geq 3(-1)+5$

$4 \geq -3+5$

$4 \geq 2$ True

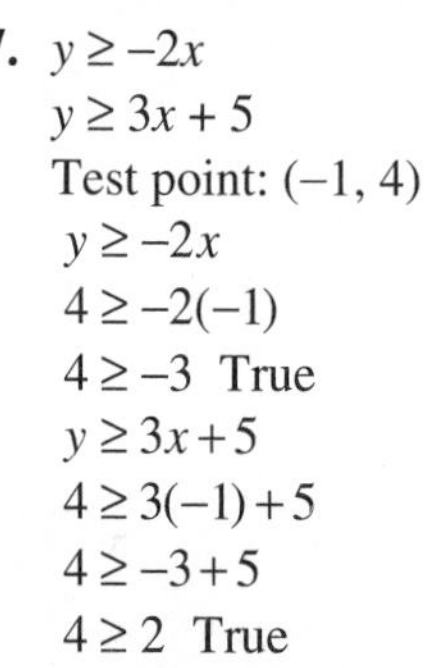

9. $y \geq 2x - 3$

Test point: (0, −1)

$-1 \geq 2(0)-3$

$-1 \geq -3$ True

$y \leq \frac{2}{3}x$

$-1 \leq \frac{2}{3}(0)$

$-1 \leq 0$ True

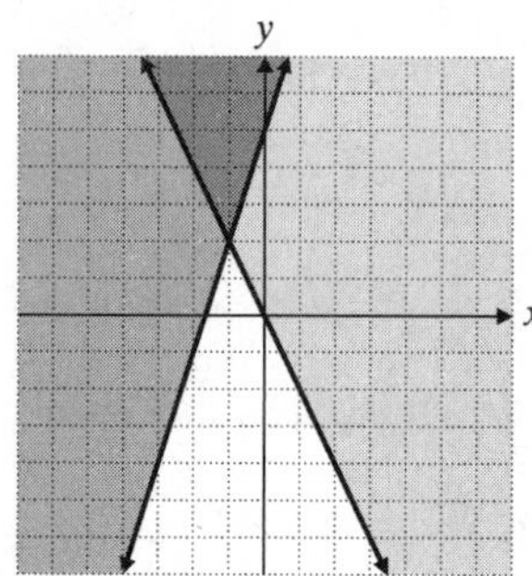

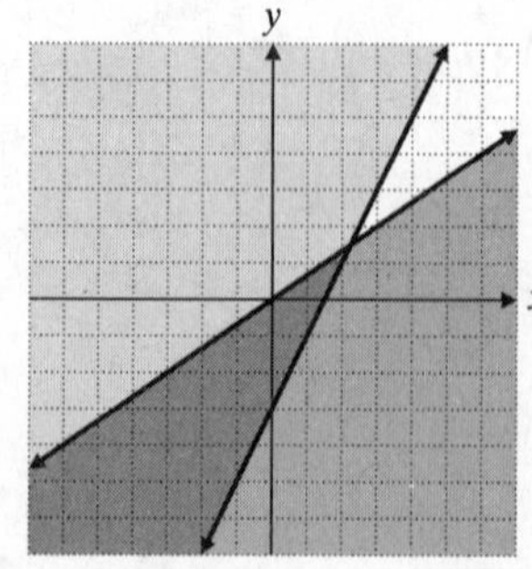

11. $x - y \geq -1$
Test point: (0, 0) $\Rightarrow$ $0 - 0 \geq -1$
$0 \geq -1$ True
$-3x - y \leq 4$, Test point: (0, 0)
$-3(0) - 0 \leq 4$
$0 \leq 4$ True

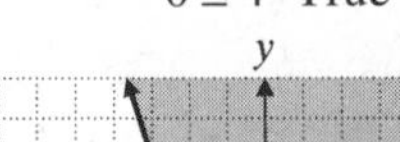

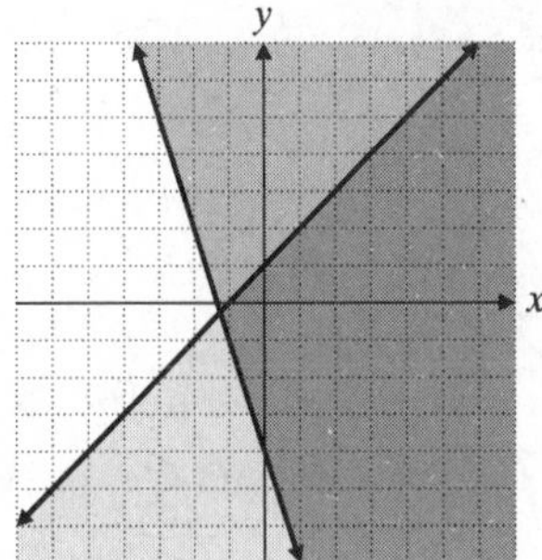

13. $x + 2y < 6$
Test point: (0, 0)
$0 + 2(0) < 6$
$0 < 6$ True
$y < 3$
Test point: (0, 0)
$0 < 3$ True

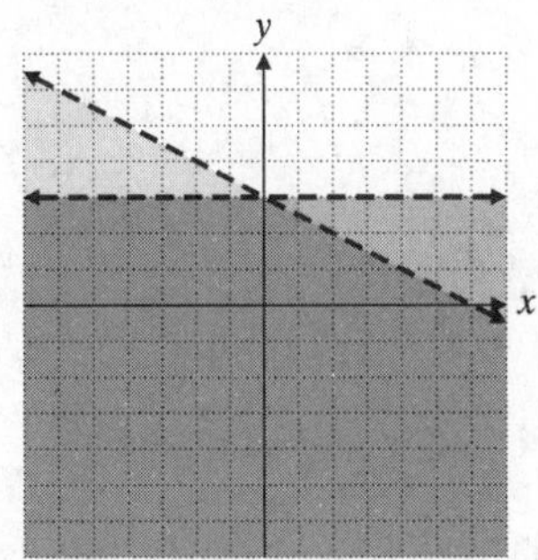

15. $y < 4, x > -2$

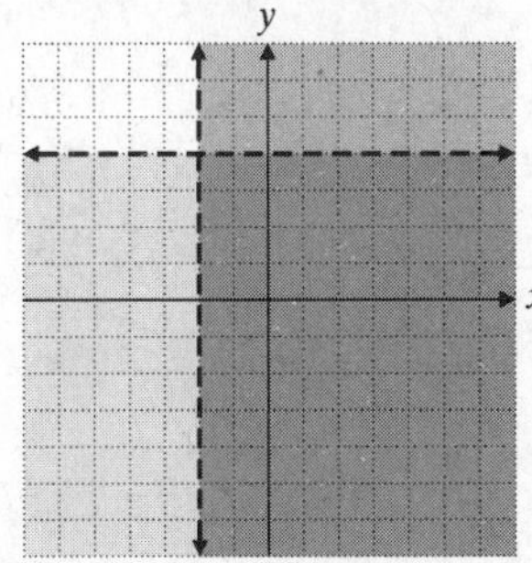

17. $x - 4y \geq -4$
Test point: (0, 0)
$0 - 4(0) \geq -4$
$0 \geq -4$ True
$3x + y \leq 3$
Test point: (0, 0)
$3(0) + 0 \leq 3$
$0 \leq 3$ True

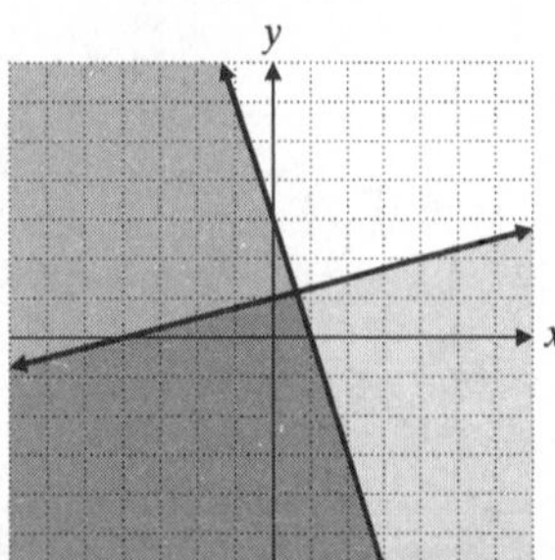

19. $3x + 2y < 6$
Test point: (0, 0)
$3(0) + 2(0) < 6$
$0 < 6$ True
$3x + 2y > -6$
Test point: (0, 0)
$3(0) + 2(0) > -6$
$0 > -6$ True

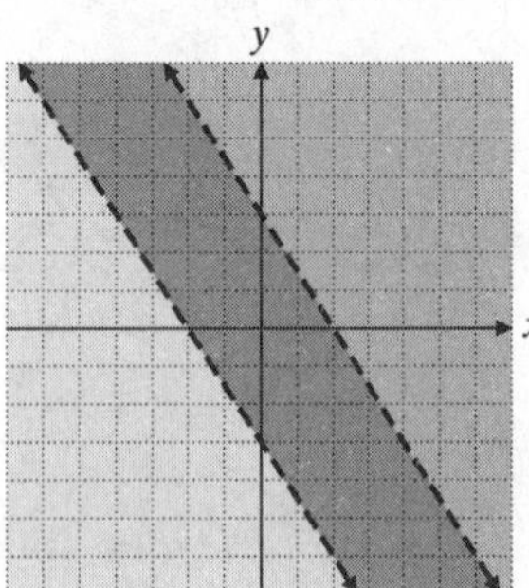

21. $x + y \leq 5$
Test point: (0, 0)
$0 + 0 \leq 5$
$0 \leq 5$ True
$2x - y \geq 1$
Test point: (0, 0)
$2(0) - 0 \geq 1$
$0 \geq 1$ False

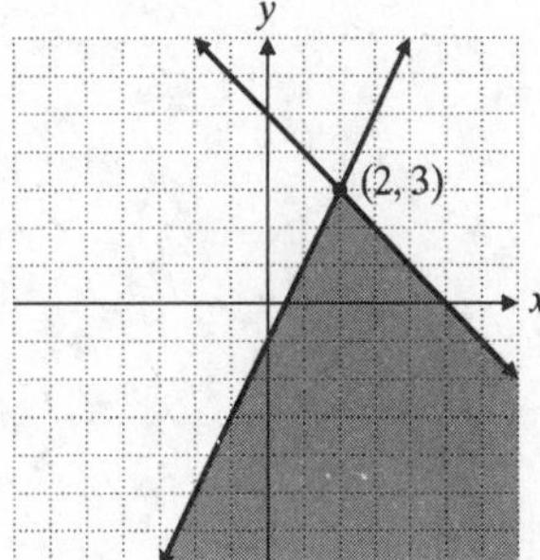

23. $x + 3y \le 12$
Test point: (0, 0)
$0+3(0)\le 12$
$0 \le 12$ True
$y < x$
Test point: (2, 0)
$0 < 2$ True

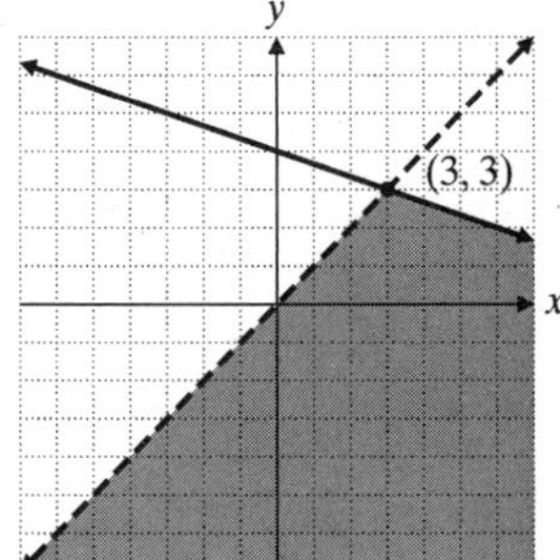

25.
$y \le x$ **(1)**
$x+y\ge 1$ **(2)**
$x\le 5$ **(3)**
Test point: (2, 1)
$y \le x$ **(1)**
$1\le 2$ True
$x+y\ge 1$ **(2)**
$1+2\ge 1$
$3\ge 1$ True
$x\le 5$ **(3)**
$2\le 5$ True

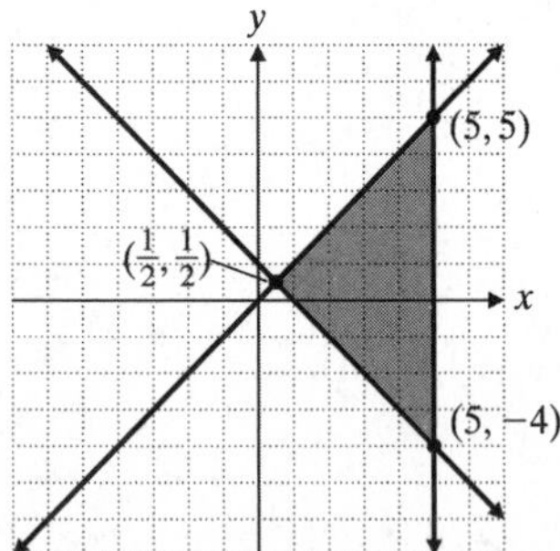

27. $y \le 3x + 6$
Test point: (0, 0)
$0\le 3(0)+6$
$0\le 6$ True
$4x + 3y \le 3$
Test point: (0, 0)
$4(0)+3(0)\le 3$
$0\le 3$ True
$x \ge -2,\ y \ge -3$

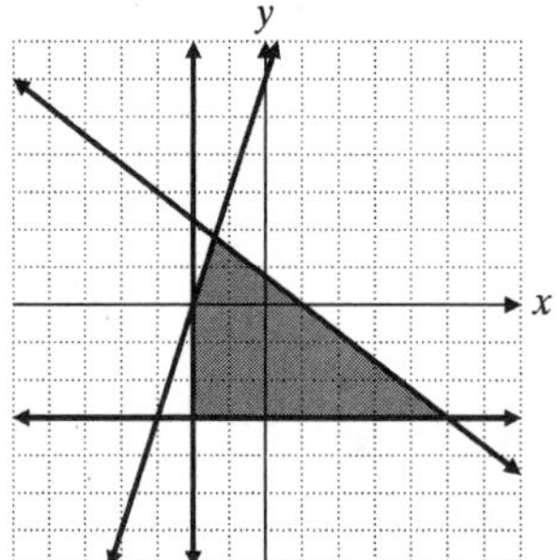

29. a. $N \le 2D$, Test point: (2, 2)
$2\le 2(2)$
$2\le 4$ True
$4N + 3D \le 20$, Test point: (2, 2)
$4(2)+3(2)\le 20$
$8+6\le 20$
$14\le 20$ True

$N\ge 0,\ D\ge 0$

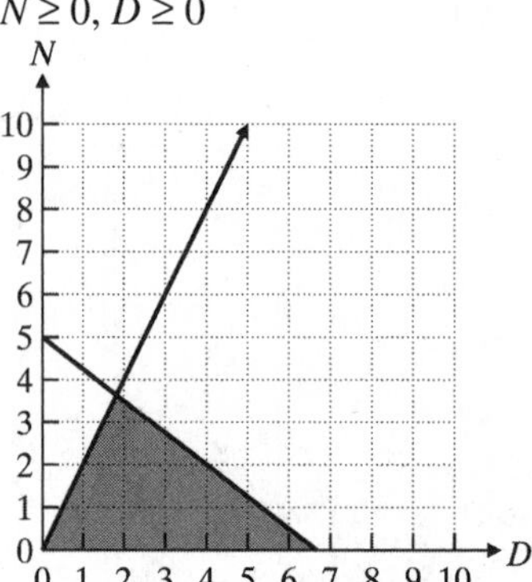

b. Yes, (3, 2) is in the shaded region.

c. No, (1, 4) is not in the shaded region.

Cumulative Review

31. $-3x^2y-x^2+5y^2 = -3(2)^2(-1)-(2)^2+5(-1)^2$
$= 12-4+5$
$= 13$

32. $2x-2[y+3(x-y)] = 2x-2[y+3x-3y]$
$= 2x-2[3x-2y]$
$= 2x-6x+4y$
$= -4x+4y$

33. x = average revenue on sunny day
y = average revenue on rainy day
$6x+y=6050$ **(1)**
$3x+4y=7400$ **(2)**
Solve **(1)** for y, substitute into **(2)**.
$y = 6050 - 6x$

$3x+4(6050-6x)=7400$
$3x+24,200-24x=7400$
$-21x=-16,800$
$x=800$
Substitute $x = 800$ into **(2)**.
$3(800)+4y=7400$
$2400+4y=7400$
$4y=5000$
$y=1250$
Average revenue on a sunny day is \$800.
Average revenue on a rainy day is \$1250.

34. Let x = price of chicken sandwich
y = price of side salad
z = price of soda
$3x+2y+3z=17.30$ **(1)**
$3x+4y+4z=23.15$ **(2)**
$4x+3y+5z=24.75$ **(3)**
Add **(1)** and –**(2)**.
$3x+2y+3z=17.30$
$-3x-4y-4z=-23.15$
$-2y-z=-5.85$
$-z=-5.85+2y$
$z=-2y+5.85$ **(4)**
Add 4 · **(2)** + –3(**3**), substitute in value of z from (**4**).
$12x+16y+16z=92.6$
$-12x-9y-15z=-74.25$
$7y+z=18.35$
$7y+(-2y+5.85)=18.35$
$7y-2y+5.85=18.35$
$5y=12.50$
$y=2.5$
Substitute $y = 2.5$ into (**4**).
$z=-2(2.5)+5.85$
$z=-5+5.85$
$z=0.85$
Substitute $z = 0.85$, $y = 2.5$ into (**1**).
$3x+2(2.5)+3(0.85)=17.30$
$3x+5+2.55=17.30$
$3x+7.55=17.30$
$3x=9.75$
$x=3.25$
One chicken sandwich costs \$3.25.
One side salad costs \$2.50.
One soda costs \$0.85.

Quick Quiz 4.4

1. Below the line since the inequality symbol used is <.

2. Solid lines since the inequality symbols used are ≥ and ≥.

3. $3x+2y>6$
$x-2y<2$

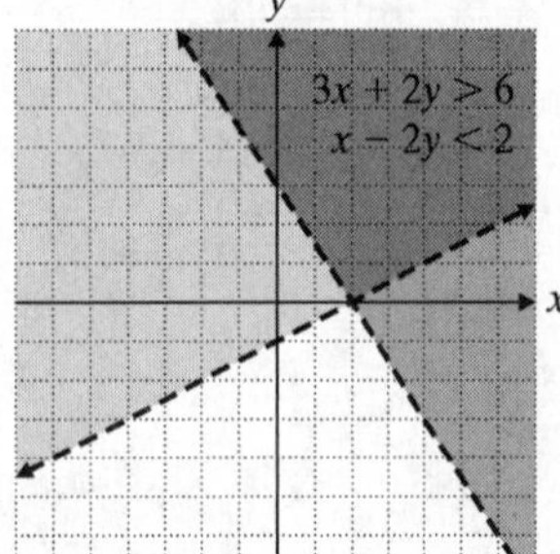

4. Answers may vary. Possible solution:
$y > x + 2$ is graphed using a dashed line and shaded above the line.
$x < 3$ is graphed using a dashed line and shaded below the line.
The region that satisfies both inequalities is the overlapping shaded regions.

Putting Your Skills to Work

1. Donivan Tech
Gross earnings per week: $\frac{\$50,310}{52} = \$967.50$
Hourly wage for 45-hour work week:
$\frac{\$967.50}{45} = \$21.50$

2. J&R Financial Group
Gross earnings per week: $6 \times 5 \times \$20.50 = \$615$
Yearly gross earnings: $\$615 \times 52 = \$31,980$

3. Based on Sharon's gross income, Donivan Tech pays the higher hourly rate. The higher rate is \$21.50 – \$20.50 = \$1.00 per hour higher.

4., 5., 6.

	Donivan Tech Salaried Full-Time Position (3 wk vacation)	J&R Financial Hourly Part-Time Position (2 wk vacation)
Total Job-Related Expenses Per Year	\$28,336.45	\$15,525.10
Net Per Year Employment Income	\$21,973.55	\$16,454.90
Net Per Week Employment Income	\$422.57	\$316.44
Net Per Hour Employment Income	\$9.39	\$10.55
Percent: Work-Related Expenses	56.32%	48.55%

7. Answers may vary.

Chapter 4 Review Problems

1. $x+2y=8$
$x-y=2$

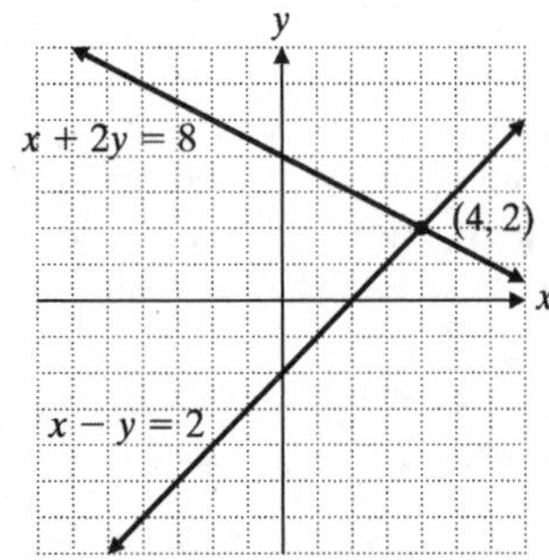

The solution is (4, 2).

2. $x+y=2$
$3x-y=6$

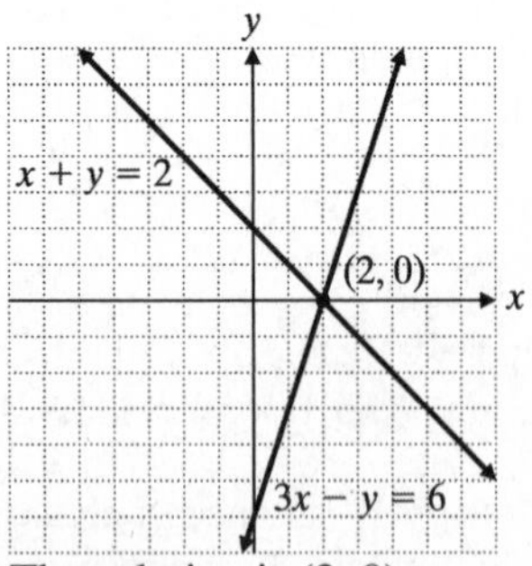

The solution is (2, 0).

3. $2x + y = 6$
$3x + 4y = 4$

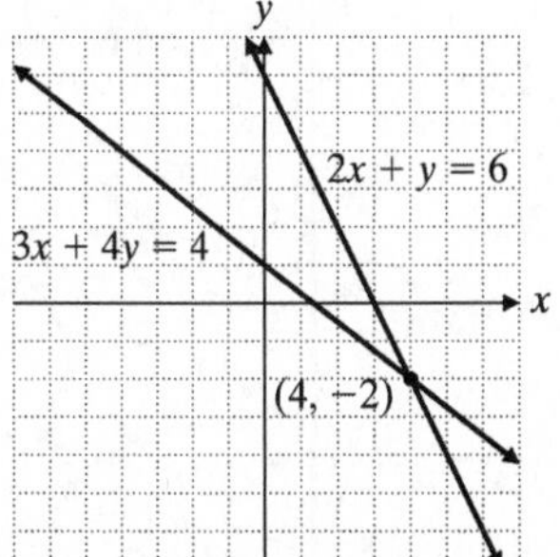

The solution is (4, –2).

4. $3x - 2y = -9,\ y = \dfrac{9+3x}{2}$

$$2x + y = 1$$
$$2x + \frac{9+3x}{2} = 1$$
$$4x + 9 + 3x = 2$$
$$7x = -7$$
$$x = -1$$
$$y = \frac{9+3x}{2}$$
$$y = \frac{9+3(-1)}{2}$$
$$y = 3$$

The solution is (–1, 3).

5. $-6x - y = 1,\ y = -6x - 1$

$$3x - 4y = 31$$
$$3x - 4(-6x - 1) = 31$$
$$3x + 24x + 4 = 31$$
$$27x = 27$$
$$x = 1$$
$$y = -6x - 1$$
$$y = -6(1) - 1$$
$$y = -7$$

The solution is (1, –7).

6. $4x + 5y = 2$ **(1)**
$3x - y = 11$ **(2)**

Solve **(1)** for *x*, then substitute into **(2)**.

$$4x = 2 - 5y$$
$$x = \frac{1}{2} - \frac{5}{4}y$$
$$3\left(\frac{1}{2} - \frac{5}{4}y\right) - y = 11$$
$$\frac{3}{2} - \frac{15}{4}y - y = 11$$
$$-\frac{19}{4}y = 11 - \frac{3}{2}$$
$$-\frac{19}{4}y = \frac{19}{2}$$
$$y = -2$$

Substitute $y = -2$ into **(2)**.

$$3x + 2 = 11$$
$$3x = 9$$
$$x = 3$$

(3, –2) is the solution.

7. $3x - 4y = -12$ **(1)**
$x + 2y = -4$ **(2)**

Solve **(2)** for *x*, substitute into **(1)**.

$$x = -4 - 2y$$
$$3(-4 - 2y) - 4y = -12$$
$$-12 - 6y - 4y = -12$$
$$-10y = 0$$
$$y = 0$$

Substitute $y = 0$ into **(2)**.

$$x + 0 = -4$$
$$x = -4$$

(–4, 0) is the solution.

8. $-2x + 5y = -12,\quad -6x + 15y = -36$
$3x + y = 1,\quad \underline{6x + 2y = 2}$

$$17y = -34$$
$$y = -2$$
$$3x + y = 1$$
$$3x - 2 = 1$$
$$3x = 3$$
$$x = 1$$

The solution is (1, –2).

9. $-3x + 4y = 9,\quad -15x + 20y = 45$
$5x + 3y = -15,\quad \underline{15x + 9y = -45}$

$$29y = 0$$
$$y = 0$$
$$-3x + 4y = 9$$
$$-3x + 4(0) = 9$$
$$-3x = 9$$
$$x = -3$$

The solution is (–3, 0).

10. $7x-4y=2,\quad -35x+20y=-10$
$6x-5y=-3,\quad \underline{24x-20y=-12}$
$-11x=-22$
$x=2$

$6x-5y=-3$
$6(2)-5y=-3$
$-5y=-15$
$y=3$
The solution is (2, 3).

11. $5x+2y=3,\quad 35x+14y=21$
$7x+5y=-20,\quad \underline{-35x-25y=100}$
$-11y=121$
$y=-11$

$5x+2y=3$
$5x+2(-11)=3$
$5x=25$
$x=5$
The solution is (5, −11).

12. $x=3-2y \Rightarrow 3x=9-6y \Rightarrow 3x+6y=2$
$3x+6y=8$
Inconsistent system; no solution

13. $x+5y=10$
$x=10-5y$

$y=2-\frac{1}{5}x$

$y=2-\frac{1}{5}(10-5y)$
$5y=10-10+5y$
$0=0$
Dependent system; infinite number of solutions

14. $4x-6y=3$ **(1)**
$8x+9y=-1$ **(2)**
Add −2**(1)** + **(2)**
$-8x+12y=-6$
$\underline{8x+9y=-1}$
$21y=-7$
$y=-\frac{1}{3}$

Substitute $y=-\frac{1}{3}$ into **(1)**.

$4x-6\left(-\frac{1}{3}\right)=3$
$4x+2=3$
$4x=1$
$x=\frac{1}{4}$

$\left(\frac{1}{4}, -\frac{1}{3}\right)$ is the solution.

15. $5x-2y=15$ **(1)**
$3x+y=-2$ **(2)**
Solve **(2)** for *y*, substitute into **(1)**.
$y=-2-3x$
$5x-2(-2-3x)=15$
$5x+4+6x=15$
$11x=11$
$x=1$
Substitute $x=1$ into **(2)**.
$3(1)+y=-2$
$3+y=-2$
$y=-5$
(1, −5) is the solution.

16. $x+\frac{1}{3}y=1,\ x=1-\frac{1}{3}y$

$\frac{1}{4}x-\frac{3}{4}y=-\frac{9}{4}$

$\frac{1}{4}\left(1-\frac{1}{3}y\right)-\frac{3}{4}y=-\frac{9}{4}$

$1-\frac{1}{3}y-3y=-9$

$-\frac{10}{3}y=-10$

$y=3$

$x=1-\frac{1}{3}y=1-\frac{1}{3}(3)=1-1$
$x=0$
The solution is (0, 3).

17. $\frac{2}{3}x+y=1$ **(1)**

$\frac{1}{3}x+y=\frac{5}{6}$ **(2)**

Add **(1)** to −2 times **(2)**.

$$\begin{aligned} \frac{2}{3}x + y &= 1 \\ -\frac{2}{3}x - 2y &= -\frac{5}{3} \\ \hline -y &= -\frac{2}{3} \\ y &= \frac{2}{3} \end{aligned}$$

$$\begin{aligned} \frac{2}{3}x + \frac{2}{3} &= 1 \\ \frac{2}{3}x &= \frac{1}{3} \\ x &= \frac{1}{2} \end{aligned}$$

The solution is $\left(\frac{1}{2}, \frac{2}{3}\right)$.

18. $3a + 8b = 0$, $9a + 2b = 11$,

$$\begin{aligned} 3a + 8b &= 0 \\ -36a - 8b &= -44 \\ \hline -33a \quad &= -44 \\ a &= \frac{4}{3} \end{aligned}$$

$$\begin{aligned} 3 \cdot \frac{4}{3} + 8b &= 0 \\ 8b &= -4 \\ b &= -\frac{1}{2} \end{aligned}$$

The solution is $\left(\frac{4}{3}, -\frac{1}{2}\right)$.

19. $3a + 5b = -2$, $10b = -6a - 4$,

$$\begin{aligned} 6a + 10b &= -4 \\ -6a - 10b &= 4 \\ \hline 0 &= 0 \end{aligned}$$

Dependent system; infinite number of solutions

20.
$$\begin{aligned} x + 3 &= 3y + 1 \\ x &= 3y - 2 \\ 1 - 2(x - 2) &= 6y + 1 \\ 1 - 2(3y - 2 - 2) &= 6y + 1 \\ 1 - 6y + 8 &= 6y + 1 \\ -12y &= -8 \\ y &= \frac{2}{3} \end{aligned}$$

$$x = 3y - 2 = 3\left(\frac{2}{3}\right) - 2 = 2 - 2$$

$$x = 0$$

The solution is $\left(0, \frac{2}{3}\right)$.

21. $10(x + 1) - 13 = -8y$, $y = \frac{10x - 3}{-8}$

$$\begin{aligned} 4(2 - y) &= 5(x + 1) \\ 8 - 4y &= 5x + 5 \\ 8 - 4\left(\frac{10x - 3}{-8}\right) &= 5x + 5 \\ 16 + 10x - 3 &= 10x + 10 \\ 13 &= 10 \end{aligned}$$

Inconsistent system; no solution

22. $0.3x - 0.2y = 0.7$, $-0.6x + 0.4y = 0.3$,

$$\begin{aligned} 0.6x - 0.4y &= 1.4 \\ -0.6x + 0.4y &= 0.3 \\ \hline 0 &= 1.7 \end{aligned}$$

Inconsistent system; no solution

23. $0.2x - 0.1y = 0.8$, $0.1x + 0.3y = 1.1$,

$$\begin{aligned} 0.6x - 0.3y &= 2.4 \\ 0.1x + 0.3y &= 1.1 \\ \hline 0.7x &= 3.5 \\ x &= 5 \end{aligned}$$

$$\begin{aligned} 0.2x - 0.1y &= 0.8 \\ 0.2(5) - 0.1y &= 0.8 \\ -0.1y &= -0.2 \\ y &= 2 \end{aligned}$$

The solution is (5, 2).

24.
$$\begin{aligned} 3x - 2y - z &= 3 \quad \textbf{(1)} \\ 2x + y + z &= 1 \quad \textbf{(2)} \\ -x - y + z &= -4 \quad \textbf{(3)} \end{aligned}$$

Add **(1)** and **(2)**.

$5x - y = 4$ **(4)**

Add **(1)** and **(3)**.

$2x - 3y = -1$ **(5)**

$$\begin{aligned} -15x + 3y &= -12 \\ 2x - 3y &= -1 \\ \hline -13x \quad &= -13 \\ x &= 1 \end{aligned}$$

Substitute $x = 1$ into **(4)**.

$$\begin{aligned} 5(1) - y &= 4 \\ -y &= -1 \\ y &= 1 \end{aligned}$$

Substitute $x = 1$, $y = 1$ into **(2)**.

$$\begin{aligned} 2(1) + 1 + z &= 1 \\ z &= -2 \end{aligned}$$

The solution is (1, 1, –2).

25.
$$\begin{aligned} -2x + y - z &= -7 \quad \textbf{(1)} \\ x - 2y - z &= 2 \quad \textbf{(2)} \\ 6x + 4y + 2z &= 4 \quad \textbf{(3)} \end{aligned}$$

Add 2 times **(2)** to **(3)**.

$$\begin{aligned} 2x-4y-2z&=4\\ 6x+4y+2z&=4\\ \hline 8x\qquad\quad &=8\\ x&=1 \end{aligned}$$

Substitute $x = 1$ into **(1)** and **(2)**.

$-2+y-z=-7 \Rightarrow \quad y-z=-5$ **(4)**

$1-2y-z=2 \Rightarrow \quad -2y-z=1$ **(5)**

Add -1 times **(5)** to **(4)**.

$$\begin{aligned} y-z&=-5\\ 2y+z&=-1\\ \hline 3y\qquad&=-6\\ y&=-2 \end{aligned}$$

Substitute $x = 1$, $y = -2$ into **(3)**.

$$\begin{aligned} 6(1)+4(-2)+2z&=4\\ 2z&=6\\ z&=3 \end{aligned}$$

The solution is (1, –2, 3).

26. $2x+5y+z=3$ **(1)**
$x+y+5z=42$ **(2)**
$2x+y=7$ **(3)**

Solve **(1)** for z and substitute into **(2)**.

$z = 3 - 2x - 5y$

$$\begin{aligned} x+y+5(3-2x-5y)&=42\\ x+y+15-10x-25y&=42\\ -9x-24y&=27 \quad \textbf{(4)} \end{aligned}$$

Solve **(3)** for y, and substitute into **(4)**.

$y = 7 - 2x$

$$\begin{aligned} -9x-24(7-2x)&=27\\ -9x-168+48x&=27\\ 39x&=195\\ x&=5 \end{aligned}$$

Substitute $x = 5$ into $y = 7 - 2x$.

$y = 7 - 2(5) = 7 - 10 = -3$

$y = -3$

Substitute $x = 5$, $y = -3$ into

$z = 3 - 2x - 5y$

$z = 3 - 2(5) - 5(-3) = 8$

$z = 8$

The solution is (5, –3, 8).

27. $x+2y+z=5$ **(1)**
$3x-8y=17$ **(2)**
$2y+z=-2$ **(3)**

Solve **(1)** for x, and substitute into **(2)**.

$x = 5 - 2y - z$

$$\begin{aligned} 3(5-2y-z)-8y&=17\\ 15-6y-3z-8y&=17\\ 3z&=-14y-2\\ z&=\frac{-14y-2}{3} \end{aligned}$$

Substitute $z=\dfrac{-14y-2}{3}$ into **(3)**.

$$\begin{aligned} 2y+\frac{-14y-2}{3}&=-2\\ 6y-14y-2&=-6\\ -8y&=-4\\ y&=\frac{1}{2} \end{aligned}$$

Substitute $y=\dfrac{1}{2}$ into **(3)**.

$$\begin{aligned} 2\left(\frac{1}{2}\right)+z&=-2\\ 1+z&=-2\\ z&=-3 \end{aligned}$$

Substitute $y=\dfrac{1}{2}$ into **(2)**.

$$\begin{aligned} 3x-8\left(\frac{1}{2}\right)&=17\\ 3x-4&=17\\ 3x&=21\\ x&=7 \end{aligned}$$

The solution is $\left(7, \frac{1}{2}, -3\right)$.

28. $3x+2y-z=-6$ **(1)**
$x-3y+4z=11$ **(2)**
$2x+5y-3z=-11$ **(3)**

Add –**(1)** and 3**(2)**.

$$\begin{aligned} -3x-2y+z&=6\\ 3x-9y+12z&=33\\ \hline -11y+13z&=39\\ -11y&=39-13z\\ y&=-\frac{39}{11}+\frac{13}{11}z \quad \textbf{(4)} \end{aligned}$$

Add 2**(2)** and –1**(3)**.

$$\begin{aligned} 2x-6y+8z&=22\\ -2x-5y+3z&=11\\ \hline -11y+11z&=33 \quad \textbf{(5)} \end{aligned}$$

Substitute **(4)** into **(5)**.

$$-11\left(-\frac{39}{11}+\frac{13}{11}z\right)+11z=33$$
$$39-13z+11z=33$$
$$-2z=-6$$
$$z=3$$

Substitute $z = 3$ into **(5)**.

$$-11y+11(3)=33$$
$$y=0$$

Substitute $y = 0$, $z = 3$ into **(1)**.

$$3x+2y-z=-6$$
$$3x+0-3=-6$$
$$3x=-3$$
$$x=-1$$

(−1, 0, 3) is the solution.

29. $-3x-4y+z=-4$ **(1)**
$x+6y+3z=-8$ **(2)**
$5x+3y-z=14$ **(3)**

Add **(1)** and 3(**2**).

$$-3x-4y+z=-4$$
$$3x+18y+9z=-24$$
$$\overline{14y+10z=-28}$$
$$10z=-28-14y$$
$$z=-\frac{28}{10}-\frac{14}{10}y \quad \textbf{(4)}$$

Add 5(**2**) and −(**3**).

$$5x+30y+15z=-40$$
$$-5x-3y+z=-14$$
$$\overline{27y+16z=-54} \quad \textbf{(5)}$$

Substitute **(4)** into **(5)**.

$$27y+16\left(-\frac{28}{10}-\frac{14}{10}y\right)=-54$$
$$27y-\frac{448}{10}-\frac{224}{10}y=-54$$
$$\frac{46}{10}y=-\frac{92}{10}$$
$$y=-2$$

Substitute $y = -2$ into **(4)**.

$$z=-\frac{28}{10}-\frac{14}{10}(-2)$$
$$z=-\frac{28}{10}+\frac{28}{10}$$
$$z=0$$

Substitute $y = -2$, $z = 0$ into **(2)**.

$$x+6(-2)+3(0)=-8$$
$$x-12=-8$$
$$x=4$$

(4, −2, 0) is the solution.

30. $3x+2y=7$ **(1)**
$2x+7z=-26$ **(2)**
$5y+z=6$ **(3)**

Solve **(3)** for z and substitute into **(2)**.

$$z = 6 - 5y$$
$$2x+7(6-5y)=-26$$
$$2x+42-35y=-26$$
$$2x=35y-68$$
$$x=\frac{35y-68}{2}$$

Substitute $x=\frac{35y-68}{2}$ into **(1)**.

$$3\left(\frac{35y-68}{2}\right)+2y=7$$
$$105y-204+4y=14$$
$$109y=218$$
$$y=2$$

Substitute $y = 2$ into **(3)**.

$$5(2)+z=6$$
$$10+z=6$$
$$z=-4$$

Substitute $y = 2$ into **(1)**.

$$3x+2(2)=7$$
$$3x+4=7$$
$$3x=3$$
$$x=1$$

The solution is (1, 2, −4).

31. $x-y=2$ **(1)**
$5x+7y-5z=2$ **(2)**
$3x-5y+2z=-2$ **(3)**

Add 2 times **(2)** to 5 times **(3)**.

$$10x+14y-10z=4$$
$$15x-25y+10z=-10$$
$$\overline{25x-11y=-6} \quad \textbf{(4)}$$

Solve **(1)** for x and substitute into **(4)**.

$$x = 2 + y$$
$$25(2+y)-11y=-6$$
$$50+25y-11y=-6$$
$$14y=-56$$
$$y=-4$$

Substitute $y = -4$ into **(1)**.

$$x-(-4)=2$$
$$x+4=2$$
$$x=-2$$

Substitute $x = -2$, $y = -4$ into **(3)**.

$3(-2)-5(-4)+2z=-2$
$-6+20+2z=-2$
$2z=-16$
$z=-8$

The solution is (–2, –4, –8).

32. v = speed of plane in still air
w = speed of wind
$720=(v-2)\cdot 3$
$720=(v+w)(2.5)$

$v-w=240$
$v+w=288$
$2v=528$
$v=264$

$w=288-v=288-264=24$
The speed of the plane in still air is 264 mph and the wind speed is 24 mph.

33. Let x = no. of touchdowns (6 points)
y = no. of field goals (3 points)
$6x+3y=63$ **(1)**
$x+y=12$ **(2)**
Solve **(2)** for x.
$x=12-y$
Substitute **(2)** into **(1)**.
$6(12-y)+3y=63$
$72-6y+3y=63$
$-3y=-9$
$y=3$
Substitute $y=3$ into **(2)**.
$x+3=12$
$x=9$
9 touchdowns and 3 field goals were scored.

34. x = number of laborers
y = number of mechanics
$70x+90y=1950$ **(1)**
$80x+100y=2200$ **(2)**
Divide both equations by 10.
$7x+9y=195$ **(1)**
$8x+10y=220$ **(2)**
Add –8 times **(1)** to 7 times **(2)**.
$-56x-72y=-1560$
$56x+70y=1540$
$-2y=-20$
$y=10$
Substitute $y=10$ into **(2)**.
$8x+10(10)=220$
$8x=120$
$x=15$
The circus hired 15 laborers and 10 mechanics.

35. Let x = number of children's tickets
y = number of adult tickets
$x+y=330$ **(1)**
$8x+13y=3215$ **(2)**
Solve **(1)** for x, substitute into **(2)**.
$x=330-y$
$8(330-y)+13y=3215$
$2640-8y+13y=3215$
$5y=575$
$y=115$
Substitute $y=115$ into **(1)**.
$x+115=330$
$x=215$
115 adult tickets sold.
215 children's tickets sold.

36. x = cost of hat
y = cost of shirt
z = cost of pants
$2x+5y+4z=129$ **(1)**
$x+y+2z=42$ **(2)**
$2x+3y+z=63$ **(3)**
Add –2 times **(2)** and **(1)**.
$3y=45$
$y=15$
Substitute $y=15$ into **(2)** and solve for x.
$x+15+2z=42$
$x=27-2z$
Substitute $x=27-2z$ and $y=15$ into **(3)**.
$2(27-2z)+3(15)+z=63$
$54-4z+45+z=63$
$-3z=-36$
$z=12$
Substitute $y=15$, $z=12$ into **(2)**.
$x+15+2(12)=42$
$x+15+24=42$
$x=3$
The hats cost \$3, shirts \$15, and pants \$12.

37. J = Jess' score
N = Nick's score
C = Chris' score
$J+C+N=249$ **(1)**
$J=C+20$ **(2)**
$2N=J+C+6$ **(3)**
Substitute J from **(2)** into **(1)** and **(3)**.
$C+20+C+N=249 \Rightarrow 2C=229-N$
$2N=C+20+C+6 \Rightarrow 2C=2N-26$
from which $229-N=2N-26$
$3N=255$
$N=85$

and $2C = 229 - 85 = 144$
$C = 72$
and $J = C + 20 = 72 + 20$
$J = 92$
Jess scored 92 points, Nick scored 85 points, and Chris scored 72 points.

38. x = cost of jelly
y = cost of peanut butter
z = cost of honey
$4x + 3y + 5z = 9.8$ **(1)**
$2x + 2y + z = 4.20$ **(2)**
$3x + 4y + 2z = 7.70$ **(3)**
Add **(1)** and –5 times **(2)**.
$-6x - 7y = -11.2$ **(4)**
Add –2 times **(2)** to **(3)**.
$-x = -0.7$
$x = 0.7$
Substitute $x = 0.7$ into **(4)**.
$-6(0.7) - 7y = -11.2$
$-7y = -7$
$y = 1$
Substitute $x = 0.7$, $y = 1$ into **(2)**.
$2(0.7) + 2(1) + z = 4.2$
$z = 0.8$
The cost of a jar of jelly is \$0.70, the cost of a jar of peanut butter is \$1, and the cost of a jar of honey is \$0.80.

39. x = number of buses
y = number of station wagons
z = number of sedans
$x + y + z = 9$ **(1)**
$40x + 8y + 5z = 127$ **(2)**
$3(3y) + 5(2z) = 126$
$24y + 10z = 126$ **(3)**
Add –40 times **(1)** to **(2)**.
$-32y - 35z = -233$ **(4)**
Add 32 times **(3)** to 24 times **(4)**.
$-520z = -1560$
$z = 3$
Substitute $z = 3$ into **(3)**.
$24y + 10(3) = 126$
$24y = 96$
$y = 3$
Substitute $y = 4$, $z = 3$ into **(1)**.
$x + 4 + 3 = 9$
$x = 2$
They should use 2 buses, 4 station wagons, and 3 sedans.

40. $-x - 5z = -5$ **(1)**
$13x + 2z = 2$ **(2)**
Solve **(1)** for x and substitute into **(2)**.
$x = 5 - 5z$
$13(5 - 5z) + 2z = 2$
$65 - 65z + 2z = 2$
$-63z = -63$
$z = 1$
Substitute $z = 1$ into **(1)**.
$-x - 5(1) = -5$
$-x = 0$
$x = 0$
The solution is (0, 1).

41. $x - y = 1$
$5x + y = 7$
Adding gives
$6x = 8$
$x = \frac{4}{3}$ and
$5 \cdot \frac{4}{3} + y = 7$
$y = \frac{1}{3}$
The solution is $\left(\frac{4}{3}, \frac{1}{3}\right)$.

42. $2x + 5y = 4$ **(1)**
$5x - 7y = -29$ **(2)**
Solve **(1)** for y and substitute into **(2)**.
$y = \frac{4 - 2x}{5}$
$5x - 7\frac{4 - 2x}{5} = -29$
$25x - 28 + 14x = -145$
$39x = -117$
$x = -3$
Substitute $x = -3$ into **(1)**.
$2(-3) + 5y = 4$
$-6 + 5y = 4$
$5y = 10$
$y = 2$
The solution is (–3, 2).

43. $\frac{x}{2} - 3y = -6$ **(1)**
$\frac{4}{3}x + 2y = 4$ **(2)**
Solve **(1)** for x and substitute into **(2)**.
$x = 6y - 12$

$\frac{4}{3}(6y-12)+2y=4$
$24y-48+6y=12$
$30y=60$
$y=2$
$x = 6y - 12 = 6(2) - 12 = 0$
The solution is (0, 2).

44. $\frac{x-2}{4}=y-2$ **(1)**
$\frac{-3y+1}{2}=x+y$ **(2)**
Multiply **(1)** by 4, solve for x.
$x-2=4y-8$
$x=4y-6$
Substitute **(1)** into **(2)**.
$\frac{-3y+1}{2}=4y-6+y$
$-3y+1=10y-12$
$-13y=-13$
$y=1$
Substitute $y = 1$ into **(1)**.
$x = 4(1) - 6$
$x = -2$
(–2, 1) is the solution.

45. $\frac{x}{2}-y=-12$ **(1)**
$x+\frac{3}{4}y=9$ **(2)**
Solve **(1)** for x.
$\frac{x}{2}=-12+y$
$x=-24+2y$
Substitute into **(2)**.
$-24+2y+\frac{3}{4}y=9$
$\frac{11}{4}y=33$
$y=12$
Substitute $y = 12$ into **(1)**.
$\frac{x}{2}-12=-12$
$x=0$
(0, 12) is the solution.

46. $3(2+x)=y+1$
$6+3x=y+1$
$y=3x+5$ **(1)**
$5(x-y)=-7-3y$
$5x-5y=-7-3y$
$5x-2y=-7$ **(2)**
Substitute y from **(1)** into **(2)**.
$5x-2(3x+5)=-7$
$5x-6x-10=-7$
$-x=3$
$x=-3$
Substitute $x = -3$ into **(1)**.
$y = 3x + 5 = 3(-3) + 5$
$y = -9 + 5 = -4$
The solution is (–3, –4).

47. $7(x+3)=2y+25$
$7x+21=2y+25$
$7x-2y=4$ **(1)**
$3(x-6)=-2(y+1)$
$3x-18=-2y-2$
$3x+2y=16$ **(2)**
Add **(1)** and **(2)**.
$10x = 20$
$x=2$
Substitute $x = 2$ into **(2)**.
$3(2)+2y=16$
$2y=10$
$y=5$
The solution is (2, 5).

48. Multiply both equations by 10 to clear decimals.
$3x-4y=9$ **(1)**
$2x-3y=4$ **(2)**
Solve **(1)** for y and substitute into **(2)**.
$y=\frac{3x-9}{4}$
$2x-3\frac{3x-9}{4}=4$
$8x-9x+27=16$
$-x=-11$
$x=11$
$y=\frac{3x-9}{4}=\frac{3(11)-9}{4}=6$
The solution is (11, 6).

49. $1.2x-y=1.6$
$x+1.5y=6$
Solve the first equation for y and substitute into the second equation.

$y = 1.2x - 1.6$
$x + 1.5(1.2x - 1.6) = 6$
$x + 1.8x - 2.4 = 6$
$2.8x = 8.4$
$x = 3$
$y = 1.2(3) - 1.6$
$y = 2$
The solution is (3, 2).

50. $x - \frac{y}{2} + \frac{1}{2}z = -1$ **(1)**
$2x + \frac{5}{2}z = -1$ **(2)**
$\frac{3}{2}y + 2z = 1$ **(3)**
Solve **(2)** for x and **(3)** for y and substitute into **(1)**.
$$x = \frac{-\frac{5}{2}z - 1}{2}$$
$$y = \frac{2(1-2z)}{3} = \frac{2-4z}{3}$$
$$\frac{-\frac{5}{2}z - 1}{2} - \frac{\frac{2-4z}{3}}{2} + \frac{1}{2}z = -1$$
$$-\frac{5}{2}z - 1 - \frac{2-4z}{3} + z = -2$$
$$-15z - 6 - 4 + 8z + 6z = -12$$
$$-z = -2$$
$$z = 2$$
$$x = \frac{-\frac{5}{2}z - 1}{2} = \frac{-\frac{5}{2}(2) - 1}{2} = -3$$
$$y = \frac{2-4z}{3} = \frac{2-4(2)}{3} = -2$$
The solution is (–3, –2, 2).

51. $2x - 3y + 2z = 0$ **(1)**
$x + 2y - z = 2$ **(2)**
$2x + y + 3z = -1$ **(3)**
Add **(1)** and 2 times **(2)**.
$4x + y = 4$ **(4)**
Add 3 times **(2)** and **(3)**.
$5x + 7y = 5$ **(5)**
Solve **(4)** for y and substitute into **(5)**.
$y = 4 - 4x$
$5x + 7(4 - 4x) = 5$
$5x + 28 - 28x = 5$
$-23x = -23$
$x = 1$
$y = 4 - 4x$
$y = 4 - 4(1) = 0$
Substitute $x = 1$, $y = 0$ into **(1)**.
$2(1) - 3(0) + 2z = 0$
$2 + 2z = 0$
$2z = -2$
$z = -1$
The solution is (1, 0, –1).

52. $x - 4y + 4z = -1$ **(1)**
$2x - y + 5z = -3$ **(2)**
$x - 3y + z = 4$ **(3)**
Add **(1)** and –4 times **(3)**.
$-3x + 8y = -17$ **(4)**
Add **(2)** and –5 times **(3)**.
$-3x + 14y = -23$ **(5)**
Subtract **(5)** from **(4)**.
$-6y = 6$
$y = -1$
Substitute $y = -1$ into **(4)**.
$-3x + 8(-1) = -17$
$-3x = -9$
$x = 3$
Substitute $x = 3$, $y = -1$ into **(3)**.
$3 - 3(-1) + z = 4$
$z = -2$
The solution is (3, –1, –2).

53. $x - 2y + z = -5$ **(1)**
$2x + z = -10$ **(2)**
$y - z = 15$ **(3)**
Solve **(3)** for y.
$y = z + 15$ **(4)**
Solve **(2)** for x.
$x = \frac{-z-10}{2}$ **(5)**
Substitute **(4)** and **(5)** into **(1)**.
$$\frac{-z-10}{2} - 2(z+15) + z = -5$$
$$-z - 10 - 4z - 60 + 2z = -10$$
$$-3z = 60$$
$$z = -20$$
Substitute $z = -20$ into **(4)** and **(5)**.
$$x = \frac{-(-20)-10}{2} = 5$$
$y = -20 + 15 = -5$
The solution is (5, –5, –20).

54. $y \geq -\frac{1}{2}x - 1$ **(1)**
$-x + y \leq 5$ **(2)**
Test point: (0, 0)

(1) $0 \geq -\frac{1}{2}(0) - 1$

$0 \geq -1$ True

(2) $0 + 0 \leq 5$ True

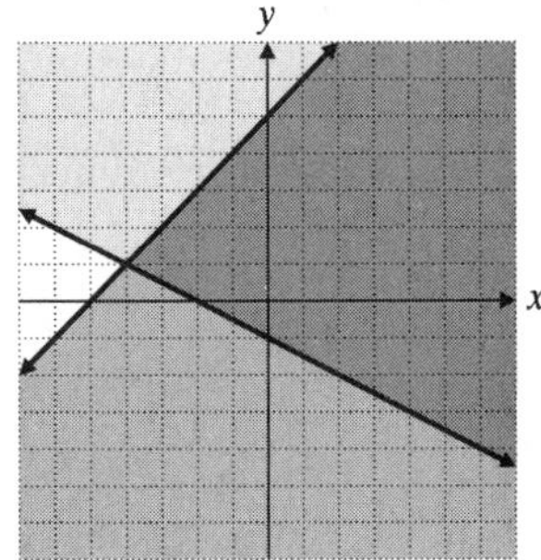

55. $-2x + 3y < 6$

Test point: (0, 0)

$-2(0) + 3(0) < 6$

$0 < 6$ True

$y > -2$

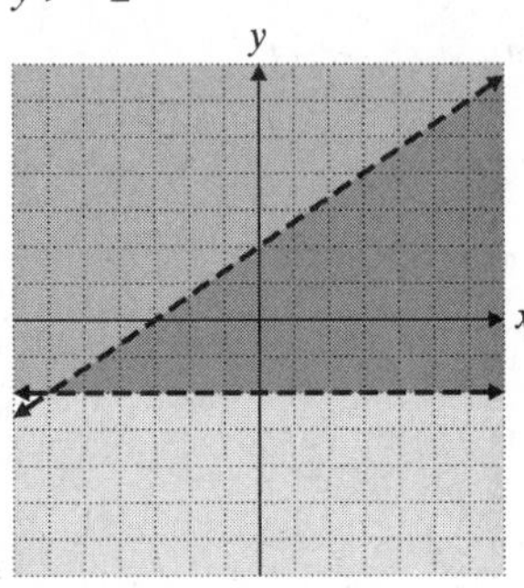

56. $x + y > 1$

Test point: (0, 0)

$0 + 0 > 1$

$0 > 1$ False

$2x - y < 5$

Test point: (0, 0)

$2(0) - 0 < 5$

$0 < 5$ True

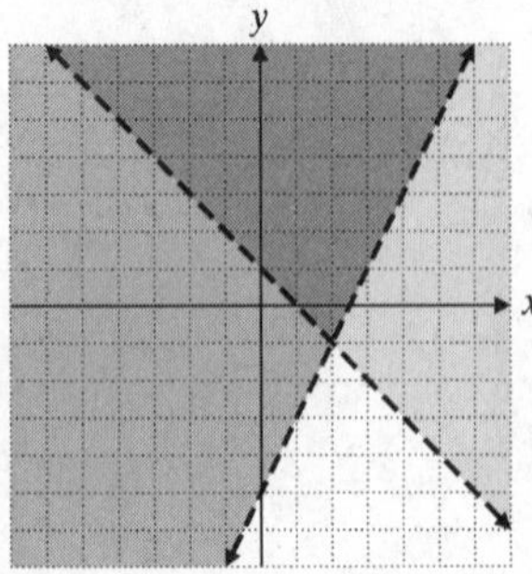

57. $x + y \geq 4$

Test point: (0, 0)

$0 + 0 \geq 4$

$0 \geq 4$ False

$y \leq x$

Test point: (2, 0)

$0 \leq 2$ True

$x \leq 6$

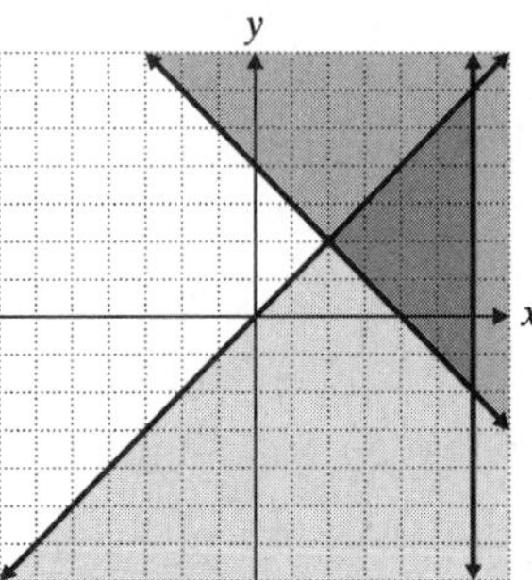

How Am I Doing? Chapter 4 Test

1. $x - y = 3$ **(1)**

$2x - 3y = -1$ **(2)**

Solve the first equation for y and substitute into the second equation.

$y = x - 3$

$2x - 3(x - 3) = -1$

$2x - 3x + 9 = -1$

$-x = -10$

$x = 10$

$y = 10 - 3$

$y = 7$

The solution is (10, 7).

2. $3x + 2y = 1$ **(1)**

$5x + 3y = 3$ **(2)**

Multiply the first equation by 3 and the second equation by –2 and add.

$$\begin{array}{r} 9x + 6y = 3 \\ \underline{-10x - 6y = -6} \\ -x \qquad = -3 \\ x = 3 \end{array}$$

Substitute $x = 3$ into the first equation.

$3(3) + 2y = 1$

$2y = -8$

$y = -4$

The solution is (3, –4).

3. $5x-3y=3$ **(1)**
$7x+y=25$ **(2)**

Add 3 times **(2)** to **(1)**.

$$\begin{aligned} 5x-3y&=3\\ 21x+3y&=75\\ \hline 26x&=78\\ x&=3 \end{aligned}$$

$$\begin{aligned} 7x+y&=25\\ 7(3)+y&=25\\ y&=4 \end{aligned}$$

The solution is (3, 4).

4. $\frac{1}{4}a-\frac{3}{4}b=-1$ **(1)**
$\frac{1}{3}a+b=\frac{5}{3}$ **(2)**

Solve **(2)** for b and substitute into **(1)**.

$$b=\frac{5}{3}-\frac{1}{3}a$$

$$\begin{aligned} \frac{1}{4}a-\frac{3}{4}\left(\frac{5}{3}-\frac{1}{3}a\right)&=-1\\ a-5+a&=-4\\ 2a&=1\\ a&=\frac{1}{2} \end{aligned}$$

$$b=\frac{5}{3}-\frac{1}{3}a$$

$$b=\frac{5}{3}-\frac{1}{3}\cdot\frac{1}{2}=\frac{3}{2}$$

The solution is $\left(\frac{1}{2},\frac{3}{2}\right)$.

5. $\frac{1}{3}x+\frac{5}{6}y=2$ **(1)**
$\frac{3}{5}x-y=-\frac{7}{5}$ **(2)**

Solve **(2)** for y and substitute into **(1)**.

$$y=\frac{3}{5}x+\frac{7}{5}$$

$$\begin{aligned} \frac{1}{3}x+\frac{5}{6}\left(\frac{3}{5}x+\frac{7}{5}\right)&=2\\ 2x+3x+7&=12\\ 5x&=5\\ x&=1 \end{aligned}$$

$$y=\frac{3}{5}x+\frac{7}{5}$$

$$y=\frac{3}{5}(1)+\frac{7}{5}$$

$$y=2$$

The solution is (1, 2).

6. $8x-3y=5$
$-16x+6y=8$

Multiply the first equation by 2 and add to the second equation.

$$\begin{aligned} 16x-6y&=10\\ -16x+6y&=8\\ \hline 0&=18 \end{aligned}$$

Inconsistent system; no solution

7. $3x+5y-2z=-5$
$2x+3y-z=-2$
$2x+4y+6z=18$

Multiply first equation by $-\frac{2}{3}$ and add to the second and third equation

$$3x+5y-2z=-5$$
$$-\frac{1}{3}y+\frac{1}{3}z=\frac{4}{3}$$
$$\frac{2}{3}y+\frac{22}{3}z=\frac{64}{3}$$

Multiply second equation by 2 and add to the third equation.

$$3x+5y-2z=-5$$
$$-\frac{1}{3}y+\frac{1}{3}z=\frac{4}{3}$$
$$8z=24$$
$$z=3$$

From the second equation

$$-\frac{1}{3}y+\frac{1}{3}(3)=\frac{4}{3}\Rightarrow y=-1$$

From the first equation

$3x+5(-1)-2(3)=-5,\ x=2$

The solution is (2, −1, 3).

8. $3x+2y=0$
$2x-y+3z=8$
$5x+3y+z=4$

Multiply the first equation by $-\frac{2}{3}$ and add to the second equation and then multiply the first equation by $-\frac{5}{3}$ and add to the third equation.

$3x+2y=0$
$-\frac{7}{3}y+3z=8$
$-\frac{1}{3}y+z=4$

Multiply the second equation by $-\frac{1}{7}$ and add to the third equation.

$3x+2y=0$
$-\frac{7}{3}y+3z=8$
$\frac{4}{7}z=\frac{20}{7}$
$z=5$

From second equation

$-\frac{7}{3}y+3(5)=8 \Rightarrow y=3$

From the first equation

$3x + 2(3) = 0, x = -2$

The solution is $(-2, 3, 5)$.

9. $x+5y+4z=-3$
$x-y-2z=-3$
$x+2y+3z=-5$

Multiply the first equation by -1 and add to the second and third equation.

$x+5y+4z=-3$
$-6y-6z=0$
$-3y-z=-2$

Multiply the second equation by $-\frac{1}{2}$ and add to the third equation.

$x+5y+4z=-3$
$-6y-6z=0$
$2z=-2$

$2z=-2, z=-1$

From the second equation

$-6y-6(-1)=0, y=1$

From the first equation

$x+5(1)+4(-1)=-3, x=-4$

The solution is $(-4, 1, -1)$.

10. v = speed of plane in still air
w = speed of wind

$$\begin{aligned} 1000=(v+w)(2) &\Rightarrow \quad v+w=500 \\ 1000=(v-w)(2.5) &\Rightarrow \quad \underline{v-w=400} \\ &\qquad\quad 2v=900 \\ &\qquad\quad v=450 \end{aligned}$$

$450+w=500$
$w=50$

The speed of the plane in still air is 450 mph.
The speed of the wind is 50 mph.

11. p = price of a pen
m = price of a mug
s = price of a shirt

$4p+m+s=20$ **(1)**
$2p+2m=11$ **(2)**
$6p+m+2s=33$ **(3)**

Add -1 times **(1)** to **(3)**.

$2p + s = 13$ **(4)**

Add -2 times **(3)** to **(2)**.

$-10p - 4s = -55$ **(5)**

Add 5 times **(4)** to **(5)**.

$s = 10$

Substitute $s = 10$ into **(4)**.

$2p+10=13$
$2p=3$
$p=1.50$

Substitute $s = 10$, $p = 1.5$ into **(1)**.

$4(1.5)+m+10=20$
$m=4$

Each pen costs \$1.50, each mug costs \$4.00, and each T-shirt costs \$10.00.

12. x = daily charge
y = mileage charge

$5x+150y=180$ **(1)**
$7x+320y=274$ **(2)**

Multiply **(1)** by $-\frac{7}{5}$ and add to **(2)**.

$110y=22$
$y=0.2$

From first equation

$5x+150(0.2)=180$
$5x=150$
$x=30$

They charge \$30 per day and \$0.20 per mile.

13. $x+2y\le 6$

Test point: (0, 0)

$0+2(0)\le 6$
$0\le 6$ True

$-2x+y\ge -2$

Test point: (0, 0)

$-2(0)+0\ge -2$
$0\ge -2$ True

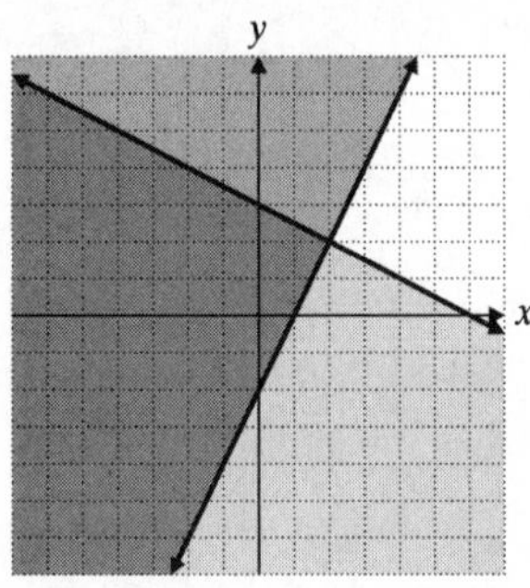

14. $3x + y > 8$
Test point: (0, 0)
$3(0) + 0 > 8$
$0 > 8$ False
$x - 2y > 5$
Test point: (0, 0)
$0 - 2(0) > 5$
$0 > 5$ False

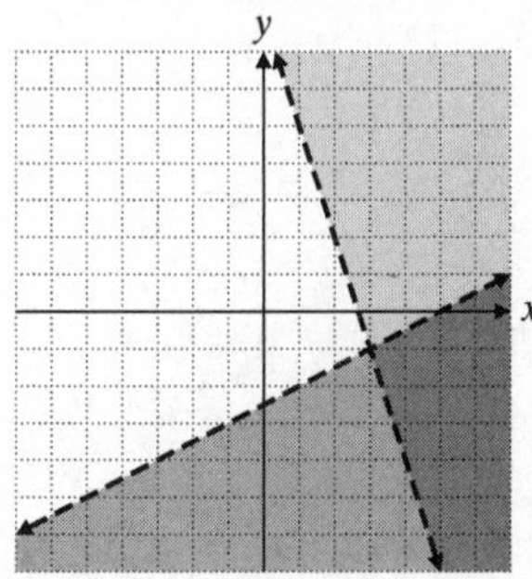

Cumulative Test for Chapters 1–4

1. $7 + 0 = 7$
Identity property of addition

2. $\sqrt{64} - (3-5)^3 + (-50) \div 5$
$= 8 - (-2)^3 - 10$
$= 8 + 8 - 10$
$= 16 - 10$
$= 6$

3. $(5x^{-2})(3x^{-4}y^2) = (5)(3)x^{-2-4}y^2$
$= 15x^{-6}y^2$
$= \dfrac{15y^2}{x^6}$

4. $2x - 4[x - 3(2x+1)] = 2x - 4[x - 6x - 3]$
$= 2x - 4[-5x - 3]$
$= 2x + 20x + 12$
$= 22x + 12$

5. $A = P(3 + 4rt)$
$\dfrac{A}{3+4rt} = \dfrac{P(3+4rt)}{3+4rt}$
$P = \dfrac{A}{3+4rt}$

6. $\dfrac{1}{4}x + 5 = \dfrac{1}{3}(x-2)$
$12\left[\dfrac{1}{4}x + 5\right] = 12\left[\dfrac{1}{3}(x-2)\right]$
$3x + 60 = 4x - 8$
$3x - 4x + 60 = 4x - 4x - 8$
$-x + 60 = -8$
$-x + 60 - 60 = -8 - 60$
$-x = -68$
$x = 68$

7. $4x - 8y = 10$ or $y = \dfrac{4x - 10}{8}$

x	$y = \frac{4x-10}{8}$	y
−2	$y = \frac{4(-2)-10}{8}$	$-\frac{9}{4}$
0	$y = \frac{4(0)-10}{8}$	$-\frac{5}{4}$
4	$y = \frac{4(4)-10}{8}$	$\frac{3}{4}$

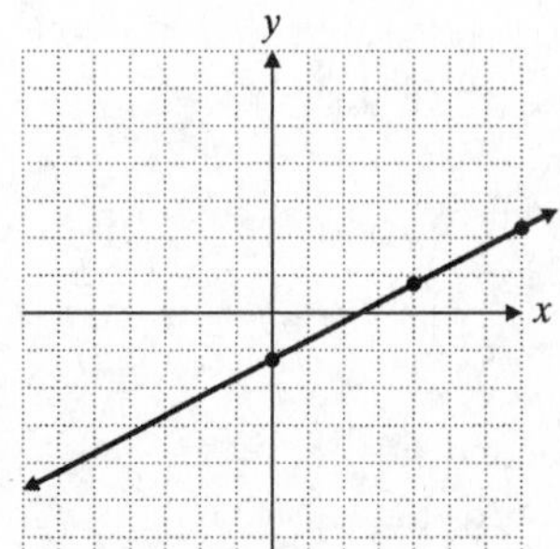

8. $m = \dfrac{y_2 - y_1}{x_2 - x_1}$
$m = \dfrac{2 - 10}{5 + 1}$
$m = \dfrac{-8}{6}$
$m = -\dfrac{4}{3}$

9. $4x+3-13x-7<2(3-4x)$
$$-9x-4<6-8x$$
$$-9x+8x<6+4$$
$$-x<10$$
$$x>-10$$

10. $\frac{2x-1}{3}\le 7$ and $2(x+1)\ge 12$
$$2x-1\le 21 \qquad 2x+2\ge 12$$
$$2x\le 22 \qquad 2x\ge 10$$
$$x\le 11 \qquad x\ge 5$$
$$5\le x\le 11$$

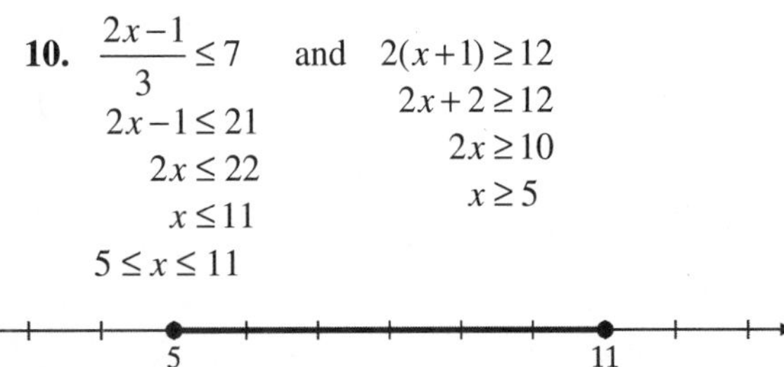

11. $5x+6y=-2$
$$6y=-5x-2$$
$$y=-\frac{5}{6}x-\frac{1}{3},\ m=-\frac{5}{6},\ m_{\perp}=\frac{6}{5}$$
$$y-y_1=m(x-x_1)$$
$$y-(-3)=\frac{6}{5}(x-2)$$
$$5(y+3)=5\cdot\frac{6}{5}(x-2)$$
$$5y+15=6x-12$$
$$6x-5y=27$$

12. x = length of first side
$x + 7$ = length of second side
$2x - 6$ = length of third side
$$x+x+7+2x-6=69$$
$$4x+1=69$$
$$4x=68$$
$$x=17$$
x = 17 m, first side
$x + 7$ = 24 m, second side
$2x - 6$ = 28 m, third side

13. x = amount invested at 7%
$6000 - x$ = amount invested at 9%
$$0.07x+0.09(6000-x)=510$$
$$0.07x+540-0.09x=510$$
$$-0.02x=-30$$
$$x=1500$$
$6000 - x = 4500$
Victor invested \$1500 at 7% and \$4500 at 9%.

14. $5x+2y=2$ **(1)**
$4x+3y=-4$ **(2)**
Solve **(1)** for y and substitute into **(2)**.
$$2y=2-5x$$
$$y=\frac{2-5x}{2}$$
$$4x+3\frac{2-5x}{2}=-4$$
$$8x+6-15x=-8$$
$$-7x=-14$$
$$x=2$$
$$y=\frac{2-5x}{2}=\frac{2-5(2)}{2}=-4$$
The solution is (2, –4).

15. $x-\frac{1}{2}y=4$ **(1)**
$2x+\frac{1}{3}y=0$ **(2)**
Solve **(1)** for x.
$$x=4+\frac{1}{2}y$$
Substitute into **(2)**.
$$2\left(4+\frac{1}{2}y\right)+\frac{1}{3}y=0$$
$$8+y+\frac{1}{3}y=0$$
$$\frac{4}{3}y=-8$$
$$y=-6$$
Substitute $y = -6$ into **(1)**.
$$x-\frac{1}{2}(-6)=4$$
$$x=1$$
(1, –6) is the solution.

16. Let x = price of T-shirts
y = price of sweatshirts
$6x+9y=156$ **(1)**
$8x+5y=124$ **(2)**
Solve **(1)** for x, substitute into **(2)**.
$$6x=156-9y$$
$$x=\frac{156}{6}-\frac{9}{6}y$$
$$8\left(\frac{156}{6}-\frac{9}{6}y\right)+5y=124$$
$$208-12y+5y=124$$
$$-7y=-84$$
$$y=12$$

Substitute $y = 12$ into **(2)**.

$8x+5(12)=124$

$8x=64$

$x=8$

T-shirt price is \$8. Sweatshirt price is \$12.

17. $7x-6y=17$ **(1)**

$3x+y=18$ **(2)**

Solve **(2)** for y and substitute into **(1)**.

$y = 18 - 3x$

$7x-6(18-3x)=17$

$7x-108+18x=17$

$25x=125$

$x=5$

$y = 18 - 3x = 18 - 3(5) = 3$

The solution is (5, 3).

18. $x+3y+z=5$

$2x-3y-2z=0$

$x-2y+3z=-9$

Multiply first equation by −2 and add to the second equation, then multiply the first equation by −1 and add to the third equation.

$x+3y+z=5$

$-9y-4z=-10$

$-5y+2z=-14$

Multiply the second equation by $-\frac{5}{9}$ and add to the third equation.

$x+3y+z=5$

$-9y-4z=-10$

$\frac{38}{9}z=-\frac{76}{9}$

From the third equation

$\frac{38}{9}z=-\frac{76}{9},\ z=-2$

From the second equation

$-9y-4(-2)=-10,\ y=2$

From the first equation

$x+3(2)+(-2)=5,\ x=1$

The solution is (1, 2, −2).

19. $-5x+6y=2$ **(1)**

$10x-12y=-4$ **(2)**

Multiplying **(1)** by 2 and adding to **(2)** gives 0 = 0. Dependent system; infinite number of solutions

20. $x-y\geq -4$

Test point: (0, 0)

$0-0\geq -4$

$0\geq -4$ True

$x+2y\geq 2$

Test point: (0, 0)

$0+2(0)\geq 2$

$0\geq 2$ False

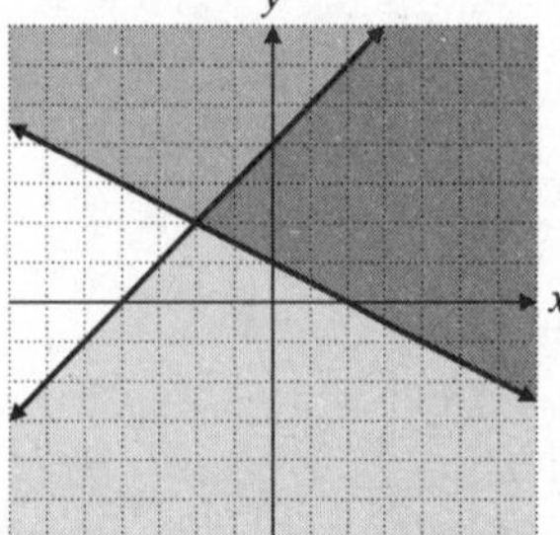

Chapter 5

5.1 Exercises

1. $2x^2 - 5x + 3$
Trinomial, 2nd degree

3. $-3.2a^4bc^3$
Monomial, 8th degree

5. $\frac{3}{5}m^3n - \frac{2}{5}mn$
Binomial, 4th degree

7. $p(x) = 5x^2 - 9x - 12$

$p(3) = 5(3)^2 - 9(3) - 12$
$p(3) = 5(9) - 9(3) - 12$
$p(3) = 45 - 27 - 12$
$p(3) = 6$

9. $g(x) = -4x^3 - x^2 + 5x - 1$

$g(-2) = -4(-2)^3 - (-2)^2 + 5(-2) - 1$
$g(-2) = -4(-8) - (4) + 5(-2) - 1$
$g(-2) = 32 - 4 - 10 - 1$
$g(-2) = 17$

11. $h(x) = 2x^4 - x^3 + 2x^2 - 4x - 3$

$h(-1) = 2(-1)^4 - (-1)^3 + 2(-1)^2 - 4(-1) - 3$
$h(-1) = 2(1) - (-1) + 2(1) - 4(-1) - 3$
$h(-1) = 2 + 1 + 2 + 4 - 3$
$h(-1) = 6$

13. $(x^2 + 3x - 2) + (-2x^2 - 5x + 1) + (x^2 - x - 5)$
$= x^2 + 3x - 2 - 2x^2 - 5x + 1 + x^2 - x - 5$
$= -3x - 6$

15. $(7m^3 + 4m^2 - m + 2.5) - (-3m^3 + 5m + 3.8)$
$= 7m^3 + 4m^2 - m + 2.5 + 3m^3 - 5m - 3.8$
$= 10m^3 + 4m^2 - 6m - 1.3$

17. $(5a^3 - 2a^2 - 6a + 8) + (5a + 6) - (-a^2 - a + 2)$
$= 5a^3 - 2a^2 - 6a + 8 + 5a + 6 + a^2 + a - 2$
$= 5a^3 - a^2 + 12$

19. $\left(\frac{2}{3}x^2 + 5x\right) + \left(\frac{1}{2}x^2 - \frac{1}{3}x\right)$
$= \frac{2}{3}x^2 + \frac{1}{2}x^2 + 5x - \frac{1}{3}x$
$= \frac{4}{6}x^2 + \frac{3}{6}x^2 + \frac{15}{3}x - \frac{1}{3}x$
$= \frac{7}{6}x^2 + \frac{14}{3}x$

21. $(2.3x^3 - 5.6x^2 - 2) - (5.5x^3 - 7.4x^2 + 2)$
$= 2.3x^3 - 5.6x^2 - 2 - 5.5x^3 + 7.4x^2 - 2$
$= -3.2x^3 + 1.8x^2 - 4$

23. $(5x + 8)(2x + 9) = 10x^2 + 45x + 16x + 72$
$= 10x^2 + 61x + 72$

25. $(5w + 2d)(3a - 4b) = 15aw - 20bw + 6ad - 8bd$

27. $(-6x + y)(2x - 5y) = -12x^2 + 30xy + 2xy - 5y^2$
$= -12x^2 + 32xy - 5y^2$

29. $(7r - s^2)(-4a - 11s^2)$
$= -28ar - 77rs^2 + 4as^2 + 11s^4$

31. $(5x - 8y)(5x + 8y) = 25x^2 - 64y^2$

33. $(5a - 2b)^2 = 25a^2 - 20ab + 4b^2$

35. $(7m - 1)^2 = 49m^2 - 14m + 1$

37. $(6 + 5x^2)(6 - 5x^2) = 36 - 25x^4$

39. $(3m^3 + 1)^2 = 9m^6 + 6m^3 + 1$

41. $2x(3x^2 - 5x + 1) = 6x^3 - 10x^2 + 2x$

43. $-\frac{1}{2}ab(4a - 5b - 10) = -2a^2b + \frac{5}{2}ab^2 + 5ab$

45. $(2x - 3)(x^2 - x + 1)$
$= 2x^3 - 2x^2 + 2x - 3x^2 + 3x - 3$
$= 2x^3 - 5x^2 + 5x - 3$

47. $(3x^2-2xy-6y^2)(2x-y)$
$=6x^3-3x^2y-4x^2y+2xy^2-12xy^2+6y^3$
$=6x^3-7x^2y-10xy^2+6y^3$

49. $(3a^3+4a^2-a-1)(a-5)$
$=3a^4-15a^3+4a^3-20a^2-a^2+5a-a+5$
$=3a^4-11a^3-21a^2+4a+5$

51. $(x+2)(x-3)(2x-5)$
$=(x^2-x-6)(2x-5)$
$=2x^3-5x^2-2x^2+5x-12x+30$
$=2x^3-7x^2-7x+30$

53. $(a-2)(5+a)(3-2a)$
$=(5a+a^2-10-2a)(3-2a)$
$=(a^2+3a-10)(3-2a)$
$=3a^2-2a^3+9a-6a^2-30+20a$
$=-2a^3-3a^2+29a-30$

55. $V=(2x^2+5x+8)(3x+5)$
$V=6x^3+10x^2+15x^2+25x+24x+40$
$V=6x^3+25x^2+49x+40$
The volume is $(6x^3+25x^2+49x+40)$ cm^3.

Cumulative Review

57. $\frac{1}{2}x+4\le\frac{2}{3}(x-3)+1$
$6\left(\frac{1}{2}x+4\right)\le 6\left[\frac{2}{3}(x-3)+1\right]$
$3x+24\le 4x-12+6$
$-x\le -30$
$x\ge 30$

58. $2(x+1)-3=4-(x+5)$
$2x+2-3=4-x-5$
$2x-1=-x-1$
$3x=0$
$x=0$

Quick Quiz 5.1

1. $(7x-4)+(5x-6)-(3x^2-9x)$
$=7x-4+5x-6-3x^2+9x$
$=-3x^2+21x-10$

2. $(5x-2y^2)^2=(5x)^2-2(5x)(2y^2)+(2y^2)^2$
$=25x^2-20xy^2+4y^4$

3. $(2x-3)(4x^2+2x-1)$
$=8x^3+4x^2-2x-12x^2-6x+3$
$=8x^3-8x^2-8x+3$

4. Answers may vary. Possible solution: First step is to use the FOIL method to multiply the first two polynomials, then combining like terms of the product. Next use the distributive property to multiply the remaining polynomial by the above product.

5.2 Exercises

1. $(24x^2-8x-44)\div 4$
$\frac{24x^2}{4}-\frac{8x}{4}-\frac{44}{4}=6x^2-2x-11$

3. $(27x^4-9x^3+63x^2)\div(-9x)$
$\frac{27x^4}{-9x}-\frac{9x^3}{-9x}+\frac{63x^2}{-9x}=-3x^3+x^2-7x$

5. $\frac{8b^4-6b^3-b^2}{2b^2}=\frac{8b^4}{2b^2}-\frac{6b^3}{2b^2}-\frac{b^2}{2b^2}$
$=4b^2-3b-\frac{1}{2}$

7. $\frac{9a^3b^3-15a^3b^2+3a^2b}{3a^2b}$
$=\frac{9a^3b^3}{3a^2b}-\frac{15a^3b^2}{3a^2b}+\frac{3a^2b}{3a^2b}$
$=3ab^2-5ab+1$

9. $(5x^2-17x+6)\div(x-3)$

$$\begin{array}{r}
5x-2 \\
x-3\overline{)5x^2-17x+6} \\
\underline{5x^2-15x} \\
-2x+6 \\
\underline{-2x+6} \\
0
\end{array}$$

$(5x^2-17x+6)\div(x-3)=5x-2$
Check: $(x-3)(5x-2)=5x^2-17x+6$

11. $(15x^2 + 23x + 4) \div (5x + 1)$

$$\begin{array}{r} 3x+4 \\ 5x+1 \overline{)\,15x^2+23x+4} \\ \underline{15x^2+\ \ 3x} \\ 20x+4 \\ \underline{20x+4} \\ 0 \end{array}$$

$(15x^2 + 23x + 4) \div (5x + 1) = 3x + 4$

Check: $(5x + 1)(3x + 4) = 15x^2 + 23x + 4$

13. $(20x^2 - 17x + 3) \div (4x - 1)$

$$\begin{array}{r} 5x-3 \\ 4x-1 \overline{)\,20x^2-17x+3} \\ \underline{20x^2-\ \ 5x} \\ -12x+3 \\ \underline{-12x+3} \\ 0 \end{array}$$

$(20x^2 - 17x + 3) \div (4x - 1) = 5x - 3$

Check: $(5x - 3)(4x - 1) = 20x^2 - 17x + 3$

15. $(x^3 - x^2 + 11x - 1) \div (x + 1)$

$$\begin{array}{r} x^2-2x+13 \\ x+1 \overline{)\,x^3-x^2\ \ +11x-1} \\ \underline{x^3+x^2} \\ -2x^2+11x \\ \underline{-2x^2\ \ -2x} \\ 13x-\ \ 1 \\ \underline{13x+13} \\ -14 \end{array}$$

$(x^3 - x^2 + 11x - 1) \div (x + 1) = x^2 - 2x + 13 - \dfrac{14}{x+1}$

Check: $(x+1)\left(x^2 - 2x + 13 - \dfrac{14}{x+1}\right)$

$= (x+1)(x^2 - 2x + 13) - (x+1)\left(\dfrac{14}{x+1}\right)$

$= x^3 - 2x^2 + 13x + x^2 - 2x + 13 - 14$

$= x^3 - x^2 + 11x - 1$

17. $(2x^3 - x^2 - 7) \div (x - 2)$

$$\begin{array}{r} 2x^2+3x+6 \\ x-2 \overline{)\,2x^3-\ \ x^2+0x\ \ -7} \\ \underline{2x^3-4x^2} \\ 3x^2+0x \\ \underline{3x^2-6x} \\ 6x-\ \ 7 \\ \underline{6x-12} \\ 5 \end{array}$$

$(2x^3 - x^2 - 7) \div (x - 2) = 2x^2 + 3x + 6 + \dfrac{5}{x-2}$

19. $\dfrac{4x^3 - 6x^2 - 3}{2x+1} = 2x^2 - 4x + 2 - \dfrac{5}{2x+1}$

$$\begin{array}{r} 2x^2-4x+2 \\ 2x+1 \overline{)\,4x^3-6x^2+0x-3} \\ \underline{4x^3+2x^2} \\ -8x^2+0x \\ \underline{-8x^2-4x} \\ 4x-3 \\ \underline{4x+2} \\ -5 \end{array}$$

21. $\dfrac{2x^4 - 3x^3 + 8x^2 - 18}{2x - 3}$

$$\begin{array}{r} x^3+4x+6 \\ 2x-3 \overline{)\,2x^4-3x^3+8x^2+\ \ 0x-18} \\ \underline{2x^4-3x^3} \\ 8x^2+\ \ 0x \\ \underline{8x^2-12x} \\ 12x-18 \\ \underline{12x-18} \\ 0 \end{array}$$

$(2x^4 - 3x^3 + 8x^2 - 18) \div (2x - 3) = x^3 + 4x + 6$

23. $\dfrac{6t^4-5t^3-8t^2+16t-8}{3t^2+2t-4}$

$$\begin{array}{r|l} & 2t^2-3t+2 \\ \hline 3t^2+2t-4 & 6t^4-5t^3-8t^2+16t-8 \\ & \underline{6t^4+4t^3-8t^2} \\ & -9t^3 \quad +16t \\ & \underline{-9t^3-6t^2+12t} \\ & 6t^2+4t-8 \\ & \underline{6t^2+4t-8} \\ & 0 \end{array}$$

$$\frac{6t^4-5t^3-8t^2+16t-8}{3t^2+2t-4}=2t^2-3t+2$$

25.

$$A = LW$$
$$18x^3-21x^2+11x-2=(6x^2-5x+2)W$$
$$W=\frac{18x^3-21x^2+11x-2}{6x^2-5x+2}$$

$$\begin{array}{r|l} & 3x-1 \\ \hline 6x^2-5x+2 & 18x^3-21x^2+11x-2 \\ & \underline{18x^3-15x^2+6x} \\ & -6x^2+5x-2 \\ & \underline{-6x^2+5x-2} \\ & 0 \end{array}$$

The width of the solar panel is $(3x-1)$ meters.

Cumulative Review

27. $m=\dfrac{y_2-y_1}{x_2-x_1}=\dfrac{0-(-1)}{-\frac{1}{3}-\frac{1}{2}}=\dfrac{1}{-\frac{5}{6}}=-\dfrac{6}{5}$

28.
$$3y-2x=7$$
$$3y=2x+7$$
$$y=\frac{2}{3}x+\frac{7}{3}$$
$$m=\frac{2}{3}$$
$$m_\perp=-\frac{3}{2}$$

29.
$$5(x-2)+3y=3x-(y-1)$$
$$5x-10+3y=3x-y+1$$
$$2x=-4y+11$$
$$x=\frac{-4y+11}{2}$$

30.
$$\frac{2x+4}{3}-\frac{y}{2}=x-2y+1$$
$$6\left(\frac{2x+4}{3}-\frac{y}{2}\right)=6(x-2y+1)$$
$$2(2x+4)-3y=6x-12y+6$$
$$4x+8-3y=6x-12y+6$$
$$2+9y=2x$$
$$x=\frac{9y+2}{2}$$

Quick Quiz 5.2

1. $(16x^3-20x^2-8x)\div(-4x)$
$$=\frac{16x^3}{-4x}-\frac{20x^2}{-4x}-\frac{8x}{-4x}$$
$$=-4x^2+5x+2$$

2. $(x^3-6x^2+7x-2)\div(x-1)$

$$\begin{array}{r|l} & x^2-5x+2 \\ \hline x-1 & x^3-6x^2+7x-2 \\ & \underline{x^3-x^2} \\ & -5x^2+7x \\ & \underline{-5x^2+5x} \\ & 2x-2 \\ & \underline{2x-2} \\ & 0 \end{array}$$

$$(x^3-6x^2+7x-2)\div(x-1)=(x^2-5x+2)$$

3. $(2x^4-x^3-12x^2-17x-12)\div(2x+3)$

$$\begin{array}{r|l} & x^3-2x^2-3x-4 \\ \hline 2x+3 & 2x^4-x^3-12x^2-17x-12 \\ & \underline{2x^4+3x^3} \\ & -4x^3-12x^2 \\ & \underline{-4x^3-6x^2} \\ & -6x^2-17x \\ & \underline{-6x^2-9x} \\ & -8x-12 \\ & \underline{-8x-12} \\ & 0 \end{array}$$

$$(2x^4-x^3-12x^2-17x-12)\div(2x+3)$$
$$=(x^3-2x^2-3x-4)$$

4. Answers may vary. Possible solution: Multiply the divisor by the quotient. The result will be the dividend if the correct quotient has been found.

5.3 Exercises

1. $(2x^2 - 11x - 8) \div (x-6)$

$$\begin{array}{r|rrr} 6 & 2 & -11 & -8 \\ & & 12 & 6 \\ \hline & 2 & 1 & \boxed{-2} \end{array}$$

$$(2x - 11x - 8) \div (x-6) = 2x + 1 + \frac{-2}{x-6}$$

3. $(3x^3 + x^2 - x + 4) \div (x+1)$

$$\begin{array}{r|rrrr} -1 & 3 & 1 & -1 & 4 \\ & & -3 & 2 & -1 \\ \hline & 3 & -2 & 1 & \boxed{3} \end{array}$$

$$(3x^3 + x^2 - x + 4) \div (x+1) = 3x^2 - 2x + 1 + \frac{3}{x+1}$$

5. $(x^3 + 7x^2 + 17x + 15) \div (x+3)$

$$\begin{array}{r|rrrr} -3 & 1 & 7 & 17 & 15 \\ & & -3 & -12 & -15 \\ \hline & 1 & 4 & 5 & \boxed{0} \end{array}$$

$$(x^3 + 7x^2 + 17x + 15) \div (x+3) = x^2 + 4x + 5$$

7. $(4x^3 - 11x^2 - 20x + 5) \div (x-4)$

$$\begin{array}{r|rrrr} 4 & 4 & -11 & -20 & 5 \\ & & 16 & 20 & 0 \\ \hline & 4 & 5 & 0 & \boxed{5} \end{array}$$

$$(4x^3 - 11x^2 - 20x + 5) \div (x-4)$$
$$= \left(4x^2 + 5x + \frac{5}{x-4}\right)$$

9. $(x^3 - 2x^2 + 8) \div (x+2)$

$$\begin{array}{r|rrrr} -2 & 1 & -2 & 0 & 8 \\ & & -2 & 8 & -16 \\ \hline & 1 & -4 & 8 & \boxed{-8} \end{array}$$

$$(x^3 - 2x^2 + 8) \div (x+2) = x^2 - 4x + 8 + \frac{-8}{x+2}$$

11. $(6x^4 + 13x^3 + 35x - 24) \div (x+3)$

$$\begin{array}{r|rrrrr} -3 & 6 & 13 & 0 & 35 & -24 \\ & & -18 & 15 & -45 & 30 \\ \hline & 6 & -5 & 15 & -10 & \boxed{6} \end{array}$$

$$(6x^4 + 13x^3 + 35x - 24) \div (x+3)$$
$$= 6x^3 - 5x^2 + 15x - 10 + \frac{6}{x+3}$$

13. $(2x^4 + 3x^3 + x^2 + 2x + 5) \div (x+1)$

$$\begin{array}{r|rrrrr} -1 & 2 & 3 & 1 & 2 & 5 \\ & & -2 & -1 & 0 & -2 \\ \hline & 2 & 1 & 0 & 2 & \boxed{3} \end{array}$$

$$(2x^4 + 3x^3 + x^2 + 2x + 5) \div (x+1)$$
$$= 2x^3 + x^2 + 2 + \frac{3}{x+1}$$

15. $(2x^4 - 5x - 3) \div (x-1)$

$$\begin{array}{r|rrrrr} 1 & 2 & 0 & 0 & -5 & -3 \\ & & 2 & 2 & 2 & -3 \\ \hline & 2 & 2 & 2 & -3 & \boxed{-6} \end{array}$$

$$(2x^4 - 5x - 3) \div (x-1)$$
$$= 2x^3 + 2x^2 + 2x - 3 - \frac{6}{x-1}$$

17. $(2x^5 - 13x^3 + 10x^2 + 6) \div (x+3)$

$$\begin{array}{r|rrrrrr} -3 & 2 & 0 & -13 & 10 & 0 & 6 \\ & & -6 & 18 & -15 & 15 & -45 \\ \hline & 2 & -6 & 5 & -5 & 15 & \boxed{-39} \end{array}$$

$$(2x^5 - 13x^3 + 10x^2 + 6) \div (x+3)$$
$$= 2x^4 - 6x^3 + 5x^2 - 5x + 15 + \frac{-39}{x+3}$$

19. $(x^6 - 5x^3 + x^2 + 12) \div (x+1)$

$$\begin{array}{r|rrrrrrr} -1 & 1 & 0 & 0 & -5 & 1 & 0 & 12 \\ & & -1 & 1 & -1 & 6 & -7 & 7 \\ \hline & 1 & -1 & 1 & -6 & 7 & -7 & \boxed{19} \end{array}$$

$$(x^6 - 5x^3 + x^2 + 12) \div (x+1)$$
$$= x^5 - x^4 + x^3 - 6x^2 + 7x - 7 + \frac{19}{x+1}$$

21. $(4x^3 - 6x^2 + 6) \div (2x+3)$

$$\begin{array}{r|rrrr} -\frac{3}{2} & 4 & -6 & 0 & 6 \\ & & -6 & 18 & -27 \\ \hline & 4 & -12 & 18 & \boxed{-21} \end{array}$$

$$(4x^3 - 6x^2 + 6) \div (2x+3)$$
$$= \frac{4x^2 - 12x + 18}{2} + \frac{-21}{2x+3}$$
$$= 2x^2 - 6x + 9 + \frac{-21}{2x+3}$$

Cumulative Review

23. $2{,}000{,}000 \text{ gallons} \cdot \frac{0.134 \text{ cubic feet}}{1 \text{ gallon}}$
$= 268{,}000$ cubic feet
There were $268{,}000 \text{ ft}^3$ of molasses.

24. $V = \pi r^2 h$
$268{,}000 = \pi(200)^2 h$
$h \approx 2.1$
The molasses was 2.1 feet deep.

25. $p(x) = 2x^4 - 3x^2 + 6x - 1$
$p(-3) = 2(-3)^4 - 3(-3)^2 + 6(-3) - 1$
$p(-3) = 2(81) - 3(9) + 6(-3) - 1$
$p(-3) = 162 - 27 - 18 - 1$
$p(-3) = 116$

Quick Quiz 5.3

1. $(x^3 - 2x^2 - 5x - 2) \div (x+1)$

```
-1 | 1  -2  -5  -2
   |    -1   3   2
   --------------------
     1  -3  -2  | 0
```

$(x^3 - 2x^2 - 5x - 2) \div (x+1) = x^2 - 3x - 2$

2. $(x^3 - 8x^2 + 24) \div (x-2)$

```
2 | 1  -8    0   24
  |     2  -12  -24
  --------------------
    1  -6  -12  | 0
```

$(x^3 - 8x^2 + 24) \div (x-2) = x^2 - 6x - 12$

3. $(x^4 - 7x^2 + 2x - 12) \div (x+3)$

```
-3 | 1   0  -7   2  -12
   |    -3   9  -6   12
   ------------------------
     1  -3   2  -4  | 0
```

$(x^4 - 7x^2 + 2x - 12) \div (x+3) = x^3 - 3x^2 + 2x - 4$

4. Answers may vary. Possible solution:
When a term x^n of the dividend is missing in the sequence of descending powers of x, a zero must be used for the missing term's coefficient. This is done to save a place in the quotient for a coefficient of the term x^{n-1}.

5.4 Exercises

1. $80 - 10y = 10(8 - y)$

3. $5a^2 - 25a = 5a(a-5)$

5. $4a^2b^3 - 8ab + 32a = 4a(ab^3 - 2b + 8)$

7. $30y^4 + 24y^3 + 18y^2 = 6y^2(5y^2 + 4y + 3)$

9. $15ab^2 + 5ab - 10a^3b = 5ab(3b + 1 - 2a^2)$

11. $10a^2b^3 - 30a^3b^3 + 10a^3b^2 - 40a^4b^2$
$= 10a^2b^2(b - 3ab + a - 4a^2)$

13. $3x(x+y) - 2(x+y) = (x+y)(3x-2)$

15. $5b(a-3b) + 8(-3b+a) = 5b(a-3b) + 8(a-3b)$
$= (a-3b)(5b+8)$

17. $3x(a+5b) + (a+5b) = 3x(a+5b) + 1(a+5b)$
$= (a+5b)(3x+1)$

19. $2a^2(3x-y) - 5b^3(3x-y) = (3x-y)(2a^2 - 5b^3)$

21. $3x(5x+y) - 8y(5x+y) - (5x+y)$
$= 3x(5x+y) - 8y(5x+y) - 1(5x+y)$
$= (5x+y)(3x - 8y - 1)$

23. $2a(a-6b) - 3b(a-6b) - 2(a-6b)$
$= (a-6b)(2a - 3b - 2)$

25. $x^3 + 5x^2 + 3x + 15 = x^2(x+5) + 3(x+5)$
$= (x+5)(x^2+3)$

27. $2x + 6 - 3ax - 9a = 2(x+3) - 3a(x+3)$
$= (x+3)(2-3a)$

29. $ab - 4a + 12 - 3b = a(b-4) - 3(-4+b)$
$= a(b-4) - 3(b-4)$
$= (b-4)(a-3)$

31. $5x-20+3xy-12y=5(x-4)+3y(x-4)$
$=(5+3y)(x-4)$

33. $9y+2x-6-3xy=9y-6-3xy+2x$
$=3(3y-2)-x(3y-2)$
$=(3y-2)(3-x)$

35. $ab^3+c+b^2+abc=ab^3+b^2+c+abc$
$=b^2(ab+1)+c(1+ab)$
$=(b^2+c)(ab+1)$

37. $\frac{1}{3}x^3+\frac{1}{2}x^2+\frac{1}{6}x=x\left(\frac{1}{3}x^2+\frac{1}{2}x+\frac{1}{6}\right)$

Cumulative Review

39. $6x-2y=-12$ or $y=3x+6$

x	$y = 3x + 6$	y
−2	$y = 3(-2) + 6$	0
−1	$y = 3(-1) + 6$	3
0	$y = 3(0) + 6$	6

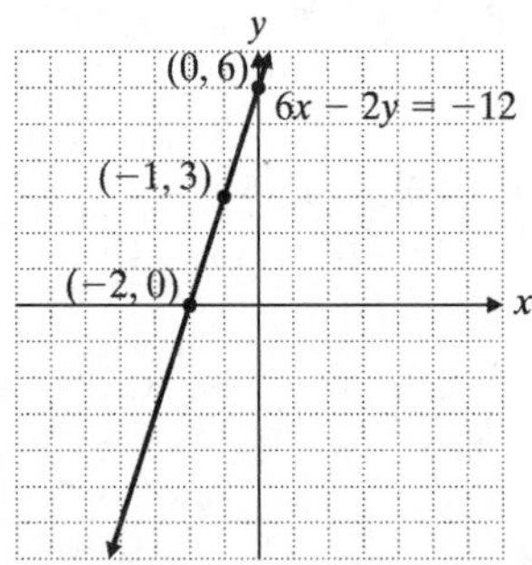

40. $y=\frac{2}{3}x-2$

x	$y=\frac{2}{3}x-2$	y
−3	$y=\frac{2}{3}(-3)-2$	−4
0	$y=\frac{2}{3}(0)-2$	−2
3	$y=\frac{2}{3}(3)-2$	0

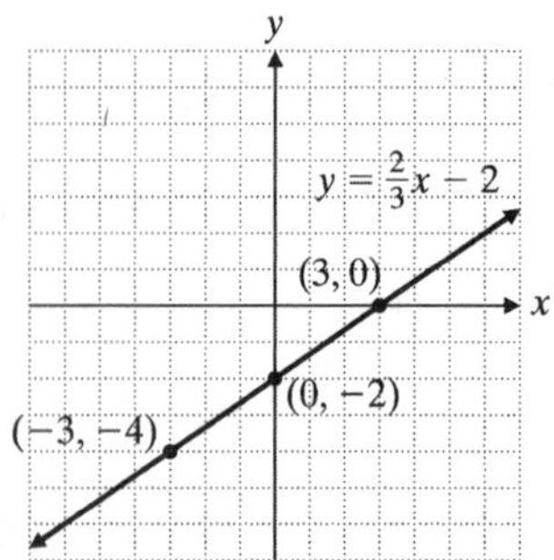

41. (5, −3) and (2, 6)
$m=\frac{y_2-y_1}{x_2-x_1}=\frac{6-(-3)}{2-5}$
$m=\frac{9}{-3}$
$m=-3$

42. $2y+6x=-3$
$2y=-6x-3$
$y=-3x-\frac{3}{2}$
$m=-3,\ b=-\frac{3}{2}$

Quick Quiz 5.4

1. $15a^2b^3-10a^3b^3-25a^2b^4=5a^2b^3(3-2a-5b)$

2. $2x(5x-3y)-5(5x-3y)=(5x-3y)(2x-5)$

3. $3ac+20b^2-4ab-15bc$
$=3ac-4ab-15bc+20b^2$
$=a(3c-4b)-5b(3c-4b)$
$=(3c-4b)(a-5b)$

4. Answers may vary. Possible solution:
Arrange the order as follows:
$8xy-10y+12x-15$.

5.5 Exercises

1. The numbers whose product is 7 and whose sum is 8 are 1 and 7.
$x^2+8x+7=(x+1)(x+7)$

3. The numbers whose product is 15 and whose sum is −8 are −5 and −3.
$x^2-8x+15=(x-5)(x-3)$

5. The numbers whose product is 24 and whose sum is –10 are –6 and –4.
$x^2 - 10x + 24 = (x-6)(x-4)$

7. The numbers whose product is –45 and whose sum is 4 are 9 and –5.
$a^2 + 4a - 45 = (a+9)(a-5)$

9. The numbers whose product is –42 and whose sum is –1 are –7 and 6.
$x^2 - xy - 42y^2 = (x-7y)(x+6y)$

11. The numbers whose product is 14 and whose sum is –15 are –14 and –1.
$x^2 - 15xy + 14y^2 = (x-14y)(x-y)$

13. The numbers whose product is –40 and whose sum is –3 are –8 and 5.
$x^4 - 3x^2 - 40 = (x^2-8)(x^2+5)$

15. The numbers whose product is 63 and whose sum is 16 are 7 and 9.
$x^4 + 16x^2y^2 + 63y^4 = (x^2+7y^2)(x^2+9y^2)$

17. $2x^2 + 26x + 44 = 2(x^2 + 13x + 22)$
$= 2(x+11)(x+2)$

19. $x^3 + x^2 - 20x = x(x^2 + x - 20) = x(x+5)(x-4)$

21. $2x^2 - x - 1 = 2x^2 + x - 2x - 1$
$= x(2x+1) - 1(2x+1)$
$= (2x+1)(x-1)$

23. $6x^2 - 7x - 5 = 6x^2 - 10x + 3x - 5$
$= 2x(3x-5) + 1(3x-5)$
$= (3x-5)(2x+1)$

25. $3a^2 - 8a + 5 = 3a^2 - 5a - 3a + 5$
$= a(3a-5) - 1(3a-5)$
$= (3a-5)(a-1)$

27. $4a^2 + a - 14 = 4a^2 - 7a + 8a - 14$
$= a(4a-7) + 2(4a-7)$
$= (4a-7)(a+2)$

29. $2x^2 + 13x + 15 = (2x+3)(x+5)$

31. $3x^4 - 8x^2 - 3 = (3x^2+1)(x^2-3)$

33. $6x^2 + 35xy + 11y^2 = (3x+y)(2x+11y)$

35. $7x^2 + 11xy - 6y^2 = (7x-3y)(x+2y)$

37. $8x^3 - 2x^2 - x = x(8x^2 - 2x - 1)$
$= x(4x+1)(2x-1)$

39. $10x^4 + 15x^3 + 5x^2 = 5x^2(2x^2 + 3x + 1)$
$= 5x^2(2x+1)(x+1)$

41. $x^2 - 2x - 63 = (x-9)(x+7)$

43. $6x^2 + x - 2 = (3x+2)(2x-1)$

45. $x^2 - 20x + 51 = (x-17)(x-3)$

47. $15x^2 + x - 2 = (5x+2)(3x-1)$

49. $2x^2 + 4x - 96 = 2(x^2 + 2x - 48)$
$= 2(x+8)(x-6)$

51. $18x^2 + 21x + 6 = 3(6x^2 + 7x + 2)$
$= 3(3x+2)(2x+1)$

53. $40ax^2 + 72ax - 16a = 8a(5x^2 + 9x - 2)$
$= 8a(5x-1)(x+2)$

55. $6x^3 + 26x^2 - 20x = 2x(3x^2 + 13x - 10)$
$= 2x(3x-2)(x+5)$

57. $7x^4 + 13x^2 - 2 = (7x^2-1)(x^2+2)$

59. $9a^2 - 18ab - 7b^2 = (3a+b)(3a-7b)$

61. $x^6 - 10x^3 - 39 = (x^3-13)(x^3+3)$

63. $10x^3y - 15x^2y - 10xy = 5xy(2x^2 - 3x - 2)$
$= 5xy(2x+1)(x-2)$

Cumulative Review

65. $A = \pi r^2 \approx 3.14(3)^2 = 28.26 \text{ in.}^2$

66. $A = \frac{1}{3}(3b + 4a)$

$3A = 3b + 4a$

$3b = 3A - 4a$

$b = \frac{3A - 4a}{3}$

67. x = number of first class seats

y = number of coach seats

$x + y = 184 \quad (1)$

$y = 6x + 16 \quad (2)$

Substitute $6x + 16$ for y in (1).

$x + 6x + 16 = 184$

$7x = 168$

$x = 24$

$y = 6x + 16 = 6(24) + 16$

$y = 160$

There are 24 first-class seats and 160 coach seats.

68. $6x + 4y = -12$

or $4y = -6x - 12$

$y = -\frac{3}{2}x - 3$

x	$y = -\frac{3}{2}x - 3$	y
−2	$y = -\frac{3}{2}(-2) - 3$	0
0	$y = -\frac{3}{2}(0) - 3$	−3
2	$y = -\frac{3}{2}(2) - 3$	−6

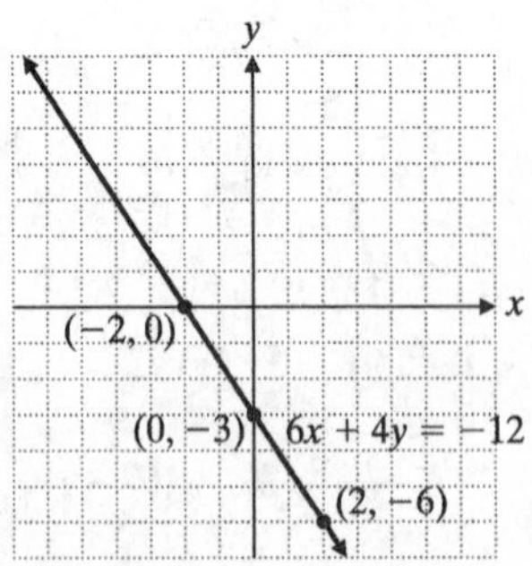

Quick Quiz 5.5

1. The numbers whose product is −84 and whose sum is 5 are −7 and 12.
   $x^2 + 5x - 84 = (x - 7)(x + 12)$

2. $6x^2 - 17x + 12 = 6x^2 - 8x - 9x + 12$
   $= 2x(3x - 4) - 3(3x - 4)$
   $= (3x - 4)(2x - 3)$

3. $4x^3 + 6x^2 - 40x = 2x(2x^2 + 3x - 20)$
   $= 2x(x + 4)(2x - 5)$

4. Answers may vary. Possible solution:
   The first step in factoring the expression $3x^2y^2 + 6xy^2 - 72y^2$ is to factor $3y^2$ out of every term leaving $3y^2(x^2 + 2x - 24)$.

How Am I Doing? Sections 5.1–5.5

1. $(5x^2 - 3x + 2) + (-3x^2 - 5x - 8) - (x^2 + 3x - 10)$
   $= 5x^2 - 3x + 2 - 3x^2 - 5x - 8 - x^2 - 3x + 10$
   $= x^2 - 11x + 4$

2. $(x^2 - 3x - 4)(2x - 3)$
   $= 2x^3 - 3x^2 - 6x^2 + 9x - 8x + 12$
   $= 2x^3 - 9x^2 + x + 12$

3. $(5a - 8)(a - 7) = 5a^2 - 35a - 8a + 56$
   $= 5a^2 - 43a + 56$

4. $(3y - 5)(3y + 5) = 9y^2 - 25$

5. $(3x^2 + 4)^2 = (3x^2)^2 + 2(3x^2)(4) + 4^2$
   $= 9x^4 + 24x^2 + 16$

6. $p(x) = 2x^3 - 5x^2 - 6x + 1$
   $p(-3) = 2(-3)^3 - 5(-3)^2 - 6(-3) + 1$
   $p(-3) = 2(-27) - 5(9) - 6(-3) + 1$
   $p(-3) = -54 - 45 + 18 + 1$
   $p(-3) = -80$

7. $\frac{25x^3y^2 - 30x^2y^3 - 50x^2y^2}{5x^2y^2}$
   $= \frac{25x^3y^2}{5x^2y^2} - \frac{30x^2y^3}{5x^2y^2} - \frac{50x^2y^2}{5x^2y^2}$
   $= 5x - 6y - 10$

8. $(3y^3-5y^2+2y-1)\div(y-2)$

$$\begin{array}{r} 3y^2+y+4 \\ y-2\overline{)3y^3-5y^2+2y-1} \\ \underline{3y^3-6y^2} \\ y^2+2y \\ \underline{y^2-2y} \\ 4y-1 \\ \underline{4y-8} \\ 7 \end{array}$$

$(3y^3-5y^2+2y-1)\div(y-2)=3y^2+y+4+\dfrac{7}{y-2}$

9. $(6x^4-x^3+11x^2-2x-2)\div(3x+1)$

$$\begin{array}{r} 2x^3-x^2+4x-2 \\ 3x+1\overline{)6x^4-x^3+11x^2-2x-2} \\ \underline{6x^4+2x^3} \\ -3x^3+11x^2 \\ \underline{-3x^3-x^2} \\ 12x^2-2x \\ \underline{12x^2+4x} \\ -6x-2 \\ \underline{-6x-2} \\ 0 \end{array}$$

$(6x^4-x^3+11x^2-2x-2)\div(3x+1)$
$=2x^3-x^2+4x-2$

10. $(2x^4+10x^3+11x^2-6x-9)\div(x+3)$

$$\begin{array}{r|rrrrr} -3 & 2 & 10 & 11 & -6 & -9 \\ & & -6 & -12 & 3 & 9 \\ \hline & 2 & 4 & -1 & -3 & 0 \end{array}$$

$(2x^4+10x^3+11x^2-6x-9)\div(x+3)$
$=2x^3+4x^2-x-3$

11. $24a^3b^2+36a^4b^2-60a^3b^3$
$=12a^3b^2(2+3a-5b)$

12. $3x(4x-3y)-2(4x-3y)=(4x-3y)(3x-2)$

13. $10wx+6xz-15yz-25wy$
$=2x(5w+3z)-5y(3z+5w)$
$=(5w+3z)(2x-5y)$

14. $18a^2+6ab-15a-5b=6a(3a+b)-5(3a+b)$
$=(3a+b)(6a-5)$

15. $x^2-7x+10=(x-5)(x-2)$

16. $4y^2-4y-15=(2y-5)(2y+3)$

17. $28x^2-19xy+3y^2=(7x-3y)(4x-y)$

18. $2x^2+21x+40=(2x+5)(x+8)$

19. $3x^2-6x-72=3(x^2-2x-24)=3(x+4)(x-6)$

20. $8x^2-18x+9=(4x-3)(2x-3)$

5.6 Exercises

1. The problem will have two terms. It will be in the form a^2-b^2. One term is positive and one term is negative. The coefficients and variables for the first and second terms are both perfect squares. So each one will be of the form 1, 4, 9, 16, 25, 36, and/or x^2, x^4, x^6, etc.

3. There will be two terms added together. It will be of the form a^3+b^3. Each term will contain a number or variable cubed or both. They will be of the form 1, 8, 27, 64, 125, and/or x^3, x^6, x^9, etc.

5. $a^2-64=a^2-8^2=(a+8)(a-8)$

7. $16x^2-81=(4x)^2-9^2=(4x+9)(4x-9)$

9. $64x^2-1=(8x)^2-1^2=(8x+1)(8x-1)$

11. $49m^2-9n^2=(7m)^2-(3n)^2$
$=(7m+3n)(7m-3n)$

13. $100y^2-81=(10y)^2-(9)^2=(10y+9)(10y-9)$

15. $1-36x^2y^2=1^2-(6xy)^2=(1+6xy)(1-6xy)$

17. $32x^2-18=2(16x^2-9)$
$=2[(4x)^2-3^2]$
$=2(4x+3)(4x-3)$

19. $5x-20x^3=5x(1-4x^2)$
$=5x[(1)^2-(2x)^2]$
$=5x(1+2x)(1-2x)$

21. $9x^2 - 6x + 1 = (3x)^2 - 2(3x)(1) + 1^2 = (3x-1)^2$

23. $49x^2 - 14x + 1 = (7x)^2 - 2(7x)(1) + 1^2 = (7x-1)^2$

25. $81w^2 + 36wt + 4t^2 = (9w)^2 + 2(9w)(2t) + (2t)^2$
$= (9w + 2t)^2$

27. $36x^2 + 60xy + 25y^2 = (6x)^2 + 2(6x)(5y) + (5y)^2$
$= (6x + 5y)^2$

29. $8x^2 + 40x + 50 = 2(4x^2 + 20x + 25) = 2(2x+5)^2$

31. $3x^3 - 24x^2 + 48x = 3x(x^2 - 8x + 16)$
$= 3x(x-4)^2$

33. $x^3 - 27 = x^3 - 3^3 = (x-3)(x^2 + 3x + 9)$

35. $x^3 + 125 = x^3 + 5^3 = (x+5)(x^2 - 5x + 25)$

37. $64x^3 - 1 = (4x)^3 - 1^3 = (4x-1)(16x^2 + 4x + 1)$

39. $8x^3 - 125 = (2x)^3 - 5^3 = (2x-5)(4x^2 + 10x + 25)$

41. $1 - 27x^3 = 1^3 - (3x)^3 = (1-3x)(1 + 3x + 9x^2)$

43. $64x^3 + 125 = (4x)^3 + 5^3$
$= (4x+5)(16x^2 - 20x + 25)$

45. $64s^6 + t^6 = (4s^2)^3 + (t^2)^3$
$= (4s^2 + t^2)(16s^4 - 4s^2t^2 + t^4)$

47. $5y^3 - 40 = 5(y^3 - 8) = 5(y-2)(y^2 + 2 + 4)$

49. $250x^3 + 2 = 2(125x^3 + 1)$
$= 2(5x+1)(25x^2 - 5x + 1)$

51. $x^5 - 8x^2y^3 = x^2(x^3 - 8y^3)$
$= x^2(x-2y)(x^2 + 2xy + 4y^2)$

53. $25w^4 - 1 = (5w^2)^2 - 1^2 = (5w^2 + 1)(5w^2 - 1)$

55. $b^4 + 6b^2 + 9 = (b^2)^2 + 2(b^2)(3) + 3^2 = (b^2 + 3)^2$

57. $9m^6 - 64 = (3m^3)^2 - 8^2 = (3m^3 + 8)(3m^3 - 8)$

59. $36y^6 - 60y^3 + 25 = (6y^3)^2 - 2(6y^3)(5) + 5^2$
$= (6y^3 - 5)^2$

61. $45z^8 - 5 = 5(9z^8 - 1) = 5(3z^4 + 1)(3z^4 - 1)$

63. $125m^3 + 8n^3 = (5m)^3 + (2n)^3$
$= (5m + 2n)(125m^2 - 10mn + 4n^2)$

65. $24a^3 - 3b^3 = 3(8a^3 - b^3)$
$= 3(2a - b)(4a^2 + 2ab + b^2)$

67. $4w^2 - 20wz + 25z^2 = (2w)^2 - 2(2w)(5z) + (5z)^2$
$= (2w - 5z)^2$

69. $36a^2 - 81b^2 = 9(4a^2 - 9b^2)$
$= 9(2a + 3b)(2a - 3b)$

71. $16x^4 - 81y^4 = (4x^2)^2 - (9y^2)^2$
$= (4x^2 + 9y^2)(4x - 9y^2)$
$= (4x^2 + 9y^2)(2x + 3y)(2x - 3y)$

73. $125m^6 + 8 = (5m^2)^3 + 2^3$
$= (5m^2 + 2)(25m^4 - 10m^2 + 4)$

75. $25x^2 + 25x + 4 = (5x)^2 + 25x + 2^2$
$25 \neq 2(5)(2) = 20$
$25x^2 + 25x + 4 = (5x + 4)(5x + 1)$

77. $49x^2 - 35x + 4 = (7x)^2 - 35x + (2)^2$
$2(7x)(2) = 28x \neq 35x$
$49x^2 - 35x + 4 = (7x - 4)(7x - 1)$

79. $A = (4x)(4x) - y^2$
$A = 16x^2 - y^2$
$A = (4x - y)(4x + y)$
The area is $(4x + y)(4x - y)$ ft^2.

Cumulative Review

81. principal salary: $y = 1.57x + 84$
teacher salary: $y = 1.04x + 44$
$1.04x + 44 + 50 = 1.57x + 84$
$0.53x = 10$
$x = 18.9$
A principal's salary will be \$50,000 per year more than a teacher's salary in $2002 + 18 = 2020$.

82. r = amount paid by Rebecca
g = amount paid by George
m = amount paid by Marcel
$g = m + 34$ (1)
$r = m - 19$ (2)
$r + g + m = 387$ (3)
Substitute $(m + 34)$ for g, and $(m - 19)$ for r in (3).

$$m-19+m+34+m=387$$
$$3m+15=387$$
$$m=124$$

Substitute 124 for m in (1) and (2).
$g = 124 + 34 = 158$
$r = 124 - 19 = 105$
Rebecca paid \$105.
Marcel paid \$124.
George paid \$158.

Quick Quiz 5.6

1. $25x^2-64y^2=(5x)^2-(8y)^2=(5x+8y)(5x-8y)$

2. $49x^2-56xy+16y^2=(7x)^2-2(7x)(4y)+(4y)^2$
$=(7x-4y)^2$

3. $64x^3-27y^3=(4x)^3-(3y)^3$
$=(4x-3y)(16x^2+12xy+9y^2)$

4. Answers may vary. Possible solution: The middle coefficient is only half of what would be required for the formula to work.

5.7 Exercises

1. In any factoring problem the first step is <u>to factor out a common factor if possible</u>.

3. You cannot factor polynomials of the form a^2+b^2. All such polynomials that are the sum of two squares are prime. You can factor polynomials of the form a^2-b^2, but you do NOT have that form in this problem.

5. $3xy - 6yz = 3y(x - 2z)$

7. $y^2+7y-18=(y+9)(y-2)$

9. $3x^2-8x+5=(3x-5)(x-1)$

11. $4x^2+20x+25=(2x)^2+2(2x)(5)+5^2$
$=(2x+5)^2$

13. $8x^3-125y^3=(2x)^3-(5y)^3$
$=(2x-5y)(4x^2+10xy+25y^2)$

15. $a^3-3ab-ac=a(a^2-3b-c)$

17. x^2+16 is prime.

19. $64y^2-25z^2=(8y)^2-(5z)^2=(8y+5z)(8y-5z)$

21. $6x^2-23x-4=(6x+1)(x-4)$

23. $3x^2-x-1$ is prime.

25. $x^3+9x^2+14x=x(x^2+9x+14)$
$=x(x+7)(x+2)$

27. $25x^2-40x+16=(5x)^2-2(5x)(4)+4^2$
$=(5x-4)^2$

29. $6a^2-6a-36=6(a^2-a-6)$
$=6(a-3)(a+2)$

31. $3x^2-3x-xy+y=3x(x-1)-y(x-1)$
$=(x-1)(3x-y)$

33. $81a^4-1=(9a^2)^2-1^2$
$=(9a^2+1)(9a^2-1)$
$=(9a^2+1)(3a+1)(3a-1)$

35. $2x^5-16x^3-18x=2x(x^4-8x^2-9)$
$=2x(x^2+1)(x^2-9)$
$=2x(x+3)(x-3)(x^2+1)$

37. $8a^3b-50ab^3=2ab(4a^2-25b^2)$
$=2ab[(2a)^2-(5b)^2]$
$=2ab(2a-5b)(2a+5b)$

39. $2ax-8xy+aw-4wy=2x(a-4y)+w(a-4y)$
$=(a-4y)(2x+w)$

41. $S=5x(x-10)+8y(x-10)$
$S=(x-10)(5x+8y)\text{ ft}^2$
$S=5x^2-50x+8xy-80y\text{ ft}^2$

Cumulative Review

43. $3x - 2 \le -5 + 2(x - 3)$
$3x - 2 \le -5 + 2x - 6$
$3x - 2x \le -5 - 6 + 2$
$x \le -9$

44. $|2 + 5x - 3| < 2$
$-2 < 2 + 5x - 3 < 2$
$-2 < 5x - 1 < 2$
$-1 < 5x < 3$
$-\frac{1}{5} < x < \frac{3}{5}$

45. $\left|\frac{1}{3}(5 - 4x)\right| > 4$
$\frac{1}{3}(5 - 4x) < -4$ or $\frac{1}{3}(5 - 4x) > 4$
$5 - 4x < -12$ $\quad$ $5 - 4x > 12$
$-4x < -17$ $\quad$ $-4x > 7$
$x > \frac{17}{4}$ $\quad$ $x < -\frac{7}{4}$

46. $x - 4 \ge 7$ or $4x + 1 \le 17$
$x \ge 11$ $\quad$ $4x \le 16$
$x \le 4$

Quick Quiz 5.7

1. $49x^3 + 84x^2y + 36xy^2$
$= x(49x^2 + 84xy + 36y^2)$
$= x[(7x)^2 + 2(7x)(6y) + (6y)^2]$
$= x(7x + 6y)^2$

2. $6x^4 + x^2 - 12 = (2x^2 + 3)(3x^2 - 4)$

3. $9x^2 + 12x - 12 = 3(3x^2 + 4x - 4)$
$= 3(x + 2)(3x - 2)$

4. Answers may vary. Possible solution: With all signs positive, the second term coefficient must equal 2 times the product of the first and third coefficient.

5.8 Exercises

1. $x^2 - x - 6 = 0$
$(x - 3)(x + 2) = 0$
$x - 3 = 0$ or $x + 2 = 0$
$x = 3$ $\quad$ $x = -2$
Check: $3^2 - 3 - 6 \stackrel{?}{=} 0,\ 0 = 0$
$(-2)^2 - (-2) - 6 \stackrel{?}{=} 0,\ 0 = 0$

3. $5x^2 - 6x = 0$
$x(5x - 6) = 0$
$x = 0$ or $5x - 6 = 0$
$x = \frac{6}{5}$
Check: $5(0)^2 - 6(0) \stackrel{?}{=} 0,\ 0 = 0$
$5\left(\frac{6}{5}\right)^2 - 6\left(\frac{6}{5}\right) \stackrel{?}{=} 0,\ 0 = 0$

5. $9x^2 - 16 = 0$
$(3x + 4)(3x - 4) = 0$
$3x + 4 = 0$ or $3x - 4 = 0$
$3x = -4$ $\quad$ $3x = 4$
$x = -\frac{4}{3}$ $\quad$ $x = \frac{4}{3}$
Check: $9\left(-\frac{4}{3}\right)^2 - 16 \stackrel{?}{=} 0,\ 0 = 0$
$9\left(\frac{4}{3}\right)^2 - 16 \stackrel{?}{=} 0,\ 0 = 0$

7. $3x^2 - 2x - 8 = 0$
$(3x + 4)(x - 2) = 0$
$3x + 4 = 0$ or $x - 2 = 0$
$x = -\frac{4}{3}$ $\quad$ $x = 2$
Check: $3\left(-\frac{4}{3}\right)^2 - 2\left(-\frac{4}{3}\right) - 8 \stackrel{?}{=} 0,\ 0 = 0$
$3(2)^2 - 2(2) - 8 \stackrel{?}{=} 0,\ 0 = 0$

9. $8x^2 - 3 = 2x$
$8x^2 - 2x - 3 = 0$
$(4x - 3)(2x + 1) = 0$
$4x - 3 = 0$ or $2x + 1 = 0$
$x = \frac{3}{4}$ $\quad$ $x = -\frac{1}{2}$
Check: $8\left(\frac{3}{4}\right)^2 - 3 \stackrel{?}{=} 2\left(\frac{3}{4}\right),\ \frac{3}{2} = \frac{3}{2}$
$8\left(-\frac{1}{2}\right)^2 - 3 \stackrel{?}{=} 2\left(-\frac{1}{2}\right),\ -1 = -1$

11. $8x^2 = 11x - 3$

$8x^2 - 11x + 3 = 0$

$(8x-3)(x-1) = 0$

$8x - 3 = 0$ or $x - 1 = 0$

$x = \frac{3}{8}$ $\quad$ $x = 1$

Check: $8\left(\frac{3}{8}\right)^2 \stackrel{?}{=} 11\left(\frac{3}{8}\right) - 3,\ \frac{9}{8} = \frac{9}{8}$

$8(1)^2 \stackrel{?}{=} 11(1) - 3,\ 8 = 8$

13. $x^2 + \frac{5}{3}x = \frac{2}{3}x$

$x^2 + x = 0$

$x(x+1) = 0$

$x = 0$ or $x + 1 = 0$

$x = -1$

Check: $0^2 + \frac{5}{3}(0) \stackrel{?}{=} \frac{2}{3}(0),\ 0 = 0$

$(-1)^2 + \frac{5}{3}(-1) \stackrel{?}{=} \frac{2}{3}(-1),\ -\frac{2}{3} = -\frac{2}{3}$

15. $25x^2 + 10x + 1 = 0$

$(5x+1)^2 = 0$

$5x + 1 = 0$

$x = -\frac{1}{5}$ double root

Check: $25\left(-\frac{1}{5}\right)^2 + 10\left(-\frac{1}{5}\right) + 1 \stackrel{?}{=} 0$

$0 = 0$

17. $x^3 + 7x^2 + 12x = 0$

$x(x^2 + 7x + 12) = 0$

$x(x+4)(x+3) = 0$

$x = 0$ or $x + 4 = 0$ or $x + 3 = 0$

$x = -4$ $\quad$ $x = -3$

Check: $0^3 + 7(0)^2 + 12(0) \stackrel{?}{=} 0,\ 0 = 0$

$(-4)^3 + 7(-4)^2 + 12(-4) \stackrel{?}{=} 0,\ 0 = 0$

$(-3)^3 + 7(-3)^2 + 12(-3) \stackrel{?}{=} 0,\ 0 = 0$

19. $\frac{x^3}{6} - 8x = \frac{x^2}{3}$

$x^3 - 2x^2 - 48x = 0$

$x(x^2 - 2x - 48) = 0$

$x(x-8)(x+6) = 0$

$x = 0$ or $x - 8 = 0$ or $x + 6 = 0$

$x = 8$ $\quad$ $x = -6$

Check: $\frac{0^3}{6} - 8(0) \stackrel{?}{=} \frac{0^2}{3},\ 0 = 0$

$\frac{(8)^3}{6} - 8(8) \stackrel{?}{=} \frac{(8)^2}{3},\ \frac{64}{3} = \frac{64}{3}$

$\frac{(-6)^3}{6} - 8(-6) \stackrel{?}{=} \frac{(-6)^2}{3},\ 12 = 12$

21. $3x^3 - 10x = 17x$

$3x^3 - 27x = 0$

$3x(x^2 - 9) = 0$

$3x(x+3)(x-3) = 0$

$3x = 0$ or $x + 3 = 0$ or $x - 3 = 0$

$x = 0$ $\quad$ $x = -3$ $\quad$ $x = 3$

Check: $3(0)^3 - 10(0) \stackrel{?}{=} 17(0),\ 0 = 0$

$3(-3)^3 - 10(-3) \stackrel{?}{=} 17(-3),\ -51 = -51$

$3(3)^3 - 10(3) \stackrel{?}{=} 17(3),\ 51 = 51$

23. $2x^3 + 4x^2 = 30x$

$2x(x^2 + 2x - 15) = 0$

$2x(x+5)(x-3) = 0$

$2x = 0$ or $x + 5 = 0$ or $x - 3 = 0$

$x = 0$ $\quad$ $x = -5$ $\quad$ $x = 3$

Check: $2(0)^3 + 4(0)^2 \stackrel{?}{=} 30(0),\ 0 = 0$

$2(-5)^3 + 4(-5)^2 \stackrel{?}{=} 30(-5),\ -150 = -150$

$2(3)^3 + 4(3)^2 \stackrel{?}{=} 30(3),\ 90 = 90$

25. $\frac{7x^2 - 3}{2} = 2x$

$7x^2 - 3 = 4x$

$7x^2 - 4x - 3 = 0$

$(7x+3)(x-1) = 0$

$7x + 3 = 0$ or $x - 1 = 0$

$x = -\frac{3}{7}$ $\quad$ $x = 1$

Check: $\dfrac{7\left(-\frac{3}{7}\right)^2-3}{2} \stackrel{?}{=} 2\left(-\dfrac{3}{7}\right), -\dfrac{6}{7}=-\dfrac{6}{7}$

$\dfrac{7(1)^2-3}{2} \stackrel{?}{=} 2(1),\ 2=2$

27.
$$2(x+3)=-3x+2(x^2-3)$$
$$2x+6=-3x+2x^2-6$$
$$2x^2-5x-12=0$$
$$(2x+3)(x-4)=0$$
$$2x+3=0 \quad\text{or}\quad x-4=0$$
$$x=-\frac{3}{2} \qquad x=4$$

Check: $2\left(-\dfrac{3}{2}+3\right) \stackrel{?}{=} -3\left(-\dfrac{3}{2}\right)+2\left(\left(-\dfrac{3}{2}\right)^2-3\right)$

$$3=3$$

$$2(4+3) \stackrel{?}{=} -3(4)+2(4^2-3)$$
$$14=14$$

29.
$$7x^2+6=2x^2+2(4x+3)$$
$$5x^2+6=8x+6$$
$$5x^2-8x=0$$
$$x(5x-8)=0$$
$$x=0 \quad\text{or}\quad 5x-8=0$$
$$x=\frac{8}{5}$$

Check: $7(0)^2+6 \stackrel{?}{=} 2(0)^2+2(4(0)+3)$

$$6=6$$

$$7\left(\frac{8}{5}\right)^2+6 \stackrel{?}{=} 2\left(\frac{8}{5}\right)^2+2\left(4\left(\frac{8}{5}\right)+3\right)$$
$$\frac{598}{25}=\frac{598}{25}$$

31.
$$2x^2-3x+c=0$$
$$2\left(-\frac{1}{2}\right)^2-3\left(-\frac{1}{2}\right)+c=0$$
$$2+c=0$$
$$c=-2$$
$$2x^2-3x-2=0$$
$$(2x+1)(x-2)=0$$
$$2x+1=0 \quad\text{or}\quad x-2=0$$
$$x=-\frac{1}{2} \qquad x=2$$

$x=2$ is the other solution.

33. $b=h+2$
$$A=\frac{1}{2}bh=\frac{1}{2}(h+2)h=180$$
$$h^2+2h-360=0$$
$$(h-18)(h+20)=0$$
$$h-18=0 \quad\text{or}\quad h+20=0$$
$$h=18 \qquad h=-20,\ \text{reject}$$
$h = 18$
$b = h + 2 = 20$
The altitude is 18 inches and the base is 20 inches.

35. $b = 3h + 2$
$$A=\frac{1}{2}bh=104$$
$$\frac{1}{2}(3h+2)h=104$$
$$3h^2+2h=208$$
$$3h^2+2h-208=0$$
$$(h-8)(3h+26)=0$$
$$h-8=0 \quad\text{or}\quad 3h+26=0$$
$$h=8 \qquad h=-\frac{26}{3}$$
reject, because $h>0$

$b = 3h + 2 = 3(8) + 2 = 26$

a. The altitude is 8 feet and the base is 26 feet.

b. $8\text{ feet}\left(\dfrac{1\text{ yard}}{3\text{ feet}}\right)=2\dfrac{2}{3}\text{ yards}$

$26\text{ feet}\left(\dfrac{1\text{ yard}}{3\text{ feet}}\right)=8\dfrac{2}{3}\text{ yards}$

The altitude is $2\frac{2}{3}$ yards and the base is $8\frac{2}{3}$ yards.

37. $L = W + 4$
$A = LW = 896$
$$W(W+4)=896$$
$$W^2+4W-896=0$$
$$(W-28)(W+32)=0$$
$$W-28=0 \quad\text{or}\quad W+32=0$$
$$W=28 \qquad W=-32$$
reject, because $W>0$

$L = W + 4 = 32$

a. The width is 28 cm and the length is 32 cm.

b. The width is 280 mm and the length is 320 mm.

39. $A = s^2$
$P = 4s$
$s^2 = 4s + 96$
$s^2 - 4s - 96 = 0$
$(s-12)(s+8) = 0$
$s - 12 = 0$ or $s + 8 = 0$
$s = 12$ $s = -8$
reject, because $s > 0$
Each side of the rug is 12 feet.

41. $W = 2$
$H = L + 2$
$V = LWH = 198$
$L \cdot 2(L+2) = 198$
$L^2 + 2L - 99 = 0$
$(L+11)(L-9) = 0$
$L + 11 = 0$ or $L - 9 = 0$
$L = -11$ $L = 9$
reject, because $L > 0$
$H = L + 2 = 11$
The length is 9 inches and the height is 11 inches.

43. $L = 2W - 3$
$A = LW = 54$
$(2W-3)W = 54$
$2W^2 - 3W - 54 = 0$
$(W-6)(2W+9) = 0$
$W - 6 = 0$ or $W + 9 = 0$
$W = 6$ $W = -9$
reject, because $W > 0$
$L = 2W - 3 = 9$
The length of the landing area is 9 miles and the width is 6 miles.

45. $N = 3.2x + 9.8$
2000: $x = 0$
$N = 3.2(0) + 9.8$
$N = 9.8$
There were 9,800,000,000 debit card transactions in year 2000.

47. $29 = 3.2x + 9.8$
$6 = x$
$2000 + 6 = 2006$
In 2006 there were 29,000,000,000 debit card transactions.

Cumulative Review

49. $(2x^3y^2)^3(5xy^2)^2 = 2^3(x^3)^3(y^2)^3 5^2 x^2 (y^2)^2$
$= 8x^{3(3)}y^{2(3)}25x^2y^{2(2)}$
$= 200x^{9+2}y^{6+4}$
$= 200x^{11}y^{10}$

50. $\dfrac{(2a^3b^2)^3}{16a^5b^8} = \dfrac{2^3a^{3\cdot3}b^{2\cdot3}}{16a^5b^8} = \dfrac{8a^9b^6}{16a^5b^8} = \dfrac{a^{9-5}}{2b^{8-6}} = \dfrac{a^4}{2b^2}$

51. $x - 2y = 8$
$x + y = -1$
Solve the second equation for y and substitute into the first equation.
$y = -x - 1$
$x - 2(-x-1) = 8$
$x + 2x + 2 = 8$
$3x = 6$
$x = 2$
$y = -x - 1$
$y = -2 - 1$
$y = -3$
The solution is $(2, -3)$.

52. $(5, 2)$ and $(6, 1)$
$m = \dfrac{y_2 - y_1}{x_2 - x_1} = \dfrac{1-2}{6-5} = -\dfrac{1}{1} = -1$
$y = mx + b$
$1 = -1(6) + b$
$b = 7$
$y = -x + 7$

Quick Quiz 5.8

1. $x^2 = -4x + 32$
$x^2 + 4x - 32 = 0$
$(x+8)(x-4) = 0$
$x + 8 = 0$ or $x - 4 = 0$
$x = -8$ $x = 4$

2. $15x^2 - 11x + 2 = 0$
$15x^2 - 5x - 6x + 2 = 0$
$5x(3x-1) - 2(3x-1) = 0$
$(3x-1)(5x-2) = 0$
$3x - 1 = 0$ or $5x - 2 = 0$
$3x = 1$ $5x = 2$
$x = \dfrac{1}{3}$ $x = \dfrac{2}{5}$

3. A = area of triangle = 160 square yards
b = base of triangle
a = altitude of triangle
$A = \frac{1}{2}ba$ (1)
$a = b - 4$ (2)
Substitute $b - 4$ for a in (1).
$160 = \frac{1}{2}b(b-4)$
$320 = b^2 - 4b$
$b^2 - 4b - 320 = 0$
$(b-20)(b+16) = 0$
$b - 20 = 0$ or $b + 16 = 0$
$b = 20$ $\quad$ $b = -16$
negative value not valid in this case
$a = b - 4 = 20 - 4 = 16$
The altitude is 16 yards. The base is 20 yards.

4. Answers may vary. Possible solution: Multiply the equation by the LCD (8) to eliminate fractions. Next move all terms to the left side of the equation, set equal to zero and factor to get $(3x - 4)(x + 2) = 0$. Set each term equal to zero and solve for x to get 2 possible solutions of $x = \frac{4}{3}, -2$. Test the validity of each possible solution in the context of the problem, discard possible solutions that make no sense.

Putting Your Skills to Work

1. \$3.00 a cup × 30 days = \$90
\$4.00 a cup × 2 times a day × 30 days = \$240
Total = \$90 + \$240 = \$330
They spend \$330.

2. (30 cups + 60 cups) ÷ 2 = 45
45 × \$1 = \$45
\$330 − \$45 = \$285
They save \$285 per month.

3. \$285 per month × 8 months = \$2280
Yes, they will have enough money for the vacation.

4. Answers may vary.

Chapter 5 Review Problems

1. $(x^2 - 3x + 5) + (-2x^2 - 7x + 8)$
$= x^2 - 3x + 5 - 2x^2 - 7x + 8$
$= -x^2 - 10x + 13$

2. $(-4x^2y - 7xy + y) + (5x^2y + 2xy - 9y)$
$= -4x^2y - 7xy + y + 5x^2y + 2xy - 9y$
$= x^2y - 5xy - 8y$

3. $(-6x^2 + 7xy - 3y^2) - (5x^2 - 3xy - 9y^2)$
$= -6x^2 + 7xy - 3y^2 - 5x^2 + 3xy + 9y^2$
$= -11x^2 + 10xy + 6y^2$

4. $(-13x^2 + 9x - 14) - (-2x^2 - 6x + 1)$
$= -13x^2 + 9x - 14 + 2x^2 + 6x - 1)$
$= -11x^2 + 15x - 15$

5. $(5x + 2) - (6 - x) + (2x + 3)$
$= 5x + 2 - 6 + x + 2x + 3$
$= 8x - 1$

6. $(4x - 5) - (3x^2 + x) + (x^2 - 2)$
$= 4x - 5 - 3x^2 - x + x^2 - 2$
$= -2x^2 + 3x - 7$

7. $p(x) = 3x^3 - 2x^2 - 6x + 1$
$p(-4) = 3(-4)^3 - 2(-4)^2 - 6(-4) + 1$
$p(-4) = -192 - 32 + 24 + 1$
$p(-4) = -199$

8. $p(x) = 3x^3 - 2x^2 - 6x + 1$
$p(-1) = 3(-1)^3 - 2(-1)^2 - 6(-1) + 1$
$p(-1) = -3 - 2 + 6 + 1$
$p(-1) = 2$

9. $p(x) = 3x^3 - 2x^2 - 6x + 1$
$p(3) = 3(3)^3 - 2(3)^2 - 6(3) + 1$
$p(3) = 81 - 18 - 18 + 1$
$p(3) = 46$

10. $g(x) = -x^4 + 2x^3 - x + 5$
$g(-1) = -(-1)^4 + 2(-1)^3 - (-1) + 5$
$g(-1) = -1 - 2 + 1 + 5$
$g(-1) = 3$

11. $g(x) = -x^4 + 2x^3 - x + 5$
$g(3) = -(3)^4 + 2(3)^3 - 3 + 5$
$g(3) = -81 + 54 - 3 + 5$
$g(3) = -25$

12. $g(x) = -x^4 + 2x^3 - x + 5$
$g(0) = -0^4 + 2(0)^3 - 0 + 5$
$g(0) = 5$

13. $h(x) = -x^3 - 6x^2 + 12x - 4$
$h(3) = -(3)^3 - 6(3)^2 + 12(3) - 4$
$h(3) = -27 - 54 + 36 - 4$
$h(3) = -49$

14. $h(x) = -x^3 - 6x^2 + 12x - 4$
$h(-2) = -(-2)^3 - 6(-2)^2 + 12(-2) - 4$
$h(-2) = 8 - 24 - 24 - 4$
$h(-2) = -44$

15. $h(x) = -x^3 - 6x^2 + 12x - 4$
$h(0) = -(0)^3 - 6(0)^2 + 12(0) - 4$
$h(0) = -4$

16. $3xy(x^2 - xy + y^2) = 3x^3y - 3x^2y^2 + 3xy^3$

17. $(3x^2 + 1)(2x - 1) = 6x^3 - 3x^2 + 2x - 1$

18. $(5x^2 + 3)^2 = (5x^2)^2 + 2(5x^2)(3) + 3^2$
$= 25x^4 + 30x^2 + 9$

19. $(x - 3)(2x - 5)(x + 2)$
$= (2x^2 - 11x + 15)(x + 2)$
$= 2x^3 + 4x^2 - 11x^2 - 22x + 15x + 30$
$= 2x^3 - 7x^2 - 7x + 30$

20. $(x^2 - 3x + 1)(-2x^2 + x - 2)$
$= -2x^4 + x^3 - 2x^2 + 6x^3 - 3x^2 + 6x - 2x^2$
$+ x - 2$
$= -2x^4 + 7x^3 - 7x^2 + 7x - 2$

21. $(3x - 5)(3x^2 + 2x - 4)$
$= 9x^3 + 6x^2 - 12x - 15x^2 - 10x + 20$
$= 9x^3 - 9x^2 - 22x + 20$

22. $(5ab - 2)(5ab + 2) = (5ab)^2 - 2^2$
$= 25a^2b^2 - 4$

23. $(3a - b^2)(6a - 5b^2) = 18a^2 - 15ab^2 - 6ab^2 + 5b^4$
$= 18a^2 - 21ab^2 + 5b^4$

24. $(25x^3y - 15x^2y - 100xy) \div (-5xy)$

$$\frac{25x^3y - 15x^2y - 100xy}{-5xy}$$
$$= \frac{25x^3y}{-5xy} - \frac{15x^2y}{-5xy} - \frac{100xy}{-5xy}$$
$$= -5x^2 + 3x + 20$$

25. $(12x^2 - 5x - 2) \div (3x - 2)$

$$\begin{array}{r} 4x + 1 \\ 3x - 2\overline{)12x^2 - 5x - 2} \\ \underline{12x^2 - 8x} \\ 3x - 2 \\ \underline{3x - 2} \\ 0 \end{array}$$

$(12x^2 - 5x - 2) \div (3x - 2) = 4x + 1$

26. $(2x^3 + x^2 - x + 1) \div (2x + 3)$

$$\begin{array}{r} x^2 - x + 1 \\ 2x + 3\overline{)2x^3 + x^2 - x + 1} \\ \underline{2x^3 + 3x^2} \\ -2x^2 - x \\ \underline{-2x^2 - 3x} \\ 2x + 1 \\ \underline{2x + 3} \\ -2 \end{array}$$

$(2x^3 + x^2 - x + 1) \div (2x + 3) = x^2 - x + 1 + \frac{-2}{2x + 3}$

27. $(3y^3 - 2y + 5) \div (y - 3)$

$$\begin{array}{r} 3y^2 + 9y + 25 \\ y - 3\overline{)3y^3 + 0y^2 - 2y + 5} \\ \underline{3y^3 - 9y^2} \\ 9y^2 - 2y \\ \underline{9y^2 - 27y} \\ 25y + 5 \\ \underline{25y - 75} \\ 80 \end{array}$$

$(3y^3 - 2y + 5) \div (y - 3) = 3y^2 + 9y + 25 + \frac{80}{y - 3}$

28. $(8a^4 + 2a^3 - 10a^2 - a + 3) \div (2a^2 - 1)$

$$\begin{array}{r}
4a^2 + a - 3 \\
2a^2 - 1 \overline{\smash{)}\, 8a^4 + 2a^3 - 10a^2 - a + 3} \\
\underline{8a^4 \qquad - 4a^2 \qquad\quad} \\
2a^3 - 6a^2 \qquad\quad \\
\underline{2a^3 \qquad\quad - a \quad} \\
-6a^2 \qquad 3 \\
\underline{-6a^2 \qquad 3} \\
0
\end{array}$$

$(8a^4 + 2a^3 - 10a^2 - a + 3) \div (2a^2 - 1)$
$= (4a^2 + a - 3)$

29. $(2x^4 - x^2 + 6x + 3) \div (x - 1)$

$$\begin{array}{r}
2x^3 + 2x^2 + x + 7 \\
x - 1 \overline{\smash{)}\, 2x^4 + 0x^3 - x^2 + 6x + 3} \\
\underline{2x^4 - 2x^3 \qquad\qquad\qquad} \\
2x^3 - x^2 \qquad\qquad \\
\underline{2x^3 - 2x^2 \qquad\qquad} \\
x^2 + 6x \qquad \\
\underline{x^2 - x \qquad} \\
7x + 3 \\
\underline{7x - 7} \\
10
\end{array}$$

$(2x^4 - x^2 + 6x + 3) \div (x - 1)$
$= 2x^3 + 2x^2 + x + 7 + \dfrac{10}{x - 1}$

30. $(2x^4 - 13x^3 + 16x^2 - 9x + 20) \div (x - 5)$

$$\begin{array}{r}
2x^3 - 3x^2 + x - 4 \\
x - 5 \overline{\smash{)}\, 2x^4 - 13x^3 + 16x^2 - 9x + 20} \\
\underline{2x^4 - 10x^3 \qquad\qquad\qquad\quad} \\
-3x^3 + 16x^2 \qquad\qquad \\
\underline{-3x^3 + 15x^2 \qquad\qquad} \\
x^2 - 9x \qquad \\
\underline{x^2 - 5x \qquad} \\
-4x + 20 \\
\underline{-4x + 20} \\
0
\end{array}$$

$(2x^4 - 13x^3 + 16x^2 - 9x + 2) \div (x - 5)$
$= 2x^3 - 3x^2 + x - 4$

31. $(3x^4 + 5x^3 - x^2 + x - 2) \div (x + 2)$

$$\begin{array}{r}
3x^3 - x^2 + x - 1 \\
x + 2 \overline{\smash{)}\, 3x^4 + 5x^3 - x^2 + x - 2} \\
\underline{3x^4 + 6x^3 \qquad\qquad\qquad} \\
-x^3 - x^2 \qquad\qquad \\
\underline{-x^3 - 2x^2 \qquad\qquad} \\
x^2 + x \qquad \\
\underline{x^2 + 2x \qquad} \\
-x - 2 \\
\underline{-x - 2} \\
0
\end{array}$$

$(3x^4 + 5x^3 - x^2 + x - 2) \div (x + 2)$
$= 3x^3 - x^2 + x - 1$

32. $15a^2b + 5ab^2 - 10ab = 5ab(3a + b - 2)$

33. $x^5 - 3x^4 + 2x^2 = x^2(x^3 - 3x^2 + 2)$

34. $12mn - 8m = 4m(3n - 2)$

35. $xy - 6y + 3x - 18 = y(x - 6) + 3(x - 6)$
$= (x - 6)(y + 3)$

36. $8x^2y + x^2b + 8y + b = x^2(8y + b) + 1(8y + b)$
$= (8y + b)(x^2 + 1)$

37. $3ab - 15a - 2b + 10 = 3a(b - 5) - 2(b - 5)$
$= (b - 5)(3a - 2)$

38. $x^2 - 9x - 22 = (x - 11)(x + 2)$

39. $4x^2 - 5x - 6 = (4x + 3)(x - 2)$

40. $6x^2 + 5x - 21 = (3x + 7)(2x - 3)$

41. $100x^2 - 49 = (10x)^2 - 7^2 = (10x + 7)(10x - 7)$

42. $4x^2 - 28x + 49 = (2x)^2 - 2(2x)(7) + 7^2$
$= (2x - 7)^2$

43. $8a^3 - 27 = (2a)^3 - 3^3 = (2a - 3)(4a^2 + 6a + 9)$

44. $9x^2 - 121 = (3x)^2 - (11)^2 = (3x + 11)(3x - 11)$

45. $5x^2 - 11x + 2 = (5x - 1)(x - 2)$

46. $x^3+8x^2+12x = x(x^2+8x+12)$
$= x(x+6)(x+2)$

47. $x^2-8wy+4wx-2xy = x^2+4wx-8wy-2xy$
$= x(x+4w)-2y(4w+x)$
$= x(x+4w)-2y(x+4w)$
$= (x+4w)(x-2y)$

48. $36x^2+25$ is prime.

49. $2x^2-7x-3$ is prime.

50. $x^2+6xy-27y^2 = (x+9y)(x-3y)$

51. $27x^4-x = x(27x^3-1)$
$= x[(3x)^3-1^3]$
$= x(3x-1)(9x^2+3x+1)$

52. $21a^2+20ab+4b^2 = (7a+2b)(3a+2b)$

53. $-3a^3b^3+2a^2b^4-a^2b^3 = -a^2b^3(3a-2b+1)$ or $a^2b^3(-3a+2b-1)$

54. $a^4b^4+a^3b^4-6a^2b^4 = a^2b^4(a^2+a-6)$
$= a^2b^4(a+3)(a-2)$

55. $3x^4-5x^2-2 = (3x^2+1)(x^2-2)$

56. $9a^2b+15ab-14b = b(9a^2+15a-14)$
$= b(3a+7)(3a-2)$

57. $2x^2+7x-6$ is prime.

58. $3x^2+5x+4$ is prime.

59. $4y^4-13y^3+9y^2 = y^2(4y^2-13y+9)$
$= y^2(4y-9)(y-1)$

60. $y^4+2y^3-35y^2 = y^2(y^2+2y-35)$
$= y^2(y+7)(y-5)$

61. $10x^2y^2-20x^2y+5x^2 = 5x^2(2y^2-4y+1)$

62. $4x^4+4x^2-15 = (2x^2-3)(2x^2+5)$

63. $a^2+5ab^3+4b^6 = (a+b^3)(a+4b^3)$

64. $3x^2-12-8x+2x^3 = 3x^2-12+2x^3-8x$
$= 3(x^2-4)+2x(x^2-4)$
$= (3+2x)(x^2+4)$
$= (2x+3)(x+2)(x-2)$

65. $2x^4-12x^2-54 = 2(x^4-6x^2-27)$
$= 2(x^2-9)(x^2+3)$
$= 2(x+3)(x-3)(x^2+3)$

66. $8a+8b-4bx-4ax = 4(2a+2b-ax-bx)$
$= 4[2(a+b)-x(a+b)]$
$= 4(a+b)(2-x)$

67. $8x^4+34x^2y^2+21y^4 = (4x^2+3y^2)(2x^2+7y^2)$

68. $4x^3+10x^2-6x = 2x(2x^2+5x-3)$
$= 2x(2x-1)(x+3)$

69. $2a^2x-15ax+7x = x(2a^2-15a+7)$
$= x(2a-1)(a-7)$

70. $9x^4y^2-30x^2y+25 = (3x^2y)^2-2(3x^2y)(5)+5^2$
$= (3x^2y-5)^2$

71. $27x^3y-3xy = 3xy(9x^2-1) = 3xy(3x+1)(3x-1)$

72. $5bx-28y+4by-35x = 5bx-35x+4by-28y$
$= 5x(b-7)+4y(b-7)$
$= (b-7)(5x+4y)$

73. $27abc^2-12ab = 3ab(9c^2-4)$
$= 3ab(3c+2)(3c-2)$

74. $5x^2-9x-2=0$
$(5x+1)(x-2)=0$
$5x+1=0$ or $x-2=0$
$x=-\frac{1}{5}$ $\quad x=2$

75. $2x^2-11x+12=0$
$(2x-3)(x-4)=0$
$2x-3=0$ or $x-4=0$
$x=\frac{3}{2}$ $\quad x=4$

76. $(6x+5)(x+2)=-2$
$6x^2+12x+5x+10=-2$
$6x^2+17x+12=0$
$(3x+4)(2x+3)=0$
$3x+4=0$ or $2x+3=0$
$3x=-4$ $2x=-3$
$x=-\frac{4}{3}$ $x=-\frac{3}{2}$

77. $6x^2=24x$
$6x^2-24x=0$
$6x(x-4)=0$
$6x=0$ or $x-4=0$
$x=0$ $x=4$

78. $3x^2+14x+3=-1+4(x+1)$
$3x^2+14x+3=-1+4x+4$
$3x^2+10x=0$
$x(3x+10)=0$
$x=0$ or $3x+10=0$
$x=0$ $x=-\frac{10}{3}$

79. $x^3+7x^2=-12x$
$x^3+7x^2+12x=0$
$x(x^2+7x+12)=0$
$x(x+4)(x+3)=0$
$x=0$ or $x+4=0$ or $x+3=0$
$x=0$ $x=-4$ $x=-3$

80. x = base
a = altitude
A = area = 75
$A=\frac{1}{2}xa$ (1)
$a=x+5$ (2)
Substitute $x + 5$ for a in (1).
$75=\frac{1}{2}x(x+5)$
$2(75)=x^2+5x$
$150=x^2+5x$
$x^2+5x-150=0$
$(x+15)(x-10)=0$
$x+15=0$ or $x-10=0$
$x=-15$ $x=10$
negative value is not valid in this case
$x = 10$
$a = x + 5 = 15$
The base = 10 meters.
The altitude = 15 meters.

81. w = width
l = length
A = area = 45
$A=wl$ (1)
$l=2w-1$ (2)
Substitute $2w - 1$ for l in (1).
$A=w(2w-1)$
$45=2w^2-w$
Rearrange.
$2w^2-w-45=0$
$(w-5)(2w+9)=0$
$w-5=0$ or $2w+9=0$
$w=5$ $2w=-9$
$w=-\frac{9}{2}$
negative not valid in this case
$w = 5$
$l = 2w - 1 = 9$
The width = 5 ft.
The length = 9 ft.

82. $P=3x^2-7x-10=30$
$3x^2-7x-40=0$
$(3x+8)(x-5)=0$
$3x+8=0$ or $x-5=0$
$x=-\frac{8}{3}$ $x=5$
reject, because $x > 0$
Five calculators should be made.

83. x = length of old side
$2x + 3$ = length of new side
$x^2+24=(2x+3)^2$
$x^2+24=4x^2+12x+9$
$3x^2+12x-15=0$
$x^2+4x-5=0$
$(x+5)(x-1)=0$
$x+5=0$ or $x-1=0$
$x=-5$ $x=1$
reject, because $x > 0$, $2x + 3 = 5$
The old side is 1 yard and the new side is 5 yards.

How Am I Doing? Chapter 5 Test

1. $(3x^2y-2xy^2-6)+(5+2xy^2-7x^2y)$
$=3x^2y-2xy^2-6+5+2xy^2-7x^2y$
$=-4x^2y-1$

2. $(5a^2-3)-(2+5a)-(4a-3)$
$=5a^2-3-2-5a-4a+3$
$=5a^2-9a-2$

3. $-2x(x+3y-4)=-2x^2-6xy+8x$

4. $(2x-3y^2)^2=(2x)^2-2(2x)(3y^2)+(3y^2)^2$
$=4x^2-12xy^2+9y^4$

5. $(x-2)(2x^2+x-1)$
$=2x^3+x^2-x-4x^2-2x+2$
$=2x^3-3x^2-3x+2$

6. $(-15x^3-12x^2+21x)\div(-3x)$
$=\frac{-15x^3}{-3x}-\frac{12x^2}{-3x}+\frac{21x}{-3x}$
$=5x^2+4x-7$

7. $(2x^4-7x^3+7x^2-9x+10)\div(2x-5)$

$$\begin{array}{r} x^3-x^2+x-2 \\ 2x-5\overline{)2x^4-7x^3+7x^2-9x+10} \\ \underline{2x^4-5x^3} \\ -2x^3+7x^2 \\ \underline{-2x^3+5x^2} \\ 2x^2-9x \\ \underline{2x^2-5x} \\ -4x+10 \\ \underline{-4x+10} \\ 0 \end{array}$$

$(2x^4-7x^3+7x^2-9x+10)\div(2x-5)$
$=x^3-x^2+x-2$

8. $(x^3-x^2-5x+2)\div(x+2)$

$$\begin{array}{r} x^2-3x+1 \\ x+2\overline{)x^3-x^2-5x+2} \\ \underline{x^3+2x^2} \\ -3x^2-5x \\ \underline{-3x^2-6x} \\ x+2 \\ \underline{x+2} \\ 0 \end{array}$$

$(x^3-x^2-5x+2)\div(x+2)=x^2-3x+1$

9. $(x^4+x^3-x-3)\div(x+1)$

$$\begin{array}{r|rrrrr} -1 & 1 & 1 & 0 & -1 & -3 \\ & & -1 & 0 & 0 & 1 \\ \hline & 1 & 0 & 0 & -1 & -2 \end{array}$$

$(x^4+x^3-x-3)\div(x+1)=x^3-1+\frac{-2}{x+1}$

10. $121x^2-25y^2=(11x)^2-(5y)^2$
$=(11x+5y)(11x-5y)$

11. $9x^2+30xy+25y^2=(3x)^2+2(3x)(5y)+(5y)^2$
$=(3x+5y)^2$

12. $x^3-26x^2+48x=x(x^2-26x+48)$
$=x(x-2)(x-24)$

13. $4x^3y+8x^2y^2+4x^2y=4x^2y(x+2y+1)$

14. $x^2-6wy+3xy-2wx=x^2+3xy-2wx-6wy$
$=x(x+3y)-2w(x+3y)$
$=(x+3y)(x-2w)$

15. $2x^2-3x+2$ is prime.

16. $18x^2+3x-15=3(6x^2+x-5)$
$=3(6x-5)(x+1)$

17. $54a^4-16a=2a(27a^3-8)$
$=2a[(3a)^3-2^3]$
$=2a(3a-2)(9a^2+6a+4)$

18. $9x^5 - 6x^3y + xy^2 = x(9x^4 - 6x^2y + y^2)$
$= x[(3x^2)^2 - 2(3x^2)(y) + y^2]$
$= x(3x^2 - y)^2$

19. $3x^4 + 17x^2 + 10 = (3x^2 + 2)(x^2 + 5)$

20. $3x - 10ay + 6y - 5ax = 3x - 5ax + 6y - 10ay$
$= x(3 - 5a) + 2y(3 - 5a)$
$= (3 - 5a)(x + 2y)$

21. $p(x) = -2x^3 - x^2 + 6x - 10$
$p(2) = -2(2)^3 - (2)^2 + 6(2) - 10$
$p(2) = -16 - 4 + 12 - 10$
$p(2) = -18$

22. $p(x) = -2x^3 - x^2 + 6x - 10$
$p(-3) = -2(-3)^3 - (-3)^3 + 6(-3) - 10$
$p(-3) = 54 - 9 - 18 - 10$
$p(-3) = 17$

23. $x^2 = 5x + 14$
$x^2 - 5x - 14 = 0$
$(x - 7)(x + 2) = 0$
$x - 7 = 0$ or $x + 2 = 0$
$x = 7$ 		$x = -2$

24. $3x^2 - 11x - 4 = 0$
$(3x + 1)(x - 4) = 0$
$3x + 1 = 0$ or $x - 4 = 0$
$x = -\frac{1}{3}$ 		$x = 4$

25. $7x^2 + 6x = 8x$
$7x^2 - 2x = 0$
$x(7x - 2) = 0$
$x = 0$ or $7x - 2 = 0$
$x = 0$ 		$x = \frac{2}{7}$

26. $h = b - 4$
$A = \frac{1}{2}bh = \frac{1}{2}b(b - 4) = 70$
$b^2 - 4b = 140$
$b^2 - 4b - 140 = 0$
$(b + 10)(b - 14) = 0$
$b + 10 = 0$ or $b - 14 = 0$
$b = -10$ 		$b = 14$
reject, because $b > 0$
$h = b - 4 = 10$
The base of the triangle is 14 inches and the altitude of the triangle is 10 inches.

Cumulative Test for Chapters 1–5

1. $3(5 \cdot 2) = (3 \cdot 5)2$ illustrates the associative property of multiplication.

2. $\frac{2 + 6(-2)}{(2 - 4)^3 + 3} = \frac{2 - 12}{(-2)^3 + 3} = \frac{-10}{-8 + 3} = \frac{-10}{-5} = 2$

3. $7x - 3[1 + 2(x - y)] = 7x - 3[1 + 2x - 2y]$
$= 7x - 3 - 6x + 6y$
$= x - 3 + 6y$

4. $5x + 7y = 2$
$5x = 2 - 7y$
$x = \frac{2 - 7y}{5}$

5. $\frac{1}{3}(x - 1) + \frac{x}{4} = x - 2$
Multiply by 12 to eliminate fractions.
$4(x - 1) + 3x = 12(x - 2)$
$4x - 4 + 3x = 12x - 24$
$-5x = -20$
$x = 4$

6. $(-2, -3)$ and $(1, 5)$
$m = \frac{y_2 - y_1}{x_2 - x_1}$
$m = \frac{-3 - 5}{-2 - 1}$
$m = \frac{8}{3}$

7. $y = -\frac{2}{3}x + 4$

x	$y = -\frac{2}{3}x + 4$	y
-3	$y = -\frac{2}{3}(-3) + 4$	6
0	$y = -\frac{2}{3}(0) + 4$	4
3	$y = -\frac{2}{3}(3) + 4$	2

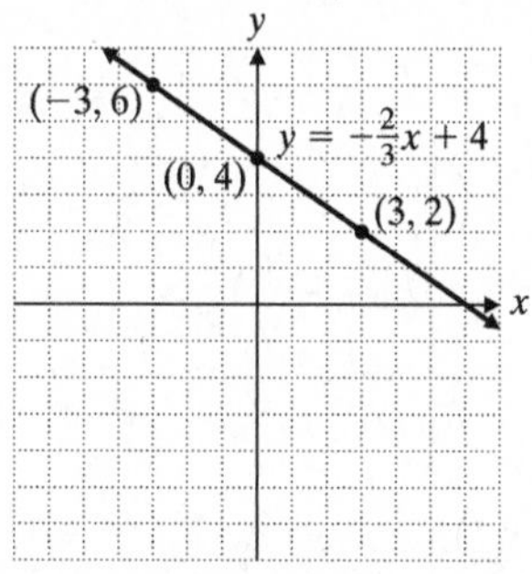

8. $3x - 4y \ge -12$
Test point: (0, 0)
$3(0) + 4(0) \ge -12$
$0 \ge -12$, true

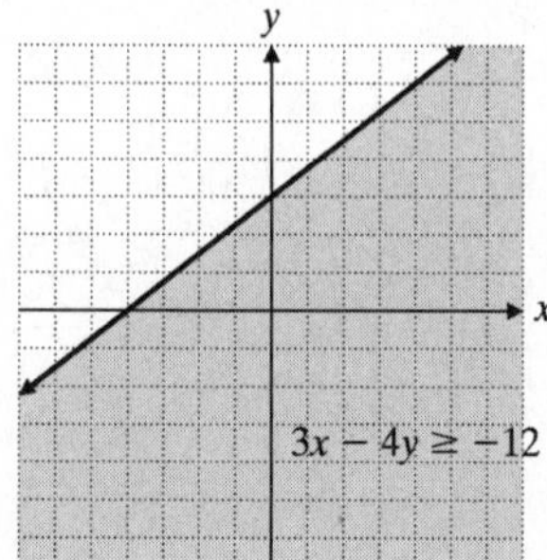

9. $-3(x+2) < 5x - 2(4+x)$
$-3x - 6 < 5x - 8 - 2x$
$-6x < -2$
$x > \frac{1}{3}$

10. $P = 2L + 2W = 46$
$2(2W+5) + 2W = 46$
$4W + 10 + 2W = 46$
$6W = 36$
$W = 6$
$L = 2W + 5 = 17$
The length of the rectangle is 17 meters and the width of the rectangle is 6 meters.

11. $x^3 - 2x^2 + 5xy - 6$
$x = -1$
$y = 0$
$(-1)^3 - 2(-1)^2 + 5(-1)(0) - 6 = -1 - 2 + 0 - 6$
$= -9$

12. $a[ab - 2b(a+4)] = a[ab - 2ab - 8b]$
$= a[-ab - 8b]$
$= -a^2b - 8ab$

13. $(3x+1)(x^2 - x + 4)$
$= 3x^3 - 3x^2 + 12x + x^2 - x + 4$
$= 3x^3 - 2x^2 + 11x + 4$

14. $(-21x^3 + 14x^2 - 28x) \div (7x)$
$$\frac{-21x^3}{7x} + \frac{14x^2}{7x} - \frac{28x}{7x} = -3x^2 + 2x - 4$$

15. $(2x^3 - 3x^2 + 3x - 4) \div (x - 2)$

$$\begin{array}{r} 2x^2 + x + 5 \\ x-2 \overline{) 2x^3 - 3x^2 + 3x - 4} \\ \underline{2x^3 - 4x^2} \qquad\qquad \\ x^2 + 3x \qquad \\ \underline{x^2 - 2x} \qquad \\ 5x - 4 \\ \underline{5x - 10} \\ 6 \end{array}$$

$(2x^3 - 3x^2 + 3x - 4) \div (x - 2)$
$= 2x^2 + x + 5 + \frac{6}{x-2}$

16. $2x^3 - 10x^2 = 2x^2(x - 5)$

17. $64x^2 - 49 = (8x)^2 - 7^2 = (8x+7)(8x-7)$

18. $3x^2 - 2x - 8 = (3x+4)(x-2)$

19. $49x^2 + 42x + 9 = (7x)^2 + 2(7x)(3) + 3^2$
$= (7x+3)^2$

20. $3x^2 - 9x - 54 = 3(x^2 - 3x - 18) = 3(x-6)(x+3)$

21. $2x^2 + 24x + 40 = 2(x^2 + 12x + 20)$
$= 2(x+10)(x+2)$

22. $16x^2 + 9$ is prime.

23. $4x^3 + 8x^2 - 5x = x(4x^2 + 8x - 5)$
$= x(2x-1)(2x+5)$

24. $27x^4 + 64x = x(27x^3 + 64)$
$= x[(3x)^3 + 4^3]$
$= x(3x+4)(9x^2 - 12x + 16)$

25. $2x-6-5xy+15y=2(x-3)-5y(x-3)$
$=(x-3)(2-5y)$

26. $3x^2-4x-4=0$
$(3x+2)(x-2)=0$
$3x+2=0$ or $x-2=0$
$x=-\frac{2}{3}$ $x=2$

27. $x^2-8x=33$
$x^2-8x-33=0$
$(x-11)(x+3)=0$
$x-11=0$ or $x+3=0$
$x=11$ $x=-3$

28. $A=\frac{1}{2}bh=\frac{1}{2}b(2b+1)=68$
$2b^2+b-136=0$
$(2b+17)(b-8)=0$
$2b+17=0$ or $b-8=0$
$b=-\frac{17}{2}$ $b=8$
reject, because $b>0$
$h=2b+1=17$
The base of the triangle is 8 meters and the altitude of the triangle is 17 meters.

Chapter 6

6.1 Exercises

1. $2x - 6 \neq 0$
$2x \neq 6$
$x \neq 3$
All real numbers except 3.

3. $g(x) = \dfrac{-3x+5}{x^2+5x-24}$
$g(x) = \dfrac{-3x+5}{(x+8)(x-3)}$
$x \neq -8, 3$
All real numbers except −8, 3

5. $\dfrac{-18x^4y}{12x^2y^6} = -\dfrac{6x^2y \cdot 3x^2}{6x^2y \cdot 2y^5} = -\dfrac{3x^2}{2y^5}$

7. $\dfrac{3x^3 - 24x^2}{6x-48} = \dfrac{3x^2(x-8)}{6(x-8)} = \dfrac{x^2}{2}$

9. $\dfrac{9x^2}{12x^2-15x} = \dfrac{3x \cdot 3x}{3x(4x-5)} = \dfrac{3x}{4x-5}$

11. $\dfrac{5x^2y^2 - 15xy^2}{10x^3y - 20x^3y^2} = \dfrac{5xy^2(x-3)}{10x^3y(1-2y)} = \dfrac{y(x-3)}{2x^2(1-2y)}$

13. $\dfrac{2x+10}{2x^2-50} = \dfrac{2(x+5)}{2(x+5)(x-5)} = \dfrac{1}{x-5}$

15. $\dfrac{3y^2-27}{3y+9} = \dfrac{3(y^2-9)}{3(y+3)} = \dfrac{3(y+3)(y-3)}{3(y+3)} = y-3$

17. $\dfrac{2x^2 - x^3 - x^4}{x^4 - x^3} = \dfrac{-x^2(x^2+x-2)}{x^3(x-1)}$
$= -\dfrac{(x+2)(x-1)}{x(x-1)}$
$= -\dfrac{x+2}{x}$

19. $\dfrac{2y^2+y-10}{4-y^2} = \dfrac{(2y+5)(y-2)}{-(y+2)(y-2)} = -\dfrac{2y+5}{y+2}$

21. $\dfrac{-8mn^5}{3m^4n^3} \cdot \dfrac{9m^3n^3}{6mn} = \dfrac{-2 \cdot 4 \cdot 3 \cdot 3m^4n^4n^4}{2 \cdot 3 \cdot 3m^4n^4m} = \dfrac{-4n^4}{m}$

23. $\dfrac{3a^2}{a^2+4a+4} \cdot \dfrac{a^2-4}{3a} = \dfrac{3a \cdot a(a+2)(a-2)}{3a \cdot (a+2)(a+2)}$
$= \dfrac{a(a-2)}{a+2}$

25. $\dfrac{x^2+5x+7}{x^2-5x+6} \cdot \dfrac{3x-6}{x^2+5x+7} = \dfrac{3(x-2)}{(x-2)(x-3)} = \dfrac{3}{x-3}$

27. $\dfrac{x^2-3xy-10y^2}{x+y} \cdot \dfrac{x^2+7xy+6y^2}{x+2y}$
$= \dfrac{(x-5y)(x+2y)}{x+y} \cdot \dfrac{(x+6y)(x+y)}{x+2y}$
$= (x-5y)(x+6y)$

29. $\dfrac{y^2-y-12}{2y^2+y-1} \cdot \dfrac{2y^2+7y-4}{2y^2-32}$
$= \dfrac{(y-4)(y+3)}{(2y-1)(y+1)} \cdot \dfrac{(2y-1)(y+4)}{2(y+4)(y-4)}$
$= \dfrac{y+3}{2(y+1)}$

31. $\dfrac{x^3-125}{x^5y} \cdot \dfrac{x^3y^2}{x^2+5x+25}$
$= \dfrac{(x-5)(x^2+5x+25)}{x^3y \cdot x^2} \cdot \dfrac{x^3y \cdot y}{(x^2+5x+25)}$
$= \dfrac{y(x-5)}{x^2}$

33. $\dfrac{2mn-m}{15m^3} \div \dfrac{2n-1}{3m^2} = \dfrac{m(2n-1)}{5m \cdot 3m^2} \cdot \dfrac{3m^2}{(2n-1)} = \dfrac{1}{5}$

35. $\dfrac{b^2-6b+9}{5b^2-16b+3} \div \dfrac{6b-3}{15b-3}$
$= \dfrac{(b-3)^2}{(b-3)(5b-1)} \cdot \dfrac{3(5b-1)}{3(2b-1)}$
$= \dfrac{b-3}{2b-1}$

37. $\frac{x^2 - xy - 6y^2}{x^2 + 2} \div (x^2 + 2xy)$
$= \frac{(x+2y)(x-3y)}{(x^2+2)} \cdot \frac{1}{x(x+2y)}$
$= \frac{x-3y}{x(x^2+2)}$

39. $\frac{7x}{y^2} \div 21x^3 = \frac{7x}{y^2} \cdot \frac{1}{7x \cdot 3x^2} = \frac{1}{3x^2y^2}$

41. $\frac{5x^2 - 2x}{15x - 6} = \frac{x(5x-2)}{3(5x-2)} = \frac{x}{3}$

43. $\frac{x^2y - 49y}{x^2y^3} \cdot \frac{3x^2y - 21xy}{x^2 - 14x + 49}$
$= \frac{y(x+7)(x-7)}{x^2y^3} \cdot \frac{3xy(x-7)}{(x-7)(x-7)}$
$= \frac{3(x+7)}{xy}$

45. $\frac{x^2 - 9x + 14}{x^3y^4} \div \frac{x^2 - 6x - 7}{x^2y^2}$
$= \frac{(x-2)(x-7)}{x^3y^4} \cdot \frac{x^2y^2}{(x+1)(x-7)}$
$= \frac{x-2}{xy^2(x+1)}$

47. $\frac{a^2 - a - 12}{2a^2 + 5a - 12} = \frac{(a-4)(a+3)}{(2a-3)(a+4)}$
Cannot be simplified.

49. Graphing $y_1 = 3.6x^2 + 1.8x - 4.3$ and using the *zero* feature to find the intercepts shows the domain of $f(x) = \frac{2x+5}{3.6x^2 + 1.8x - 4.3}$ is all real number except $x \approx -1.4$ and $x \approx 0.9$.

51. $P(x) = \frac{90(1+1.5x)}{1+0.5x}$
$P(1) = \frac{90(1+1.5(1))}{1+0.5(1)}$
$P(1) = 150$
The total number of fish is 150.

53. $P(x) = \frac{90(1+1.5x)}{1+0.5x}$
$P(6) = \frac{90(1+1.5(6))}{1+0.5(6)}$
$P(6) = 225$
The total number of fish is 225.

55.

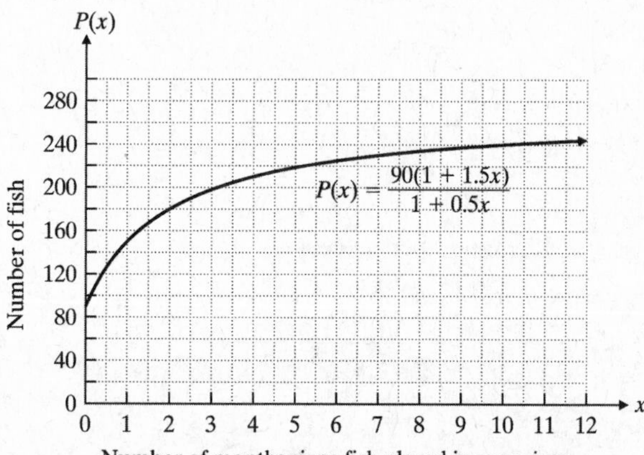

Cumulative Review

57. $2x - y = 8$
$x + 3y = 4$
Solve the first equation for y and substitute into the second equation.
$y = 2x - 8$
$x + 3(2x - 8) = 4$
$x + 6x - 24 = 4$
$7x = 28$
$x = 4$
$y = 2x - 8$
$y = 2(4) - 4 = 0$
The solution is (4, 0).

58. $3x - (x+5) < 7x + 10$
$3x - x - 5 < 7x + 10$
$2x - 5 < 7x + 10$
$-5x < 15$
$x > -3$

59. 1996 to 2006
$\frac{4810 - 2160}{2160} \cdot 100\% \approx 123\%$
x = number of riders killed not wearing helmets
$x = 4810(0.42)$
$x = 2020$ people
The percent increase in the number of deaths between 1996 and 2006 was 123%. Of the number of people killed in 2006, 2020 people were not wearing helmets.

Quick Quiz 6.1

1. $$\frac{x^3+11x^2+30x}{3x^3+17x^2-6x}=\frac{x(x^2+11x+30)}{x(3x^2+17x-6)}=\frac{x(x+6)(x+5)}{x(x+6)(3x-1)}=\frac{x+5}{3x-1}$$

2. $$\frac{2x^2-128}{x^2+16x+64}\cdot\frac{3x^2+30x+48}{x^2-6x-16}=\frac{2(x^2-64)}{(x+8)(x+8)}\cdot\frac{3(x^2+10x+16)}{(x-8)(x+2)}=\frac{2(x+8)(x-8)}{(x+8)(x+8)}\cdot\frac{3(x+8)(x+2)}{(x-8)(x+2)}=6$$

3. $$\frac{x^2+x-6}{x-5}\div\frac{x^2+8x+15}{15-3x}=\frac{(x+3)(x-2)}{(x-5)}\cdot\frac{3(5-x)}{(x+5)(x+3)}=\frac{(x+3)(x-2)}{(x-5)}\cdot\frac{-3(x-5)}{(x+5)(x+3)}=\frac{-3(x-2)}{x+5}$$

4. Answers may vary. Possible solution:
$$\frac{9-x^2}{x^2-7x+12}$$
The first step is to factor −1 out of the numerator.
$$\frac{-(x^2-9)}{x^2-7x+12}$$
The next step is to factor the numerator and denominator.
$$\frac{-(x+3)(x-3)}{(x-4)(x-3)}$$
Lastly, cancel like terms.
$$-\frac{x+3}{x-4}$$

6.2 Exercises

1. The factors are 5, x, and y. The factor y is repeated in one fraction three times. Since the highest power of y is 3, the LCD is $5xy^3$.

3. $x-1=x-1$
$x^2-2x+1=(x-1)^2$
$\text{LCD}=(x-1)^2$

5. $2m^3n$ and $2mn^2$
$\text{LCD}=2m^3n^2$

7. $$\frac{3x}{(2x+5)^3},\ \frac{x-2}{(x+1)(2x+5)^2}$$
$\text{LCD}=(x+1)(2x+5)^3$

9. $3x^2+2x=x(3x+2)$
$18x^2+9x-2=(6x-1)(3x+2)$
$\text{LCD}=x(6x-1)(3x+2)$

11. $$\frac{3}{x+4}+\frac{2}{x^2-16}=\frac{3(x-4)}{(x+4)(x-4)}+\frac{2}{(x+4)(x-4)}=\frac{3x-12+2}{(x+4)(x-4)}=\frac{3x-10}{(x+4)(x-4)}$$

13. $$\frac{9}{4xy}+\frac{3}{4y^2}=\frac{9}{4xy}\cdot\frac{y}{y}+\frac{3}{4y^2}\cdot\frac{x}{x}=\frac{9y+3x}{4xy^2}=\frac{3(3y+x)}{4xy^2}$$

15. $$\frac{3}{x^2-7x+12}+\frac{5}{x^2-4x}=\frac{3x}{(x-4)(x-3)x}+\frac{5(x-3)}{x(x-4)(x-3)}=\frac{3x+5x-15}{x(x-4)(x-3)}=\frac{8x-15}{x(x-4)(x-3)}$$

17. $$\frac{6x}{2x-5}+4=\frac{6x+4(2x-5)}{2x-5}=\frac{6x+8x-20}{2x-5}=\frac{14x-20}{2x-5}$$

19. $\dfrac{-5y}{y^2-1}+\dfrac{6}{y^2-2y+1}$
$=\dfrac{-5y(y-1)}{(y-1)^2(y+1)}+\dfrac{6(y+1)}{(y-1)^2(y+1)}$
$=\dfrac{-5y^2+5y+6y+6}{(y-1)^2(y+1)}$
$=\dfrac{-5y^2+11y+6}{(y-1)^2(y+1)}$

21. $\dfrac{a+4}{3a-6}+\dfrac{a-1}{a^2-4}$
$=\dfrac{a+4}{3(a-2)}+\dfrac{a-1}{(a+2)(a-2)}$
$=\dfrac{a+4}{3(a-2)}\cdot\dfrac{(a+2)}{(a+2)}+\dfrac{a-1}{(a+2)(a-2)}\cdot\dfrac{3}{3}$
$=\dfrac{(a+4)(a+2)+(3a-3)}{3(a+2)(a-2)}$
$=\dfrac{a^2+2a+4a+8+3a-3}{3(a+2)(a-2)}$
$=\dfrac{a^2+9a+5}{3(a+2)(a-2)}$

23. $\dfrac{5}{x-4}-\dfrac{3}{x+1}=\dfrac{5(x+1)-3(x-4)}{(x-4)(x+1)}$
$=\dfrac{5x+5-3x+12}{(x-4)(x+1)}$
$=\dfrac{2x+17}{(x-4)(x+1)}$

25. $\dfrac{1}{x^2-x-2}-\dfrac{3}{x^2+2x+1}$
$=\dfrac{1(x+1)}{(x-2)(x+1)(x+1)}-\dfrac{3(x-2)}{(x+1)(x+1)(x-2)}$
$=\dfrac{x+1-3x+6}{(x-2)(x+1)^2}$
$=\dfrac{-2x+7}{(x-2)(x+1)^2}$

27. $\dfrac{4y}{y^2+3y+2}-\dfrac{y-3}{y+2}$
$=\dfrac{4y}{(y+2)(y+1)}-\dfrac{(y-3)(y+1)}{(y+2)(y+1)}$
$=\dfrac{4y-y^2+2y+3}{(y+2)(y+1)}$
$=\dfrac{-y^2+6y+3}{(y+2)(y+1)}$

29. $a+2+\dfrac{3}{2a-5}=\dfrac{a}{1}\cdot\dfrac{2a-5}{2a-5}+\dfrac{2}{1}\cdot\dfrac{2a-5}{2a-5}+\dfrac{3}{2a-5}$
$=\dfrac{2a^2-5a+4a-10+3}{2a-5}$
$=\dfrac{2a^2-a-7}{2a-5}$

31. $P(x)=R(x)-C(x)$
$P(x)=\dfrac{80-24x}{2-x}-\dfrac{60-12x}{3-x}$
$=\dfrac{8(10-3x)(3-x)-12(5-x)(2-x)}{(2-x)(3-x)}$
$=\dfrac{8(30-19x+3x^2)-12(10-7x+x^2)}{x^2-5x+6}$
$=\dfrac{240-152x+24x^2-120+84x-12x^2}{x^2-5x+6}$
$=\dfrac{12x^2-68x+120}{x^2-5x+6}$

33. $P(x)=R(x)-C(x)$
$P(x)=\dfrac{12x^2-68x+120}{x^2-5x+6}$
$P(10)=\dfrac{12(10)^2-68(10)+120}{(10)^2-5(10)+6}$
$P(10)=11.4285714286...$
If ten machines per day are manufactured, the daily profit will be \$11,429.

Cumulative Review

35. $$\frac{-6\cdot 9\div|1-10|}{4^2-5\cdot 2+3\sqrt{25}}=\frac{-54\div 9}{16-10+3(5)}=\frac{-6}{6+15}=-\frac{6}{21}=-\frac{2}{7}$$

36. $0.000351=3.51\times 10^{-4}$

37. j = cost of Jamie's car
d = cost of Denise's car
a = cost of Amanda's car
$$j+d+a=24,000 \quad (1)$$
$$j=2a \quad (2)$$
$$d=a+2000 \quad (3)$$
Substitute $2a=j$ into (1), and $a+2000=d$ into (1).
$$2a+a+2000+a=24,000$$
$$4a=22,000$$
$$a=5500$$
$j=2a=11,000$
$d=a+2000=7500$
Amanda's car cost \$5500.
Jamie's car cost \$11,000.
Denise's car cost \$7500.

38. x = number of liters of 15% acid
$60-x$ = number of liters of 30% acid
$$0.15x+0.30(60-x)=0.20(60)$$
$$0.15x+18-0.3x=12$$
$$-0.15x=-6$$
$$x=40$$
$60-x=20$
He should use 40 liters of 15% acid and 20 liters of 30% acid.

Quick Quiz 6.2

1. $$\frac{5}{x}-\frac{4}{x+3}=\frac{5}{x}\cdot\frac{x+3}{x+3}-\frac{4}{x+3}\cdot\frac{x}{x}=\frac{5(x+3)-4x}{x(x+3)}=\frac{5x+15-4x}{x(x+3)}=\frac{x+15}{x(x+3)}$$

2. $$\frac{4}{x^2+3x+2}+\frac{3}{x^2+6x+8}$$
$$=\frac{4}{(x+2)(x+1)}+\frac{3}{(x+4)(x+2)}$$
$$=\frac{4}{(x+2)(x+1)}\cdot\frac{x+4}{x+4}+\frac{3}{(x+4)(x+2)}\cdot\frac{x+1}{x+1}$$
$$=\frac{4(x+4)+3(x+1)}{(x+1)(x+2)(x+4)}$$
$$=\frac{4x+16+3x+3}{(x+1)(x+2)(x+4)}$$
$$=\frac{7x+19}{(x+1)(x+2)(x+4)}$$

3. $$\frac{6x}{3x-7}+5=\frac{6x}{3x-7}+\frac{5}{1}\cdot\frac{3x-7}{3x-7}=\frac{6x+5(3x-7)}{3x-7}=\frac{6x+15x-35}{3x-7}=\frac{21x-35}{3x-7}=\frac{7(3x-5)}{3x-7}$$

4. Answers may vary. Possible solution:
Factor both denominators into prime factors. List all the different prime factors. The LCD is the product of these factors, each of which is raised to the highest power that appears on that factor in the denominators.

6.3 Exercises

1. $$\frac{\frac{7}{x}}{\frac{3}{xy}}=\frac{7}{x}\div\frac{3}{xy}=\frac{7}{x}\cdot\frac{xy}{3}=\frac{7y}{3}$$

3. $$\frac{\frac{2x}{x+5}}{\frac{x^2}{x-1}}=\frac{2x}{x+5}\div\frac{x^2}{x-1}=\frac{2x}{x+5}\cdot\frac{x-1}{x^2}=\frac{2(x-1)}{x(x+5)}$$

5. $$\frac{1-\frac{4}{3y}}{\frac{2}{y}+1}=\frac{\frac{1}{1}-\frac{4}{3y}}{\frac{2}{y}+\frac{1}{1}}\cdot\frac{3y}{3y}=\frac{\frac{3y}{1}-\frac{12y}{3y}}{\frac{6y}{y}+\frac{3y}{1}}=\frac{3y-4}{6+3y}=\frac{3y-4}{3(2+y)}$$

7. $$\frac{\frac{y}{6}-\frac{1}{2y}}{\frac{3}{2y}-\frac{1}{y}}=\frac{\frac{y}{6}-\frac{1}{2y}}{\frac{3}{2y}-\frac{1}{y}}\cdot\frac{6y}{6y}=\frac{y^2-3}{9-6}=\frac{y^2-3}{3}$$

9. $$\frac{\frac{2}{y^2-9}}{\frac{3}{y+3}+1}=\frac{\frac{2}{y^2-9}}{\frac{3}{y+3}+1}\cdot\frac{(y+3)(y-3)}{(y+3)(y-3)}=\frac{2}{3(y-3)+y^2-9}=\frac{2}{3y-9+y^2-9}=\frac{2}{y^2+3y-18}=\frac{2}{(y+6)(y-3)}$$

11. $$\frac{\frac{3}{2x+4}+\frac{1}{2}}{\frac{2}{x^2-4}+\frac{x}{x+2}}=\frac{\frac{6+2x+4}{2(2x+4)}}{\frac{2+x(x-2)}{(x+2)(x-2)}}=\frac{6+2x+4}{2(2x+4)}\cdot\frac{(x+2)(x-2)}{2+x(x-2)}=\frac{2x+10}{4(x+2)}\cdot\frac{(x-2)(x+2)}{2+x(x-2)}=\frac{2(x+5)}{4}\cdot\frac{(x-2)}{2+x^2-2x}=\frac{(x+5)(x-2)}{2(x^2-2x+2)}$$

13. $$\frac{-8}{\frac{6x}{x-1}-4}=\frac{-8}{\frac{6x}{x-1}-4}\cdot\frac{(x-1)}{(x-1)}=\frac{-8(x-1)}{6x-4(x-1)}=\frac{-8(x-1)}{6x-4x+4}=\frac{-8(x-1)}{2x+4}=\frac{-8(x-1)}{2(x+2)}=-\frac{4(x-1)}{x+2}$$

15. $$\frac{\frac{1}{2x+1}+\frac{4}{4x^2+4x+1}}{\frac{6x}{2x^2+x}}=\frac{\frac{1}{2x+1}+\frac{4}{(2x+1)(2x+1)}}{\frac{6}{x(2x+1)}}\cdot\frac{x(2x+1)(2x+1)}{x(2x+1)(2x+1)}=\frac{\frac{x(2x+1)(2x+1)}{(2x+1)}+\frac{4x(2x+1)(2x+1)}{(2x+1)(2x+1)}}{\frac{6x(2x+1)(2x+1)}{x(2x+1)}}=\frac{x(2x+1)+4x}{6(2x+1)}=\frac{x(2x+5)}{6(2x+1)}$$

17. $$\frac{\frac{4}{x+y}+\frac{1}{3}}{\frac{x}{x+y}-1}=\frac{\frac{4}{x+y}+\frac{1}{3}}{\frac{x}{x+y}-\frac{1}{1}}\cdot\frac{3(x+y)}{3(x+y)}=\frac{\frac{12(x+y)}{x+y}+\frac{3(x+y)}{3}}{\frac{3x(x+y)}{x+y}-\frac{3(x+y)}{1}}=\frac{12+x+y}{3x-3x-3y}=-\frac{12+x+y}{3y}$$

19. $\dfrac{\frac{1}{x-a}-\frac{1}{x}}{a}=\dfrac{\frac{1}{x-a}-\frac{1}{x}}{a}\cdot\dfrac{x(x-a)}{x(x-a)}$
$=\dfrac{x-(x-a)}{ax(x-a)}$
$=\dfrac{x-x+a}{ax(x-a)}$
$=\dfrac{a}{ax(x-a)}$
$=\dfrac{1}{x(x-a)}$

Cumulative Review

21. $|2-3x|=4$
$2-3x=4$ or $2-3x=-4$
$3x=-2$ $-3x=-6$
$x=-\dfrac{2}{3}$ $x=2$

22. $\left|\dfrac{1}{2}(5-x)\right|=5$
$\dfrac{1}{2}(5-x)=5$ or $\dfrac{1}{2}(5-x)=-5$
$5-x=10$ $5-x=-10$
$x=-5$ $x=15$

23. $|7x-3-2x|<6$
$|5x-3|<6$
$-6<5x-3<6$
$-3<5x<9$
$-\dfrac{3}{5}<x<\dfrac{9}{5}$

24. $|0.6x+0.3|\ge 1.2$
$0.6x+0.3\le -1.2$ or $0.6x+0.3\ge 1.2$
$0.6x\le -1.5$ $0.6x\ge 0.9$
$x\le -2.5$ $x\ge 1.5$

Quick Quiz 6.3

1. $\dfrac{\frac{1}{4x}+\frac{1}{2x}}{\frac{1}{3y}+\frac{5}{6y}}=\dfrac{\frac{1}{4x}+\frac{1}{2x}}{\frac{1}{3y}+\frac{5}{6y}}\cdot\dfrac{12xy}{12xy}$
$=\dfrac{\frac{12xy}{4x}+\frac{12xy}{2x}}{\frac{12xy}{3y}+\frac{60xy}{6y}}$
$=\dfrac{3y+6y}{4x+10x}$
$=\dfrac{9y}{14x}$

2. $\dfrac{\frac{3}{x+4}-2}{4-\frac{3}{x+4}}=\dfrac{\frac{3}{x+4}-\frac{2}{1}}{\frac{4}{1}-\frac{3}{x+4}}\cdot\dfrac{x+4}{x+4}$
$=\dfrac{\frac{3(x+4)}{x+4}-\frac{2(x+4)}{1}}{\frac{4(x+4)}{1}-\frac{3(x+4)}{x+4}}$
$=\dfrac{3-2x-8}{4x+16-3}$
$=\dfrac{-2x-5}{4x+13}$

3. $\dfrac{\frac{3}{x+3}-\frac{1}{x}}{\frac{5}{x^2+3x}}=\dfrac{\frac{3}{x+3}-\frac{1}{x}}{\frac{5}{x(x+3)}}\cdot\dfrac{x(x+3)}{x(x+3)}$
$=\dfrac{\frac{3x(x+3)}{x+3}-\frac{x(x+3)}{x}}{\frac{5x(x+3)}{x(x+3)}}$
$=\dfrac{3x-x-3}{5}$
$=\dfrac{2x-3}{5}$

4. Answers may vary. Possible solution:

1. Find the LCD of all the fractions in the numerator and denominator. The LCD is *xy*.
2. Multiply the numerator and denominator by the LCD. Use the distributive property.
3. The result is simplified, but should be written in factored form.

How Am I Doing? Sections 6.1–6.3

1. $\dfrac{49x^2-9}{7x^2+4x-3}=\dfrac{(7x-3)(7x+3)}{(x+1)(7x-3)}=\dfrac{7x+3}{x+1}$

2. $\frac{x^2-4x-21}{x^2+x-56}=\frac{(x-7)(x+3)}{(x-7)(x+8)}=\frac{x+3}{x+8}$

3. $\frac{2x^3-5x^2-3x}{x^3-8x^2+15x}=\frac{x(2x^2-5x-3)}{x(x^2-8+15)}$

$=\frac{(2x+1)(x-3)}{(x-5)(x-3)}$

$=\frac{2x+1}{x-5}$

4. $\frac{6a-30}{3a+3}\cdot\frac{9a^2+a-8}{2a^2-15a+25}$

$=\frac{6(a-5)}{3(a+1)}\cdot\frac{(9a-8)(a+1)}{(2a-5)(a-5)}$

$=\frac{2(9a-8)}{2a-5}$

5. $\frac{5x^3y^2}{x^2y+10xy^2+25y^3}\div\frac{2x^4y^5}{3x^3-75xy^2}$

$=\frac{5x^3y^2}{y(x^2+10xy+5y^2)}\cdot\frac{3x(x^2-25y^2)}{2x^4y^5}$

$=\frac{15}{2y^4}\cdot\frac{(x-5y)(x+5y)}{(x+5y)(x+5y)}$

$=\frac{15(x-5y)}{2y^4(x+5y)}$

6. $\frac{8x^3+1}{4x^2+4x+1}\cdot\frac{6x+3}{4x^2-2x+1}$

$=\frac{(2x+1)(4x^2-2x+1)}{4x^2+4x+1}\cdot\frac{3(2x+1)}{4x^2-2x+1}$

$=\frac{3(2x+1)^2}{4x^2+4x+1}$

$=\frac{3(4x^2+4x+1)}{(4x^2+4x+1)}$

$=3$

7. $\frac{x}{2x-4}-\frac{5}{2x}=\frac{x}{2(x-2)}\cdot\frac{x}{x}-\frac{5}{2x}\cdot\frac{x-2}{x-2}$

$=\frac{x^2-5(x-2)}{2x(x-2)}$

$=\frac{x^2-5x+10}{2x(x-2)}$

8. $\frac{2}{x+5}+\frac{3}{x-5}+\frac{7x}{x^2-25}$

$=\frac{2(x-5)}{(x+5)(x-5)}+\frac{3(x+5)}{(x-5)(x+5)}+\frac{7x}{(x-5)(x+5)}$

$=\frac{2x-10+3x+15+7x}{(x-5)(x+5)}$

$=\frac{12x+5}{(x-5)(x+5)}$

9. $\frac{y-1}{y^2-2y-8}-\frac{y+2}{y^2+6y+8}$

$=\frac{y-1}{(y-4)(y+2)}-\frac{y+2}{(y+4)(y+2)}$

$=\frac{(y-1)}{(y-4)(y+2)}\cdot\frac{y+4}{y+4}-\frac{y+2}{(y+4)(y+2)}\cdot\frac{y-4}{y-4}$

$=\frac{(y-1)(y+4)-(y+2)(y-4)}{(y-4)(y+2)(y+4)}$

$=\frac{y^2+4y-y-4-y^2+4y-2y+8}{(y-4)(y+2)(y+4)}$

$=\frac{5y+4}{(y-4)(y+2)(y+4)}$

10. $\frac{x+1}{x+4}+\frac{4-x^2}{x^2-16}=\frac{x+1}{x+4}-\frac{x^2-4}{(x+4)(x-4)}$

$=\frac{(x+1)(x-4)-(x^2-4)}{(x+4)(x-4)}$

$=\frac{x^2-3x-4-x^2+4}{(x+4)(x-4)}$

$=\frac{-3x}{(x+4)(x-4)}$

11. $\frac{\frac{1}{12x}+\frac{5}{3x}}{\frac{2}{3x^2}}=\frac{\frac{1}{12x}+\frac{5}{3x}}{\frac{2}{3x^2}}\cdot\frac{12x^2}{12x^2}=\frac{x+20x}{8}=\frac{21x}{8}$

12. $\frac{\frac{x}{4x^2-1}}{3-\frac{2}{2x+1}}=\frac{\frac{x}{(2x+1)(2x-1)}}{\frac{3(2x+1)-2}{2x+1}}$

$=\frac{x}{(2x+1)(2x-1)}\cdot\frac{2x+1}{6x+3-2}$

$=\frac{x}{(2x-1)(6x+1)}$

13. $\frac{\frac{5}{x}+3}{\frac{6}{x}-2}=\frac{\frac{5}{x}+3}{\frac{6}{x}-2}\cdot\frac{x}{x}=\frac{5+3x}{6-2x}$

14. $\dfrac{\frac{1}{x}+\frac{x}{x+3}}{\frac{x+1}{x+3}+\frac{4}{x}}=\dfrac{\frac{1}{x}+\frac{x}{x+3}}{\frac{x+1}{x+3}+\frac{4}{x}}\cdot\dfrac{x(x+3)}{x(x+3)}$

$=\dfrac{\frac{x(x+3)}{x}+\frac{x^2(x+3)}{x+3}}{\frac{x(x+1)(x+3)}{x+3}+\frac{4x(x+3)}{x}}$

$=\dfrac{x+3+x^2}{x^2+x+4x+12}$

$=\dfrac{x^2+x+3}{x^2+5x+12}$

6.4 Exercises

1. $\dfrac{2}{x}+\dfrac{3}{2x}=\dfrac{7}{6}$

$6x\left(\dfrac{2}{x}+\dfrac{3}{2x}\right)=6x\left(\dfrac{7}{6}\right)$

$12+9=7x$

$7x=21$

$x=3$

Check: $\dfrac{2}{3}+\dfrac{3}{2(3)}\overset{?}{=}\dfrac{7}{6},\ \dfrac{7}{6}=\dfrac{7}{6}$

3. $2-\dfrac{1}{x}=\dfrac{1}{5x}$

$5x\left(2-\dfrac{1}{x}\right)=5x\cdot\dfrac{1}{5x}$

$10x-5=1$

$10x=6$

$x=\dfrac{3}{5}$ or 0.6

Check: $2-\dfrac{1}{\frac{3}{5}}\overset{?}{=}\dfrac{1}{5\cdot\frac{3}{5}},\ \dfrac{1}{3}=\dfrac{1}{3}$

5. $\dfrac{5}{2x+3}+\dfrac{1}{x}=\dfrac{3}{x}$

$x(2x+3)\dfrac{5}{2x+3}+x(2x+3)\cdot\dfrac{1}{x}=x(2x+3)\cdot\dfrac{3}{x}$

$5x+2x+3=3(2x+3)$

$7x+3=6x+9$

$x=6$

Check: $\dfrac{5}{2(6)+3}+\dfrac{1}{6}\overset{?}{=}\dfrac{3}{6},\ \dfrac{1}{2}=\dfrac{1}{2}$

7. $\dfrac{2}{y}=\dfrac{5}{y-3}$

$2y-6=5y$

$3y=-6$

$y=-2$

Check: $\dfrac{2}{-2}\overset{?}{=}\dfrac{5}{-2-3},\ -1=-1$

9. $\dfrac{y+6}{y+3}-2=\dfrac{3}{y+3}$

$y+3\left(\dfrac{y+6}{y+3}-2\right)=(y+3)\cdot\dfrac{3}{y+3}$

$y+6-2y-6=3$

$-y=3$

$y=-3$

Check: $\dfrac{-3+6}{-3+3}-2\overset{?}{=}\dfrac{3}{-3+3}$

$\dfrac{3}{0}-2\overset{?}{=}\dfrac{3}{0}$

No solution

11. $\dfrac{1}{3x}-\dfrac{2}{x}=\dfrac{-5}{x+4}$

$3x(x+4)\left(\dfrac{1}{3x}-\dfrac{2}{x}\right)=3x(x+4)\left(\dfrac{-5}{x+4}\right)$

$x+4-2(3(x+4))=-5(3x)$

$x+4-6x-24=-15x$

$10x=20$

$x=2$

Check: $\dfrac{1}{3(2)}-\dfrac{2}{2}\overset{?}{=}\dfrac{-5}{2+4},\ -\dfrac{5}{6}=-\dfrac{5}{6}$

13. $\dfrac{2x+3}{x+3}=\dfrac{2x}{x+1}$

$(2x+3)(x+1)=2x(x+3)$

$2x^2+5x+3=2x^2+6x$

$5x+3=6x$

$x=3$

Check: $\dfrac{2\cdot3+3}{3+3}\overset{?}{=}\dfrac{2\cdot3}{3+1},\ \dfrac{3}{2}=\dfrac{3}{2}$

15.

$$\frac{3}{x^2-4}=\frac{2}{2x^2+4x}$$
$$\frac{3}{(x+2)(x-2)}=\frac{2}{2x(x+2)}$$
$$2x(x+2)(x-2)\cdot\frac{3}{(x+2)(x-2)}=2x(x+2)(x-2)\cdot\frac{2}{2x(x+2)}$$
$$6x=2(x-2)$$
$$6x=2x-4$$
$$4x=-4$$
$$x=-1$$

Check: $\frac{3}{(-1)^2-4}\stackrel{?}{=}\frac{2}{2(-1)^2+4(-1)}$

$$-1=-1$$

17.

$$\frac{1}{x^2+x}+\frac{5}{x}=\frac{3}{x+1}$$
$$\frac{1}{x(x+1)}+\frac{5}{x}=\frac{3}{x+1}$$
$$x(x+1)\cdot\frac{1}{x(x+1)}+x(x+1)\cdot\frac{5}{x}=x(x+1)\cdot\frac{3}{x+1}$$
$$1+5(x+1)=3x$$
$$1+5x+5=3x$$
$$2x=-6$$
$$x=-3$$

Check: $\frac{1}{(-3)^2+(-3)}+\frac{5}{-3}\stackrel{?}{=}\frac{3}{-3+1}$

$$-\frac{3}{2}=-\frac{3}{2}$$

19.

$$\frac{5}{y-3}+2=\frac{3}{3y-9}$$
$$3(y-3)\cdot\frac{5}{y-3}+3(y-3)\cdot 2=3(y-3)\cdot\frac{3}{3(y-3)}$$
$$15+6(y-3)=3$$
$$12+6y-18=0$$
$$6y=6$$
$$y=1$$

Check: $\frac{5}{1-3}+2\stackrel{?}{=}\frac{3}{3(1)-9},\ -\frac{1}{2}=-\frac{1}{2}$

21.
$$1-\frac{10}{z-3}=\frac{-5}{3z-9}$$
$$3(z-3)\cdot 1-3(z-3)\cdot\frac{10}{z-3}=3(z-3)\cdot\frac{-5}{3(z-3)}$$
$$3(z-3)-30=-5$$
$$3z-9=25$$
$$3z=34$$
$$z=\frac{34}{3}\text{ or }11\frac{1}{3}$$

Check: $1-\frac{10}{\frac{34}{3}-3}\stackrel{?}{=}\frac{-5}{3\cdot\frac{34}{3}-9},\ -\frac{1}{5}=-\frac{1}{5}$

23.
$$\frac{8}{3x+2}-\frac{7x+4}{3x^2+5x+2}=\frac{2}{x+1}$$
$$\frac{8}{3x+2}-\frac{7x+4}{(3x+2)(x+1)}=\frac{2}{x+1}$$
$$(3x+2)(x+1)\cdot\frac{8}{3x+2}-(3x+2)(x+1)\cdot\frac{7x+4}{(3x+2)(x+1)}=(3x+2)(x+1)\cdot\frac{2}{x+1}$$
$$8(x+1)-(7x+4)=2(3x+2)$$
$$8x+8-7x-4=6x+4$$
$$5x=0$$
$$x=0$$

Check: $\frac{8}{3(0)+2}-\frac{7(0)+4}{3(0)^2+5(0)+2}\stackrel{?}{=}\frac{2}{(0)+1}$
$$2=2$$

25.
$$\frac{4}{z^2-9}=\frac{2}{z^2-3z}$$
$$\frac{4}{(z-3)(z+3)}=\frac{2}{z(z-3)}$$
$$z(z-3)(z+3)\cdot\frac{4}{(z-3)(z+3)}=z(z-3)(z+3)\cdot\frac{2}{z(z-3)}$$
$$4z=2(z+3)$$
$$4z=2z+6$$
$$2z=6$$
$z=3$ which gives division by zero, no solution.

27.
$$\frac{3x+1}{3}+\frac{1}{x+2}=x$$
$$3(x+2)\cdot\frac{3x+1}{3}+3(x+2)\cdot\frac{1}{x+2}=3(x+2)\cdot x$$
$$(3x+1)(x+2)+3=3x(x+2)$$
$$3x^2+6x+x+2+3=3x^2+6x$$
$$x=-5$$

Check: $\frac{3(-5)}{3}+\frac{1}{-5+2}\stackrel{?}{=}-5$
$$-5=-5$$

29. When the solved-for value of the variable causes the denominator of any fraction to equal to 0, or when the variable drops out and leaves a false statement.

Cumulative Review

31. $7x^2-63=7(x^2-9)=7(x+3)(x-3)$

32. $2x^2+20x+50=2(x^2+10x+25)=2(x+5)^2$

33. $$\begin{aligned}64x^3-27y^3&=(4x)^3-(3y)^3\\&=(4x-3y)(16x^2+12xy+9y^2)\end{aligned}$$

34. $3x^2-13x+14=(3x-7)(x-2)$

Quick Quiz 6.4

1. $$\begin{aligned}2+\frac{x}{x+3}&=\frac{3x}{x-3}\\(x+3)(x-3)\cdot 2+(x+3)(x-3)\cdot\frac{x}{x+3}&=(x+3)(x-3)\cdot\frac{3x}{x-3}\\2(x+3)(x-3)+x(x-3)&=3x(x+3)\\2x^2-18+x^2-3x&=3x^2+9x\\12x+18&=0\\12x&=-18\\x&=-\frac{18}{12}\\x&=-\frac{3}{2}\end{aligned}$$

2. $$\begin{aligned}\frac{1}{x+2}-\frac{1}{3}&=\frac{-2}{3x+6}\\3(x+2)\cdot\frac{1}{x+2}-3(x+2)\cdot\frac{1}{3}&=3(x+2)\cdot\frac{-2}{3(x+2)}\\3-x-2&=-2\\x&=3\end{aligned}$$

3. $$\begin{aligned}\frac{1}{3x-2}+\frac{2x}{x+1}&=2\\(3x-2)(x+1)\cdot\frac{1}{3x-2}+(3x-2)(x+1)\cdot\frac{2x}{x+1}&=(3x-2)(x+1)\cdot 2\\(x+1)+2x(3x-2)&=2(3x^2+x-2)\\x+1+6x^2-4x&=6x^2+2x-4\\-5x&=-5\\x&=1\end{aligned}$$

4. Answers may vary. Possible solution: The left side of the equation equals zero. It cannot therefore equal $\frac{3}{2}$. It has no solution.

6.5 Exercises

1.
$$x = \frac{y-b}{m}$$
$$m \cdot x = m \cdot \frac{y-b}{m}$$
$$mx = y-b$$
$$m = \frac{y-b}{x}$$

3.
$$\frac{1}{f} = \frac{1}{a} + \frac{1}{b}$$
$$abf \cdot \frac{1}{f} = abf\left(\frac{1}{a} + \frac{1}{b}\right)$$
$$ab = bf + af$$
$$ab - bf = af$$
$$b(a-f) = af$$
$$b = \frac{af}{a-f}$$

5.
$$\frac{V}{lh} = w$$
$$lh = \frac{V}{w}$$
$$h = \frac{V}{lw}$$

7.
$$G = \frac{ab+ac}{3}$$
$$3G = ab + ac$$
$$3G = a(b+c)$$
$$\frac{3G}{b+c} = a$$

9.
$$\frac{r^3}{V} = \frac{3}{4\pi}$$
$$3V = 4\pi r^3$$
$$V = \frac{4}{3}\pi r^3$$

11.
$$\frac{E}{e} = \frac{R+r}{r}$$
$$e(R+r) = Er$$
$$e = \frac{Er}{R+r}$$

13.
$$\frac{QR_1}{S_1} = \frac{QR_2}{S_2}$$
$$QR_1S_2 = QR_2S_1$$
$$S_1 = \frac{R_1S_2}{R_2}$$

15.
$$\frac{S-2lw}{2w+2l} = h$$
$$S - 2lw = 2wh + 2lh$$
$$2wh + 2lw = S - 2lh$$
$$w(2h+2l) = S - 2lh$$
$$w = \frac{S-2lh}{2h+2l}$$

17.
$$E = T_1 - \frac{T_1}{T_2}$$
$$ET_2 = T_1T_2 - T_1$$
$$T_1(T_2 - 1) = ET_2$$
$$T_1 = \frac{ET_2}{T_2 - 1}$$

19.
$$m = \frac{y_2 - y_1}{x_2 - x_1}$$
$$m(x_2 - x_1) = y_2 - y_1$$
$$mx_2 - mx_1 = y_2 - y_1$$
$$mx_1 = mx_2 + y_1 - y_2$$
$$x_1 = \frac{mx_2 + y_1 - y_2}{m}$$

21.
$$\frac{2D - at^2}{2t} = V$$
$$2D - at^2 = 2Vt$$
$$2D = 2Vt + at^2$$
$$D = \frac{2Vt + at^2}{2}$$

23. $Q = \dfrac{kA(t_1 - t_2)}{L}$
$QL = kAt_1 - kAt_2$
$kAt_2 = kAt_1 - QL$
$t_2 = \dfrac{kAt_1 - QL}{kA}$

25. $\dfrac{T_2W}{T_2 - T_1} = q$
$T_2W = qT_2 - qT_1$
$qT_2 - T_2W = qT_1$
$T_2(q - W) = qT_1$
$T_2 = \dfrac{qT_1}{q - W}$

27. $\dfrac{s - s_0}{v_0 + gt} = t$
$s - s_0 = v_0t + gt^2$
$v_0t = s - s_0 - gt^2$
$v_0 = \dfrac{s - s_0 - gt^2}{t}$

29. $\dfrac{1.98V}{1.96V_0} = 0.983 + 5.936(T - T_0)$
$\dfrac{1.98V}{1.96V_0} = 0.983 + 5.936T - 5.936T_0$
$5.936T = 5.936T_0 + \dfrac{1.98V}{1.96V_0} - 0.983$
$T \approx T_0 + 0.1702\left(\dfrac{V}{V_0}\right) - 0.1656$

31. $\dfrac{3}{55} = \dfrac{6.5}{x}$
$3x = 357.5$
$x = 119.1\overline{6}$
The two cities are approximately 119.2 kilometers apart.

33. $\dfrac{60}{88} = \dfrac{x}{80}$
$88x = 4800$
$x = 54.\overline{54}$
The speed limit is approximately 54.55 mph.

35. $\dfrac{35}{x} = \dfrac{22}{50}$
$22x = 1750$
$x = 79.\overline{54}$
The number of grizzly bears, to the nearest whole number, is 80.

37. x = number of officers
$\dfrac{2}{7} = \dfrac{x}{117 - x}$
$234 - 2x = 7x$
$9x = 234$
$x = 26$ officers
$117 - 26 = 91$ seamen
There are 26 officers and 91 seamen on the ship.

39. x = number of people in marketing
$\dfrac{4}{13} = \dfrac{x}{187 - x}$
$748 - 4x = 13x$
$17x = 748$
$x = 44$
$187 - 44 = 143$
There are 44 people in marketing and 143 people in sales.

41. $\dfrac{L}{W} = \dfrac{10}{8} \Rightarrow L = \dfrac{10}{8}W$
$2L + 2W = 54$
$\dfrac{10}{8}W + W = 27$
$\dfrac{18}{8}W = 27$
$W = 12$ in. for the width
$L = \dfrac{10}{8}(12) = 15$ in. for the length
The frame is 12 inches wide and 15 inches long.

43. x = number preferring new software
$\dfrac{3}{11} = \dfrac{x}{280 - x}$
$840 - 3x = 11x$
$14x = 840$
$x = 60$
Sixty people prefer the new software.

45. $\dfrac{x}{8} = \dfrac{12}{15}$
$15x = 96$
$x = 6.4$
The wall is 6.4 feet high.

47. $\frac{t}{6}+\frac{t}{9}=1$

$\frac{5}{18}t=1$

$t=\frac{18}{5}=3.6$

It will take them 3.6 hours to do the work together.

49. $\frac{1}{t}\cdot 2+\frac{1}{3}\cdot 2=1$

$6+2t=3t$

$t=6$

Using just the hot water pipe the pool will fill in 6 hours.

51. $\frac{b}{a}=\frac{x+8}{c}$

$\frac{5}{2}=\frac{x+8}{116}$

$2x+16=580$

$2x=564$

$x=282$

The width of the river is 282 feet.

Cumulative Review

53. $7x-3y=8$

$3y=7x-8$

$y=\frac{7}{3}x-\frac{8}{3}\Rightarrow m=\frac{7}{3},\ b=-\frac{8}{3}$

54. $m=\frac{y_2-y_1}{x_2-x_1}=\frac{2.5-2}{-1-1.5}=\frac{0.5}{-2.5}=-\frac{1}{5}$

55. A line perpendicular to $y=\frac{1}{3}x-4$ passing through (0, 6).

$m=\frac{1}{3}$, so $m_\perp=-3$

$b=6$

$y=-3x+6$

56. $x-5y=-9$ (1)

$3x+2y=-10$ (2)

Solve (1) for x.

$x=-9+5y$ (3)

Substitute $-9+5y$ for x in (2).

$3(-9+5y)+2y=-10$

$-27+15y+2y=-10$

$17y=17$

$y=1$

Substitute 1 for y in (3).

$x=-9+5(1)$

$x=-4$

$(-4, 1)$

Quick Quiz 6.5

1. $\frac{3W+2AH}{4W-8H}=A$

Solving for H

$3W+2AH=A(4W-8H)$

$3W+2AH=4AW-8AH$

$10AH=4AW-3W$

$H=\frac{4AW-3W}{10A}$

2. x = number of students accepted

$\frac{3}{22+3}=\frac{x}{2400}$

$\frac{3}{25}=\frac{x}{2400}$

$25x=3(2400)$

$x=288$

288 students were accepted.

3. w = new width

l = new length

$\frac{w}{l}=\frac{22}{35}$ (1)

$2w+2l=342$ (2)

Solve (1) for w.

$w=\frac{22}{35}l$ (3)

Substitute $\frac{22}{25}l$ for w in (2).

$2\left(\frac{22}{35}l\right)+2l=342$

$l=105$

$w=\frac{22}{35}l=66$

Window dimensions are 105 inches long, 66 inches wide.

4. Answers may vary. Possible solution:
To eliminate fractions, multiply both sides of the equation by the denominator $B - H$.
To isolate H, first subtract $5B$ from both sides of the equation, then divide both sides by -5.

Putting Your Skills to Work

1. Option A: $\frac{15,000}{21} = 714.29$ gallons

Option B: $\frac{15,000}{32} = 468.75$ gallons

Option C: $\frac{15,000}{60} = 250$ gallons

2. Option A: $714.29 \times 4.15 = \$2964.30$
Option B: $468.75 \times 4.15 = \$1945.31$
Option C: $250 \times 4.15 = \$1037.50$

3. $\text{Option A} - \text{Option B} = \$2964.30 - \$1945.31$
$= \$1018.99$
She would save \$1018.99.

4. $\frac{6610}{1018.99} \approx 6.49$
It will take her 6.49 years.

5. $5 \times \$1018.99 = \$5094.95$
$10 \times \$1018.99 = \$10,189.90$
After 5 years, she would save \$5094.95 and after 10 years she would save \$10,189.90.

6. $\text{Option A} - \text{Option C} = \$2964.30 - \$1037.50$
$= \$1926.80$
She will save \$1926.80.

7. $\frac{2230}{1926.80} \approx 1.16$
It would take her 1.16 years.

8. $5 \times \$1926.80 = \$9634$
$10 \times \$1926.80 = \$19,268$
After 5 years, she would save \$9634 and in 10 years she would save \$19,268.

9. More miles per year would increase the gallons of gas, money spent on gas, and savings compared with Option A, per year. It would also decrease the number of years necessary to make up the higher cost in purchase price. Fewer miles per year would decrease the gallons of gas, money spent on gas, and savings compared with Option A, per year. It would also increase the number of years necessary to make up the higher cost in purchase price. More miles driven per year makes MPG an even more critical issue, while fewer miles driven per year does the opposite.

10. Higher gas prices would increase the money spent on gas and savings compared with Option A, per year. This would also decrease the number of years necessary to make up the higher cost in purchase price. Lower gas prices would decrease the money spent on gas and savings compared with Option A, per year. This would also increase the number of years necessary to make up the higher cost in purchase price. Higher gas prices make MPG an even more critical issue, while lower gas prices do the opposite.

Chapter 6 Review Problems

1. $\frac{6x^3 - 9x^2}{12x^2 - 18x} = \frac{3x^2(2x-3)}{6x(2x-3)} = \frac{x}{2}$

2. $\frac{15x^4}{5x^2 - 20x} = \frac{15x^4}{5x(x-4)} = \frac{3x^3}{x-4}$

3. $\frac{26x^3y^2}{39xy^4} = \frac{13 \cdot 2x \cdot x^2y^2}{13 \cdot 3xy^2y^2} = \frac{2x^2}{3y^2}$

4. $\frac{42a^4bc^3}{24a^7b} = \frac{7c^3}{4a^{7-4}} = \frac{7c^3}{4a^3}$

5. $\frac{2x^2 - 5x + 3}{3x^2 + 2x - 5} = \frac{(2x-3)(x-1)}{(3x+5)(x-1)} = \frac{2x-3}{3x+5}$

6. $\frac{ax + 2a - bx - 2b}{3x^2 - 12} = \frac{a(x+2) - b(x+2)}{3(x^2-4)}$
$= \frac{(a-b)\cancel{(x+2)}}{3(x-2)\cancel{(x+2)}}$
$= \frac{a-b}{3(x-2)}$

7. $\frac{4x^2 - 1}{x^2 - 4} \cdot \frac{2x^2 + 4x}{4x + 2} = \frac{(2x-1)(2x+1)}{(x+2)(x-2)} \cdot \frac{2x(x+2)}{2(2x+1)}$
$= \frac{x(2x-1)}{x-2}$

8. $\frac{3y}{4xy-6y^2}\cdot\frac{2x-3y}{12xy}=\frac{3y}{2y(2x-3y)}\cdot\frac{(2x-3y)}{12xy}$
$=\frac{1}{8xy}$

9. $\frac{y^2+8y-20}{y^2+6y-16}\cdot\frac{y^2+3y-40}{y^2+6y-40}$
$=\frac{(y+10)(y-2)}{(y+8)(y-2)}\cdot\frac{(y+8)(y-5)}{(y+10)(y-4)}$
$=\frac{y-5}{y-4}$

10. $\frac{3x^3y}{x^2+7x+12}\cdot\frac{x^2+8x+15}{6xy^2}$
$=\frac{3xx^2y}{(x+4)(x+3)}\cdot\frac{(x+5)(x+3)}{3\cdot 2xyy}$
$=\frac{x^2(x+5)}{2y(x+4)}$

11. $\frac{3x+9}{5x-20}\div\frac{3x^2-3x-36}{x^2-8x+16}$
$=\frac{3(x+3)}{5(x-4)}\cdot\frac{(x-4)(x-4)}{3(x+3)(x-4)}$
$=\frac{1}{5}$

12. $\frac{4a^3b^2}{4x^2-4x-3}\div\frac{8ab^4}{6x^2-11x+3}$
$=\frac{4a^3b^2}{(2x-3)(2x+1)}\cdot\frac{(2x-3)(3x-1)}{8ab^4}$
$=\frac{a^2(3x-1)}{2b^2(2x+1)}$

13. $\frac{9y^2-3y-2}{6y^2-13y-5}\div\frac{3y^2+10y-8}{2y^2+13y+20}$
$=\frac{(3y+1)(3y-2)}{(3y+1)(2y-5)}\cdot\frac{(y+4)(2y+5)}{(y+4)(3y-2)}$
$=\frac{2y+5}{2y-5}$

14. $\frac{4a^2+12a+5}{2a^2-7a-13}\div(4a^2+2a)$
$=\frac{(2a+1)(2a+5)}{2a^2-7a-13}\cdot\frac{1}{2a(2a+1)}$
$=\frac{2a+5}{2a(2a^2-7a-13)}$

15. $\frac{x-5}{2x+1}-\frac{x+1}{x-2}$
$=\frac{(x-5)(x-2)-(2x+1)(x+1)}{(2x+1)(x-2)}$
$=\frac{x^2-7x+10-2x^2-3x-1}{(2x+1)(x-2)}$
$=\frac{-x^2-10x+9}{(2x+1)(x-2)}$
$=-\frac{x^2+10x-9}{(2x+1)(x-2)}$

16. $\frac{5}{4x}+\frac{-3}{x+4}=\frac{5(x+4)-3(4x)}{4x(x+4)}$
$=\frac{5x+20-12x}{4x(x+4)}$
$=\frac{-7x+20}{4x(x+4)}$

17. $\frac{2y-1}{12y}-\frac{3y+2}{9y}=\frac{(2y-1)(3)}{12y(3)}-\frac{(3y+2)(4)}{9y(4)}$
$=\frac{6y-3-12y-8}{36y}$
$=\frac{-6y-11}{36y}$ or $-\frac{6y+11}{36y}$

18. $\frac{4}{y+5}+\frac{3y+2}{y^2-25}=\frac{4(y-5)}{(y+5)(y-5)}+\frac{3y+2}{(y+5)(y-5)}$
$=\frac{4y-20+3y+2}{(y+5)(y-5)}$
$=\frac{7y-18}{(y+5)(y-5)}$

19. $\frac{2}{y^2-4}+\frac{y}{y^2+4y+4}$
$=\frac{2}{(y+2)(y-2)}+\frac{y}{(y+2)(y+2)}$
$=\frac{2}{(y+2)(y-2)}\cdot\frac{y+2}{y+2}+\frac{y}{(y+2)(y+2)}\cdot\frac{y-2}{y-2}$
$=\frac{2(y+2)+y(y-2)}{(y+2)(y+2)(y-2)}$
$=\frac{2y+4+y^2-2y}{(y+2)^2(y-2)}$
$=\frac{y^2+4}{(y+2)^2(y-2)}$

20. $\frac{y^2+2y+20}{y^2+8y+12}-\frac{2y+1}{y+6}$
$=\frac{y^2+2y+20}{(y+6)(y+2)}-\frac{2y+1}{(y+6)}\cdot\frac{y+2}{y+2}$
$=\frac{y^2+2y+20-(y+2)(2y+1)}{(y+6)(y+2)}$
$=\frac{y^2+2y+20-2y^2-y-4y-2}{(y+6)(y+2)}$
$=\frac{-y^2-3y+18}{(y+6)(y+2)}$
$=\frac{-(y+6)(y-3)}{(y+6)(y+2)}$
$=-\frac{y-3}{y+2}$

21. $\frac{a}{5-a}-\frac{2}{a+3}+\frac{2a^2-2a}{a^2-2a-15}$
$=\frac{-a(a+3)}{(a-5)(a+3)}-\frac{2(a-5)}{(a+3)(a-5)}+\frac{2a(a-1)}{(a-5)(a+3)}$
$=\frac{-a^2-3a-2a+10+2a^2-2a}{(a+3)(a-5)}$
$=\frac{a^2-7a+10}{(a+3)(a-5)}$
$=\frac{(a-2)(a-5)}{(a+3)(a-5)}$
$=\frac{a-2}{a+3}$

22. $\frac{2}{x^2+8x+16}-\frac{x}{2x^2+9x+4}$
$=\frac{2}{(x+4)^2}-\frac{x}{(2x+1)(x+4)}$
$=\frac{2(2x+1)-x(x+4)}{(x+4)^2(2x+1)}$
$=\frac{4x+2-x^2-4x}{(x+4)^2(2x+1)}$
$=\frac{-x^2+2}{(x+4)^2(2x+1)}$

23. $5b-1-\frac{b+2}{b+3}=\frac{5b(b+3)-(b+3)-(b+2)}{b+3}$
$=\frac{5b^2+15b-b-3-b-2}{b+3}$
$=\frac{5b^2+13b-5}{b+3}$

24. $\frac{1}{x}+\frac{3}{2x}+3+2x=\frac{2}{2x}+\frac{3}{2x}+\frac{6x}{2x}+\frac{4x^2}{2x}$
$=\frac{2+3+6x+4x^2}{2x}$
$=\frac{4x^2+6x+5}{2x}$

25. $\frac{\frac{5}{x}+1}{1-\frac{25}{x^2}}=\frac{\frac{5}{x}+1}{1-\frac{25}{x^2}}\cdot\frac{x^2}{x^2}$
$=\frac{5x+x^2}{x^2-25}$
$=\frac{x(x+5)}{(x-5)(x+5)}$
$=\frac{x}{x-5}$

26. $\frac{\frac{4}{x+3}}{\frac{2}{x-2}-\frac{1}{x^2+x-6}}$
$=\frac{\frac{4}{x+3}}{\frac{2}{(x-2)}-\frac{1}{(x+3)(x-2)}}\cdot\frac{(x+3)(x-2)}{(x+3)(x-2)}$
$=\frac{4(x-2)}{2(x+3)-1}$
$=\frac{4(x-2)}{2x+6-1}$
$=\frac{4(x-2)}{2x+5}$

27. $$\frac{\frac{8}{y+2}-4}{\frac{2}{y+2}-1}=\frac{\frac{8}{y+2}-\frac{4}{1}}{\frac{2}{y+2}-\frac{1}{1}}\cdot\frac{y+2}{y+2}$$
$$=\frac{\frac{8(y+2)}{y+2}-\frac{4(y+2)}{1}}{\frac{2(y+2)}{y+2}-\frac{y+2}{1}}$$
$$=\frac{8-4y-8}{2-y-2}$$
$$=4$$

28. $$\frac{\frac{1}{a}+\frac{a}{a+1}}{\frac{a}{a+1}-\frac{1}{a}}=\frac{\frac{1}{a}+\frac{a}{a+1}}{\frac{a}{a+1}-\frac{1}{a}}\cdot\frac{a(a+1)}{a(a+1)}$$
$$=\frac{\frac{a(a+1)}{a}+\frac{a^2(a+1)}{a+1}}{\frac{a^2(a+1)}{a+1}-\frac{a(a+1)}{a}}$$
$$=\frac{a^2+a+1}{a^2-a-1}$$

29. $$\frac{\frac{2}{x+4}-\frac{1}{x^2+4x}}{\frac{3}{2x+8}}=\frac{\frac{2}{x+4}-\frac{1}{x(x+4)}}{\frac{3}{2(x+4)}}\cdot\frac{2x(x+4)}{2x(x+4)}$$
$$=\frac{4x-2}{3x}$$
$$=\frac{2(2x-1)}{3x}$$

30. $$\frac{\frac{y^2}{y^2-x^2}-1}{x+\frac{xy}{x-y}}=\frac{\frac{-y^2}{(x-y)(x+y)}-1}{x+\frac{xy}{x-y}}\cdot\frac{(x-y)(x+y)}{(x-y)(x+y)}$$
$$=\frac{-y^2-x^2+y^2}{x(x-y)(x+y)+xy(x+y)}$$
$$=\frac{-x^2}{x(x+y)(x-y+y)}$$
$$=\frac{-x}{(x+y)x}$$
$$=\frac{-1}{x+y}\text{ or }-\frac{1}{x+y}$$

31. $$\frac{\frac{2x+1}{x-1}}{1+\frac{x}{x+1}}=\frac{\frac{2x+1}{x-1}}{1+\frac{x}{x+1}}\cdot\frac{(x+1)(x-1)}{(x+1)(x-1)}$$
$$=\frac{(2x+1)(x+1)}{(x+1)(x-1)+x(x-1)}$$
$$=\frac{2x^2+3x+1}{x^2-1+x^2-x}$$
$$=\frac{(2x+1)(x+1)}{2x^2-x-1}$$
$$=\frac{(2x+1)(x+1)}{(2x+1)(x-1)}$$
$$=\frac{x+1}{x-1}$$

32. $$\frac{\frac{3}{x}-\frac{2}{x+1}}{\frac{5}{x^2+5x+4}-\frac{1}{x+4}}$$
$$=\frac{\frac{3}{x}-\frac{2}{x+1}}{\frac{5}{(x+4)(x+1)}-\frac{1}{x+4}}\cdot\frac{x(x+1)(x+4)}{x(x+1)(x+4)}$$
$$=\frac{3(x+1)(x+4)-2x(x+4)}{5x-x(x+1)}$$
$$=\frac{3x^2+15x+12-2x^2-8x}{5x-x^2-x}$$
$$=\frac{x^2+7x+12}{-x^2+4x}$$
$$=\frac{(x+4)(x+3)}{-x(x-4)}$$
$$=\frac{-(x+3)(x+4)}{x(x-4)}$$

33. $$\frac{3}{2}=1-\frac{1}{x-1}$$
$$2(x-1)\cdot\frac{3}{2}=2(x-1)\left[1-\frac{1}{x-1}\right]$$
$$3(x-1)=2(x-1)-2$$
$$3x-3=2x-2-2$$
$$x=-1$$
Check: $\frac{3}{2}\stackrel{?}{=}1-\frac{1}{-1-1},\ \frac{3}{2}=\frac{3}{2}$

34.

$$\frac{3}{7}+\frac{4}{x+1}=1$$
$$7(x+1)\cdot\frac{3}{7}+7(x+1)\cdot\frac{4}{x+1}=7(x+1)\cdot 1$$
$$3(x+1)+28=7(x+1)$$
$$3x+3+28=7x+7$$
$$4x=24$$
$$x=6$$

Check: $\frac{3}{7}+\frac{4}{6+1}\stackrel{?}{=}1,\ 1=1$

35.

$$\frac{7}{2x-3}+\frac{3}{x+2}=\frac{6}{2x-3}$$
$$(2x-3)(x+2)\cdot\frac{7}{2x-3}+(2x-3)(x+2)\cdot\frac{3}{x+2}=(2x-3)(x+2)\cdot\frac{6}{2x-3}$$
$$7(x+2)+3(2x-3)=6(x+2)$$
$$7x+14+6x-9=6x+12$$
$$7x=7$$
$$x=1$$

Check: $\frac{7}{2(1)-3}+\frac{3}{1+2}\stackrel{?}{=}\frac{6}{2(1)-3},\ -6=-6$

36.

$$\frac{9}{5x-2}-\frac{2}{x}=\frac{1}{x}$$
$$x(5x-2)\left(\frac{9}{5x-2}-\frac{2}{x}\right)=x(5x-2)\cdot\frac{1}{x}$$
$$9x-2(5x-2)=5x-2$$
$$9x-10x+4=5x-2$$
$$-6x=-6$$
$$x=1$$

Check: $\frac{9}{5(1)-2}-\frac{2}{1}\stackrel{?}{=}\frac{1}{1},\ 1=1$

37.

$$\frac{5}{2a}=\frac{2}{a}-\frac{1}{12}$$
$$12a\left(\frac{5}{2a}\right)=12a\left(\frac{2}{a}-\frac{1}{12}\right)$$
$$30=24-a$$
$$a=-6$$

Check: $\frac{5}{2(-6)}\stackrel{?}{=}\frac{2}{-6}-\frac{1}{12},\ -\frac{5}{12}=-\frac{5}{12}$

38. $$\frac{1}{2a} = \frac{2}{a} - \frac{3}{10}$$

$$10a\left(\frac{1}{2a}\right) = 10a\left(\frac{2}{9} - \frac{3}{10}\right)$$

$$5 = 20 - 3a$$

$$3a = 15$$

$$a = 5$$

Check: $\frac{1}{2(5)} \stackrel{?}{=} \frac{2}{5} - \frac{3}{10}, \frac{1}{10} = \frac{1}{10}$

39. $$\frac{1}{y} + \frac{1}{2y} = 2$$

$$2y\left(\frac{1}{y} + \frac{1}{2y}\right) = 2y(2)$$

$$2 + 1 = 4y$$

$$y = \frac{3}{4}$$

Check: $\frac{1}{\frac{3}{4}} + \frac{1}{2 \cdot \frac{3}{4}} \stackrel{?}{=} 2, 2 = 2$

40. $$\frac{5}{y^2} + \frac{7}{y} = \frac{6}{y^2}$$

$$y^2\left(\frac{5}{y^2} + \frac{7}{y}\right) = y^2\left(\frac{6}{y^2}\right)$$

$$5 + 7y = 6$$

$$7y = 1$$

$$y = \frac{1}{7}$$

Check: $\frac{5}{\left(\frac{1}{7}\right)^2} + \frac{7}{\frac{1}{7}} \stackrel{?}{=} \frac{6}{\left(\frac{1}{7}\right)^2}, 294 = 294$

41.
$$\frac{3}{a+4}+\frac{a+1}{3a+12}=\frac{5}{3}$$
$$\frac{3}{a+4}+\frac{a+1}{3(a+4)}=\frac{5}{3}$$
$$3(a+4)\cdot\frac{3}{a+4}+3(a+4)\cdot\frac{a+1}{3(a+4)}=3(a+4)\cdot\frac{5}{3}$$
$$9+a+1=5(a+4)$$
$$10+a=5a+20$$
$$-4a=10$$
$$a=-\frac{5}{2}$$

Check: $\frac{3}{-\frac{5}{2}+4}+\frac{-\frac{5}{2}+1}{3\left(-\frac{5}{2}\right)+12}\stackrel{?}{=}\frac{5}{3}$

$$\frac{5}{3}=\frac{5}{3}$$

42.
$$\frac{a+3}{2a+12}=\frac{3}{2}-\frac{5}{a+6}$$
$$2(a+6)\left(\frac{a+3}{2(a+6)}\right)=2(a+6)\left(\frac{3}{2}-\frac{5}{a+6}\right)$$
$$a+3=3(a+6)-10$$
$$a+3=3a+18-10$$
$$-2a=5$$
$$a=-\frac{5}{2}$$

Check: $\frac{-\frac{5}{2}+3}{2\left(-\frac{5}{2}\right)+12}\stackrel{?}{=}\frac{3}{2}-\frac{5}{-\frac{5}{2}+6}$

$$\frac{1}{14}=\frac{1}{14}$$

43.
$$\frac{3x-23}{2x^2-5x-3}+\frac{2}{x-3}=\frac{5}{2x+1}$$
$$\frac{3x-23}{(2x+1)(x-3)}+\frac{2}{x-3}=\frac{5}{2x+1}$$
$$(2x+1)(x-3)\left[\frac{3x-23}{(2x+1)(x-3)}+\frac{2}{x-3}\right]=(2x+1)(x-3)\cdot\frac{5}{2x+1}$$
$$3x-23+2(2x+1)=5(x-3)$$
$$3x-23+4x+2=5x-15$$
$$2x=6$$
$$x=3$$

$x = 3$ results in division by 0.
No solution

44.
$$\frac{2x-10}{2x^2-5x-3}+\frac{1}{x-3}=\frac{3}{2x+1}$$
$$\frac{2x-10}{(2x+1)(x-3)}+\frac{1}{x-3}=\frac{3}{2x+1}$$
$$(2x+1)(x-3)\left[\frac{2x-10}{(2x+1)(x-3)}+\frac{1}{x-3}\right]=(2x+1)(x-3)\cdot\frac{3}{2x+1}$$
$$2x-10+2x+1=3(x-3)$$
$$4x-9=3x-9$$
$$x=0$$

Check: $\frac{2(0)-10}{2(0)^2-5(0)-3}+\frac{1}{0-3}\stackrel{?}{=}\frac{3}{2(0)+1}$
$$3=3$$

45.
$$\frac{N}{V}=\frac{m}{M+N}$$
$$MN+N^2=mV$$
$$MN=mV-N^2$$
$$M=\frac{mV-N^2}{N}$$
$$M=\frac{mV}{N}-N$$

46.
$$m=\frac{y-y_0}{x-x_0}$$
$$mx-mx_0=y-y_0$$
$$mx=y-y_0+mx_0$$
$$x=\frac{y-y_0+mx_0}{m}$$

47.
$$\frac{1}{f}=\frac{1}{a}+\frac{1}{b}$$
$$abf\cdot\frac{1}{f}=abf\left(\frac{1}{a}+\frac{1}{b}\right)$$
$$ab=fb+fa$$
$$ab-fa=fb$$
$$a(b-f)=fb$$
$$a=\frac{fb}{b-f}$$

48.
$$S=\frac{V_1t+V_2t}{2}$$
$$t(V_1+V_2)=2S$$
$$t=\frac{2S}{V_1+V_2}$$

49. $d = \frac{LR_2}{R_2 + R_1}$

$dR_2 + dR_1 = LR_2$

$R_2(L - d) = dR_1$

$R_2 = \frac{dR_1}{L - d}$

50. $\frac{S - P}{Pr} = t$

$Prt = S - P$

$r = \frac{S - P}{Pt}$

51. $\frac{x^2 - x - 42}{x^2 - 2x - 35} = \frac{(x-7)(x+6)}{(x-7)(x+5)} = \frac{x+6}{x+5}$

52. $\frac{2x^2 - 5x - 3}{x^2 - 9} \cdot \frac{2x^2 + 5x - 3}{2x^2 + 5x + 2}$

$= \frac{(2x+1)(x-3)(2x-1)(x+3)}{(x+3)(x-3)(2x+1)(x+2)}$

$= \frac{2x-1}{x+2}$

53. $\frac{-2x-1}{x+4} + 4x + 3 = \frac{-2x - 1 + (4x+3)(x+4)}{x+4}$

$= \frac{-2x - 1 + 4x^2 + 19x + 12}{x+4}$

$= \frac{4x^2 + 17x + 11}{x+4}$

54. $\frac{\frac{1}{x^2-3x+2}}{\frac{3}{x-2} - \frac{2}{x-1}} = \frac{\frac{1}{(x-2)(x-1)}}{\frac{3}{x-2} - \frac{2}{x-1}}$

$= \frac{\frac{1}{(x-2)(x-1)}}{\frac{3}{x-2} - \frac{2}{x-1}} \cdot \frac{(x-2)(x-1)}{(x-2)(x-1)}$

$= \frac{1}{3(x-1) - 2(x-2)}$

$= \frac{1}{3x - 3 - 2x + 4}$

$= \frac{1}{x+1}$

55. $\frac{5}{2} - \frac{3}{x+1} = \frac{2-x}{x+1}$

$2(x+1) \cdot \left[\frac{5}{2} - \frac{3}{x+1}\right] = 2(x+1)\left(\frac{2-x}{x+1}\right)$

$5(x+1) - 6 = 2(2-x)$

$5x + 5 - 6 = 4 - 2x$

$7x = 5$

$x = \frac{5}{7}$

56. x = graphing calculators ordered
y = scientific calculators ordered

$x + y = 320$ (1)

$\frac{x}{y} = \frac{7}{3}$ (2)

Solve (2) for x.

$x = \frac{7}{3}y$ (3)

Substitute $\frac{7}{3}y$ for x in (1).

$\frac{7}{3}y + y = 320$

$10y = 960$

$y = 96$

Substitute 96 for y in (3).

$x = \frac{7}{3}(96) = 224$

224 graphing calculators ordered.
96 scientific calculators ordered.

57. x = number of one-story homes

$\frac{3}{13} = \frac{x}{112 - x}$

$336 - 3x = 13x$

$16x = 336$

$x = 21$

$112 - x = 91$

Walter Johnson built 21 one-story homes and 91 two-story homes.

58. $\frac{5}{7} = \frac{w}{l}$

$\frac{5}{7} = \frac{w}{\frac{168-2w}{2}}$

$\frac{5}{7} = \frac{2w}{168 - 2w}$

$840 - 10w = 14w$

$24w = 840$

$w = 35$

$l = \frac{168-2w}{2} = 49$

The enlarged photograph will be 35 inches wide and 49 inches long.

59. $\frac{4}{3500} = \frac{x}{4900}$

$3500x = 19{,}600$

$x = 5.6$

The pump can empty the 4900-gallon pool in 5.6 hours.

60. $\frac{100}{P} = \frac{8}{40}$

$8P = 4000$

$P = 500$

The population is 500 rabbits.

61. x = number of officers

$\frac{2}{9} = \frac{x}{154-x}$

$308 - 2x = 9x$

$11x = 308$

$x = 28$

There are 28 officers.

62. x = number of nautical miles

$\frac{7}{2} = \frac{x}{3.5}$

$2x = 24.5$

$x = 12.25$

The course is 12.25 nautical miles.

63. x = height of building

$\frac{7}{6} = \frac{x}{156}$

$6x = 1092$

$x = 182$

The building is 182 feet tall.

64. x = time to plant flowers if Marianne and her son work together

$\frac{1}{10} + \frac{1}{15} = \frac{1}{x}$

$15x\left(\frac{1}{10} + \frac{1}{15}\right) = 15x\left(\frac{1}{x}\right)$

$15x + 10x = 150$

$25x = 150$

$x = 6$

It will take 6 hours to plant flowers when working together.

65. t = time if both faucets are open

$\frac{t}{15} + \frac{t}{10} = 1$

$\frac{5t}{30} = 1$

$\frac{t}{6} = 1$

$t = 6$

It would take 6 minutes to fill if both faucets are left open.

66. $\text{Rate} = \frac{\text{Rise}}{\text{Run}} = \frac{170.9-115.6}{4} = 13.825$

2013: 170.9 + 13.825(2) = 199

$199 billion is projected for 2013.

67. x = projected amount spent online in 2013

$\frac{170.9}{144.8} = \frac{x}{170.9}$

$(170.9)^2 = 144.8x$

$202 \approx x$

$202 billion is projected for 2013.

68. 2001–2005 increase:

$\frac{83.7-31}{4} = 13.175$ billion dollars/year

2011–2013 increase:

$\frac{13.175}{2} = 6.5875$ billion dollars/year

Expected sales 2013:

170.9 + 6.5875(2) = $184 billion

69. 2005–2007 increase:

$\frac{115.6-83.7}{2} = 15.95$ billion dollars/year

2011–2013 increase:

2(15.95) = 31.9 billion dollars/year

Expected sales 2013:

170.9 + 31.9(2) = 235 billion dollars

How Am I Doing? Chapter 6 Test

1. $\frac{x^3+3x^2+2x}{x^3-2x^2-3x} = \frac{x(x^2+3x+2)}{x(x^2-2x-3)}$

$= \frac{(x+2)(x+1)}{(x-3)(x+1)}$

$= \frac{x+2}{x-3}$

2. $$\frac{-25p^4qr^3}{45pqr^6}=-\frac{5p^{4-1}}{9r^{6-3}}=-\frac{5p^3}{9r^3}$$

3. $$\frac{2y^2+7y-4}{y^2+2y-8}\cdot\frac{2y^2-8}{3y^2+11y+10}$$
$$=\frac{(2y-1)(y+4)}{(y+4)(y-2)}\cdot\frac{2(y+2)(y-2)}{(3y+5)(y+2)}$$
$$=\frac{2(2y-1)}{3y+5}$$

4. $$\frac{4-2x}{3x^2-2x-8}\div\frac{2x^2+x-1}{9x+12}$$
$$=\frac{-2(x-2)}{(3x+4)(x-2)}\cdot\frac{3(3x+4)}{(2x-1)(x+1)}$$
$$=-\frac{6}{(2x-1)(x+1)}$$

5. $$\frac{3}{x}-\frac{2}{x+1}=\frac{3(x+1)-2x}{x(x+1)}=\frac{3x+3-2x}{x(x+1)}=\frac{x+3}{x(x+1)}$$

6. $$\frac{2}{x^2+5x+6}+\frac{3x}{x^2+6x+9}$$
$$=\frac{2(x+3)}{(x+3)(x+2)(x+3)}+\frac{3x(x+2)}{(x+3)(x+3)(x+2)}$$
$$=\frac{2x+6+3x^2+6x}{(x+3)(x+2)(x+3)}$$
$$=\frac{3x^2+8x+6}{(x+3)^2(x+2)}$$

7. $$\frac{\frac{4}{y+2}-2}{5-\frac{10}{y+2}}=\frac{\frac{4}{y+2}-2}{5-\frac{10}{y+2}}\cdot\frac{y+2}{y+2}$$
$$=\frac{4-2(y+2)}{5(y+2)-10}$$
$$=\frac{4-2y-4}{5y+10-10}$$
$$=\frac{-2y}{5y}$$
$$=-\frac{2}{5}$$

8. $$\frac{\frac{1}{x}-\frac{3}{x+2}}{\frac{2}{x^2+2x}}=\frac{\frac{1}{x}-\frac{3}{x+2}}{\frac{2}{x(x+2)}}\cdot\frac{x(x+2)}{x(x+2)}$$
$$=\frac{x+2-3x}{2}$$
$$=\frac{-2x+2}{2}$$
$$=\frac{2(-x+1)}{2}$$
$$=-x+1$$

9. $$\frac{7}{4}=\frac{x+4}{x}$$
$$7x=4x+16$$
$$3x=16$$
$$x=\frac{16}{3}\text{ or }5\frac{1}{3}$$
Check: $\frac{7}{4}\stackrel{?}{=}\frac{\frac{16}{3}+4}{\frac{16}{3}}$, $\frac{7}{4}=\frac{7}{4}$

10. $$2+\frac{x}{x+4}=\frac{3x}{x-4}$$
$$(x+4)(x-4)\left(2+\frac{x}{x+4}\right)=(x+4)(x-4)\left(\frac{3}{x-4}\right)$$
$$2(x+4)(x-4)+x(x-4)=3x(x+4)$$
$$2x^2-32+x^2-4x=3x^2+12x$$
$$16x=-32$$
$$x=-2$$
Check: $2+\frac{-2}{-2+4}\stackrel{?}{=}\frac{3(-2)}{-2-4}$, $1=1$

11. $$\frac{1}{2y+4}-\frac{1}{6}=\frac{-2}{3y+6}$$
$$\frac{1}{2(y+2)}-\frac{1}{6}=\frac{-2}{3(y+2)}$$
$$6(y+2)\left(\frac{1}{2(y+2)}-\frac{1}{6}\right)=6(y+2)\left(\frac{-2}{3(y+2)}\right)$$
$$3-(y+2)=-4$$
$$3-y-2=-4$$
$$y=5$$
Check: $\frac{1}{2(5)+4}-\frac{1}{6}\stackrel{?}{=}\frac{-2}{3(5)+6}$, $-\frac{2}{21}=-\frac{2}{21}$

12.
$$2+\frac{3}{x}=\frac{2x}{x-1}$$
$$x(x-1)\left[2+\frac{3}{x}\right]=x(x-1)\left(\frac{2x}{x-1}\right)$$
$$2x(x-1)+3(x-1)=2x^2$$
$$2x^2-2x+3x-3=2x^2$$
$$x=3$$
Check: $2+\frac{3}{3}\stackrel{?}{=}\frac{2(3)}{3-1},\ 3=3$

13.
$$h=\frac{S-2WL}{2W+2L}$$
$$2hW+2hL=S-2WL$$
$$(2h+2L)W=S-2hL$$
$$W=\frac{S-2hL}{2h+2L}$$

14. $\frac{3V}{\pi h}=r^2\Rightarrow \pi r^2h=3V\Rightarrow h=\frac{3V}{\pi r^2}$

15.
$$\frac{3}{19}=\frac{x}{286-x}$$
$$858-3x=19x$$
$$22x=858$$
$$x=39$$
$286-39=247$

39 employees got the bonus, 247 did not.

16.
$$\frac{500}{850}=\frac{W}{\frac{8100-2W}{2}}$$
$$\frac{500}{850}=\frac{2W}{8100-2W}$$
$$4,050,000-1000W=1700W$$
$$2700W=4,050,000$$
$$W=1500$$
$$\frac{8100-2W}{2}=2550$$
The width is 1500 feet and the length is 2550 feet.

Cumulative Test for Chapters 1–6

1.
$$(3x^{-2}y)^2(x^4y^{-3})=3^2x^{-4}y^2x^4y^{-3}$$
$$=9x^{-4+4}y^{2-3}$$
$$=9x^0y^{-1}$$
$$=\frac{9}{y}$$

2.
$$\frac{3}{4}(x+2)=\frac{1}{3}x-2$$
$$12\left[\frac{3}{4}(x+2)\right]=12\left[\frac{1}{3}x-2\right]$$
$$9(x+2)=4x-24$$
$$9x+18=4x-24$$
$$5x=-42$$
$$x=-\frac{42}{5}$$

3. $-6x+2y=-12$

Let $x=0$: $-6(0)+2y=-12$
$$y=-6$$
Let $x=1$: $-6(1)+2y=-12$
$$y=-3$$
Let $y=0$: $-6x+2(0)=-12$
$$x=2$$

x	y
0	–6
1	–3
2	0

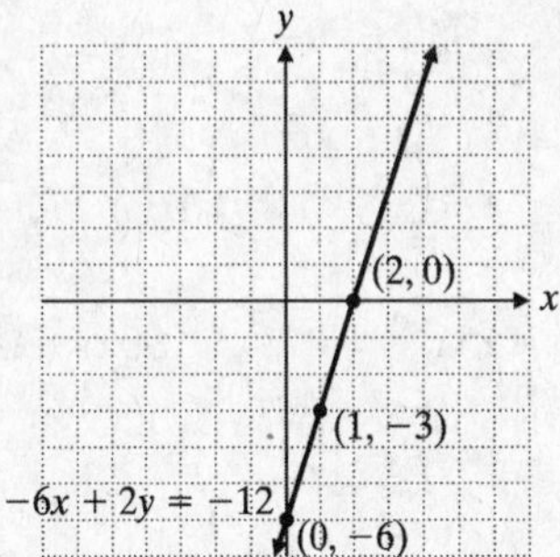

4.
$$5x-6y=8$$
$$6y=5x-8$$
$$y=\frac{5}{6}x-\frac{4}{3}\Rightarrow m=\frac{5}{6},\ m_{\|}=\frac{5}{6}$$
$$y-y_1=m(x-x_1)$$
$$y-(-3)=\frac{5}{6}(x-(-1))$$
$$6y+18=5x+5$$
$$5x-6y=13$$

5. x = amount at 5%
$7000 - x$ = amount at 8%

$$0.05x + 0.08(7000 - x) = 539$$
$$0.05x + 560 - 0.08x = 539$$
$$-0.03x = -21$$
$$x = \$700 \text{ at } 5\%$$

$7000 - x = \$6300$ at 8%
She invested \$700 at 5% and \$6300 at 8%.

6.
$$3(2-6x) > 4(x+1)+24$$
$$6-18x > 4x+4+24$$
$$-22x > 22$$
$$x < -1$$

−3 −2 −1 0 1 2 3

7.
$$2x^2 - 3x - 4y^2 = 2(-2)^2 - 3(-2) - 4(3)^2$$
$$= 2(4) + 6 - 4(9)$$
$$= 8 + 6 - 36$$
$$= 14 - 36$$
$$= -22$$

8.
$$|2x+1| \le 8$$
$$-8 \le 2x+1 \le 8$$
$$-9 \le 2x \le 7$$
$$-\frac{9}{2} \le x \le \frac{7}{2}$$

9.
$$8x^3 - 125y^3 = (2x)^3 - (5y)^3$$
$$= (2x-5y)(4x^2 + 10xy + 25y^2)$$

10.
$$81x^3 - 90x^2y + 25xy^2 = x(81x^2 - 90xy + 25y^2)$$
$$= x(9x-5y)^2$$

11.
$$x^2 + 20x + 36 = 0$$
$$(x+18)(x+2) = 0$$
$$x+18 = 0 \quad \text{or} \quad x+2 = 0$$
$$x = -18 \qquad x = -2$$

12.
$$3x^2 - 11x - 4 = 0$$
$$(3x+1)(x-4) = 0$$
$$3x+1 = 0 \quad \text{or} \quad x-4 = 0$$
$$x = -\frac{1}{3} \qquad x = 4$$

13.
$$\frac{7x^2-28}{x^2+6x+8} = \frac{7(x+2)(x-2)}{(x+4)(x+2)} = \frac{7(x-2)}{x+4}$$

14.
$$\frac{2x^2+x-1}{2x^2-9x+4} \cdot \frac{3x^2-12x}{6x+15}$$
$$= \frac{(2x-1)(x+1)}{(2x-1)(x-4)} \cdot \frac{3x(x-4)}{3(2x+5)}$$
$$= \frac{x(x+1)}{2x+5}$$

15.
$$\frac{x^3+2x^2}{3x-21} \div \frac{2x^3+5x^2+2x}{x-7}$$
$$= \frac{x^2(x+2)}{3(x-7)} \cdot \frac{x-7}{x(2x^2+5x+2)}$$
$$= \frac{x(x+2)}{3} \cdot \frac{1}{(2x+1)(x+2)}$$
$$= \frac{x}{3(2x+1)}$$

16.
$$\frac{5}{2x-4} - \frac{x+1}{x^2-4}$$
$$= \frac{5}{2(x-2)} - \frac{x+1}{(x-2)(x+2)}$$
$$= \frac{5}{2(x-2)} \cdot \frac{x+2}{x+2} - \frac{x+1}{(x-20(x+2)} \cdot \frac{2}{2}$$
$$= \frac{5(x+2) - 2(x+1)}{2(x-2)(x+2)}$$
$$= \frac{5x+10-2x-2}{2(x-2)(x+2)}$$
$$= \frac{3x+8}{2(x-2)(x+2)}$$

17.
$$\frac{\frac{1}{2x+1}+1}{4-\frac{3}{4x^2-1}} = \frac{\frac{1}{2x+1}+1}{4-\frac{3}{(2x-1)(2x+1)}} \cdot \frac{(2x-1)(2x+1)}{(2x-1)(2x+1)}$$
$$= \frac{2x-1+(2x-1)(2x+1)}{4(2x-1)(2x+1)-3}$$
$$= \frac{(2x-1)(1+2x+1)}{16x^2-4-3}$$
$$= \frac{(2x-1)(2x+2)}{16x^2-7}$$
$$= \frac{2(x+1)(2x-1)}{16x^2-7}$$

18.
$$\frac{3}{x-6} + \frac{4}{x+4} = \frac{3(x+4)+4(x-6)}{(x-6)(x+4)}$$
$$= \frac{3x+12+4x-24}{(x-6)(x+4)}$$
$$= \frac{7x-12}{(x-6)(x+4)}$$

19. $$\frac{1}{2x+3}-\frac{4}{4x^2-9}=\frac{3}{2x-3}$$
$$(2x+3)(2x-3)\left[\frac{1}{2x+3}-\frac{4}{(2x+3)(2x-3)}\right]=(2x+3)(2x-3)\frac{3}{2x-3}$$
$$2x-3-4=3(2x+3)$$
$$2x-7=6x+9$$
$$4x=-16$$
$$x=-4$$

Check: $\frac{1}{2(-4)+3}-\frac{4}{4(-4)^2-9}\stackrel{?}{=}\frac{3}{2(-4)-3}$
$$-\frac{3}{11}=-\frac{3}{11}$$

20. $$\frac{2}{5x}-\frac{4}{x}=\frac{3}{10}$$
$$10x\left(\frac{2}{5x}\right)-10x\left(\frac{4}{x}\right)=10x\left(\frac{3}{10}\right)$$
$$4-40=3x$$
$$-36=3x$$
$$-12=x$$

Check: $\frac{2}{5(-12)}-\frac{4}{-12}\stackrel{?}{=}\frac{3}{10}$
$$\frac{3}{10}=\frac{3}{10}$$

21. $$H=\frac{3b+2x}{5-4b}$$
$$5H-4bH=3b+2x$$
$$3b+4bH=5H-2x$$
$$b(3+4H)=5H-2x$$
$$b=\frac{5H-2x}{3+4H}$$

22. x = number patrolling on foot
$3234-x$ = number of patrolling in squad cars
$$\frac{3}{11}=\frac{x}{3234-x}$$
$$9702-3x=11x$$
$$14x=9702$$
$$x=693$$
$$3234-x=2541$$
693 are patrolling on foot and 2541 are patrolling in squad cars.

Chapter 7

7.1 Exercises

1. $\left(\frac{3xy^{-1}}{z^2}\right)^4 = \frac{3^4x^4y^{-4}}{z^8} = \frac{81x^4}{y^4z^8}$

3. $\left(\frac{4ab^{-2}}{3b}\right)^2 = \frac{16a^2b^{-4}}{9b^2} = \frac{16a^2}{9b^{2+4}} = \frac{16a^2}{9b^6}$

5. $\left(\frac{2x^2}{y}\right)^{-3} = \frac{2^{-3}x^{-6}}{y^{-3}} = \frac{y^3}{8x^6}$

7. $\left(\frac{3xy^{-2}}{y^3}\right)^{-2} = \frac{3^{-2}x^{-2}y^4}{y^{-6}} = \frac{y^{4+6}}{3^2x^2} = \frac{y^{10}}{9x^2}$

9. $(x^{3/4})^2 = x^{6/4} = x^{3/2}$

11. $(y^{12})^{2/3} = y^{12\cdot\frac{2}{3}} = y^8$

13. $\frac{x^{3/5}}{x^{1/5}} = x^{\frac{3}{5}-\frac{1}{5}} = x^{2/5}$

15. $\frac{x^{8/9}}{x^{2/9}} = x^{\frac{8}{9}-\frac{2}{9}} = x^{6/9} = x^{2/3}$

17. $\frac{a^2}{a^{1/4}} = a^{\frac{8}{4}-\frac{1}{4}} = a^{7/4}$

19. $x^{1/7}\cdot x^{3/7} = x^{\frac{1}{7}+\frac{3}{7}} = x^{4/7}$

21. $a^{3/8}\cdot a^{1/2} = a^{\frac{3}{8}+\frac{4}{8}} = a^{7/8}$

23. $y^{3/5}\cdot y^{-1/10} = y^{\frac{6}{10}-\frac{1}{10}} = y^{5/10} = y^{1/2}$

25. $x^{-3/4} = \frac{1}{x^{3/4}}$

27. $a^{-5/6}b^{1/3} = \frac{b^{1/3}}{a^{5/6}}$

29. $6^{-1/2} = \frac{1}{6^{1/2}}$

31. $3a^{-1/3} = \frac{3}{a^{1/3}}$

33. $(27)^{5/3} = (3^3)^{5/3} = 3^{3\cdot\frac{5}{3}} = 3^5 = 243$

35. $(4)^{3/2} = (2^2)^{3/2} = 2^{2\cdot 3/2} = 2^3 = 8$

37. $(-8)^{5/3} = ((-2)^3)^{5/3} = (-2)^{3\cdot 5/3} = (-2)^5 = -32$

39. $(-27)^{2/3} = ((-3)^3)^{2/3} = (-3)^{3\cdot 2/3} = (-3)^2 = 9$

41. $(x^{1/4}y^{-1/3})(x^{3/4}y^{1/2}) = x^{\frac{1}{4}+\frac{3}{4}}y^{-\frac{2}{6}+\frac{3}{6}} = xy^{1/6}$

43. $(7x^{1/3}y^{1/4})(-2x^{1/4}y^{-1/6}) = 7(-2)x^{\frac{1}{3}+\frac{1}{4}}y^{\frac{1}{4}-\frac{1}{6}}$
$= -14x^{\frac{3}{12}+\frac{4}{12}}y^{\frac{3}{12}-\frac{2}{12}}$
$= -14x^{7/12}y^{1/12}$

45. $6^2\cdot 6^{-2/3} = 6^{\frac{6}{3}-\frac{2}{3}} = 6^{4/3}$

47. $\frac{2x^{1/5}}{x^{-1/2}} = 2x^{\frac{2}{10}+\frac{5}{10}} = 2x^{7/10}$

49. $\frac{-20x^2y^{-1/5}}{5x^{-1/2}y} = -\frac{4x^{2+\frac{1}{2}}}{y^{1+\frac{1}{5}}} = -\frac{4x^{5/2}}{y^{6/5}}$

51. $\left(\frac{8a^2b^6}{a^{-1}b^3}\right)^{1/3} = (8a^3b^3)^{1/3} = 8^{1/3}a^{3\cdot\frac{1}{3}}b^{3\cdot\frac{1}{3}} = 2ab$

53. $(-4x^{1/4}y^{5/2}z^{1/2})^2 = 16x^{2/4}y^{10/2}z^{2/2}$
$= 16x^{1/2}y^5z$

55. $x^{2/3}(x^{4/3} - x^{1/5}) = x^{\frac{2}{3}+\frac{4}{3}} - x^{\frac{2}{3}+\frac{1}{5}}$
$= x^{6/3} - x^{\frac{10}{15}+\frac{3}{15}}$
$= x^2 - x^{13/15}$

57. $m^{7/8}(m^{-1/2} + 2m) = m^{\frac{7}{8}-\frac{4}{8}} + 2m^{\frac{7}{8}+\frac{8}{8}}$
$= m^{3/8} + 2m^{15/8}$

59. $(8)^{-1/3} = (2^3)^{-1/3} = 2^{3(-1/3)} = 2^{-1} = \frac{1}{2}$

61. $(25)^{-3/2} = (5^2)^{-3/2} = 5^{2(-3/2)} = 5^{-3} = \frac{1}{5^3} = \frac{1}{125}$

63. $$\begin{aligned}(81)^{3/4} + (25)^{1/2} &= (3^4)^{3/4} + (5^2)^{1/2}\\ &= 3^{4\cdot 3/4} + 5^{2\cdot 1/2}\\ &= 3^3 + 5\\ &= 27 + 5\\ &= 32\end{aligned}$$

65. $$\begin{aligned}3y^{1/2} + y^{-1/2} &= 3y^{1/2} \cdot \frac{y^{1/2}}{y^{1/2}} + \frac{1}{y^{1/2}}\\ &= \frac{3y}{y^{1/2}} + \frac{1}{y^{1/2}}\\ &= \frac{3y+1}{y^{1/2}}\end{aligned}$$

67. $$\begin{aligned}x^{-1/3} + 6^{4/3} &= \frac{1}{x^{1/3}} + 6^{4/3}\\ &= \frac{1}{x^{1/3}} + \frac{6^{4/3}x^{1/3}}{x^{1/3}}\\ &= \frac{1+6^{4/3}x^{1/3}}{x^{1/3}}\end{aligned}$$

69. $$\begin{aligned}10a^{5/4} - 4a^{8/5} &= 2a^{4/4} \cdot 5a^{1/4} - 2a^{5/5} \cdot 2a^{3/5}\\ &= 2a(5a^{1/4} - 2^{3/5})\end{aligned}$$

71. $$\begin{aligned}12x^{4/3} - 3x^{5/2} &= 3x^{3/3} \cdot 4x^{1/3} - 3x^{2/2} \cdot x^{3/2}\\ &= 3x(4x^{1/3} - x^{3/2})\end{aligned}$$

73. $x^a \cdot x^{1/4} = x^{-1/8} \Rightarrow x^{a+\frac{1}{4}} = x^{-1/8}$

$$a + \frac{1}{4} = -\frac{1}{8}$$
$$a = -\frac{3}{8}$$

75. $r = 0.62(V)^{1/3}$

$r = 0.62(27)^{1/3}$

$r = 0.62(3) = 1.86$

The radius is 1.86 meters.

77. $r = \left(\frac{3V}{\pi h}\right)^{1/2}$

$r = \left(\frac{3(314)}{3.14(12)}\right)^{1/2} = (25)^{1/2} = 5$

The radius is 5 feet.

Cumulative Review

79. $$\begin{aligned}-4(x+1) &= \frac{1}{3}(3-2x)\\ -12(x+1) &= 3-2x\\ -12x-12 &= 3-2x\\ -10x &= 15\\ x &= -\frac{3}{2}\end{aligned}$$

80. $$\begin{aligned}A &= \frac{h}{2}(a+b)\\ 2A &= h(a+b)\\ 2A &= ha+hb\\ ha+hb &= 2A\\ hb &= 2A-ha\\ b &= \frac{2A-ha}{h} = \frac{2A}{h} - \frac{ha}{h} = \frac{2A}{h} - a\end{aligned}$$

Quick Quiz 7.1

1. $$\begin{aligned}(-4x^{2/3}y^{1/4})(3x^{1/6}y^{1/2}) &= (-4)(3)x^{\frac{2}{3}+\frac{1}{6}}y^{\frac{1}{4}+\frac{1}{2}}\\ &= -12x^{\frac{4}{6}+\frac{1}{6}}y^{\frac{1}{4}+\frac{2}{4}}\\ &= -12x^{5/6}y^{3/4}\end{aligned}$$

2. $\frac{16x^4}{8x^{2/3}} = 2x^{4-\frac{2}{3}} = 2x^{\frac{12}{3}-\frac{2}{3}} = 2x^{10/3}$

3. $$\begin{aligned}(25x^{1/4})^{3/2} &= (5^2x^{2/8})^{3/2}\\ &= (5x^{1/8})^{2(3/2)}\\ &= (5x^{1/8})^3\\ &= 5^3x^{3\cdot\frac{1}{8}}\\ &= 125x^{3/8}\end{aligned}$$

4. Answers may vary. Possible answer: Change the exponents to have equal denominators, then add the numerators over the common denominator. This is the combined, simplified exponent for x.

7.2 Exercises

1. A square root of a number is a value that when multiplied by itself is equal to the original number.

3. One answer is $\sqrt[3]{-8} = -2$ because $(-2)(-2)(-2) = -8$.

5. $\sqrt{100} = 10$ because $10^2 = 100$.

7. $\sqrt{16} + \sqrt{81} = 4 + 9 = 13$

9. $-\sqrt{\frac{1}{9}} = -\frac{1}{3}$

11. $\sqrt{-36}$ is not a real number.

13. $\sqrt{0.04} = 0.2$ because $0.2^2 = 0.04$.

15. $f(x) = \sqrt{3x+21}$

$f(0) = \sqrt{3(0)+21} = \sqrt{21} \approx 4.6$
$f(1) = \sqrt{3(1)+21} = \sqrt{24} \approx 4.9$
$f(5) = \sqrt{3(5)+21} = \sqrt{36} = 6$
$f(-4) = \sqrt{3(-4)+21} = \sqrt{9} = 3$

The domain is $3x + 21 \geq 0$
$3x \geq -21$
$x \geq -7$

The domain is all real numbers x where $x \geq -7$.

17. $f(x) = \sqrt{0.5x-5}$

$f(10) = \sqrt{0.5(10)-5} = \sqrt{0} = 0$
$f(12) = \sqrt{0.5(12)-5} = \sqrt{1} = 1$
$f(18) = \sqrt{0.5(18)-5} = \sqrt{4} = 2$
$f(20) = \sqrt{0.5(20)-5} = \sqrt{5} \approx 2.2$

The domain is $0.5x - 5 \geq 0$
$0.5x \geq 5$
$x \geq 10$

The domain is all real numbers x where $x \geq 10$.

19. $f(x) = \sqrt{x-1}$

$f(1) = \sqrt{1-1} = \sqrt{0} = 0$
$f(2) = \sqrt{2-1} = \sqrt{1} = 1$
$f(5) = \sqrt{5-1} = \sqrt{4} = 2$
$f(10) = \sqrt{10-1} = \sqrt{9} = 3$

21. $f(x) = \sqrt{3x+9}$

$f(-3) = \sqrt{3(-3)+9} = \sqrt{0} = 0$
$f\left(-\frac{8}{3}\right) = \sqrt{3\left(-\frac{8}{3}\right)+9} = \sqrt{1} = 1$
$f\left(-\frac{5}{3}\right) = \sqrt{3\left(-\frac{5}{3}\right)+9} = \sqrt{4} = 2$
$f(0) = \sqrt{3(0)+9} = \sqrt{9} = 3$

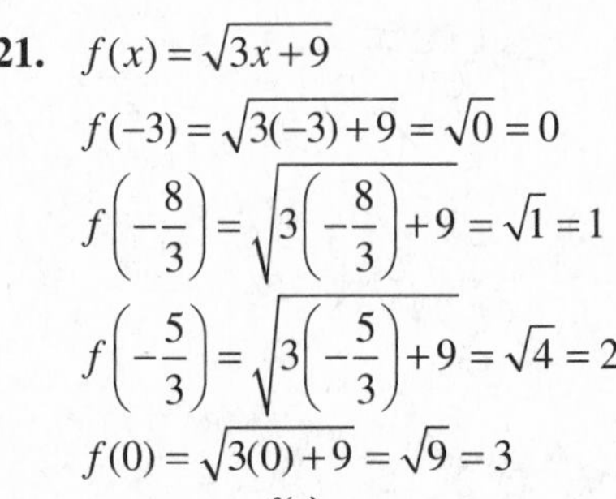

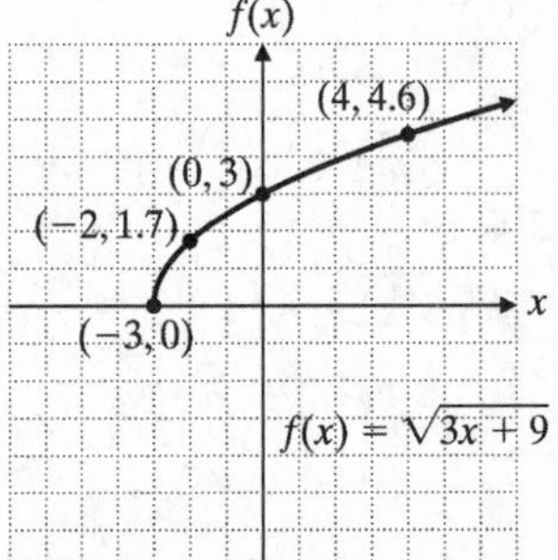

23. $\sqrt[3]{64} = 4$ because $4^3 = 64$.

25. $\sqrt[3]{-1000} = -10$ because $(-10)^3 = -1000$.

27. $\sqrt[4]{16} = \sqrt[4]{2^4} = 2$

29. $\sqrt[4]{81} = \sqrt[4]{3^4} = 3$

31. $\sqrt[5]{(8)^5} = 8$

33. $\sqrt[8]{(5)^8} = 5$

35. $\sqrt[3]{-\frac{1}{8}} = \sqrt[3]{\left(-\frac{1}{2}\right)^3} = -\frac{1}{2}$

37. $\sqrt[3]{y} = y^{1/3}$

39. $\sqrt[5]{m^3} = m^{3/5}$

41. $\sqrt[4]{2a} = (2a)^{1/4}$

43. $\sqrt[7]{(a+b)^3} = (a+b)^{3/7}$

45. $\sqrt{\sqrt[3]{x}} = (x^{1/3})^{1/2} = x^{1/6}$

47. $\left(\sqrt[6]{3x}\right)^5 = ((3x)^{1/6})^5 = (3x)^{5/6}$

49. $\sqrt[6]{(12)^6} = 12$

51. $\sqrt[3]{x^{12}y^3} = \sqrt[3]{(x^4y)^3} = x^4y$

53. $\sqrt{36x^8y^4} = \sqrt{(6x^4y^2)^2} = 6x^4y^2$

55. $\sqrt[4]{16a^8b^4} = \sqrt[4]{(2a^2b)^4} = 2a^2b$

57. $y^{4/7} = \sqrt[7]{y^4}$

59. $7^{-2/3} = \frac{1}{7^{2/3}} = \frac{1}{\sqrt[3]{7^2}} = \frac{1}{\sqrt[3]{49}}$

61. $(a+5b)^{3/4} = \sqrt[4]{(a+5b)^3}$

63. $(-x)^{3/5} = \left(\sqrt[5]{(-x)^3}\right)^3$

65. $(2xy)^{3/5} = \sqrt[5]{(2xy)^3} = \sqrt[5]{8x^3y^3}$

67. $9^{3/2} = (3)^{2(3/2)} = 3^3 = 27$

69. $\left(\frac{4}{25}\right)^{1/2} = \sqrt{\frac{4}{25}} = \frac{2}{5}$

71. $\left(\frac{1}{8}\right)^{-1/3} = (8^{-1})^{-1/3} = 8^{-1(-1/3)} = 8^{1/3} = \sqrt[3]{8} = 2$

73. $(64x^4)^{-1/2} = (8x^2)^{2(-1/2)} = (8x^2)^{-1} = \frac{1}{8x^2}$

75. $\sqrt{121x^4} = \sqrt{(11x^2)^2} = 11x^2$

77. $\sqrt{144a^6b^{24}} = \sqrt{(12a^3b^{12})^2} = 12a^3b^{12}$

79. $\sqrt{25x^2} = 5|x|$

81. $\sqrt[3]{-8x^6} = -2x^2$

83. $\sqrt[4]{x^8y^{16}} = x^2y^4$

85. $\sqrt[4]{a^{12}b^4} = \sqrt[4]{(a^3b)^4} = |a^3b|$

87. $\sqrt{25x^{12}y^4} = 5x^6y^2$

Cumulative Review

89.
$$\begin{aligned} -5x+2y &= 6 \\ 5x-2y &= -6 \\ 5x &= 2y-6 \\ x &= \frac{2y-6}{5} \end{aligned}$$

90.
$$\begin{aligned} x &= \frac{2}{3}y+4 \\ 3(x) &= 3\left(\frac{2}{3}y+4\right) \\ 3x &= 2y+12 \\ 3x-12 &= 2y \\ \frac{3x-12}{2} &= y \text{ or } y = \frac{3x}{2}-6 \end{aligned}$$

Quick Quiz 7.2

1. $\left(\frac{4}{25}\right)^{3/2} = \left(\frac{2^2}{5^2}\right)^{3/2} = \left(\frac{2}{5}\right)^{2(3/2)} = \left(\frac{2}{5}\right)^3 = \frac{8}{125}$

2. $\sqrt[3]{-64} = -4$ because $(-4)^3 = -64$.

3. $\sqrt{121x^{10}y^{12}} = \sqrt{(11x^5y^6)^2} = 11x^5y^6$

4. Answers may vary. Possible solution: Factor the coefficient completely to identify the fourth root. Remove the fourth root from under the radical. Divide the exponents of the variables by 4 to remove the variables from under the radical.

7.3 Exercises

1. $\sqrt{8} = \sqrt{4\cdot 2} = \sqrt{4}\sqrt{2} = 2\sqrt{2}$

3. $\sqrt{18} = \sqrt{9\cdot 2} = \sqrt{9}\sqrt{2} = 3\sqrt{2}$

5. $\sqrt{28} = \sqrt{4\cdot 7} = \sqrt{4}\sqrt{7} = 2\sqrt{7}$

7. $\sqrt{50} = \sqrt{25}\sqrt{2} = 5\sqrt{2}$

9. $\sqrt{9x^3} = \sqrt{9}\sqrt{x^2}\sqrt{x} = 3x\sqrt{x}$

11. $\sqrt{40a^6b^7} = \sqrt{4\cdot 10a^6b^6b}$
$= \sqrt{4a^6b^6}\sqrt{10b}$
$= 2a^3b^3\sqrt{10b}$

13. $\sqrt{90x^3yz^4} = \sqrt{9\cdot 10x^2xyz^4}$
$= \sqrt{9x^2z^4}\sqrt{10xy}$
$= 3xz^2\sqrt{10xy}$

15. $\sqrt[3]{8} = \sqrt[3]{2^3} = 2$

17. $\sqrt[3]{40} = \sqrt[3]{8\cdot 5} = 2\sqrt[3]{5}$

19. $\sqrt[3]{54a^2} = \sqrt[3]{27\cdot 2a^2} = 3\sqrt[3]{2a^2}$

21. $\sqrt[3]{27a^5b^9} = \sqrt[3]{27a^3b^9}\sqrt[3]{a^2} = 3ab^3\sqrt[3]{a^2}$

23. $\sqrt[3]{24x^6y^{11}} = \sqrt[3]{8\cdot 3x^6y^9y^2}$
$= \sqrt[3]{8x^6y^9}\sqrt[3]{3y^2}$
$= 2x^2y^3\sqrt[3]{3y^2}$

25. $\sqrt[4]{81kp^{23}} = \sqrt[4]{3^4kp^{20}p^3}$
$= \sqrt[4]{3^4p^{20}}\sqrt[4]{kp^3}$
$= 3p^5\sqrt[4]{kp^3}$

27. $\sqrt[5]{-32x^5y^6} = \sqrt[5]{(-2)^5x^5y^5y}$
$= \sqrt[5]{(-2)^5x^5y^5}\sqrt[5]{y}$
$= -2xy\sqrt[5]{y}$

29. $\sqrt[4]{1792} = a\sqrt[4]{7}$
$\sqrt[4]{256\cdot 7} = a\sqrt[4]{7}$
$\sqrt[4]{4^4\cdot 7} = a\sqrt[4]{7}$
$4\sqrt[4]{7} = a\sqrt[4]{7}$
$a = 4$

31. $4\sqrt{5} + 8\sqrt{5} = 12\sqrt{5}$

33. $4\sqrt{3} + \sqrt{7} - 5\sqrt{7} = 4\sqrt{3} - 4\sqrt{7}$

35. $3\sqrt{32} - \sqrt{2} = 3\sqrt{16\cdot 2} - \sqrt{2} = 12\sqrt{2} - \sqrt{2} = 11\sqrt{2}$

37. $4\sqrt{12} + \sqrt{27} = 4\sqrt{4\cdot 3} + \sqrt{9\cdot 3}$
$= 8\sqrt{3} + 3\sqrt{3}$
$= 11\sqrt{3}$

39. $\sqrt{8} + \sqrt{50} - 2\sqrt{72} = \sqrt{4\cdot 2} + \sqrt{25\cdot 2} - 2\sqrt{36\cdot 2}$
$= 2\sqrt{2} + 5\sqrt{2} - 12\sqrt{2}$
$= -5\sqrt{2}$

41. $\sqrt{48} - 2\sqrt{27} + \sqrt{12} = \sqrt{16\cdot 3} - 2\sqrt{9\cdot 3} + \sqrt{4\cdot 3}$
$= 4\sqrt{3} - 6\sqrt{3} + 2\sqrt{3}$
$= 0$

43. $-5\sqrt{45} + 6\sqrt{20} + 3\sqrt{5} = -5\sqrt{9\cdot 5} + 6\sqrt{4\cdot 5} + 3\sqrt{5}$
$= -15\sqrt{5} + 12\sqrt{5} + 3\sqrt{5}$
$= 0$

45. $3\sqrt{48x} - 2\sqrt{12x} = 3\sqrt{16\cdot 3x} - 2\sqrt{4\cdot 3x}$
$= 12\sqrt{3x} - 4\sqrt{3x}$
$= 8\sqrt{3x}$

47. $5\sqrt{2x} + 2\sqrt{18x} + 2\sqrt{32x}$
$= 5\sqrt{2x} + 2\sqrt{9\cdot 2x} + 2\sqrt{16\cdot 2x}$
$= 5\sqrt{2x} + 6\sqrt{2x} + 8\sqrt{2x}$
$= 19\sqrt{2x}$

49. $\sqrt{44} - 3\sqrt{63x} + 4\sqrt{28x}$
$= \sqrt{4\cdot 11} - 3\sqrt{9\cdot 7x} + 4\sqrt{4\cdot 7x}$
$= 2\sqrt{11} - 9\sqrt{7x} + 8\sqrt{7x}$
$= 2\sqrt{11} - \sqrt{7x}$

51. $\sqrt{200x^3} - x\sqrt{32x} = \sqrt{100x^2\cdot 2x} - x\sqrt{16\cdot 2x}$
$= 10x\sqrt{2x} - 4x\sqrt{2x}$
$= 6x\sqrt{2x}$

53. $\sqrt[3]{16} + 3\sqrt[3]{54} = \sqrt[3]{2^3\cdot 2} + 3\sqrt[3]{3^3\cdot 2}$
$= 2\sqrt[3]{2} + 9\sqrt[3]{2}$
$= 11\sqrt[3]{2}$

55. $4\sqrt[3]{x^4y^3} - 3\sqrt[3]{xy^5} = 4\sqrt[3]{x^3y^3\cdot x} - 3\sqrt[3]{y^3\cdot xy^2}$
$= 4xy\sqrt[3]{x} - 3y\sqrt[3]{xy^2}$

57. $\sqrt{48}+\sqrt{27}+\sqrt{75}$
$= 6.92820323+5.196152423+8.660254038$
$= 20.78460969$
$12\sqrt{3} = 20.78460969$ which shows
$\sqrt{48}+\sqrt{27}+\sqrt{75} = 12\sqrt{3}.$

59. $I = \sqrt{\frac{P}{R}} = \sqrt{\frac{500}{10}} = \sqrt{50} \approx 7.071$
The current is approximately 7.071 amps.

61. $T = 2\pi\sqrt{\frac{L}{32}} = 2(3.14)\sqrt{\frac{8}{32}} \approx 3.14$
The period of the pendulum is approximately 3.14 seconds.

Cumulative Review

63. $16x^3 - 56x^2y + 49xy^2$
$= x(16x^2 - 56xy + 49y^2)$
$= x[(4x)^2 - 2(4x)(7y) + (7y)^2]$
$= x(4x - 7y)^2$

64. $81x^2y - 25y = y(81x^2 - 25)$
$= y[(9x)^2 - 5^2]$
$= y(9x+5)(9x-5)$

65. $S + M = 4.5 \Rightarrow S = 4.5 - M$
$0.2S + 0.25M = 1$
$0.2(4.5 - M) + 0.25M = 1$
$0.9 - 0.2M + 0.25M = 1$
$0.05M = 0.1$
$M = 2$
$S = 4.5 - M = 2.5$
2.5 small servings of scallops (fifteen scallops) and two small servings of skim milk (2 cups) would meet the requirement.

66. $0.2S = 1$
$2S = 10$
$S = 5$
If you eat only scallops, 5 servings (30 scallops) would meet the requirement.
$0.25M = 1$
$25M = 100$
$M = 4$
If you drink only skim milk, 4 servings (4 cups) would meet the requirement.

Quick Quiz 7.3

1. $\sqrt{120x^7y^8} = \sqrt{4\cdot 30xx^6y^8}$
$= \sqrt{4x^6y^8}\sqrt{30x}$
$= 2x^3y^4\sqrt{30x}$

2. $\sqrt[3]{16x^{15}y^{10}} = \sqrt[3]{2y\cdot 2^3x^{15}y^9}$
$= \sqrt[3]{2^3x^{15}y^9}\sqrt[3]{2y}$
$= 2x^5y^3\sqrt[3]{2y}$

3. $2\sqrt{75}+3\sqrt{48}-4\sqrt{27}$
$= 2\sqrt{25\cdot 3}+3\sqrt{16\cdot 3}-4\sqrt{9\cdot 3}$
$= 10\sqrt{3}+12\sqrt{3}-12\sqrt{3}$
$= 10\sqrt{3}$

4. Answers may vary. Possible solution: Completely factor the coefficient of the radicand, and express as its primes to their respective powers. Divide exponents of the radicand coefficient primes and variables by the index. The results are moved outside the radical. The remainders remain inside.

7.4 Exercises

1. $\sqrt{5}\sqrt{7} = \sqrt{5\cdot 7} = \sqrt{35}$

3. $(5\sqrt{2})(-6\sqrt{5}) = -30\sqrt{2\cdot 5} = -30\sqrt{10}$

5. $(3\sqrt{10})(-4\sqrt{2}) = -12\sqrt{20} = -12\sqrt{4\cdot 5} = -24\sqrt{5}$

7. $(-3\sqrt{y})(\sqrt{5x}) = -3\sqrt{5xy}$

9. $(3x\sqrt{2x})(-2\sqrt{10xy}) = -6x\sqrt{20x^2y}$
$= -6x\sqrt{4x^2\cdot 5y}$
$= -12x^2\sqrt{5y}$

11. $5\sqrt{a}(3\sqrt{b}-5) = 15\sqrt{ab} - 25\sqrt{a}$

13. $-3\sqrt{a}(\sqrt{2b}+2\sqrt{5}) = -3\sqrt{2ab} - 6\sqrt{5a}$

15. $-\sqrt{a}(\sqrt{a}-2\sqrt{b}) = -a + 2\sqrt{ab}$

17. $7\sqrt{x}(2\sqrt{3}-5\sqrt{x}) = 14\sqrt{3x} - 35x$

19. $(3-\sqrt{2})(8+\sqrt{2}) = 24+3\sqrt{2}-8\sqrt{2}-2$
$= 24 - 5\sqrt{2} - 2$
$= 22 - 5\sqrt{2}$

21. $\left(2\sqrt{3}+\sqrt{2}\right)\left(2\sqrt{3}-4\sqrt{2}\right)$
$= 4\cdot 3 - 8\sqrt{6} + 2\sqrt{6} - 4\cdot 2$
$= 12 - 6\sqrt{6} - 8$
$= 4 - 6\sqrt{6}$

23. $\left(\sqrt{7}+4\sqrt{5x}\right)\left(2\sqrt{7}+3\sqrt{5x}\right)$
$= 2\cdot 7 + 3\sqrt{35x} + 8\sqrt{35x} + 12(5x)$
$= 14 + 11\sqrt{35x} + 60x$

25. $\left(\sqrt{3}+2\sqrt{2}\right)\left(\sqrt{5}+\sqrt{3}\right) = \sqrt{15} + 3 + 2\sqrt{10} + 2\sqrt{6}$

27. $\left(\sqrt{5}-2\sqrt{6}\right)^2 = \left(\sqrt{5}\right)^2 - 2\cdot 2\sqrt{5}\sqrt{6} + \left(2\sqrt{6}\right)^2$
$= 5 - 4\sqrt{30} + 24$
$= 29 - 4\sqrt{30}$

29. $\left(9-2\sqrt{b}\right)^2 = 9^2 - 2(9)2\sqrt{b} + 2^2\cdot b$
$= 81 - 36\sqrt{b} + 4b$

31. $\left(\sqrt{3x+4}+3\right)^2 = \left(\sqrt{3x+4}\right)^2 + 2\cdot 3\sqrt{3x+4} + 3^2$
$= 3x + 4 + 6\sqrt{3x+4} + 9$
$= 3x + 13 + 6\sqrt{3x+4}$

33. $\left(\sqrt[3]{x^2}\right)\left(3\sqrt[3]{4x} - 4\sqrt[3]{x^5}\right) = 3\sqrt[3]{4x^3} - 4\sqrt[3]{x^7}$
$= 3x\sqrt[3]{4} - 4\sqrt[3]{x^6\cdot x}$
$= 3x\sqrt[3]{4} - 4x^2\sqrt[3]{x}$

35. $\left(\sqrt[3]{3}+\sqrt[3]{2}\right)\left(\sqrt[3]{9}-\sqrt[3]{4}\right) = \sqrt[3]{27} - \sqrt[3]{12} + \sqrt[3]{18} - \sqrt[3]{8}$
$= 3 - \sqrt[3]{12} + \sqrt[3]{18} - 2$
$= 1 + \sqrt[3]{18} - \sqrt[3]{12}$

37. $\sqrt{\dfrac{49}{25}} = \dfrac{\sqrt{49}}{\sqrt{25}} = \dfrac{7}{5}$

39. $\sqrt{\dfrac{12x}{49y^6}} = \dfrac{\sqrt{4\cdot 3x}}{\sqrt{49y^6}} = \dfrac{2\sqrt{3x}}{7y^3}$

41. $\sqrt[3]{\dfrac{8x^5y^6}{27}} = \dfrac{\sqrt[3]{8x^3y^6\cdot x^2}}{\sqrt[3]{27}} = \dfrac{2xy^2\sqrt[3]{x^2}}{3}$

43. $\dfrac{\sqrt[3]{5y^8}}{\sqrt[3]{27x^3}} = \dfrac{\sqrt[3]{5y^6y^2}}{3x} = \dfrac{y^2\sqrt[3]{5y^2}}{3x}$

45. $\dfrac{3}{\sqrt{2}} = \dfrac{3}{\sqrt{2}}\cdot\dfrac{\sqrt{2}}{\sqrt{2}} = \dfrac{3\sqrt{2}}{2}$

47. $\sqrt{\dfrac{4}{3}} = \dfrac{\sqrt{4}}{\sqrt{3}} = \dfrac{\sqrt{4}}{\sqrt{3}}\cdot\dfrac{\sqrt{3}}{\sqrt{3}} = \dfrac{2\sqrt{3}}{3}$

49. $\dfrac{1}{\sqrt{5y}} = \dfrac{1}{\sqrt{5y}}\cdot\dfrac{\sqrt{5y}}{\sqrt{5y}} = \dfrac{\sqrt{5y}}{5y}$

51. $\dfrac{\sqrt{14a}}{\sqrt{2y}} = \dfrac{\sqrt{14a}}{\sqrt{2y}}\cdot\dfrac{\sqrt{2y}}{\sqrt{2y}}$
$= \dfrac{\sqrt{28ay}}{2y}$
$= \dfrac{\sqrt{4\cdot 7ay}}{2y}$
$= \dfrac{2\sqrt{7ay}}{2y}$
$= \dfrac{\sqrt{7ay}}{y}$

53. $\dfrac{\sqrt{2}}{\sqrt{6x}} = \dfrac{\sqrt{2}}{\sqrt{6x}}\cdot\dfrac{\sqrt{6x}}{\sqrt{6x}}$
$= \dfrac{\sqrt{12x}}{6x}$
$= \dfrac{\sqrt{4\cdot 3x}}{6x}$
$= \dfrac{2\sqrt{3x}}{6x}$
$= \dfrac{\sqrt{3x}}{3x}$

55. $\dfrac{x}{\sqrt{5}-\sqrt{2}} = \dfrac{x}{\sqrt{5}-\sqrt{2}}\cdot\dfrac{\sqrt{5}+\sqrt{2}}{\sqrt{5}+\sqrt{2}}$
$= \dfrac{x\left(\sqrt{5}+\sqrt{2}\right)}{5-2}$
$= \dfrac{x\left(\sqrt{5}+\sqrt{2}\right)}{3}$

57. $\dfrac{2y}{\sqrt{6}+\sqrt{5}} = \dfrac{2y}{\sqrt{6}+\sqrt{5}}\cdot\dfrac{\sqrt{6}-\sqrt{5}}{\sqrt{6}-\sqrt{5}}$
$= \dfrac{2y\left(\sqrt{6}-\sqrt{5}\right)}{6-5}$
$= 2y\left(\sqrt{6}-\sqrt{5}\right)$

59. $\dfrac{\sqrt{y}}{\sqrt{6}+\sqrt{2y}} = \dfrac{\sqrt{y}}{\sqrt{6}+\sqrt{2y}} \cdot \dfrac{\sqrt{6}-\sqrt{2y}}{\sqrt{6}-\sqrt{2y}}$

$= \dfrac{\sqrt{6y}-\sqrt{2y^2}}{6-2y}$

$= \dfrac{\sqrt{6y}-y\sqrt{2}}{6-2y}$

61. $\dfrac{\sqrt{5}+\sqrt{3}}{\sqrt{5}-\sqrt{3}} = \dfrac{\sqrt{5}+\sqrt{3}}{\sqrt{5}-\sqrt{3}} \cdot \dfrac{\sqrt{5}+\sqrt{3}}{\sqrt{5}+\sqrt{3}}$

$= \dfrac{5+2\sqrt{15}+3}{5-3}$

$= \dfrac{8+2\sqrt{15}}{2}$

$= \dfrac{2\left(4+\sqrt{15}\right)}{2}$

$= 4+\sqrt{15}$

63. $\dfrac{\sqrt{3x}-2\sqrt{y}}{\sqrt{3x}+\sqrt{y}} = \dfrac{\sqrt{3x}-2\sqrt{y}}{\sqrt{3x}+\sqrt{y}} \cdot \dfrac{\sqrt{3x}-\sqrt{y}}{\sqrt{3x}-\sqrt{y}}$

$= \dfrac{3x-3\sqrt{3xy}+2y}{3x-y}$

65. $2\sqrt{32}-\sqrt{72}+3\sqrt{18}$

$= 2\sqrt{16\cdot 2}-\sqrt{36\cdot 2}+3\sqrt{9\cdot 2}$

$= 8\sqrt{2}-6\sqrt{2}+9\sqrt{2}$

$= 11\sqrt{2}$

67. $\left(3\sqrt{2}-5\sqrt{3}\right)\left(\sqrt{2}+2\sqrt{3}\right)$

$= 3\cdot 2+6\sqrt{6}-5\sqrt{6}-10\cdot 3$

$= 6+\sqrt{6}-30$

$= -24+\sqrt{6}$

69. $\dfrac{9}{\sqrt{8x}} = \dfrac{9}{\sqrt{8x}} \cdot \dfrac{\sqrt{8x}}{\sqrt{8x}} = \dfrac{9\sqrt{4\cdot 2x}}{8x} = \dfrac{18\sqrt{2x}}{8x} = \dfrac{9\sqrt{2x}}{4x}$

71. $\dfrac{\sqrt{5}+1}{\sqrt{5}+2} = \dfrac{\sqrt{5}+1}{\sqrt{5}+2} \cdot \dfrac{\sqrt{5}-2}{\sqrt{5}-2}$

$= \dfrac{5-2\sqrt{5}+\sqrt{5}-2}{5-4}$

$= 5-2\sqrt{5}+\sqrt{5}-2$

$= 3-\sqrt{5}$

73. $\dfrac{\sqrt{6}}{2\sqrt{3}-\sqrt{2}} = 1.194938299...$

$\dfrac{\sqrt{3}+3\sqrt{2}}{5} = 1.194938299...$

The decimal approximations are the same. The student worked correctly.

75. $\dfrac{\sqrt{2}+3\sqrt{5}}{4} = \dfrac{\sqrt{2}+3\sqrt{5}}{4} \cdot \dfrac{\sqrt{2}-3\sqrt{5}}{\sqrt{2}-3\sqrt{5}}$

$= \dfrac{2-9(5)}{4\left(\sqrt{2}-3\sqrt{5}\right)}$

$= \dfrac{-43}{4\left(\sqrt{2}-3\sqrt{5}\right)}$

77. $C = 0.25\dfrac{\left(10+\sqrt{15}\right)\left(\sqrt{60}\right)}{2} = 13.43$

Cost to fertilize lawn is \$13.43.

79. $A = LW = \left(\sqrt{x}+5\right)\left(\sqrt{x}+3\right)$

$A = x+8\sqrt{x}+15$

The area is $\left(x+8\sqrt{x}+15\right)$ mm^2.

Cumulative Review

81. $-2x+3y=21$ (1)

$3x+2y=1$ (2)

Multiply (1) by 3, (2) by 2 to eliminate x.

$-6x+9y=63$

$\underline{6x+4y=2}$

$13y=65$

$y=5$

Substitute 5 for y in (2).

$3x+2(5)=1$

$3x=-9$

$x=-3$

$(-3, 5)$

82. $2x+3y-z=8$ (1)

$-x+2y+3z=-14$ (2)

$3x-y-z=10$ (3)

Solve (3) for z.

$z = 3x - y - 10$ (4)

Substitute $3x - y - 10$ for z in (1), (2).

$2x+3y-(3x-y-10)=8$

$-x+2y+3(3x-y-10)=-14$

Simplify above equations.

$-x+4y=-2$ (5)
$8x-y=16$ (6)
Multiply top equation by 8 to eliminate x.

$$\begin{array}{r} -8x+32y=-16 \\ 8x-y=16 \\ \hline 31y=0 \\ y=0 \end{array}$$

Substitute 0 for y in (5).
$-x+4(0)=-2$
$x=2$
Substitute 2 for x, 0 for y in (4).
$z=(3)(2)-0-10=-4$
(2, 0, −4)

83. 15% + 31% + 26% = 72%
72% of the rings sold for $23,000 or less.

84. 0.85(85,000) = 72,250
72,250 cost more than $5000.

Quick Quiz 7.4

1. $\left(2\sqrt{3}-\sqrt{5}\right)\left(3\sqrt{3}+2\sqrt{5}\right)$
$=6(3)+4\sqrt{15}-3\sqrt{15}-2(5)$
$=18+\sqrt{15}-10$
$=8+\sqrt{15}$

2. $\dfrac{9}{\sqrt{3x}}=\dfrac{9}{\sqrt{3x}}\cdot\dfrac{\sqrt{3x}}{\sqrt{3x}}=\dfrac{9\sqrt{3x}}{3x}=\dfrac{3\sqrt{3x}}{x}$

3. $\dfrac{1+2\sqrt{5}}{4-\sqrt{5}}=\dfrac{1+2\sqrt{5}}{4-\sqrt{5}}\cdot\dfrac{4+\sqrt{5}}{4+\sqrt{5}}$
$=\dfrac{4+\sqrt{5}+8\sqrt{5}+2(5)}{16-5}$
$=\dfrac{14+9\sqrt{5}}{11}$

4. Answers may vary. Possible solution:
To rationalize the denominator, the numerator and the denominator must be multiplied by the conjugate of the denominator. In this case, the conjugate of the denominator is $3\sqrt{2}+2\sqrt{3}$. The product will not contain a radical in the denominator.

How Am I Doing? Sections 7.1–7.4

1. $(-3x^{1/4}y^{1/2})(-2x^{-1/2}y^{1/3})=6x^{\frac{1}{4}-\frac{1}{2}}y^{\frac{1}{2}+\frac{1}{3}}$
$=6x^{-1/4}y^{5/6}$
$=\dfrac{6y^{5/6}}{x^{1/4}}$

2. $(-4x^{-1/4}y^{1/3})^3=(-4)^3x^{-3/4}y^{3/3}=-\dfrac{64y}{x^{3/4}}$

3. $\dfrac{-18x^{-2}y^2}{-3x^{-5}y^{1/3}}=6x^{-2+5}y^{2-\frac{1}{3}}=6x^3y^{5/3}$

4. $\left(\dfrac{27x^2y^{-5}}{x^{-4}y^4}\right)^{2/3}=\left(\dfrac{3^3x^6}{y^9}\right)^{2/3}$
$=\left[\left(\dfrac{3^3x^6}{y^9}\right)^{1/3}\right]^2$
$=\left(\dfrac{3x^2}{y^3}\right)^2$
$=\dfrac{3^2x^4}{y^6}$
$=\dfrac{9x^4}{y^6}$

5. $27^{-4/3}=\dfrac{1}{(3^3)^{4/3}}=\dfrac{1}{3^4}=\dfrac{1}{81}$

6. $\sqrt[5]{-243}=\sqrt[5]{(-3)^5}=-3$

7. $\sqrt{169}+\sqrt[3]{-64}=13-4=9$

8. $\sqrt{64a^8y^{16}}=\sqrt{(8a^4y^8)^2}=8a^4y^8$

9. $\sqrt[3]{27a^{12}b^6c^{15}}=\sqrt[3]{27}\sqrt[3]{a^{12}}\sqrt[3]{b^6}\sqrt[3]{c^{15}}=3a^4b^2c^5$

10. $\left(\sqrt[6]{4x}\right)^5=[(4x)^{1/6}]^5=(4x)^{5/6}$

11. $\sqrt[4]{16x^{20}y^{28}}=\sqrt[4]{(2x^5y^7)^4}=2x^5y^7$

12. $\sqrt[3]{32x^8y^{15}} = \sqrt[3]{8 \cdot 4x^2x^6y^{15}}$
$= \sqrt[3]{8}\sqrt[3]{x^6}\sqrt[3]{y^{15}}\sqrt[3]{4x^2}$
$= 2x^2y^5\sqrt[3]{4x^2}$

13. $\sqrt{44} - 2\sqrt{99} + 7\sqrt{11} = \sqrt{4 \cdot 11} - 2\sqrt{9 \cdot 11} + 7\sqrt{11}$
$= 2\sqrt{11} - 2 \cdot 3\sqrt{11} + 7\sqrt{11}$
$= 2\sqrt{11} - 6\sqrt{11} + 7\sqrt{11}$
$= 3\sqrt{11}$

14. $3\sqrt{48y^3} - 2\sqrt[3]{16} + 3\sqrt[3]{54} - 5y\sqrt{12y}$
$= 3\sqrt{16y^2 \cdot 3y} - 2\sqrt[3]{8 \cdot 2} + 3\sqrt[3]{27 \cdot 2} - 5y\sqrt{4 \cdot 3y}$
$= 12y\sqrt{3y} - 4\sqrt[3]{2} + 9\sqrt[3]{2} - 10y\sqrt{3y}$
$= 2y\sqrt{3y} + 5\sqrt[3]{2}$

15. $\left(5\sqrt{2} - 3\sqrt{5}\right)\left(\sqrt{8} + 3\sqrt{5}\right)$
$= 5\sqrt{16} + 15\sqrt{10} - 3\sqrt{40} - 9(5)$
$= 20 + 15\sqrt{10} - 6\sqrt{10} - 45$
$= -25 + 9\sqrt{10}$

16. $\frac{5}{\sqrt{18x}} = \frac{5}{\sqrt{18x}} \cdot \frac{\sqrt{18x}}{\sqrt{18x}}$
$= \frac{5\sqrt{18x}}{18x}$
$= \frac{5\sqrt{9 \cdot 2x}}{18x}$
$= \frac{15\sqrt{2x}}{18x}$
$= \frac{5\sqrt{2x}}{6x}$

17. $\frac{\sqrt{2}+\sqrt{3}}{\sqrt{2}-\sqrt{3}} = \frac{\sqrt{2}+\sqrt{3}}{\sqrt{2}-\sqrt{3}} \cdot \frac{\sqrt{2}+\sqrt{3}}{\sqrt{2}+\sqrt{3}}$
$= \frac{2 + 2\sqrt{6} + 3}{2 - 3}$
$= \frac{5 + 2\sqrt{6}}{-1}$
$= -5 - 2\sqrt{6}$

7.5 Exercises

1. Isolate one of the radicals on one side of the equation.

3. $\sqrt{8x+1} = 5$
$\left(\sqrt{8x+1}\right)^2 = 5^2$
$8x + 1 = 25$
$8x = 24$
$x = 3$
Check: $\sqrt{8(3)+1} \stackrel{?}{=} 5,\ 5 = 5$
$x = 3$ is the solution.

5. $\sqrt{7x-3} - 2 = 0$
$\sqrt{7x-3} = 2$
$\left(\sqrt{7x-3}\right)^2 = 2^2$
$7x - 3 = 4$
$7x = 7$
$x = 1$
Check: $\sqrt{7(1)-3} - 2 \stackrel{?}{=} 0,\ 0 = 0$
$x = 1$ is the solution.

7. $y + 1 = \sqrt{5y-1}$
$(y+1)^2 = \left(\sqrt{5y-1}\right)^2$
$y^2 + 2y + 1 = 5y - 1$
$y^2 - 3y + 2 = 0$
$(y-2)(y-1) = 0$
$y = 2 \qquad y = 1$
Check: $2 + 1 \stackrel{?}{=} \sqrt{5(2)-1},\ 3 = 3$
$1 + 1 \stackrel{?}{=} \sqrt{5(1)-1},\ 2 = 2$
$y = 2,\ y = 1$ is the solution.

9. $2x = \sqrt{11x+3}$
$(2x)^2 = \left(\sqrt{11x+3}\right)^2$
$4x^2 = 11x + 3$
$4x^2 - 11x - 3 = 0$
$(x-3)(4x+1) = 0$
$x - 3 = 0 \qquad 4x + 1 = 0$
$x = 3 \qquad 4x = -1$
$x = -\frac{1}{4}$

Check:
$2(3) \stackrel{?}{=} \sqrt{11 \cdot 3 + 3}$
$6 = 6$
$2\left(-\frac{1}{4}\right) \stackrel{?}{=} \sqrt{11 \cdot \left(-\frac{1}{4}\right) + 3}$
$-\frac{1}{2} \neq \frac{1}{2}$
$x = 3$ only

11. $2 = 5 + \sqrt{2x+1} \Rightarrow \sqrt{2x+1} = -3$
No solution since $\sqrt{2x+1} \ge 0$.

13. $y - \sqrt{y-3} = 5$
$y - 5 = \sqrt{y-3}$
$(y-5)^2 = \left(\sqrt{y-3}\right)^2$
$y^2 - 10y + 25 = y - 3$
$y^2 - 11y + 28 = 0$
$(y-7)(y-4) = 0$
$y = 7, y = -4$
Check: $4 - \sqrt{4-3} \stackrel{?}{=} 5, 3 \ne 5$
$7 - \sqrt{7-3} \stackrel{?}{=} 5, 5 = 5$
$x = 7$ is the only solution.

15. $\sqrt{y+1} - 1 = y$
$\sqrt{y+1} = y + 1$
$\left(\sqrt{y+1}\right)^2 = (y+1)^2$
$y + 1 = y^2 + 2y + 1$
$y^2 + y = 0$
$y(y+1) = 0$
$y = 0, y = -1$
Check: $\sqrt{0+1} - 1 \stackrel{?}{=} 0, 0 = 0$
$\sqrt{-1+1} - 1 \stackrel{?}{=} -1, -1 = -1$
$y = 0, y = -1$ is the solution.

17. $x - 2\sqrt{x-3} = 3$
$x - 3 = 2\sqrt{x-3}$
$(x-3)^2 = \left(2\sqrt{x-3}\right)^2$
$x^2 - 6x + 9 = 4x - 12$
$x^2 - 10x + 21 = 0$
$(x-7)(x-3) = 0$
$x = 7, x = 3$
Check: $7 - 2\sqrt{7-3} \stackrel{?}{=} 3, 3 = 3$
$3 - 2\sqrt{3-3} \stackrel{?}{=} 3, 3 = 3$
$x = 7, x = 3$ is the solution.

19. $\sqrt{3x^2 - x} = x$
$\left(\sqrt{3x^2 - x}\right)^2 = x^2$
$3x^2 - x = x^2$
$2x^2 - x = 0$
$x(2x-1) = 0$
$x = 0, x = \frac{1}{2}$
Check: $\sqrt{3(0)^2 - 0} \stackrel{?}{=} 0, 0 = 0$
$\sqrt{3\left(\frac{1}{2}\right)^2 - \frac{1}{2}} \stackrel{?}{=} \frac{1}{2}, \frac{1}{2} = \frac{1}{2}$
$x = 0, x = \frac{1}{2}$ is the solution.

21. $\sqrt[3]{2x+3} = 2$
$\left(\sqrt[3]{2x+3}\right)^3 = 2^3$
$2x + 3 = 8$
$2x = 5$
$x = \frac{5}{2}$
Check: $\sqrt[3]{2\left(\frac{5}{2}\right) + 3} \stackrel{?}{=} 2$
$2 = 2$

23. $\sqrt[3]{4x-1} = 3$
$\left(\sqrt[3]{4x-1}\right)^3 = 3^3$
$4x - 1 = 27$
$4x = 28$
$x = 7$
Check: $\sqrt[3]{4(7)-1} \stackrel{?}{=} 3, 3 = 3$
$x = 7$ is the solution.

25. $\sqrt{x+4} = 1 + \sqrt{x-3}$
$\left(\sqrt{x+4}\right)^2 = \left(1 + \sqrt{x-3}\right)^2$
$x + 4 = 1 + 2\sqrt{x-3} + x - 3$
$6 = 2\sqrt{x-3}$
$3 = \sqrt{x-3}$
$3^2 = \left(\sqrt{x-3}\right)^2$
$9 = x - 3$
$x = 12$
Check: $\sqrt{12+4} \stackrel{?}{=} 1 + \sqrt{12-3}, 4 = 4$
$x = 12$ is the solution.

27. $\sqrt{7x+1}=1+\sqrt{5x}$

$$\left(\sqrt{7x+1}\right)^2=\left(1+\sqrt{5x}\right)^2$$
$$7x+1=1+2\sqrt{5x}+5x$$
$$2x=2\sqrt{5x}$$
$$x=\sqrt{5x}$$
$$x^2=\left(\sqrt{5x}\right)^2$$
$$x^2=5x$$
$$x^2-5x=0$$
$$x(x-5)=0$$
$x=0,\ x=5$

Check:

$$\sqrt{7(0)+1}\overset{?}{=}1+\sqrt{5\cdot 0}$$
$$1=1$$
$$\sqrt{7\cdot 5+1}\overset{?}{=}1+\sqrt{5\cdot 5}$$
$$6=6$$

29. $\sqrt{x+6}=1+\sqrt{x+2}$

$$\left(\sqrt{x+6}\right)^2=\left(1+\sqrt{x+2}\right)^2$$
$$x+6=1+2\sqrt{x+2}+x+2$$
$$9=4(x+2)$$
$$4x+8=9$$
$$4x=1$$
$$x=\frac{1}{4}$$

Check: $\sqrt{\frac{1}{4}+6}\overset{?}{=}1+\sqrt{\frac{1}{4}+2},\ \frac{5}{2}=\frac{5}{2}$

$x=\frac{1}{4}$ is the solution.

31. $\sqrt{6x+6}=1+\sqrt{4x+5}$

$$\left(\sqrt{6x+6}\right)^2=\left(1+\sqrt{4x+5}\right)^2$$
$$6x+6=1+2\sqrt{4x+5}+4x+5$$
$$x^2=4x+5$$
$$x^2-4x-5=0$$
$$(x-5)(x+1)=0$$
$x=5,\ x=-1$

Check: $\sqrt{6(5)+6}\overset{?}{=}1+\sqrt{4(5)+5},\ 6=6$

$\sqrt{6(-1)+6}\overset{?}{=}1+\sqrt{4(-1)+5},\ 0\neq 2$

$x=5$ is the only solution.

33. $\sqrt{2x+9}-\sqrt{x+1}=2$

$$\sqrt{2x+9}=2+\sqrt{x+1}$$
$$\left(\sqrt{2x+9}\right)^2=\left(2+\sqrt{x+1}\right)^2$$
$$2x+9=4+4\sqrt{x+1}+x+1$$
$$x+4=4\sqrt{x+1}$$
$$(x+4)^2=\left(4\sqrt{x+1}\right)^2$$
$$x^2+8x+16=16x+16$$
$$x^2-8x=0$$
$$x(x-8)=0$$
$x=0,\ x=8$

Check: $\sqrt{2(0)+9}-\sqrt{0+1}\overset{?}{=}2,\ 2=2$

$\sqrt{2(8)+9}-\sqrt{8+1}\overset{?}{=}2,\ 2=2$

$x=0,\ x=8$ is the solution.

35. $\sqrt{4x+6}=\sqrt{x+1}-\sqrt{x+5}$

$$\left(\sqrt{4x+6}\right)^2=\left(\sqrt{x+1}-\sqrt{x+5}\right)^2$$
$$4x+6=x+1-2\sqrt{x+1}\sqrt{x+5}+x+5$$
$$2x=2\sqrt{x^2+6x+5}$$
$$x=\sqrt{x^2+6x+5}$$
$$x^2=\left(\sqrt{x^2+6x+5}\right)^2$$
$$x^2=x^2+6x+5$$
$$6x+5=0$$
$$x=-\frac{5}{6}$$

Check: $\sqrt{4\left(-\frac{5}{6}\right)+6}\overset{?}{=}\sqrt{-\frac{5}{6}+1}-\sqrt{-\frac{5}{6}+5}$

$$\frac{4\sqrt{6}}{6}\neq\frac{-4\sqrt{6}}{6}$$

No solution

37. $2\sqrt{x}-\sqrt{x-5}=\sqrt{2x-2}$

$$\left(2\sqrt{x}-\sqrt{x-5}\right)^2=\left(\sqrt{2x-2}\right)^2$$
$$4x-4\sqrt{x}\sqrt{x-5}+x-5=2x-2$$
$$4\sqrt{x}\sqrt{x-5}=3x-3$$
$$\left(4\sqrt{x}\sqrt{x-5}\right)^2=(3x-3)^2$$
$$16x^2-80x=9x^2-18x+9$$
$$7x^2-62x-9=0$$
$$(7x+1)(x-9)=0$$
$$x=-\frac{1}{7},\ x=9$$

$x=-\frac{1}{7}$ does not check since it gives the square root of a negative.

Check: $2\sqrt{9}-\sqrt{9-5} \stackrel{?}{=} \sqrt{2(9)-2},\ 4=4$

$x=9$ is the solution.

39.
$$x=\sqrt{5.326x-1.983}$$
$$x^2=\left(\sqrt{5.326x-1.983}\right)^2$$
$$x^2=5.326x-1.983$$

$x^2-5.326x+1.983=0$

$x=0.40279$ or $x=4.92321$

Check:

$0.40279 \stackrel{?}{=} \sqrt{5.326(0.40279)-1.983}$

$0.4028=0.4028$

$4.92321 \stackrel{?}{=} \sqrt{5.326(4.92321)-1.983}$

$4.9232=4.9232$

$x=4.9232,\ x=0.4028$ is the solution to four decimal places.

41. a. $V=2\sqrt{3S} \Rightarrow V^2=4(3S) \Rightarrow S=\frac{V^2}{12}$

b. $S=\frac{V^2}{12}=\frac{(30)^2}{12}=75$

The skid mark is 75 feet.

43.
$$0.11y+1.25=\sqrt{3.7625+0.22x}$$
$$(0.11y+1.25)^2=\left(\sqrt{3.7625+0.22x}\right)^2$$
$$0.0121y^2+0.275y+1.5625=3.7625+0.22x$$
$$0.22x=0.0121y^2+0.275y-2.2$$
$$x=0.055y^2+1.25y-10$$

45. $\sqrt{x^2-3x+c}=x-2$

$x=3$: $\sqrt{3^2-3(3)+c}=3-2$

$\sqrt{c}=1$

$c=1$

Cumulative Review

47. $(4^3x^6)^{2/3}=4^{3\cdot2/3}x^{6\cdot2/3}=4^2x^4=16x^4$

48. $(2^{-3}x^{-6})^{1/3}=2^{-3\cdot\frac{1}{3}}x^{-6\cdot\frac{1}{3}}=2^{-1}x^{-2}=\frac{1}{2x^2}$

49. $\sqrt[3]{-216x^6y^9}=\sqrt[3]{(-6x^2y^3)^3}=-6x^2y^3$

50. $\sqrt[5]{-32x^{15}y^5}=\sqrt[5]{(-2)^5(x^3)^5y^5}=-2x^3y$

51. w = speed of current

$$(12+w)\cdot 4=(12-w)\cdot 5$$
$$48+4w=60-5w$$
$$9w=12$$
$$w=\frac{12}{9}\approx 1.33$$

The speed of the current is approximately 1.33 mph.

52. d = dogs
c = cats

$$c+d=28 \quad (1)$$
$$55c+68d=1748 \quad (2)$$

Solve (1) for c.
$c=28-d$ (3)
Substitute $28-d$ for c in (2).

$$55(28-d)+68d=1748$$
$$13d=208$$
$$d=16$$

Substitute 16 for d in (3).
$c=28-16=12$
There were 16 dogs and 12 cats examined.

Quick Quiz 7.5

1.
$$\sqrt{5x-4}=x$$
$$\left(\sqrt{5x-4}\right)^2=x^2$$
$$5x-4=x^2$$
$$x^2-5x+4=0$$
$$(x-4)(x-1)=0$$
$$x-4=0 \qquad x-1=0$$
$$x=4 \qquad x=1$$

Check:
$$\sqrt{5(4)-4}\stackrel{?}{=}4$$
$$4=4$$
$$\sqrt{5(1)-4}\stackrel{?}{=}1$$
$$1=1$$

2.
$$x=3-\sqrt{2x-3}$$
$$x-3=-\sqrt{2x-3}$$
$$3-x=\sqrt{2x-3}$$
$$(3-x)^2=\left(\sqrt{2x-3}\right)^2$$
$$9-6x+x^2=2x-3$$
$$x^2-8x+12=0$$
$$(x-2)(x-6)=0$$
$$x-2=0 \qquad x-6=0$$
$$x=2 \qquad x=6$$

Check:
$$2\stackrel{?}{=}3-\sqrt{2(2)-3}$$
$$2\stackrel{?}{=}3-1$$
$$2=2$$
$$6\stackrel{?}{=}3-\sqrt{2(6)-3}$$
$$6\stackrel{?}{=}3-3$$
$$6\neq 0$$

$x=2$ is the only solution.

3.
$$4-\sqrt{x-4}=\sqrt{2x-1}$$
$$\left(4-\sqrt{x-4}\right)^2=\left(\sqrt{2x-1}\right)^2$$
$$16-8\sqrt{x-4}+x-4=2x-1$$
$$8\sqrt{x-4}=13-x$$
$$\left(8\sqrt{x-4}\right)^2=(13-x)^2$$
$$64(x-4)=169-26x+x^2$$
$$64x-256=169-26x+x^2$$
$$x^2-90x+425=0$$
$$(x-85)(x-5)=0$$
$$x-85=0 \qquad x-5=0$$
$$x=85 \qquad x=5$$

Check:
$$4-\sqrt{85-4}\stackrel{?}{=}\sqrt{2(85)-1}$$
$$4-9\stackrel{?}{=}\sqrt{169}$$
$$-5\neq 13$$
$$4-\sqrt{5-4}\stackrel{?}{=}\sqrt{2(5)-1}$$
$$4-1\stackrel{?}{=}\sqrt{9}$$
$$3=3$$

$x=5$ is the only solution.

4. Answers may vary. Possible solution: Substitute the found values for x back into the original equation to test for validity.

7.6 Exercises

1. No; there is no real number that, when squared, will equal –9.

3. No; to be equal, the real number parts must be equal, and the imaginary number parts must be equal. $2\neq 3$ and $3i\neq 2i$.

5. $\sqrt{-25}=\sqrt{25}\sqrt{-1}=5i$

7. $\sqrt{-50}=\sqrt{25\cdot 2}\sqrt{-1}=5i\sqrt{2}$

9. $\sqrt{-\frac{25}{4}}=\sqrt{\frac{25}{4}}\sqrt{-1}=\frac{5}{2}i$

11. $-\sqrt{-81} = -\sqrt{81}\sqrt{-1} = -9i$

13. $2+\sqrt{-3} = 2+\sqrt{3}\sqrt{-1} = 2+i\sqrt{3}$

15. $-2.8+\sqrt{-16} = -2.8+\sqrt{16}\sqrt{-1} = -2.8+4i$

17. $-3+\sqrt{-24} = -3+\sqrt{4\cdot 6}\sqrt{-1} = -3+2i\sqrt{6}$

19. $$\begin{aligned}\left(\sqrt{-5}\right)\left(\sqrt{-2}\right) &= \sqrt{5}\sqrt{-1}\sqrt{2}\sqrt{-1}\\ &= i\sqrt{5}\cdot i\sqrt{2}\\ &= i^2\sqrt{10}\\ &= -\sqrt{10}\end{aligned}$$

21. $$\begin{aligned}\left(\sqrt{-36}\right)\left(\sqrt{-4}\right) &= \left(\sqrt{36}\sqrt{-1}\right)\left(\sqrt{4}\sqrt{-1}\right)\\ &= (6i)(2i)\\ &= 12i^2\\ &= -12\end{aligned}$$

23. $x-3i = 5+yi,\ x=5$
$y = -3$

25. $1.3-2.5yi = x-5i,\ x = 1.3$
$-2.5y = -5$
$y = 2$

27. $23+yi = 17-x+3i$
$23 = 17-x$
$x = -6$
$y = 3$

29. $(1+8i)+(-6+3i) = 1-6+(8+3)i = -5+11i$

31. $$\left(-\frac{3}{2}+\frac{1}{2}i\right)+\left(\frac{5}{2}-\frac{3}{2}i\right) = -\frac{3}{2}+\frac{5}{2}+\left(\frac{1}{2}-\frac{3}{2}\right)i = 1-i$$

33. $$\begin{aligned}&(2.8-0.7i)-(1.6-2.8i)\\ &= 2.8-1.6+(-0.7+2.8)i\\ &= 1.2+2.1i\end{aligned}$$

35. $(2i)(7i) = 14i^2 = -14$

37. $(-7i)(6i) = -42i^2 = (-42)(-1) = 42$

39. $$\begin{aligned}(2+3i)(2-i) &= 4-2i+6i-3i^2\\ &= 4+4i-3(-1)\\ &= 4+4i+3\\ &= 7+4i\end{aligned}$$

41. $9i-3(-2+i) = 9i+6-3i = 6+6i$

43. $2i(5i-6) = 10i^2-12i = -10-12i$

45. $$\left(\frac{1}{2}+i\right)^2 = \frac{1}{4}+\frac{1}{2}i+\frac{1}{2}i+i^2 = \frac{1}{4}+i-1 = -\frac{3}{4}+i$$

47. $\left(i\sqrt{3}\right)\left(i\sqrt{7}\right) = i^2\sqrt{21} = -\sqrt{21}$

49. $$\begin{aligned}\left(3+\sqrt{-2}\right)\left(4+\sqrt{-5}\right) &= \left(3+i\sqrt{2}\right)\left(4+i\sqrt{5}\right)\\ &= 12+3i\sqrt{5}+4i\sqrt{2}+i^2\sqrt{10}\\ &= 12+3i\sqrt{5}+4i\sqrt{2}-\sqrt{10}\\ &= 12-\sqrt{10}+\left(3\sqrt{5}+4\sqrt{2}\right)i\end{aligned}$$

51. $i^{17} = (i^4)^4\cdot i = 1^4\cdot i = i$

53. $i^{24} = (i^4)^6 = 1^6 = 1$

55. $i^{46} = (i^4)^{11}\cdot i^2 = 1^{11}(-1) = -1$

57. $i^{37} = i^{36}i = (i^4)^9 i = 1^9 i = i$

59. $$\begin{aligned}i^{30}+i^{28} &= (i^4)^7\cdot i^2+(i^4)^7\\ &= 1^7(-1)+1^7\\ &= -1+1\\ &= 0\end{aligned}$$

61. $i^{100}-i^7 = (i^4)^{25}-i^4i^3 = 1^{25}-1\cdot(-i) = 1+i$

63. $$\begin{aligned}\frac{2+i}{3-i} &= \frac{2+i}{3-i}\cdot\frac{3+i}{3+i}\\ &= \frac{6+5i+i^2}{9+1}\\ &= \frac{6+5i-1}{10}\\ &= \frac{5(1+i)}{10}\\ &= \frac{1+i}{2}\end{aligned}$$

65. $\frac{2i}{3+3i}=\frac{2i}{3+3i}\cdot\frac{3-3i}{3-3i}$
$=\frac{6i-6i^2}{9-9i^2}$
$=\frac{6i+6}{9+9}$
$=\frac{6+6i}{18}$
$=\frac{1+i}{3}$

67. $\frac{5-2i}{6i}=\frac{5-2i}{6i}\cdot\frac{-6i}{-6i}$
$=\frac{-30i+12i^2}{-36i^2}$
$=\frac{-12-30i}{36}$
$=-\frac{2+5i}{6}$

69. $\frac{2}{i}=\frac{2}{i}\cdot\frac{i}{i}=\frac{2i}{i^2}=\frac{2i}{-1}=-2i$

71. $\frac{7}{5-6i}=\frac{7}{5-6i}\cdot\frac{5+6i}{5+6i}=\frac{35+42i}{25+36}=\frac{35+42i}{61}$

73. $\frac{5-2i}{3+2i}=\frac{5-2i}{3+2i}\cdot\frac{3-2i}{3-2i}$
$=\frac{15-10i-6i+4i^2}{9+4}$
$=\frac{15-16i-4}{13}$
$=\frac{11-16i}{13}$

75. $\sqrt{-98}=\sqrt{98}\sqrt{-1}=\sqrt{49}\sqrt{2}\sqrt{-1}=7i\sqrt{2}$

77. $(8-5i)-(-1+3i)=8-5i+1-3i=9-8i$

79. $(3i-1)(5i-3)=15i^2-9i-5i+3$
$=15i^2-14i+3$
$=-15-14i+3$
$=-12-14i$

81. $\frac{2-3i}{2+i}=\frac{2-3i}{2+i}\cdot\frac{2-i}{2-i}$
$=\frac{4-2i-6i+3i^2}{4+1}$
$=\frac{4-8i-3}{5}$
$=\frac{1-8i}{5}$

83. $Z=\frac{V}{I}$
$=\frac{3+2i}{3i}$
$=\frac{3+2i}{3i}\cdot\frac{-3i}{-3i}$
$=\frac{-9i-6i^2}{9}$
$=\frac{6-9i}{9}$
$=\frac{2-3i}{3}$

Cumulative Review

85. $x+3+2x-5+4x+2=105$
$7x=105$
$x=15$

$x + 3 = 18$
$2x - 5 = 25$
$4x + 2 = 62$
18 hours producing juice in glass bottles, 25 hours producing juice in cans, and 62 hours producing juice in plastic bottles.

86. $C = 60[1850(0.93) - 120] = 96{,}030$
The net cost to the bank is $96,030.

Quick Quiz 7.6

1. $(6-7i)(3+2i)=18+12i-21i-14i^2$
$=18-9i+14$
$=32-9i$

2. $\frac{4+3i}{1-2i}=\frac{4+3i}{1-2i}\cdot\frac{1+2i}{1+2i}$
$=\frac{4+8i+3i+6i^2}{1-4i^2}$
$=\frac{4+11i-6}{1+4}$
$=\frac{-2+11i}{5}$

3. $i^{33} = i \cdot i^{32} = i \cdot (i^4)^8 = i(1)^8 = i \cdot (1) = i$

4. Answers may vary. Possible solution:
Multiply by FOIL method. Replace i^2 with -1.
Combine like terms.

7.7 Exercises

1. Answers may vary. A person's weekly paycheck varies as the number of hours worked, $y = kx$ where y is the weekly salary, k is the hourly salary, and x is the number of hours worked.

3. $y = \frac{k}{x}$

5. $y = kx,\ 15 = k \cdot 40,\ k = \frac{3}{8}$

$y = \frac{3}{8}x,\ y = \frac{3}{8} \cdot 64 = 24$

7. $p = kd,\ 21 = k \cdot 50,\ k = \frac{21}{50}$

$p = \frac{21}{50}d,\ p = \frac{21}{50} \cdot 170 = 71.4$

The pressure would be 71.4 pounds per square inch.

9. $d = ks^2,\ 40 = k \cdot 30^2,\ k = \frac{2}{45}$

$d = \frac{2}{45}s^2,\ d = \frac{2}{45} \cdot 60^2 = 160$

It will take 160 feet to stop.

11. $y = \frac{k}{x^2},\ 10 = \frac{k}{2^2},\ k = 40$

$y = \frac{40}{x^2},\ y = \frac{40}{0.5^2} = 160$

13. g = gallons sold
p = price in dollars
k = constant

$g = \frac{k}{p}$

When $g = 2800$, $p = \$4.10$.
$k = gp = 2800(4.10) = 11{,}480$
When $p = \$3.90$, $k = 11{,}480$.

$g = \frac{11{,}480}{3.90} \approx 2944$

He could expect to sell approximately 2944 gallons.

15. v = number of volunteers
t = time in hours
k = constant

$t = \frac{k}{v}$

When $v = 39$, $t = 6$.
$k = tv = 39(6) = 234$
When $v = 60$, $k = 234$.

$t = \frac{234}{60} = 3.9$

Cleanup will take 3.9 hours.

17. $w = \frac{k}{l},\ 900 = \frac{k}{8},\ k = 7200$

$w = \frac{7200}{l},\ w = \frac{7200}{18} = 400$

The beam can safely support 400 pounds.

19. t = time in minutes
r = radius of pipe in inches
k = constant

$t = \frac{k}{r^2}$

When $r = 2.5$, $t = 6$.
$k = tr^2 = 6(2.5)^2 = 37.5$
When $r = 3.5$, $k = 37.5$.

$t = \frac{k}{r^2} = \frac{37.5}{(3.5)^2} \approx 3.1$

The tub would fill in approximately 3.1 minutes.

Cumulative Review

21.
$$3x^2 - 8x + 4 = 0$$
$$(3x - 2)(x - 2) = 0$$
$$3x - 2 = 0 \quad \text{or} \quad x - 2 = 0$$
$$x = \frac{2}{3} \qquad\qquad x = 2$$

22.
$$4x^2 = -28x + 32$$
$$4x^2 + 28x - 32 = 0$$
$$x^2 + 7x - 8 = 0$$
$$(x + 8)(x - 1) = 0$$
$$x + 8 = 0 \quad \text{or} \quad x - 1 = 0$$
$$x = -8 \qquad\qquad x = 1$$

23. $488.75 = 1.0625p$
$p = 460$
The original price was \$460.

24. $\frac{7.5}{3}=\frac{x}{22}$
$3x=165$
$x=55$
It will take 55 gallons of paint.

Quick Quiz 7.7

1. $y=\frac{k}{x}$
When $y = 9$, $x = 3$.
$k = yx = 9(3) = 27$
When $x = 6$, $k = 27$.
$y=\frac{k}{x}=\frac{27}{6}=4.5$

2. $y=\frac{kx}{z^2}$
When $x = 3$, $z = 5$, $y = 6$.
$k=\frac{yz^2}{x}=\frac{6(5)^2}{3}=50$
When $x = 6$, $z = 10$, $k = 25$.
$y=\frac{kx}{z^2}=\frac{50(6)}{(10)^2}=3$

3. d = distance to stop in feet
v = speed traveling in miles per hour
$d = kv^2$
When $v = 50$, $d = 80$.
$k=\frac{d}{v^2}=\frac{80}{(50)^2}=0.032$
When $v = 65$, $k = 0.032$.
$d = kv^2 = 0.032(65)^2 = 135.2$
Distance to stop will be 135.2 feet.

4. Answers may vary. Possible solution:
Writing the equation is always the first step. In this case:
$y=\sqrt{xk}$
Next solve the equation for k, and substitute known values of x and y to find the numerical value of k.

Putting Your Skills to Work

1. Total deposits:
200 + 150.50 + 120.25 + 50 + 25 = 545.75
The total of her deposits is \$545.75.

2. Total checks:
238.50 + 75 + 200 + 28.56 + 36 = 578.06
The total of her checks is \$578.06.

3. $578.06 \ge 545.75$
She spent more than she deposited, but the \$300.50 would help to cover her expenses.

4. \$300.50 + \$200 + \$150.50 + \$120.25 + \$50 − (\$238.50 + \$75.00 + \$200.00)
= \$821.25 − \$513.50
= \$307.75
Her balance was \$307.75.

5. \$300.50 + \$545.75 − \$578.06 = \$268.19
Her balance is assumed to be \$268.19.

6. Eventually she will be in debt.

Chapter 7 Review Problems

1. $(3xy^{1/2})(5x^2y^{-3})=(3\cdot 5)x^{1+3}y^{\frac{1}{2}-3}$
$=15x^3y^{-5/2}$
$=\frac{15x^3}{y^{5/2}}$

2. $\frac{3x^{2/3}}{6x^{1/6}}=\frac{x^{\frac{2}{3}-\frac{1}{6}}}{2}=\frac{x^{1/2}}{2}$

3. $(16a^6b^5)^{1/2}=(4a^3b^{5/2})^{2(1/2)}=4a^3b^{5/2}$

4. $3^{1/2}\cdot 3^{1/6}=3^{\frac{3}{6}+\frac{1}{6}}=3^{4/6}=3^{2/3}$

5. $(2a^{1/3}b^{1/4})(-3a^{1/2}b^{1/2})=-6a^{\frac{1}{3}+\frac{1}{2}}b^{\frac{1}{4}+\frac{1}{2}}$
$=-6a^{5/6}b^{3/4}$

6. $\frac{6x^{2/3}y^{1/10}}{12x^{1/6}y^{-1/5}}=\frac{x^{\frac{2}{3}-\frac{1}{6}}y^{\frac{1}{10}+\frac{1}{5}}}{2}=\frac{x^{1/2}y^{3/10}}{2}$

7. $(2x^{-1/5}y^{1/10}z^{4/5})^{-5}=2^{-5}x^{-\frac{1}{5}(-5)}y^{\frac{1}{10}(-5)}z^{\frac{4}{5}(-5)}$
$=\frac{xy^{-1/2}z^{-4}}{2^5}$
$=\frac{x}{32y^{1/2}z^4}$

8. $\left(\dfrac{49a^3b^6}{a^{-7}b^4}\right)^{1/2} = (49a^{10}b^2)^{1/2}$
$= 49^{1/2}a^{10(1/2)}b^{2(1/2)}$
$= 7a^5b$

9. $\dfrac{(x^{3/4}y^{2/5})^{1/2}}{x^{-1/8}} = \dfrac{x^{\frac{3}{4}\cdot\frac{1}{2}}y^{\frac{2}{5}\cdot\frac{1}{2}}}{x^{-1/8}} = x^{\frac{3}{8}+\frac{1}{8}}y^{1/5} = x^{1/2}y^{1/5}$

10. $\left(\dfrac{8a^4}{a^{-2}}\right)^{1/3} = (8a^{(4+2)})^{1/3}$
$= (8a^6)^{1/3}$
$= 8^{1/3}a^{6\cdot\frac{1}{3}}$
$= 2a^2$

11. $(4^{5/3})^{6/5} = 4^{\frac{5\cdot 6}{3\cdot 5}} = 4^{30/15} = 4^2 = 16$

12. $2x^{1/3} + x^{-2/3} = 2x^{1/3} + \dfrac{1}{x^{2/3}}$
$= \dfrac{2x^{1/3}}{1}\cdot\dfrac{x^{2/3}}{x^{2/3}} + \dfrac{1}{x^{2/3}}$
$= \dfrac{2x^{\frac{1}{3}+\frac{2}{3}}+1}{x^{2/3}}$
$= \dfrac{2x+1}{x^{2/3}}$

13. $6x^{3/2} - 9x^{1/2} = 3x\cdot 2x^{1/2} - 3x\cdot 3x^{-1/2}$
$= 3x(2x^{1/2} - 3x^{-1/2})$

14. $-\sqrt{16} = -4$

15. $\sqrt[5]{-32} = \sqrt[5]{(-2)^5} = -2$

16. $\sqrt[6]{-20}$ is not a real number.

17. $-\sqrt{\dfrac{1}{25}} = -\dfrac{1}{5}$

18. $\sqrt{0.04} = 0.2$ because $0.2^2 = 0.04$.

19. $\sqrt[4]{-256}$ is not a real number.

20. $\sqrt[3]{-\dfrac{1}{8}} = -\dfrac{1}{2}$ because $\left(-\dfrac{1}{2}\right)^3 = -\dfrac{1}{8}$.

21. $\sqrt[3]{\dfrac{27}{64}} = \dfrac{3}{4}$ because $\left(\dfrac{3}{4}\right)^3 = \dfrac{27}{64}$.

22. $64^{2/3} = (4^3)^{2/3} = 4^{3\cdot\frac{2}{3}} = 4^2 = 16$

23. $125^{4/3} = (5^3)^{4/3} = 5^{3\cdot\frac{4}{3}} = 5^4 = 625$

24. $\sqrt{49x^4y^{10}z^2} = \sqrt{7^2(x^2)^2(y^5)^2(z)^2} = 7x^2y^5z$

25. $\sqrt[3]{64a^{12}b^{30}} = \sqrt[3]{4^3(a^4)^3(b^{10})^3} = 4a^4b^{10}$

26. $\sqrt[3]{-8a^{12}b^{15}c^{21}} = \sqrt[3]{(-2)^3(a^4)^3(b^5)^3(c^7)^3}$
$= -2a^4b^5c^7$

27. $\sqrt{49x^{22}y^2} = \sqrt{7^2(x^{11})^2y^2} = 7x^{11}y$

28. $\sqrt[5]{a^2} = a^{2/5}$

29. $\sqrt[4]{y^3} = y^{3/4}$

30. $\sqrt{2b} = (2b)^{1/2}$

31. $\sqrt[3]{5a} = (5a)^{1/3}$

32. $\left(\sqrt[5]{xy}\right)^7 = ((xy)^{1/5})^7 = (xy)^{7/5}$

33. $m^{1/2} = \sqrt[2]{m^1} = \sqrt{m}$

34. $n^{1/4} = \sqrt[4]{n^1} = \sqrt[4]{n}$

35. $y^{3/5} = \sqrt[5]{y^3}$

36. $(3z)^{2/3} = \sqrt[3]{(3z)^2} = \sqrt[3]{9z^2}$

37. $(2x)^{3/7} = \sqrt[7]{(2x)^3} = \sqrt[7]{8x^3}$

38. $16^{3/4} = (2^4)^{3/4} = 2^{4\cdot\frac{3}{4}} = 2^3 = 8$

39. $64^{5/6} = (2^6)^{5/6} = 2^{6\cdot\frac{5}{6}} = 2^5 = 32$

40. $(-27)^{2/3} = ((-3)^3)^{2/3} = (-3)^{3\cdot\frac{2}{3}} = (-3)^2 = 9$

41. $(-8)^{1/3}=((-2)^3)^{1/3}=(-2)^{3\cdot\frac{1}{3}}=(-2)^1=-2$

42. $\left(\frac{1}{9}\right)^{1/2}=(3^{-2})^{1/2}=3^{-2\cdot\frac{1}{2}}=3^{-1}=\frac{1}{3^1}=\frac{1}{3}$

43. $(0.49)^{1/2}=((0.7)^2)^{1/2}=0.7^{2\cdot\frac{1}{2}}=0.7^1=0.7$

44. $\left(\frac{1}{16}\right)^{-1/4}=(2^{-4})^{-1/4}=2^{-4\cdot\frac{-1}{4}}=2^1=2$

45. $\left(\frac{1}{36}\right)^{-1/2}=(6^{-2})^{-1/2}=6^{-2\cdot\frac{-1}{2}}=6^1=6$

46.
$$\begin{aligned}(25a^2b^4)^{3/2}&=(5^2)^{3/2}(a^2)^{3/2}(b^4)^{3/2}\\&=5^{2\cdot\frac{3}{2}}a^{2\cdot\frac{3}{2}}b^{4\cdot\frac{3}{2}}\\&=5^3a^3b^6\\&=125a^3b^6\end{aligned}$$

47.
$$\begin{aligned}(4a^6b^2)^{5/2}&=(2^2)^{5/2}(a^6)^{5/2}(b^2)^{5/2}\\&=2^5a^{6\cdot\frac{5}{2}}b^{2\cdot\frac{5}{2}}\\&=32a^{15}b^5\end{aligned}$$

48.
$$\begin{aligned}\sqrt{50}+2\sqrt{32}-\sqrt{8}&=\sqrt{25\cdot 2}+2\sqrt{16\cdot 2}-\sqrt{4\cdot 2}\\&=5\sqrt{2}+8\sqrt{2}-2\sqrt{2}\\&=11\sqrt{2}\end{aligned}$$

49.
$$\begin{aligned}\sqrt{28}-4\sqrt{7}+5\sqrt{63}&=\sqrt{4\cdot 7}-4\sqrt{7}+5\sqrt{9\cdot 7}\\&=2\sqrt{7}-4\sqrt{7}+15\sqrt{7}\\&=13\sqrt{7}\end{aligned}$$

50.
$$\begin{aligned}2\sqrt{12}-\sqrt{48}+5\sqrt{75}&=2\sqrt{4\cdot 3}-\sqrt{16\cdot 3}+5\sqrt{25\cdot 3}\\&=2\cdot 2\sqrt{3}-4\sqrt{3}+5\cdot 5\sqrt{3}\\&=4\sqrt{3}-4\sqrt{3}+25\sqrt{3}\\&=25\sqrt{3}\end{aligned}$$

51.
$$\begin{aligned}\sqrt{125x^3}+x\sqrt{45x}&=\sqrt{25x^2\cdot 5x}+x\sqrt{9\cdot 5x}\\&=5x\sqrt{5x}+3x\sqrt{5x}\\&=8x\sqrt{5x}\end{aligned}$$

52.
$$\begin{aligned}&2\sqrt{32x}-5x\sqrt{2}+\sqrt{18x}\\&=2\sqrt{16\cdot 2x}-5x\sqrt{2}+\sqrt{9\cdot 2x}\\&=8\sqrt{2x}-5x\sqrt{2}+3\sqrt{2x}\\&=11\sqrt{2x}-5x\sqrt{2}\end{aligned}$$

53.
$$\begin{aligned}3\sqrt[3]{16}-4\sqrt[3]{54}&=3\sqrt[3]{2^3\cdot 2}-4\sqrt[3]{3^3\cdot 2}\\&=3\cdot 2\sqrt[3]{2}-4\cdot 3\sqrt[3]{2}\\&=6\sqrt[3]{2}-12\sqrt[3]{2}\\&=-6\sqrt[3]{2}\end{aligned}$$

54.
$$\begin{aligned}\left(5\sqrt{12}\right)\left(3\sqrt{6}\right)&=15\sqrt{72}\\&=15\sqrt{36\cdot 2}\\&=15\cdot 6\sqrt{2}\\&=90\sqrt{2}\end{aligned}$$

55.
$$\begin{aligned}\left(-2\sqrt{15}\right)\left(4x\sqrt{3}\right)&=-8x\sqrt{45}\\&=-8x\sqrt{9\cdot 5}\\&=-24x\sqrt{5}\end{aligned}$$

56.
$$\begin{aligned}3\sqrt{x}\left(2\sqrt{8x}-3\sqrt{48}\right)&=3\sqrt{x}\left(4\sqrt{2x}-12\sqrt{3}\right)\\&=12x\sqrt{2}-36\sqrt{3x}\end{aligned}$$

57.
$$\begin{aligned}\sqrt{3a}\left(4-\sqrt{21a}\right)&=4\sqrt{3a}-\sqrt{3\cdot 21a^2}\\&=4\sqrt{3a}-\sqrt{3^2a^2\cdot 7}\\&=4\sqrt{3a}-3a\sqrt{7}\end{aligned}$$

58.
$$\begin{aligned}-\sqrt{5xy}\left(\sqrt{3x}-\sqrt{5y}\right)&=-\sqrt{15x^2y}+\sqrt{25xy^2}\\&=-x\sqrt{15y}+5y\sqrt{x}\end{aligned}$$

59.
$$\begin{aligned}2\sqrt{7b}\left(\sqrt{ab}-b\sqrt{3bc}\right)&=2\sqrt{7ab^2}-2b\sqrt{21b^2c}\\&=2b\sqrt{7a}-2b^2\sqrt{21c}\end{aligned}$$

60.
$$\begin{aligned}&\left(5\sqrt{2}+\sqrt{3}\right)\left(\sqrt{2}-2\sqrt{3}\right)\\&=5\cdot 2-10\sqrt{6}+\sqrt{6}-2\cdot 3\\&=10-9\sqrt{6}-6\\&=4-9\sqrt{6}\end{aligned}$$

61.
$$\begin{aligned}\left(5\sqrt{6}-2\sqrt{2}\right)\left(\sqrt{6}-\sqrt{2}\right)&=30-7\sqrt{12}+4\\&=34-7\sqrt{4\cdot 3}\\&=34-14\sqrt{3}\end{aligned}$$

62. $\left(2\sqrt{5}-3\sqrt{6}\right)^2=20-12\sqrt{30}+54=74-12\sqrt{30}$

63. $\left(\sqrt[3]{2x}+\sqrt[3]{6}\right)\left(\sqrt[3]{4x^2}-\sqrt[3]{y}\right)$
$= \sqrt[3]{8x^3} - \sqrt[3]{2xy} + \sqrt[3]{24x^2} - \sqrt[3]{6y}$
$= 2x - \sqrt[3]{2xy} + 2\sqrt[3]{3x^2} - \sqrt[3]{6y}$

64. $f(x) = \sqrt{4x+16}$

a. $f(12) = \sqrt{4(12)+16} = \sqrt{64} = 8$

b. Domain is all real numbers x where $4x + 16 \ge 0$ or $x \ge -4$.

65. $f(x) = \sqrt{36-3x}$

a. $f(9) = \sqrt{36-3(9)} = \sqrt{9} = 3$

b. Domain is all real numbers x where $36 - 3x \ge 0$ or $x \le 12$.

66. $f(x) = \sqrt{\frac{3}{4}x - \frac{1}{2}}$

a. $f(1) = \sqrt{\frac{3}{4}\cdot 1 - \frac{1}{2}} = \sqrt{\frac{1}{4}} = \frac{1}{2}$

b. domain: $\frac{3}{4}x - \frac{1}{2} \ge 0$
$3x - 2 \ge 0$
$3x \ge 2$
$x \ge \frac{2}{3}$
The domain is all real numbers where $x \ge \frac{2}{3}$.

67. $\sqrt{\frac{6y^2}{x}} = \frac{\sqrt{6y^2}}{\sqrt{x}} \cdot \frac{\sqrt{x}}{\sqrt{x}} = \frac{\sqrt{6xy^2}}{\sqrt{x^2}} = \frac{y\sqrt{6x}}{x}$

68. $\frac{3}{\sqrt{5y}} = \frac{3}{\sqrt{5y}} \cdot \frac{\sqrt{5y}}{\sqrt{5y}} = \frac{3\sqrt{5y}}{5y}$

69. $\frac{3\sqrt{7x}}{\sqrt{21x}} = \frac{3\sqrt{7x}}{\sqrt{3}\cdot\sqrt{7x}} = \frac{3}{\sqrt{3}} \cdot \frac{\sqrt{3}}{\sqrt{3}} = \frac{3\sqrt{3}}{3} = \sqrt{3}$

70. $\frac{2}{\sqrt{6}-\sqrt{5}} = \frac{2}{\sqrt{6}-\sqrt{5}} \cdot \frac{\sqrt{6}+\sqrt{5}}{\sqrt{6}+\sqrt{5}}$
$= \frac{2\sqrt{6}+2\sqrt{5}}{6-5}$
$= 2\sqrt{6}+2\sqrt{5}$

71. $\frac{\sqrt{x}}{3\sqrt{x}+\sqrt{y}} = \frac{\sqrt{x}}{3\sqrt{x}+\sqrt{y}} \cdot \frac{3\sqrt{x}-\sqrt{y}}{3\sqrt{x}-\sqrt{y}} = \frac{3x-\sqrt{xy}}{9x-y}$

72. $\frac{\sqrt{5}}{\sqrt{7}-3} = \frac{\sqrt{5}}{\sqrt{7}-3} \cdot \frac{\sqrt{7}+3}{\sqrt{7}+3}$
$= \frac{\sqrt{35}+3\sqrt{5}}{7-9}$
$= -\frac{\sqrt{35}+3\sqrt{5}}{2}$

73. $\frac{2\sqrt{3}+\sqrt{6}}{\sqrt{3}+2\sqrt{6}} = \frac{2\sqrt{3}+\sqrt{6}}{\sqrt{3}+2\sqrt{6}} \cdot \frac{\sqrt{3}-2\sqrt{6}}{\sqrt{3}-2\sqrt{6}}$
$= \frac{6-4\sqrt{18}+\sqrt{18}-12}{3-24}$
$= \frac{-6-3\sqrt{9\cdot 2}}{-21}$
$= \frac{-6-9\sqrt{2}}{-21}$
$= \frac{2+3\sqrt{2}}{7}$

74. $\frac{5\sqrt{2}-\sqrt{3}}{\sqrt{6}-\sqrt{3}} = \frac{5\sqrt{2}-\sqrt{3}}{\sqrt{6}-\sqrt{3}} \cdot \frac{\sqrt{6}+\sqrt{3}}{\sqrt{6}+\sqrt{3}}$
$= \frac{5\sqrt{12}+5\sqrt{6}-\sqrt{18}-3}{6-3}$
$= \frac{5\sqrt{4\cdot 3}+5\sqrt{6}-\sqrt{9\cdot 2}-3}{3}$
$= \frac{10\sqrt{3}+5\sqrt{6}-3\sqrt{2}-3}{3}$

75. $\frac{3\sqrt{x}+\sqrt{y}}{\sqrt{x}-\sqrt{y}} = \frac{3\sqrt{x}+\sqrt{y}}{\sqrt{x}-\sqrt{y}} \cdot \frac{\sqrt{x}+\sqrt{y}}{\sqrt{x}+\sqrt{y}}$
$= \frac{3x+4\sqrt{xy}+y}{x-y}$

76. $\frac{2xy}{\sqrt[3]{16xy^5}} = \frac{2xy}{\sqrt[3]{16xy^5}} \cdot \frac{\sqrt[3]{4x^2y}}{\sqrt[3]{4x^2y}}$
$= \frac{2xy\sqrt[3]{4x^2y}}{\sqrt[3]{4^3x^3y^6}}$
$= \frac{2xy\sqrt[3]{4x^2y}}{4xy^2}$
$= \frac{\sqrt[3]{4x^2y}}{2y}$

77. $\sqrt{-16}+\sqrt{-45} = \sqrt{16}\sqrt{-1}+\sqrt{45}\sqrt{-1}$
$= 4i + i\sqrt{9\cdot 5}$
$= 4i + 3i\sqrt{5}$

78. $2x-3i+5 = yi-2+\sqrt{6}$
$2x+5 = -2+\sqrt{6}$
$2x = -7+\sqrt{6}$
$x = \frac{-7+\sqrt{6}}{2}$
$y = -3$

79. $(-12-6i)+(3-5i) = -12-6i+3-5i$
$= -12+3+(-6-5)i$
$= -9-11i$

80. $(2-i)-(12-3i) = 2-i-12+3i = -10+2i$

81. $(5-2i)(3+3i) = 15+15i-6i-6i^2$
$= 15+15i-6i+6$
$= 21+9i$

82. $(6-2i)^2 = (6-2i)(6-2i)$
$= 36-12i-12i+4i^2$
$= 36-24i-4$
$= 32-24i$

83. $2i(3+4i) = 6i+8i^2 = -8+6i$

84. $3-4(2+i) = 3-8-4i = -5-4i$

85. $i^{34} = (i^4)^8 \cdot i^2 = 1^8(-1) = -1$

86. $i^{65} = (i^4)^{16} \cdot i = 1^{16} \cdot i = i$

87. $\frac{7-2i}{3+4i} = \frac{7-2i}{3+4i} \cdot \frac{3-4i}{3-4i}$
$= \frac{21-6i-28i+8i^2}{9+16}$
$= \frac{21-34i-8}{25}$
$= \frac{13-34i}{25}$

88. $\frac{5-2i}{1-3i} = \frac{5-2i}{1-3i} \cdot \frac{1+3i}{1+3i}$
$= \frac{5-2i+15i-6i^2}{1+9}$
$= \frac{5+13i+6}{10}$
$= \frac{11+13i}{10}$

89. $\frac{4-3i}{5i} = \frac{4-3i}{5i} \cdot \frac{-5i}{-5i}$
$= \frac{-20i+15i^2}{-25i^2}$
$= \frac{-20i-15}{-25(-1)}$
$= \frac{-15-20i}{25}$
$= -\frac{5(3+4i)}{5\cdot 5}$
$= -\frac{3+4i}{5}$

90. $\frac{12}{3-5i} = \frac{12}{3-5i} \cdot \frac{3+5i}{3+5i}$
$= \frac{36+60i}{9+25}$
$= \frac{36+60i}{34}$
$= \frac{18+30i}{17}$

91. $\frac{10-4i}{2+5i} = \frac{10-4i}{2+5i} \cdot \frac{2-5i}{2-5i}$
$= \frac{20-58i+20i^2}{4+25}$
$= \frac{20-58i-20}{29}$
$= \frac{-58i}{29}$
$= -2i$

92. $\sqrt{3x-2}=5$

$\left(\sqrt{3x-2}\right)^2=5^2$

$3x-2=25$

$3x=27$

$x=9$

Check: $\sqrt{3(9)-2}\stackrel{?}{=}5,\ 5=5$

$x=9$ is the solution.

93. $\sqrt[3]{3x-1}=2$

$\left(\sqrt[3]{3x-1}\right)^3=2^3$

$3x-1=8$

$3x=9$

$x=3$

Check: $\sqrt[3]{3(3)-1}\stackrel{?}{=}2,\ 2=2$

$x=3$ is the solution.

94. $\sqrt{2x+1}=2x-5$

$\left(\sqrt{2x+1}\right)^2=(2x-5)^2$

$2x+1=4x^2-20x+25$

$4x^2-22x+24=0$

$2x-11x+12=0$

$(x-4)(2x-3)=0$

$x=4,\ x=\frac{3}{2}$

Check: $\sqrt{2(4)+1}\stackrel{?}{=}2(4)-5,\ 3=3$

$\sqrt{2\left(\frac{3}{2}\right)+1}\stackrel{?}{=}2\left(\frac{3}{2}\right)-5,\ \sqrt{4}\neq -2$

$x=4$ is the solution.

95. $1+\sqrt{3x+1}=x$

$\sqrt{3x+1}=x-1$

$\left(\sqrt{3x+1}\right)^2=(x-1)^2$

$3x+1=x^2-2x+1$

$x^2-5x=0$

$x(x-5)=0$

$x=0, x=5$

Check: $1+\sqrt{3(0)+1}\stackrel{?}{=}0,\ 2\neq 0$

$1+\sqrt{3(5)+1}\stackrel{?}{=}5,\ 5=5$

$x=5$ is the solution.

96. $\sqrt{3x+1}-\sqrt{2x-1}=1$

$\sqrt{3x+1}=\sqrt{2x-1}+1$

$\left(\sqrt{3x+1}\right)^2=\left(\sqrt{2x-1}+1\right)^2$

$3x+1=2x-1+2\sqrt{2x-1}+1$

$x+1=2\sqrt{2x-1}$

$(x+1)^2=\left(2\sqrt{2x-1}\right)^2$

$x^2+2x+1=8x-4$

$x^2-6x+5=0$

$(x-5)(x-1)=0$

$x=5, x=1$

Check: $\sqrt{3(5)+1}-\sqrt{2(5)-1}\stackrel{?}{=}1,\ 1=1$

$\sqrt{3(1)+1}-\sqrt{2(1)-1}\stackrel{?}{=}1,\ 1=1$

$x=5, x=1$ is the solution.

97. $\sqrt{7x+2}=\sqrt{x+3}+\sqrt{2x-1}$

$\left(\sqrt{7x+2}\right)^2=\left[\sqrt{x+3}+\sqrt{2x-1}\right]^2$

$7x+2=x+3+2\sqrt{x+3}\sqrt{2x-1}+2x-1$

$4x=2\sqrt{x+3}\sqrt{2x-1}$

$2x=\sqrt{x+3}\sqrt{2x-1}$

$(2x)^2=\left(\sqrt{x+3}\sqrt{2x-1}\right)^2$

$4x^2=2x^2+5x-3$

$2x^2-5x+3=0$

$(x-1)(2x-3)=0$

$x=1,\ x=\frac{3}{2}$

Check:

$\sqrt{7(1)+2}\stackrel{?}{=}\sqrt{1+3}+\sqrt{2(1)-1},\ 3=3$

$\sqrt{7\left(\frac{3}{2}\right)+2}\stackrel{?}{=}\sqrt{\left(\frac{3}{2}\right)+3}+\sqrt{2\left(\frac{3}{2}\right)-1}$

$\frac{5}{\sqrt{2}}=\frac{5}{\sqrt{2}}$

$x=1,\ x=\frac{3}{2}$ is the solution.

98. $y=kx$

When $y=11, x=4,\ k=\frac{y}{x}=\frac{11}{4}.$

When $x=6,\ k=\frac{11}{4},\ y=\frac{11}{4}(6)=16.5.$

99. $y = kx,\ 5 = k \cdot 20,\ k = \frac{1}{4}$

$k = \frac{1}{4}x,\ y = \frac{1}{4} \cdot 50 = \frac{50}{4} = 12.5$

100. c = calories consumed
g = fat grams
k = constant
$g = kc$

When $c = 2000$, $g = 18$, $k = \frac{g}{c} = \frac{18}{2000} = 0.009$.
When $c = 2500$, $k = 0.009$,
$g = 0.009(2500) = 22.5$.
She can consume a maximum of 22.5 grams of fat.

101. $t = k\sqrt{d},\ 2 = k\sqrt{64},\ k = \frac{1}{4}$

$t = \frac{1}{4}\sqrt{d},\ t = \frac{1}{4}\sqrt{196} = \frac{14}{4} = 3.5$

It will take 3.5 seconds.

102. $y = \frac{k}{x},\ 8 = \frac{k}{3},\ k = 24$

$y = \frac{24}{x},\ y = \frac{24}{48} = 0.5$

103. $V = \frac{k}{P},\ 70 = \frac{k}{24},\ k = 1680$

$V = \frac{1680}{P},\ 100 = \frac{1680}{P},\ P = 16.8$

A pressure of 16.8 lb/in.^{2} corresponds to a volume of 100 in.^{3}.

104. $y = \frac{kx}{(z)^2}$

When $x = 10$, $z = 5$, $y = 20$:

$k = \frac{y(z)^2}{x} = \frac{20(5)^2}{10} = 50$

When $x = 8$, $z = 2$, $k = 50$:

$y = \frac{kx}{(z)^2} = \frac{50(8)}{(2)^2} = 100$

105. $V = khr^2,\ 50 = k(5)(3)^2,\ k = \frac{10}{9}$

$V = \frac{10}{9}r^2h,\ V = \frac{10}{9}(4)^2(9) = 160$

The capacity is 160 cm^3.

How Am I Doing? Chapter 7 Test

1. $(2x^{1/2}y^{1/3})(-3x^{1/3}y^{1/6}) = -6x^{\frac{1}{2}+\frac{1}{3}}y^{\frac{1}{3}+\frac{1}{6}}$
$= -6x^{5/6}y^{1/2}$

2. $\frac{7x^3}{4x^{3/4}} = \frac{7x^{3-\frac{3}{4}}}{4} = \frac{7x^{9/4}}{4}$

3. $(8x^{1/3})^{3/2} = 8^{3/2}x^{\frac{1}{3}\cdot\frac{3}{2}} = 8^{3/2}x^{1/2}$

4. $\left(\frac{4}{9}\right)^{3/2} = \left(\left(\frac{2}{3}\right)^2\right)^{3/2} = \left(\frac{2}{3}\right)^{2\cdot\frac{3}{2}} = \left(\frac{2}{3}\right)^3 = \frac{8}{27}$

5. $\sqrt[5]{-32} = \sqrt[5]{(-2)^5} = -2$

6. $8^{-2/3} = \frac{1}{(8^{1/3})^2} = \frac{1}{2^2} = \frac{1}{4}$

7. $16^{5/4} = (16^{1/4})^5 = 2^5 = 32$

8. $\sqrt{75a^4b^9} = \sqrt{25a^4b^8 \cdot 3b} = 5a^2b^4\sqrt{3b}$

9. $\sqrt{49a^4b^{10}} = \sqrt{7^2(a^2)^2(b^5)^2} = 7a^2b^5$

10. $\sqrt[3]{54m^3n^5} = \sqrt[3]{3^3m^3n^3 \cdot 2n^2} = 3mn\sqrt[3]{2n^2}$

11. $3\sqrt{24} - \sqrt{18} + \sqrt{50} = 3\sqrt{4 \cdot 6} - \sqrt{9 \cdot 2} + \sqrt{25 \cdot 2}$
$= 6\sqrt{6} - 3\sqrt{2} + 5\sqrt{2}$
$= 6\sqrt{6} + 2\sqrt{2}$

12. $\sqrt{40x} - \sqrt{27x} + 2\sqrt{12x}$
$= \sqrt{4 \cdot 10x} - \sqrt{9 \cdot 3x} + 2\sqrt{4 \cdot 3x}$
$= 2\sqrt{10x} - 3\sqrt{3x} + 4\sqrt{3x}$
$= 2\sqrt{10x} + \sqrt{3x}$

13. $\left(-3\sqrt{2y}\right)\left(5\sqrt{10xy}\right) = -15\sqrt{20xy^2}$
$= -15\sqrt{4y^2 \cdot 5x}$
$= -30y\sqrt{5x}$

14. $2\sqrt{3}\left(3\sqrt{6} - 5\sqrt{2}\right) = 6\sqrt{18} - 10\sqrt{6}$
$= 6\sqrt{9 \cdot 2} - 10\sqrt{6}$
$= 18\sqrt{2} - 10\sqrt{6}$

15. $(5\sqrt{3}-\sqrt{6})(2\sqrt{3}+3\sqrt{6})$
$=5\cdot 2\cdot 3+5\cdot 3\sqrt{3\cdot 6}-2\sqrt{3\cdot 6}-3\cdot 6$
$=30+15\sqrt{18}-2\sqrt{18}-18$
$=12+13\sqrt{9\cdot 2}$
$=12+39\sqrt{2}$

16. $\dfrac{30}{\sqrt{5x}}=\dfrac{30}{\sqrt{5x}}\cdot\dfrac{\sqrt{5x}}{\sqrt{5x}}=\dfrac{30\sqrt{5x}}{5x}=\dfrac{6\sqrt{5x}}{x}$

17. $\sqrt{\dfrac{xy}{3}}=\dfrac{\sqrt{xy}}{\sqrt{3}}\cdot\dfrac{\sqrt{3}}{\sqrt{3}}=\dfrac{\sqrt{3xy}}{3}$

18. $\dfrac{1+2\sqrt{3}}{3-\sqrt{3}}=\dfrac{1+2\sqrt{3}}{3-\sqrt{3}}\cdot\dfrac{3+\sqrt{3}}{3+\sqrt{3}}$
$=\dfrac{3+\sqrt{3}+6\sqrt{3}+2(3)}{9-3}$
$=\dfrac{9+7\sqrt{3}}{6}$

19. $\sqrt{3x-2}=x$
$\left(\sqrt{3x-2}\right)^2=x^2$
$3x-2=x^2$
$x^2-3x+2=0$
$(x-2)(x-1)=0$
$x=2,\ x=1$
Check: $\sqrt{3(2)-2}\stackrel{?}{=}2,\ 2=2$
$\sqrt{3(1)-2}\stackrel{?}{=}1,\ 1=1$
$x=2,\ x=1$ is the solution.

20. $5+\sqrt{x+15}=x$
$\sqrt{x+15}=x-5$
$\left(\sqrt{x+15}\right)^2=(x-5)^2$
$x+15=x^2-10x+25$
$x^2-11x+10=0$
$(x-10)(x-11)=0$
$x=10,\ x=11$
Check: $5+\sqrt{10+15}\stackrel{?}{=}10,\ 10=10$
$5+\sqrt{11+15}\stackrel{?}{=}11,\ 5+\sqrt{26}\neq 11$
$x=10$ is the solution.

21. $5-\sqrt{x-2}=\sqrt{x+3}$
$\left(5-\sqrt{x-2}\right)^2=\left(\sqrt{x+3}\right)^2$
$25-10\sqrt{x-2}+x-2=x+3$
$20=10\sqrt{x-2}$
$2=\sqrt{x-2}$
$2^2=\left(\sqrt{x-2}\right)^2$
$4=x-2$
$x=6$
Check: $5-\sqrt{6-2}\stackrel{?}{=}\sqrt{6+3},\ 3=3$
$x=6$ is the solution.

22. $(8+2i)-3(2-4i)=8+2i-6+12i=2+14i$

23. $i^{18}+\sqrt{-16}=(i^4)^4\cdot i^2+\sqrt{16}\sqrt{-1}$
$=1^4(-1)+4i$
$=-1+4i$

24. $(3-2i)(4+3i)=12+i-6i^2=12+i+6=18+i$

25. $\dfrac{2+5i}{1-3i}=\dfrac{2+5i}{1-3i}\cdot\dfrac{1+3i}{1+3i}$
$=\dfrac{2+11i+15i^2}{1+9}$
$=\dfrac{2+11i-15}{10}$
$=\dfrac{-13+11i}{10}$

26. $(6+3i)^2=36+36i+9i^2=36+36i-9=27+36i$

27. $i^{43}=(i^4)^{10}\cdot i^3=1^{10}\cdot(-i)=-i$

28. $y=\dfrac{k}{x},\ 9=\dfrac{k}{2},\ k=18$
$y=\dfrac{18}{x},\ y=\dfrac{18}{6}=3$

29. $y=k\cdot\dfrac{x}{z^2},\ 3=k\cdot\dfrac{8}{4^2},\ k=6$
$y=6\cdot\dfrac{x}{z^2},\ y=6\cdot\dfrac{5}{6^2}=\dfrac{5}{6}$

30. $d = kv^2$, $30 = k \cdot 30^2$, $k = \frac{1}{30}$

$d = \frac{1}{30} \cdot v^2$, $d = \frac{1}{30} \cdot 50^2 \approx 83.3$

The car's stopping distance is about 83.3 feet.

Cumulative Test for Chapters 1–7

1. $7 + (2 + 3) = (7 + 2) + 3$
Associative property of addition

2. $-3a(2ab - a^3) + b(ab^2 + 4a^2)$
$= -6a^2b + 3a^4 + ab^3 + 4a^2b$
$= -2a^2b + 3a^4 + ab^3$

3. $7(12-14)^3 - 7 + 3 \div (-3) = 7(-2)^3 - 7 - 1$
$= 7(-8) - 7 - 1$
$= -56 - 7 - 1$
$= -63 - 1$
$= -64$

4. $y = -\frac{2}{5}x + 3$
$-5y = 2x - 15$
$2x = -5y + 15$
$x = \frac{-5y+15}{2}$ or $x = -\frac{5y-15}{2}$

5. $3x - 5y = 15$ or $y = \frac{3x-15}{5}$

x	$y = \frac{3x-15}{5}$	y
5	$y = \frac{3(5)-15}{5}$	0
0	$y = \frac{3(0)-15}{5}$	−3

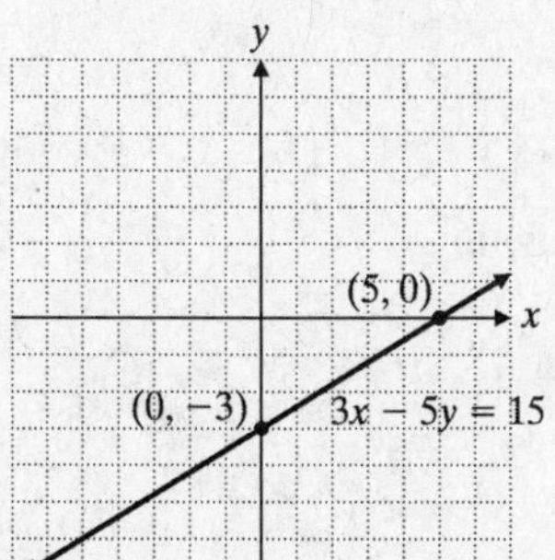

6. $16x^2 - 16x - 12 = 4(4x^2 - 4x - 3)$
$= 4(2x+1)(2x-3)$

7. $x + 4y - z = 10$
$3x + 2y + z = 4$
$2x - 3y + 2z = -7$

Multiply first equation by −3 and add to second equation; then multiply first equation by −2 and add to third equation.

$x + 4y - z = 10$
$-10y + 4z = -26$
$-11y + 4z = -27$

Multiply the second equation by $-\frac{11}{10}$ and add to the third equation.

$x + 4y - z = 10$
$-10y + 4z = -26$
$-\frac{2}{5}z = \frac{8}{5} \Rightarrow z = -4$

$z = -4$, substitute in second equation.
$-10y + 4(-4) = -26$
$-10y - 16 = -26$
$-10y = -10$

$y = 1$, substitute in first equation.
$x + 4(1) - (-4) = 10$
$x = 2$

$x = 2$, $y = 1$, $z = -4$ is the solution.

8. $\frac{7x}{x^2 - 2x - 15} - \frac{2}{x-5}$
$= \frac{7x}{(x-5)(x+3)} - \frac{2(x+3)}{(x-5)(x+3)}$
$= \frac{7x - 2x - 6}{(x-5)(x+3)}$
$= \frac{5x-6}{(x-5)(x+3)}$

9. W = Width
$2W + 3$ = Length
$P = 2L + 2W$
$48 = 2(2W + 3) + 2W$
$24 = 2W + 3 + W$
$3W = 21$
$W = 7$
$L = 2W + 3 = 17$
The width is 7 meters and the length is 17 meters.

10.
$$2ax - 3b = y - 5ax$$
$$2ax + 5ax = y + 3b$$
$$7ax = y + 3b$$
$$a = \frac{y+3b}{7x}$$

11.
$$\frac{3x^3y^{-2}}{9x^{-1/2}y^{5/2}} = \frac{1}{3}x^{3-(-1/2)}y^{-2-(5/2)} = \frac{1}{3}x^{7/2}y^{-9/2} = \frac{x^{7/2}}{3y^{9/2}}$$

12.
$$(3x^{-1/2}y^2)^{-1/3} = 3^{-1/3}x^{-1/2(-1/3)}y^{2\cdot(-1/3)} = 3^{-1/3}x^{1/6}y^{-2/3} = \frac{x^{1/6}}{3^{1/3}y^{2/3}}$$

13. $64^{-1/3} = \frac{1}{64^{1/3}} = \frac{1}{4}$

14. $\sqrt[3]{40x^5y^9} = \sqrt[3]{8\cdot 5x^3y^9x^2} = 2xy^3\sqrt[3]{5x^2}$

15.
$$\sqrt{80x} + 2\sqrt{45x} - 3\sqrt{20x}$$
$$= \sqrt{16\cdot 5x} + 2\sqrt{9\cdot 5x} - 3\sqrt{4\cdot 5x}$$
$$= 4\sqrt{5x} + 6\sqrt{5x} - 6\sqrt{5x}$$
$$= 4\sqrt{5x}$$

16.
$$\left(2\sqrt{3} - 5\sqrt{2}\right)\left(\sqrt{3} + 4\sqrt{2}\right)$$
$$= 2\cdot 3 + 8\sqrt{6} - 5\sqrt{6} - 20\cdot 2$$
$$= 6 + 3\sqrt{6} - 40$$
$$= -34 + 3\sqrt{6}$$

17.
$$\frac{\sqrt{3}+2}{2\sqrt{3}-5} = \frac{\sqrt{3}+2}{2\sqrt{3}-5}\cdot\frac{2\sqrt{3}+5}{2\sqrt{3}+5} = \frac{6+9\sqrt{3}+10}{12-25} = -\frac{16+9\sqrt{3}}{13}$$

18.
$$i^{21} + \sqrt{-16} + \sqrt{-49}$$
$$= (i^4)^5\cdot i + \sqrt{16}\sqrt{-1} + \sqrt{49}\sqrt{-1}$$
$$= 1^5\cdot i + 4i + 7i$$
$$= 12i$$

19.
$$(3-4i)^2 = 3^2 - 2(3)(4i) + (4i)^2 = 9 - 24i + 16i^2 = 9 - 24i - 16 = -7 - 24i$$

20.
$$\frac{1+2i}{1-3i} = \frac{1+2i}{1-3i}\cdot\frac{1+3i}{1+3i} = \frac{1+3i+2i+6i^2}{1-9i^2} = \frac{1+5i-6}{1+9} = \frac{-5+5i}{10} = \frac{-1+i}{2}$$

21.
$$x - 4 = \sqrt{3x+6}$$
$$(x-4)^2 = \left(\sqrt{3x+6}\right)^2$$
$$x^2 - 2(4)x + 16 = 3x + 6$$
$$x^2 - 8x - 3x + 10 = 0$$
$$x^2 - 11x + 10 = 0$$
$$(x-10)(x-1) = 0$$
$$x - 10 = 0 \qquad x - 1 = 0$$
$$x = 10 \qquad x = 1$$

Check:
$$10 - 4 \stackrel{?}{=} \sqrt{3\cdot 10 + 6}$$
$$6 = 6$$
$$1 - 4 \stackrel{?}{=} \sqrt{3\cdot 1 + 6}$$
$$-3 \neq 3$$

$x = 10$ only

22.
$$1 + \sqrt{x+1} = \sqrt{x+2}$$
$$\left(1 + \sqrt{x+1}\right)^2 = \left(\sqrt{x+2}\right)^2$$
$$1 + 2\sqrt{x+1} + x + 1 = x + 2$$
$$\sqrt{x+1} = 0$$
$$x + 1 = 0$$
$$x = -1$$

Check: $1 + \sqrt{-1+1} \stackrel{?}{=} \sqrt{-1+2},\ 1 = 1$

$x = -1$ is the solution.

23. $y = kx^2,\ 12 = k\cdot 2^2,\ k = 3$

$y = 3x^2,\ y = 3\cdot 5^2 = 75$

24. $I = \frac{k}{d^2}$, $120 = \frac{k}{10^2}$, $k = 12,000$

$I = \frac{12,000}{d^2}$, $I = \frac{12,000}{15^2} \approx 53.3$

There are about 53.3 lumens provided.

Chapter 8

8.1 Exercises

1. $x^2 = 100$
$x = \pm\sqrt{100}$
$x = \pm 10$

3. $3x^2 - 45 = 0$
$3x^2 = 45$
$x^2 = 15$
$x = \pm\sqrt{15}$

5. $2x^2 - 80 = 0$
$2x^2 = 80$
$x^2 = 40$
$x = \pm\sqrt{40} = \pm\sqrt{4 \cdot 10}$
$x = \pm 2\sqrt{10}$

7. $x^2 = -81$
$x = \pm\sqrt{-81}$
$x = \pm 9i$

9. $x^2 + 16 = 0$
$x^2 = -16$
$x = \pm\sqrt{-16}$
$x = \pm 4i$

11. $(x-3)^2 = 12$
$x - 3 = \pm\sqrt{12} = \pm\sqrt{4 \cdot 3}$
$x - 3 = \pm 2\sqrt{3}$
$x = 3 \pm 2\sqrt{3}$

13. $(x+9)^2 = 21$
$x + 9 = \pm\sqrt{21}$
$x = -9 \pm \sqrt{21}$

15. $(2x+1)^2 = 7$
$2x + 1 = \pm\sqrt{7}$
$2x = -1 \pm \sqrt{7}$
$x = \dfrac{-1 \pm \sqrt{7}}{2}$

17. $(4x-3)^2 = 36$
$4x - 3 = \pm 6$
$4x = 3 \pm 6$
$x = \dfrac{3 \pm 6}{4}$
$x = \dfrac{9}{4},\ x = -\dfrac{3}{4}$

19. $(2x+5)^2 = 49$
$2x + 5 = \pm 7$
$2x = -5 \pm 7$
$x = \dfrac{-5 \pm 7}{2}$
$x = 1, x = -6$

21. $2x^2 - 9 = 0$
$2x^2 = 9$
$x^2 = \dfrac{9}{2}$
$x = \pm\sqrt{\dfrac{9}{2}}$
$x = \pm\dfrac{3}{\sqrt{2}}$
$x = \pm\dfrac{3\sqrt{2}}{2}$

23. $x^2 + 10x + 5 = 0$
$x^2 + 10x + 25 = -5 + 25$
$(x+5)^2 = 20$
$x + 5 = \pm\sqrt{20} = \pm 2\sqrt{5}$
$x = -5 \pm 2\sqrt{5}$

25. $x^2 - 8x = 17$
$x^2 - 8x + 16 = 17 + 16$
$(x-4)^2 = 33$
$x - 4 = \pm\sqrt{33}$
$x = 4 \pm \sqrt{33}$

27. $x - 14x = -48$
$x^2 - 14x + 49 = -48 + 49$
$(x-7)^2 = 1$
$x - 7 = \pm 1$
$x = 7 \pm 1$
$x = 6, x = 8$

29. $\frac{x^2}{2}+\frac{3}{2}x=4$

$x^2+3x=8$

$x^2+3x+\frac{9}{4}=8+\left(\frac{3}{2}\right)^2$

$\left(x+\frac{3}{2}\right)^2=\frac{41}{4}$

$x+\frac{3}{2}=\pm\frac{\sqrt{41}}{2}$

$x=-\frac{3}{2}\pm\frac{\sqrt{41}}{2}$

$x=\frac{-3\pm\sqrt{41}}{2}$

31. $2y^2+10y=-11$

$y^2+5y=-\frac{11}{2}$

$y^2+5y+\frac{25}{4}=-\frac{11}{2}+\frac{25}{4}$

$\left(y+\frac{5}{2}\right)^2=\frac{3}{4}$

$y+\frac{5}{2}=\pm\frac{\sqrt{3}}{2}$

$y=-\frac{5}{2}\pm\frac{\sqrt{3}}{2}$

$y=\frac{-5\pm\sqrt{3}}{2}$

33. $3x^2+10x-2=0$

$3x^2+10x=2$

$x^2+\frac{10}{3}x=\frac{2}{3}$

$x^2+\frac{10}{3}x+\frac{25}{9}=\frac{2}{3}+\frac{25}{9}$

$\left(x+\frac{5}{3}\right)^2=\frac{31}{9}$

$x+\frac{5}{3}=\pm\frac{\sqrt{31}}{3}$

$x=-\frac{5}{3}\pm\frac{\sqrt{31}}{3}$

$x=\frac{-5\pm\sqrt{31}}{3}$

35. $x^2+4x-6=0$

$x^2+4x=6$

$x^2+4x+4=6+4$

$(x+2)^2=10$

$x+2=\pm\sqrt{10}$

$x=-2\pm\sqrt{10}$

37. $\frac{x^2}{2}-x=4$

$x^2-2x=8$

$x^2-2x+1=8+1$

$(x-1)^2=9$

$x-1=\pm 3$

$x=1\pm 3$

$x=4, x=-2$

39. $3x^2+1=x$

$3x^2-x=-1$

$x^2-\frac{1}{3}x=-\frac{1}{3}$

$x^2-\frac{1}{3}x+\frac{1}{36}=-\frac{1}{3}+\frac{1}{36}$

$\left(x-\frac{1}{6}\right)^2=-\frac{11}{36}$

$x-\frac{1}{6}=\pm\frac{i\sqrt{11}}{6}$

$x=\frac{1}{6}\pm\frac{i\sqrt{11}}{6}$

$x=\frac{1\pm i\sqrt{11}}{6}$

41. $x^2+2=x$

$x^2-x=-2$

$x^2-x+\frac{1}{4}=-2+\frac{1}{4}$

$\left(x-\frac{1}{2}\right)^2=-\frac{7}{4}$

$x-\frac{1}{2}=\pm\frac{i\sqrt{7}}{2}$

$x=\frac{1}{2}\pm\frac{i\sqrt{7}}{2}$

$x=\frac{1\pm i\sqrt{7}}{2}$

43.
$$\begin{aligned} 2x^2+2&=3x\\ 2x^2-3x&=-2\\ x^2-\frac{3}{2}x&=-1\\ x^2-\frac{3}{2}x+\frac{9}{16}&=-1+\frac{9}{16}\\ \left(x-\frac{3}{4}\right)^2&=-\frac{7}{16}\\ x-\frac{3}{4}&=\pm\frac{i\sqrt{7}}{4}\\ x&=\frac{3}{4}\pm\frac{i\sqrt{7}}{4}\\ x&=\frac{3\pm i\sqrt{7}}{4} \end{aligned}$$

45.
$$\begin{aligned} x^2+2x-5&=0\\ \left(-1+\sqrt{6}\right)^2+2\left(-1+\sqrt{6}\right)-5&\stackrel{?}{=}0\\ 1-2\sqrt{6}+6-2+2\sqrt{6}-5&\stackrel{?}{=}0\\ 0&=0\ \checkmark \end{aligned}$$

47.
$$\begin{aligned} (x-7)^2(8)&=648\\ (x-7)^2&=81\\ x-7&=\pm 9\\ x&=7\pm 9 \end{aligned}$$
Since the value of x must be greater than 7, $x = 16$.

49.
$$\begin{aligned} 4t^2&=L\\ 4t^2&=5\\ t^2&=1.25\\ t&=\pm\sqrt{1.25} \end{aligned}$$
Since the time must be positive, $t=\sqrt{1.25}\approx 1.12$. The hang time is approximately 1.12 seconds.

Cumulative Review

51.
$$\begin{aligned} \sqrt{b^2-4ac}&=\sqrt{4^2-4(3)(-4)}\\ &=\sqrt{16+48}\\ &=\sqrt{64}\\ &=8 \end{aligned}$$

52.
$$\begin{aligned} \sqrt{b^2-4ac}&=\sqrt{(-5)^2-4(2)(-3)}\\ &=\sqrt{25+24}\\ &=\sqrt{49}\\ &=7 \end{aligned}$$

53.
$$\begin{aligned} 5x^2-6x+8&=5(-2)^2-6(-2)+8\\ &=5(4)-6(-2)+8\\ &=20+12+8\\ &=40 \end{aligned}$$

54.
$$\begin{aligned} 2x^2+3x-5&=2(-3)^2+3(-3)-5\\ &=2(9)+3(-3)-5\\ &=18-9-5\\ &=4 \end{aligned}$$

Quick Quiz 8.1

1.
$$\begin{aligned} (4x-3)^2&=12\\ 4x-3&=\pm\sqrt{12}=\pm 2\sqrt{3}\\ 4x&=3\pm 2\sqrt{3}\\ x&=\frac{3\pm 2\sqrt{3}}{4} \end{aligned}$$

2.
$$\begin{aligned} x^2-8x&=28\\ x^2-8x+16&=28+16\\ (x-4)^2&=44\\ x-4&=\pm\sqrt{44}=\pm 2\sqrt{11}\\ x&=4\pm 2\sqrt{11} \end{aligned}$$

3.
$$\begin{aligned} 2x^2+10x&=-11\\ x^2+5x&=-\frac{11}{2}\\ x^2+5x+\frac{25}{4}&=-\frac{11}{2}+\frac{25}{4}\\ \left(x+\frac{5}{2}\right)^2&=\frac{3}{4}\\ x+\frac{5}{2}&=\pm\frac{\sqrt{3}}{2}\\ x&=-\frac{5}{2}\pm\frac{\sqrt{3}}{2}\\ x&=\frac{-5\pm\sqrt{3}}{2} \end{aligned}$$

4. Answers may vary. Possible solution: Divide the coefficient of x (1) by 2 to get $\frac{1}{2}$. Since $\left(\frac{1}{2}\right)^2=\frac{1}{4}$, add $\frac{1}{4}$ to both sides of the equation.

8.2 Exercises

1. Place the quadratic in standard form. Find *a*, *b*, and *c*. Substitute these values into the quadratic formula.

3. If the discriminant in the quadratic formula is zero, then the quadratic equation will have one real solution.

5. $x^2+x-5=0$
$a=1, b=1, c=-5$
$$x=\frac{-b\pm\sqrt{b^2-4ac}}{2a}$$
$$x=\frac{-1\pm\sqrt{(1)^2-4(1)(-5)}}{2(1)}$$
$$x=\frac{-1\pm\sqrt{21}}{2}$$

7. $3x^2-x-1=0$
$a=3, b=-1, c=-1$
$$x=\frac{-b\pm\sqrt{b^2-4ac}}{2a}$$
$$x=\frac{-(-1)\pm\sqrt{(-1)^2-4(3)(-1)}}{2(3)}$$
$$x=\frac{1\pm\sqrt{13}}{6}$$

9. $x^2=\frac{2}{3}x$
$$x^2-\frac{2}{3}x=0$$
$$3x^2-2x=0$$
$a=3, b=-2, c=0$
$$x=\frac{-b\pm\sqrt{b^2-4ac}}{2a}$$
$$x=\frac{-(-2)\pm\sqrt{(-2)^2-4(3)(0)}}{2(3)}$$
$$x=\frac{2\pm\sqrt{4}}{6}=\frac{2\pm2}{6}$$
$$x=0,\ x=\frac{2}{3}$$

11. $3x^2-x-2=0$
$a=3, b=-1, c=-2$
$$x=\frac{-b\pm\sqrt{b^2-4ac}}{2a}$$
$$x=\frac{-(-1)\pm\sqrt{(-1)^2-4(3)(-2)}}{2(3)}$$
$$x=\frac{1\pm\sqrt{25}}{6}=\frac{1\pm5}{6}$$
$$x=1,\ x=-\frac{2}{3}$$

13. $4x^2+3x-2=0$
$a=4, b=3, c=-2$
$$x=\frac{-b\pm\sqrt{b^2-4ac}}{2a}$$
$$x=\frac{-3\pm\sqrt{(3)^2-4(4)(-2)}}{2(4)}$$
$$x=\frac{-3\pm\sqrt{41}}{8}$$

15. $4x^2+1=7$
$$4x^2-6=0$$
$a=4, b=0, c=-6$
$$x=\frac{-b\pm\sqrt{b^2-4ac}}{2a}$$
$$x=\frac{-0\pm\sqrt{(0)^2-4(4)(-6)}}{2(4)}$$
$$x=\pm\frac{\sqrt{96}}{8}=\pm\frac{4\sqrt{6}}{8}$$
$$x=\pm\frac{\sqrt{6}}{2}$$

17. $2x(x+3)-3=4x-2$
$$2x^2+6x-4x-1=0$$
$$2x^2+2x-1=0$$
$a=2, b=2, c=-1$
$$x=\frac{-b\pm\sqrt{b^2-4ac}}{2a}$$
$$x=\frac{-2\pm\sqrt{(2)^2-4(2)(-1)}}{2(2)}$$
$$x=\frac{-2\pm\sqrt{12}}{4}=\frac{-2\pm2\sqrt{3}}{4}$$
$$x=\frac{2\left(-1\pm\sqrt{3}\right)}{4}=\frac{-1\pm\sqrt{3}}{2}$$

19. $x(x+3)-2=3x+7$
$$x^2+3x-2=3x+7$$
$$x^2-9=0$$
$a=1, b=0, c=-9$
$$x=\frac{-b\pm\sqrt{b^2-4ac}}{2a}$$
$$x=\frac{-0\pm\sqrt{(0)^2-4(1)(-9)}}{2(1)}$$
$$x=\pm\frac{\sqrt{36}}{2}=\pm\frac{6}{2}=\pm 3$$

21. $(x-2)(x+1)=\dfrac{2x+3}{2}$
$$x^2-x-2=\frac{2x+3}{2}$$
$$2x^2-2x-4=2x+3$$
$$2x^2-4x-7=0$$
$a=2, b=-4, c=-7$
$$x=\frac{-b\pm\sqrt{b^2-4ac}}{2a}$$
$$x=\frac{-(-4)\pm\sqrt{(-4)^2-4(2)(-7)}}{2(2)}$$
$$x=\frac{4\pm\sqrt{72}}{4}=\frac{4\pm 6\sqrt{2}}{4}$$
$$x=\frac{2\left(2\pm 3\sqrt{2}\right)}{4}=\frac{2\pm 3\sqrt{2}}{2}$$

23. The LCD is $3x(x + 2)$.
$$\frac{1}{x+2}+\frac{1}{x}=\frac{1}{3}$$
$$\frac{1}{x+2}(3x)(x+2)+\frac{1}{x}(3x)(x+2)=\frac{1}{3}(3x)(x+2)$$
$$3x+3(x+2)=x(x+2)$$
$$3x+3x+6=x^2+2x$$
$$x^2-4x-6=0$$
$a=1, b=-4, c=-6$
$$x=\frac{-b\pm\sqrt{b^2-4ac}}{2a}$$
$$x=\frac{-(-4)\pm\sqrt{(-4)^2-4(1)(-6)}}{2(1)}$$
$$x=\frac{4\pm\sqrt{40}}{2}=\frac{4\pm 2\sqrt{10}}{2}$$
$$x=\frac{2\left(2\pm\sqrt{10}\right)}{2}=2\pm\sqrt{10}$$

25. The LCD is $12y(y + 2)$.

$$\frac{1}{12}(12y)(y+2)+\frac{1}{y}(12y)(y+2)=\frac{2}{y+2}(12y)(y+2)$$
$$y(y+2)+12(y+2)=2(12y)$$
$$y^2+2y+12y+24=24y$$
$$y^2-10y+24=0$$

$a = 1, b = -10, c = 24$

$$y=\frac{-b\pm\sqrt{b^2-4ac}}{2a}$$
$$y=\frac{-(-10)\pm\sqrt{(-10)^2-4(1)(24)}}{2(1)}$$
$$y=\frac{10\pm\sqrt{4}}{2}$$
$$y=\frac{10\pm 2}{2}$$
$$y=6,\ y=4$$

27. $x(x+4)=-12$

$$x^2+4x+12=0$$

$a = 1, b = 4, c = 12$

$$x=\frac{-b\pm\sqrt{b^2-4ac}}{2a}$$
$$x=\frac{-4\pm\sqrt{(4)^2-4(1)(12)}}{2(1)}$$
$$x=\frac{-4\pm\sqrt{-32}}{2}=\frac{-4\pm 4i\sqrt{2}}{2}$$
$$x=\frac{2\left(-2\pm 2i\sqrt{2}\right)}{2}=-2\pm 2i\sqrt{2}$$

29. $2x^2+11=0$

$a = 2, b = 0, c = 11$

$$x=\frac{-b\pm\sqrt{b^2-4ac}}{2a}$$
$$x=\frac{-0\pm\sqrt{(0)^2-4(2)(11)}}{2(2)}$$
$$x=\pm\frac{\sqrt{-88}}{4}$$
$$x=\pm\frac{2i\sqrt{22}}{4}$$
$$x=\pm\frac{i\sqrt{22}}{2}$$

31. $3x^2 - 8x + 7 = 0$
$a = 3, b = -8, c = 7$
$$x = \frac{-b \pm \sqrt{b^2 - 4ac}}{2a}$$
$$x = \frac{-(-8) \pm \sqrt{(-8)^2 - 4(3)(7)}}{2(3)}$$
$$x = \frac{8 \pm \sqrt{-20}}{6} = \frac{8 \pm 2i\sqrt{5}}{6}$$
$$x = \frac{2\left(4 \pm i\sqrt{5}\right)}{6} = \frac{4 \pm i\sqrt{5}}{3}$$

33. $3x^2 + 4x = 2$
$3x^2 + 4x - 2 = 0$
$a = 3, b = 4, c = -2$
$b^2 - 4ac = 4^2 - 4(3)(-2) = 40$
2 irrational roots

35. $2x^2 + 10x + 8 = 0$
$a = 2, b = 10, c = 8$
$b^2 - 4ac = 10^2 - 4(2)(8) = 36 = 6^2$
2 rational roots

37. $9x^2 + 4 = 12x$
$9x^2 - 12x + 4 = 0$
$a = 9, b = -12, c = 4$
$b^2 - 4ac = (-12)^2 - 4(9)(4) = 0$
1 rational root

39. $7x(x-1) + 15 = 10$
$7x^2 - 7x + 5 = 0$
$a = 7, b = -7, c = 5$
$b^2 - 4ac = (-7)^2 - 4(7)(5) = -91$
2 nonreal complex roots

41. 13, −2
$x = 13$ $\quad$ $x = -2$
$x - 13 = 0$ $\quad$ $x + 2 = 0$
$(x-13)(x+2) = 0$
$x^2 - 11x - 26 = 0$

43. −3, −9
$x = -3$ $\quad$ $x = -9$
$x + 3 = 0$ $\quad$ $x + 9 = 0$
$(x+3)(x+9) = 0$
$x^2 + 12x + 27 = 0$

45. $4i, -4i$
$x = 4i$ $\quad$ $x = -4i$
$x - 4i = 0$ $\quad$ $x + 4i = 0$
$(x-4i)(x+4i) = 0$
$x^2 + 16 = 0$

47. $3, -\frac{5}{2}$
$x = 3$ $\quad$ $x = -\frac{5}{2}$
$x - 3 = 0$ $\quad$ $2x + 5 = 0$
$(x-3)(2x+5) = 0$
$2x^2 - x - 15 = 0$

49. $0.162x^2 + 0.094x - 0.485 = 0$
$a = 0.162, b = 0.094, c = -0.485$
$$x = \frac{-0.094 \pm \sqrt{0.094^2 - 4(0.162)(-0.485)}}{2(0.162)}$$
$$x = \frac{-0.094 \pm \sqrt{0.323116}}{0.324}$$
$x \approx 1.4643, x \approx -2.0445$

51. $p = -100x^2 + 4800x - 54{,}351 = 0$
$a = -100, b = 4800, c = -54{,}351$
$$x = \frac{-4800 \pm \sqrt{4800^2 - 4(-100)(-54{,}351)}}{2(-100)}$$
$$x = \frac{-4800 \pm \sqrt{1{,}299{,}600}}{-200}$$
$x = 18.3, x = 29.7$
Eighteen or thirty bikes per day will produce a zero profit.

Cumulative Review

53. $9x^2 - 6x + 3 - 4x - 12x^2 + 8 = -3x^2 - 10x + 11$

54. $3y(2-y) + \frac{1}{5}(10y^2 - 15y) = 6y - 3y^2 + 2y^2 - 3y$
$= -y^2 + 3y$

Quick Quiz 8.2

1. $11x^2 - 9x - 1 = 0$
$a = 11, b = -9, c = -1$
$x = \frac{-(-9) \pm \sqrt{(-9)^2 - 4(11)(-1)}}{2(11)}$
$x = \frac{9 \pm \sqrt{125}}{22}$
$x = \frac{9 \pm 5\sqrt{5}}{22}$

2. The LCD is $4x^2$.
$\frac{3}{4} + \frac{5}{4x} = \frac{2}{x^2}$
$\frac{3}{4}(4x^2) + \frac{5}{4x}(4x^2) = \frac{2}{x^2}(4x^2)$
$3x^2 + 5x = 8$
$3x^2 + 5x - 8 = 0$
$a = 3, b = 5, c = -8$
$x = \frac{-5 \pm \sqrt{(5)^2 - 4(3)(-8)}}{2(3)}$
$x = \frac{-5 \pm \sqrt{121}}{6}$
$x = \frac{-5 \pm 11}{6}$
$x = 1, x = -\frac{8}{3}$

3. $(x+2)(x+1) + (x-4)^2 = 9$
$x^2 + 3x + 2 + x^2 - 8x + 16 = 9$
$2x^2 - 5x + 18 = 9$
$2x^2 - 5x + 9 = 0$
$a = 2, b = -5, c = 9$
$x = \frac{-(-5) \pm \sqrt{(-5)^2 - 4(2)(9)}}{2(2)}$
$x = \frac{5 \pm \sqrt{-47}}{4}$
$x = \frac{5 \pm i\sqrt{47}}{4}$

4. Answers may vary. Possible solution: Subtract 3 from both sides to put the equation in standard form. Identify the values of a, b, and c, and find the value of the discriminant, $b^2 - 4ac$. If the discriminant is a perfect square, there are two rational solutions; if it is a positive number that is not a perfect square, there are two irrational solutions; if it is zero, there is one rational solution; if it is negative, there are two nonreal complex solutions.

8.3 Exercises

1. $x^4 - 9x^2 + 20 = 0$
$y = x^2$
$y^2 - 9y + 20 = 0$
$(y-5)(y-4) = 0$
$y = 5, y = 4$
$x^2 = 5$ $\quad$ $x^2 = 4$
$x = \pm\sqrt{5}$ $\quad$ $x = \pm 2$

3. $x^4 + x^2 - 12 = 0$
$y = x^2$
$y^2 + y - 12 = 0$
$(y-3)(y+4) = 0$
$y = 3, y = -4$
$x^2 = 3$ $\quad$ $x^2 = -4$
$x = \pm\sqrt{3}$ $\quad$ $x = \pm 2i$

5. $4x^4 - x^2 - 3 = 0$
$y = x^2$
$4y^2 - y - 3 = 0$
$(y-1)(4y+3) = 0$
$y = 1, -\frac{3}{4}$
$x^2 = 1$ $\quad$ $x^2 = -\frac{3}{4}$
$x = \pm 1$ $\quad$ $x = \pm\sqrt{-\frac{3}{4}} = \pm i\sqrt{\frac{3}{4}}$
$x = \pm\frac{i\sqrt{3}}{2}$

7. $x^6 - 7x^3 - 8 = 0$
$y = x^3$
$y^2 - 7y - 8 = 0$
$(y-8)(y+1) = 0$
$y = 8, y = -1$
$x^3 = 8$ $\quad$ $x^3 = -1$
$x = 2$ $\quad$ $x = -1$

9. $x^6 - 3x^3 = 0$

$y = x^3$

$y^2 - 3y = 0$

$y(y-3) = 0$

$y = 0, y = 3$

$x^3 = 0$ $\quad x^3 = 3$

$x = 0$ $\quad x = \sqrt[3]{3}$

11. $x^8 = 17x^4 - 16$

$x^8 - 17x^4 + 16 = 0$

$y = x^4$

$y^2 - 17y + 16 = 0$

$(y-16)(y-1) = 0$

$y = 16, y = 1$

$x^4 = 16$ $\quad x^4 = 1$

$x = \pm 2$ $\quad x = \pm 1$

13. $3x^8 + 13x^4 = 10$

$3x^8 + 13x^4 - 10 = 0$

$y = x^4$

$3y^2 + 13y - 10 = 0$

$(3y-2)(y+5) = 0$

$y = \frac{2}{3}, y = -5$

$x^4 = \frac{2}{3}$ $\quad x^4 = -5$

$x = \pm\sqrt[4]{\frac{2}{3}} = \pm\frac{\sqrt[4]{54}}{3}$ $\quad$ no real roots

15. $x^{2/3} + x^{1/3} - 12 = 0$

$y = x^{1/3}$

$y^2 + y - 12 = 0$

$(y+4)(y-3) = 0$

$y = -4, y = 3$

$x^{1/3} = -4$ $\quad x^{1/3} = 3$

$x = (-4)^3$ $\quad x = 3^3$

$x = -64$ $\quad x = 27$

17. $12x^{2/3} + 5x^{1/3} - 2 = 0$

$y = x^{1/3}$

$12y^2 + 5y - 2 = 0$

$(4y-1)(3y+2) = 0$

$y = \frac{1}{4}, y = -\frac{2}{3}$

$x^{1/3} = \frac{1}{4}$ $\quad x^{1/3} = -\frac{2}{3}$

$x = \left(\frac{1}{4}\right)^3$ $\quad x = \left(-\frac{2}{3}\right)^3$

$x = \frac{1}{64}$ $\quad x = -\frac{8}{27}$

19. $2x^{1/2} - 5x^{1/4} - 3 = 0$

$y = x^{1/4}$

$2y^2 - 5y - 3 = 0$

$(2y+1)(y-3) = 0$

$y = -\frac{1}{2}, y = 3$

$x^{1/4} = -\frac{1}{2}$ $\quad x^{1/4} = 3$

$x = \left(-\frac{1}{2}\right)^4 = \frac{1}{16}$, extraneous $\quad x = 3^4 = 81$

21. $2x^{1/2} - x^{1/4} - 6 = 0$

$y = x^{1/4}$

$2y^2 - y - 6 = 0$

$(y-2)(2y+3) = 0$

$y = 2, y = -\frac{3}{2}$

$x^{1/4} = 2$ $\quad x^{1/4} = -\frac{3}{2}$

$x = 2^4 = 16$ $\quad x = \left(-\frac{3}{2}\right)^4 = \frac{81}{16}$, extraneous

23. $x^{2/5} - x^{1/5} - 2 = 0$

$y = x^{1/5}$

$y^2 - y - 2 = 0$

$(y-2)(y+1) = 0$

$y = 2, y = -1$

$x^{1/5} = 2$ $\quad x^{1/5} = -1$

$x = 2^5$ $\quad x = (-1)^5$

$x = 32$ $\quad x = -1$

25. $x^6 - 5x^3 = 14$
$x^6 - 5x^3 - 14 = 0$
$y = x^3$
$y^2 - 5y - 14 = 0$
$(y-7)(y+2) = 0$
$y = 7, y = -2$
$x^3 = 7 \qquad x^3 = -2$
$x = \sqrt[3]{7} \qquad x = \sqrt[3]{-2}$

27. $(x^2 - x)^2 - 10(x^2 - x) + 24 = 0$
$y = x^2 - x$
$y^2 - 10y + 24 = 0$
$(y-6)(y-4) = 0$
$y = 6, y = 4$
$y = 6$
$x^2 - x = 6$
$x^2 - x - 6 = 0$
$(x-3)(x+2) = 0$
$x - 3 = 0 \qquad x + 2 = 0$
$x = 3 \qquad x = -2$

$y = 4$
$x^2 - x = 4$
$x^2 - x - 4 = 0$
$$x = \frac{-(-1) \pm \sqrt{(-1)^2 - 4(1)(-4)}}{2(1)}$$
$$x = \frac{1 \pm \sqrt{17}}{2}$$

29. $x - 5x^{1/2} + 6 = 0$
$y = x^{1/2}$
$y^2 - 5y + 6 = 0$
$(y-3)(y-2) = 0$
$y = 3, y = 2$
$x^{1/2} = 3 \qquad x^{1/2} = 2$
$x = 3^2 = 9 \qquad x = 2^2 = 4$

31. $x^{-2} + 3x^{-1} = 0$
$y = x^{-1}$
$y^2 + 3y = 0$
$y(y+3) = 0$
$y = 0, y = -3$
$x^{-1} = -3 \qquad x^{-1} = 0$, extraneous
$x = -\frac{1}{3}$

33.
$$15 - \frac{2x}{x-1} = \frac{x^2}{x^2 - 2x + 1}$$
$$15 - \frac{2x}{x-1} = \frac{x^2}{(x-1)(x-1)}$$
$15(x-1)^2 - 2x(x-1) = x^2$
$15x^2 - 30x + 15 - 2x^2 + 2x = x^2$
$12x^2 - 28x + 15 = 0$
$(6x-5)(2x-3) = 0$
$x = \frac{5}{6}, x = \frac{3}{2}$

Cumulative Review

35. $2x + 3y = 5$ **(1)**
$-5x - 2y = 4$ **(2)**
Multiply equation **(1)** by 2 and equation **(2)** by 3.
$4x + 6y = 10$
$-15x - 6y = 12$
$-11x \quad = 22$
$x = -2$
Replace x with -2 in equation **(1)**.
$2(-2) + 3y = 5$
$-4 + 3y = 5$
$3y = 9$
$y = 3$
The solution is $(-2, 3)$.

36. $$\frac{5 + \frac{2}{x}}{\frac{7}{3x} - 1} = \frac{5 + \frac{2}{x}}{\frac{7}{3x} - 1} \cdot \frac{3x}{3x} = \frac{15x + 6}{7 - x}$$

37. $3\sqrt{2}\left(\sqrt{5} - 2\sqrt{6}\right) = 3\sqrt{10} - 6\sqrt{12}$
$= 3\sqrt{10} - 6\sqrt{4 \cdot 3}$
$= 3\sqrt{10} - 12\sqrt{3}$

38. $\left(\sqrt{2} + \sqrt{6}\right)\left(3\sqrt{2} - 2\sqrt{5}\right)$
$= 6 - 2\sqrt{10} + 3\sqrt{12} - 2\sqrt{30}$
$= 6 - 2\sqrt{10} + 3\sqrt{4 \cdot 3} - 2\sqrt{30}$
$= 6 - 2\sqrt{10} + 6\sqrt{3} - 2\sqrt{30}$

39. $\frac{35.2-22.2}{22.2}\cdot 100\% = 58.6\%$

High school graduate: 58.6%

$\frac{46.2-30.9}{30.9}\cdot 100\% = 49.5\%$

Associate's degree: 49.5%

$\frac{68-40.7}{40.7}\cdot 100\% = 67.1\%$

Bachelor's degree: 67.1%

$\frac{105.2-66.4}{66.4}\cdot 100\% = 58.4\%$

Doctorate: 58.4%

40. $(30.9)x = 30(46.2)$

$x \approx 44.9$

The woman would have to work an additional 14.9 years.

Quick Quiz 8.3

1. $x^4 - 18x^2 + 32 = 0$

$y = x^2$

$y^2 - 18y + 32 = 0$

$(y-16)(y-2) = 0$

$y = 16,\ y = 2$

$x^2 = 16 \qquad x^2 = 2$

$x = \pm 4 \qquad x = \pm\sqrt{2}$

2. $2x^{-2} - 3x^{-1} - 20 = 0$

$y = x^{-1}$

$2y^2 - 3y - 20 = 0$

$(2y+5)(y-4) = 0$

$y = -\frac{5}{2},\ y = 4$

$x^{-1} = -\frac{5}{2} \qquad x^{-1} = 4$

$x = -\frac{2}{5} \qquad x = \frac{1}{4}$

3. $x^{2/3} - 4x^{1/3} - 5 = 0$

$y = x^{1/3}$

$y^2 - 4y - 5 = 0$

$(y-5)(y+1) = 0$

$y = 5,\ y = -1$

$x^{1/3} = 5 \qquad x^{1/3} = -1$

$x = 5^3 \qquad x = (-1)^3$

$x = 125 \qquad x = -1$

4. Answers may vary. Possible solution: Substitute y for x^4, y^2 for x^8 which yields $y^2 - 6y = 0$. Next factor out y from the left side of the equation yielding $y(y-6) = 0$. Set each term equal to zero and solve for y yielding $y = 0$ and $y = 6$. Substitute x^4 for y and solve for x yielding $x = 0$ and $x = \pm\sqrt[4]{6}$.

How Am I Doing? Sections 8.1–8.3

1. $2x^2 + 3 = 39$

$2x^2 = 36$

$x^2 = 18$

$x = \pm\sqrt{18}$

$x = \pm\sqrt{9\cdot 2}$

$x = \pm 3\sqrt{2}$

2. $(4x+1)^2 = 24$

$4x + 1 = \pm\sqrt{24} = \pm 2\sqrt{6}$

$4x = -1 \pm 2\sqrt{6}$

$x = \frac{-1 \pm 2\sqrt{6}}{4}$

3. $x^2 - 8x = -12$

$x^2 - 8x + 16 = -12 + 16$

$(x-4)^2 = 4$

$x - 4 = \pm 2$

$x = 4 \pm 2$

$x = 2,\ x = 6$

4. $2x^2 - 4x - 3 = 0$

$$2x^2 - 4x = 3$$
$$x^2 - 2x = \frac{3}{2}$$
$$x^2 - 2x + 1 = \frac{3}{2} + 1$$
$$(x-1)^2 = \frac{5}{2}$$
$$x - 1 = \pm\sqrt{\frac{5}{2}}$$
$$x - 1 = \pm\frac{\sqrt{10}}{2}$$
$$x = 1 \pm \frac{\sqrt{10}}{2}$$
$$x = \frac{2 \pm \sqrt{10}}{2}$$

5. $8x^2 - 2x - 7 = 0$

$a = 8,\ b = -2,\ c = -7$

$$x = \frac{-b \pm \sqrt{b^2 - 4ac}}{2a}$$
$$x = \frac{-(-2) \pm \sqrt{(-2)^2 - 4(8)(-7)}}{2(8)}$$
$$x = \frac{2 \pm \sqrt{228}}{16}$$
$$x = \frac{2 \pm 2\sqrt{57}}{16}$$
$$x = \frac{2\left(1 \pm \sqrt{57}\right)}{16}$$
$$x = \frac{1 \pm \sqrt{57}}{8}$$

6. $(x-1)(x+5) = 2$

$$x^2 + 4x - 5 = 2$$
$$x^2 + 4x - 7 = 0$$

$a = 1,\ b = 4,\ c = -7$

$$x = \frac{-b \pm \sqrt{b^2 - 4ac}}{2a}$$
$$x = \frac{-4 \pm \sqrt{(4)^2 - 4(1)(-7)}}{2(1)}$$
$$x = \frac{-4 \pm \sqrt{44}}{2}$$
$$x = \frac{-4 \pm 2\sqrt{11}}{2}$$
$$x = \frac{2\left(-2 \pm \sqrt{11}\right)}{2}$$
$$x = -2 \pm \sqrt{11}$$

7. $4x^2 = -12x - 17$

$$4x^2 + 12x + 17 = 0$$

$a = 4,\ b = 12,\ c = 17$

$$x = \frac{-b \pm \sqrt{b^2 - 4ac}}{2a}$$
$$x = \frac{-12 \pm \sqrt{(12)^2 - 4(4)(17)}}{2(4)}$$
$$x = \frac{-12 \pm \sqrt{-128}}{8}$$
$$x = \frac{-12 \pm 8i\sqrt{2}}{8}$$
$$x = \frac{4\left(-3 \pm 2i\sqrt{2}\right)}{8}$$
$$x = \frac{-3 \pm 2i\sqrt{2}}{2}$$

8. $5x^2 + 4x - 12 = 0$

$a = 5,\ b = 4,\ c = -12$

$$x = \frac{-b \pm \sqrt{b^2 - 4ac}}{2a}$$
$$x = \frac{-4 \pm \sqrt{(4)^2 - 4(5)(-12)}}{2(5)}$$
$$x = \frac{-4 \pm \sqrt{256}}{10}$$
$$x = \frac{-4 \pm 16}{10}$$
$$x = -2,\ x = \frac{6}{5}$$

9. $8x^2+3x=3x^2+4x$

$5x^2-x=0$

$a=5, b=-1, c=0$

$x=\dfrac{-b\pm\sqrt{b^2-4ac}}{2a}$

$x=\dfrac{-(-1)\pm\sqrt{(-1)^2-4(5)(0)}}{2(5)}$

$x=\dfrac{1\pm1}{10}$

$x=0,\ x=\dfrac{1}{5}$

10. $4x^2-3x=-6$

$4x^2-3x+6=0$

$a=4, b=-3, c=6$

$x=\dfrac{-b\pm\sqrt{b^2-4ac}}{2a}$

$x=\dfrac{-(-3)\pm\sqrt{(-3)^2-4(4)(6)}}{2(4)}$

$x=\dfrac{3\pm\sqrt{-87}}{8}$

$x=\dfrac{3\pm i\sqrt{87}}{8}$

11. The LCD is $x(x+1)$.

$\dfrac{18}{x}+\dfrac{12}{x+1}=9$

$\dfrac{18}{x}(x)(x+1)+\dfrac{12}{x+1}(x)(x+1)=9(x)(x+1)$

$18(x+1)+12x=9x(x+1)$

$18x+18+12x=9x^2+9x$

$9x^2-21x-18=0$

$3x^2-7x-6=0$

$a=3, b=-7, c=-6$

$x=\dfrac{-b\pm\sqrt{b^2-4ac}}{2a}$

$x=\dfrac{-(-7)\pm\sqrt{(-7)^2-4(3)(-6)}}{2(3)}$

$x=\dfrac{7\pm\sqrt{121}}{6}$

$x=\dfrac{7\pm11}{6}$

$x=3,\ x=-\dfrac{2}{3}$

12. $x^6-7x^3-8=0$

$y=x^3$

$y^2-7y-8=0$

$(y-8)(y+1)=0$

$y=8, y=-1$

$x^3=8 \qquad x^3=-1$

$x=2 \qquad x=-1$

13. $w^{4/3}-6w^{2/3}+8=0$

$y=w^{2/3}$

$y^2-6y+8=0$

$(y-4)(y-2)=0$

$y=4, y=2$

$w^{2/3}=4 \qquad w^{2/3}=2$

$w^2=64 \qquad w^2=8$

$w=\pm8 \qquad w=\pm2\sqrt{2}$

14. $x^8=5x^4-6$

$x^8-5x^4+6=0$

$y=x^4$

$y^2-5y+6=0$

$(y-2)(y-3)=0$

$y=2, y=3$

$x^4=2 \qquad x^4=3$

$x=\pm\sqrt[4]{2} \qquad x=\pm\sqrt[4]{3}$

15. $2x^{2/5}=7x^{1/5}-3$

$2x^{2/5}-7x^{1/5}+3=0$

$y=x^{1/5}$

$2y^2-7y+3=0$

$(2y-1)(y-3)=0$

$y=\dfrac{1}{2},\ y=3$

$x^{1/5}=\dfrac{1}{2} \qquad x^{1/5}=3$

$x=\dfrac{1}{32} \qquad x=243$

8.4 Exercises

1. $S = 16t^2$

$t^2 = \frac{S}{16}$

$t = \pm\sqrt{\frac{S}{16}} = \pm\frac{\sqrt{S}}{4}$

3. $S = 9\pi r^2$

$r^2 = \frac{S}{9\pi}$

$r = \pm\sqrt{\frac{S}{9\pi}} = \pm\frac{1}{3}\sqrt{\frac{S}{\pi}}$

5. $5M = \frac{1}{2}by^2$

$y^2 = \frac{10M}{b}$

$y = \pm\sqrt{\frac{10M}{b}}$

7. $4(y^2 + w) - 5 = 7R$

$4(y^2 + w) = 5 + 7R$

$4y^2 + 4w = 5 + 7R$

$y^2 = \frac{7R - 4w + 5}{4}$

$y = \pm\sqrt{\frac{7R - 4w + 5}{4}}$

$y = \pm\frac{\sqrt{7R - 4w + 5}}{2}$

9. $Q = \frac{3mwM^2}{2c}$

$M^2 = \frac{2Qc}{3mw}$

$M = \pm\sqrt{\frac{2Qc}{3mw}}$

11. $V = \pi(r^2 + R^2)h$

$V = \pi r^2 h + \pi R^2 h$

$\pi r^2 h = V - \pi R^2 h$

$r^2 = \frac{V - \pi R^2 h}{\pi h}$

$r = \pm\sqrt{\frac{V - \pi R^2 h}{\pi h}}$

13. $7bx^2 - 3ax = 0$

$x(7bx - 3a) = 0$

$x = 0,\ x = \frac{3a}{7b}$

15. $P = EI - RI^2$

$RI^2 - EI + P = 0$

$a = R, b = -E, c = P$

$I = \frac{-(-E) \pm \sqrt{(-E)^2 - 4RP}}{2R}$

$I = \frac{E \pm \sqrt{E^2 - 4RP}}{2R}$

17. $9w^2 + 5tw - 2 = 0$

$a = 9, b = 5t, c = -2$

$w = \frac{-(5t) \pm \sqrt{(5t)^2 - 4(9)(-2)}}{2(9)}$

$w = \frac{-5t \pm \sqrt{25t^2 + 72}}{18}$

19. $S = 2\pi rh + \pi r^2$

$\pi r^2 + 2\pi hr - S = 0$

$a = \pi, b = 2\pi h, c = -S$

$r = \frac{-2\pi h \pm \sqrt{(2\pi h)^2 - 4\pi(-S)}}{2\pi}$

$r = \frac{-2\pi h \pm \sqrt{4\pi^2 h^2 + 4\pi S}}{2\pi}$

$r = \frac{-2\pi h \pm \sqrt{4(\pi^2 h^2 + \pi S)}}{2\pi}$

$r = \frac{-2\pi h \pm 2\sqrt{\pi^2 h^2 + \pi S}}{2\pi}$

$r = \frac{-\pi h \pm \sqrt{\pi^2 h^2 + \pi S}}{\pi}$

21. $(a+1)x^2 + 5x + 2w = 0$

$a = a + 1, b = 5, c = 2w$

$x = \frac{-5 \pm \sqrt{5^2 - 4(a+1)(2w)}}{2(a+1)}$

$x = \frac{-5 \pm \sqrt{25 - 8aw - 8w}}{2a + 2}$

23. $c^2 = a^2 + b^2$
$6^2 = 4^2 + b^2$
$36 = 16 + b^2$
$b^2 = 20$
$b = \sqrt{20} = 2\sqrt{5}$

25. $c^2 = a^2 + b^2$
$\left(\sqrt{34}\right)^2 = a^2 + \left(\sqrt{19}\right)^2$
$34 = a^2 + 19$
$a^2 = 15$
$a = \sqrt{15}$

27. $c^2 = a^2 + b^2$
$12^2 = a^2 + (2a)^2$
$144 = 5a^2$
$a^2 = \frac{144}{5}$
$a = \sqrt{\frac{144}{5}}$
$a = \frac{12}{\sqrt{5}}$
$a = \frac{12\sqrt{5}}{5}$
$b = 2a = \frac{24\sqrt{5}}{5}$

29. $c^2 = a^2 + b^2$
$10^2 = x^2 + x^2$
$100 = 2x^2$
$x^2 = 50$
$x = \sqrt{50} = 5\sqrt{2}$
Each leg is $5\sqrt{2}$ inches long.

31. Let x = the width of the parking lot. Then the length is $x + 0.07$. The area of a rectangle is $A = LW$.
$x(x + 0.07) = 0.026$
$x^2 + 0.07x - 0.026 = 0$
$x = \frac{-0.07 \pm \sqrt{(0.07)^2 - 4(1)(-0.026)}}{2(1)}$
$x = \frac{-0.07 \pm \sqrt{0.1089}}{2} = \frac{-0.07 \pm 0.33}{2}$
$x = 0.13,\ x = -0.2$
Since the width must be positive, $x = 0.13$. The width is 0.13 mile and the length is $0.13 + 0.07 = 0.2$ mile.

33. If W is the width of the barn, then $2W + 4$ is the length.
$(2W + 4)W = 126$
$2W^2 + 4W - 126 = 0$
$W^2 + 2W - 63 = 0$
$(W - 7)(W + 9) = 0$
$W = 7,\ W = -9$
Since the width must be positive, $W = 7$.
The width of the barn is 7 feet and the length is $2(7) + 4 = 18$ feet.

35. If b = base of triangle, then $2b + 2$ = the altitude.
$A = \frac{1}{2}bh$
$\frac{1}{2}b(2b + 2) = 72$
$2b^2 + 2b = 144$
$b^2 + b - 72 = 0$
$(b - 8)(b + 9) = 0$
$b = 8,\ b = -9$
Since the base must have a positive length, $b = 8$.
The base is 8 cm and the altitude is $2(8) + 2 = 18$ cm.

37. If v = his speed in the rain then $v + 5$ is his speed after the rain stopped.
$\text{Time} = \frac{\text{distance}}{\text{rate}}$
$\frac{225}{v} + \frac{150}{v+5} = 8$
$225(v + 5) + 150v = 8v(v + 5)$
$225v + 1125 + 150v = 8v^2 + 40v$
$8v^2 - 335v - 1125 = 0$
$(v - 45)(8v + 25) = 0$
$v = 45,\ v = -3{,}125$
Since his speed is positive, $v = 45$. His speed in the rain was 45 mph and his speed after the rain stopped was 50 mph.

39. t = time from home to work

$1 \text{ hr } 16 \text{ min} = 1 + \frac{16}{60} = \frac{19}{15} \text{ hr}$

$\frac{19}{15} - t$ = time from work to home

The distance from home to work and the distance from work to home are the same.

Distance = (Rate)(Time)

$$50t = 45\left(\frac{19}{15} - t\right)$$
$$50t = 57 - 45t$$
$$95t = 57$$
$$t = \frac{57}{95} = \frac{3}{5}$$

$\frac{50 \text{ mi}}{\text{hr}} \cdot \frac{3}{5} \text{ hr} = 30 \text{ mi}$

Bob lives 30 miles from his job.

41. $N = -2.06x^2 + 77.82x + 743$

In the year 2000, $x = 10$.

$N = -2.06(10)^2 + 77.82(10) + 743$

$N = 1315.2$

There were approximately 1,315,200 inmates in the year 2000.

43. $1349.76 = -2.06x^2 + 77.82x + 743$

$2.06x^2 - 77.82x + 606.76 = 0$

$$x = \frac{-(-77.82) \pm \sqrt{(-77.82)^2 - 4(2.06)(606.76)}}{2(2.06)}$$

$$x = \frac{77.82 \pm 32.5}{4.12}$$

$x \approx 26.78,\ x = 11$

The number of inmates was expected to be 1,349,760 in 2001 (and again in 2016).

Cumulative Review

45. $\frac{4}{\sqrt{3x}} = \frac{4}{\sqrt{3x}} \cdot \frac{\sqrt{3x}}{\sqrt{3x}} = \frac{4\sqrt{3x}}{3x}$

46. $\frac{5\sqrt{6}}{2\sqrt{5}} = \frac{5\sqrt{6}}{2\sqrt{5}} \cdot \frac{\sqrt{5}}{\sqrt{5}} = \frac{5\sqrt{30}}{10} = \frac{\sqrt{30}}{2}$

47. $\frac{3}{\sqrt{x}+\sqrt{y}} = \frac{3}{\sqrt{x}+\sqrt{y}} \cdot \frac{\sqrt{x}-\sqrt{y}}{\sqrt{x}-\sqrt{y}} = \frac{3\left(\sqrt{x}-\sqrt{y}\right)}{x-y}$

48.
$$\frac{2\sqrt{3}}{\sqrt{3}-\sqrt{6}} = \frac{2\sqrt{3}}{\sqrt{3}-\sqrt{6}} \cdot \frac{\sqrt{3}+\sqrt{6}}{\sqrt{3}+\sqrt{6}}$$
$$= \frac{6+2\sqrt{18}}{3-6}$$
$$= \frac{6+6\sqrt{2}}{3-6}$$
$$= \frac{-3\left(-2-2\sqrt{2}\right)}{-3}$$
$$= -2-2\sqrt{2}$$

Quick Quiz 8.4

1. $H = \frac{5ab}{y^2}$

$y^2 = \frac{5ab}{H}$

$y = \pm\sqrt{\frac{5ab}{H}}$

2. $6z^2 + 7yz - 5w = 0$

$a = 6,\ b = 7y,\ c = -5w$

$$z = \frac{-7y \pm \sqrt{(7y)^2 - 4(6)(-5w)}}{2(6)}$$
$$z = \frac{-7y \pm \sqrt{49y^2 + 120w}}{12}$$

3. If x = the width, then $3x + 4$ = the length.

$A = \text{length} \cdot \text{width}$

$175 = (3x + 4)x$

$3x^2 + 4x - 175 = 0$

$(3x + 25)(x - 7) = 0$

$x = -\frac{25}{3},\ x = 7$

Since the width must be positive, $x = 7$. The width is 7 yards and the length is $3(7) + 4 = 25$ yards.

4. Answers may vary. Possible solution:

Set one leg's length = x, then the other leg's length = $3x$.

Use $c^2 = a^2 + b^2$ with $c = 12$, $a = x$, and $b = 3x$.

Solve for x to find one leg length then multiply the found value of x by 3 to find the other leg's length.

8.5 Exercises

1. $f(x) = x^2 - 2x - 8$

$x_{\text{vertex}} = \dfrac{-b}{2a}$

$x_{\text{vertex}} = \dfrac{-(-2)}{2(1)}$

$x_{\text{vertex}} = 1$

$f(x_{\text{vertex}}) = 1^2 - 2(1) - 8$

$y_{\text{vertex}} = -9$

$V(1, -9)$

$f(0) = 0^2 - 2(0) - 8$

$f(0) = -8$

$x^2 - 2x - 8 = 0$

$(x-4)(x+2) = 0$

$x = 4, x = -2$

$V(1, -9); (0, -8); (4, 0), (-2, 0)$

3. $g(x) = -x^2 - 8x + 9$

$x_{\text{vertex}} = \dfrac{-b}{2a}$

$x_{\text{vertex}} = \dfrac{-(-8)}{2(-1)}$

$x_{\text{vertex}} = -4$

$g(x_{\text{vertex}}) = -(-4)^2 - 8(-4) + 9$

$g(x_{\text{vertex}}) = 25$

$V(-4, 25)$

$g(0) = 0^2 - 8(0) + 9$

$g(0) = 9$

$-x^2 - 8x + 9 = 0$

$x^2 + 8x - 9 = 0$

$(x+9)(x-1) = 0$

$x = -9, x = 1$

$V(-4, 25); (0, 9); (-9, 0), (1, 0)$

5. $p(x) = 3x^2 + 12x + 3$

$x_{\text{vertex}} = \dfrac{-b}{2a}$

$x_{\text{vertex}} = \dfrac{-(12)}{2(3)}$

$x_{\text{vertex}} = -2$

$p(x_{\text{vertex}}) = 3(-2)^2 + 12(-2) + 3$

$p(x_{\text{vertex}}) = -9$

$V(-2, -9)$

$p(0) = 3(0)^2 + 12(0) + 3$

$p(0) = 3$

$3x^2 + 12x + 3 = 0$

$x^2 + 4x + 1 = 0$

$a = 1, b = 4, c = 1$

$x = \dfrac{-4 \pm \sqrt{4^2 - 4(1)(1)}}{2(1)}$

$x \approx -0.3, x \approx -3.7$

$V(-2, -9); (0, 3); (-0.3, 0), (-3.7, 0)$

7. $r(x) = -3x^2 - 2x - 6$

$x_{\text{vertex}} = \dfrac{-b}{2a}$

$x_{\text{vertex}} = \dfrac{-(-2)}{2(-3)}$

$x_{\text{vertex}} = -\dfrac{1}{3}$

$r(x_{\text{vertex}}) = -3\left(-\dfrac{1}{3}\right)^2 - 2\left(-\dfrac{1}{3}\right) - 6$

$r(x_{\text{vertex}}) = -\dfrac{17}{3}$

$V\left(-\dfrac{1}{3}, -\dfrac{17}{3}\right)$

$r(0) = -3(0)^2 - 2(0) - 6$

$r(0) = -6$

$-3x^2 - 2x - 6 = 0$

$a = -3, b = -2, c = -6$

$b^2 - 4ac = -68 < 0 \Rightarrow$ no x-intercepts

$V\left(-\dfrac{1}{3}, -\dfrac{17}{3}\right)$; $(0, -6)$

9. $f(x) = 2x^2 + 2x - 4$

$x_{\text{vertex}} = \dfrac{-b}{2a}$

$x_{\text{vertex}} = \dfrac{-(2)}{2(2)}$

$x_{\text{vertex}} = -\dfrac{1}{2}$

$f(x_{\text{vertex}}) = 2\left(-\dfrac{1}{2}\right)^2 + 2\left(-\dfrac{1}{2}\right) - 4$

$f(x_{\text{vertex}}) = -\dfrac{9}{2}$

$V\left(-\dfrac{1}{2}, -\dfrac{9}{2}\right)$

$f(0) = 2(0)^2 + 2(0) - 4$
$f(0) = -4$

$2x^2 + 2x - 4 = 0$
$x^2 + x - 2 = 0$
$(x-1)(x+2) = 0$
$x = 1, x = -2$
$V\left(-\frac{1}{2}, -\frac{9}{2}\right)$; (0, –4); (1, 0), (–2, 0)

11. $f(x) = x^2 - 6x + 8$

$\frac{-b}{2a} = \frac{-(-6)}{2(1)} = 3$

$f(3) = 3^2 - 6(3) + 8 = -1$
$V(3, -1)$
$f(0) = 0^2 - 6(0) + 8 = 8$
y-intercept: (0, 8)
$x^2 - 6x + 8 = 0$
$(x-4)(x-2) = 0$
$x = 4, x = 2$
x-intercepts: (4, 0), (2, 0)

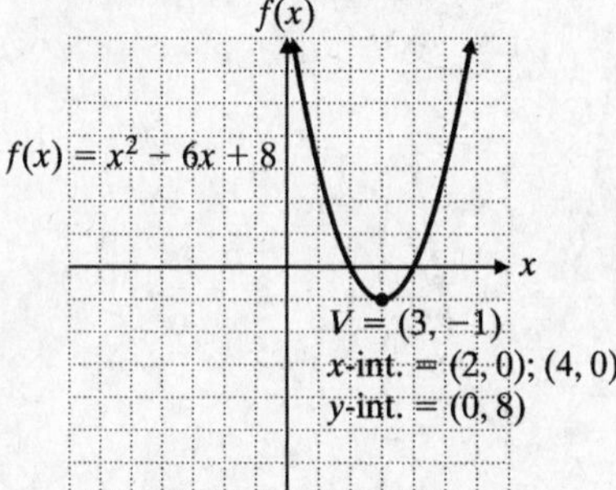

13. $g(x) = x^2 + 2x - 8$

$\frac{-b}{2a} = \frac{-2}{2} = -1$

$g(-1) = (-1)^2 + 2(-1) - 8 = -9$
$V(-1, -9)$
$g(0) = 0^2 + 2(0) - 8 = -8$
y-intercept: (0, –8)
$x^2 + 2x - 8 = 0$
$(x+4)(x-2) = 0$
$x = -4, x = 2$
x-intercepts: (–4, 0), (2, 0)

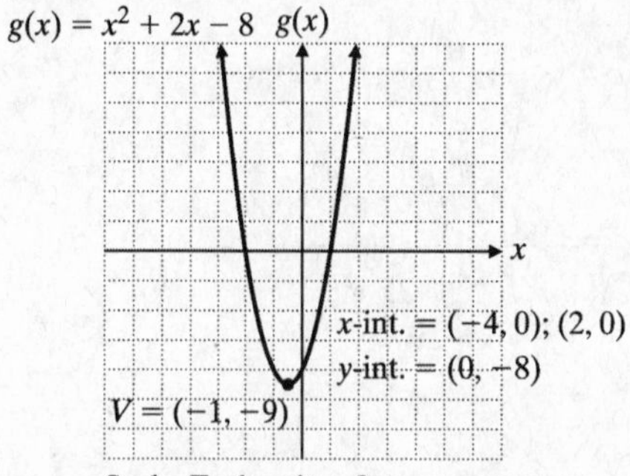

Scale: Each unit = 2

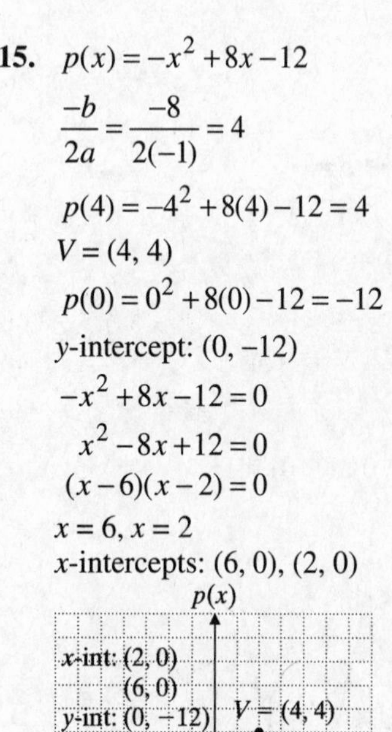

15. $p(x) = -x^2 + 8x - 12$

$\frac{-b}{2a} = \frac{-8}{2(-1)} = 4$

$p(4) = -4^2 + 8(4) - 12 = 4$
$V = (4, 4)$
$p(0) = 0^2 + 8(0) - 12 = -12$
y-intercept: (0, –12)
$-x^2 + 8x - 12 = 0$
$x^2 - 8x + 12 = 0$
$(x-6)(x-2) = 0$
$x = 6, x = 2$
x-intercepts: (6, 0), (2, 0)

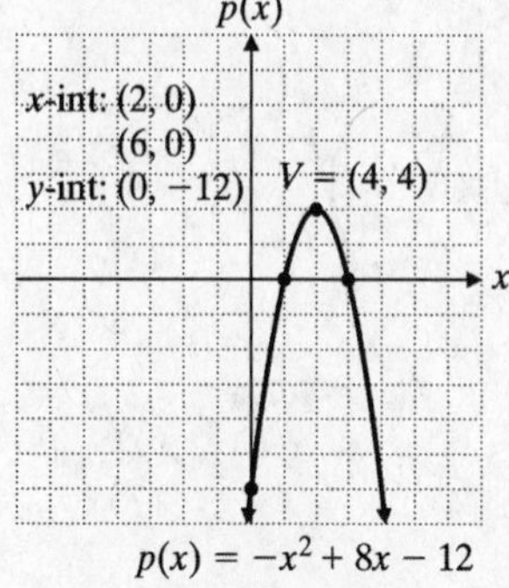

Scale: Each unit = 2

17. $r(x) = 3x^2 + 6x + 4$

$\frac{-b}{2a} = \frac{-6}{2(3)} = -1$

$r(-1) = 3(-1)^2 + 6(-1) + 4 = 1$
$V(-1, 1)$
$r(0) = 3 \cdot 0^2 + 6(0) + 4 = 4$
y-intercept: (0, 4)
$3x^2 + 6x + 4 = 0$
$a = 3, b = 6, c = 4$
$b^2 - 4ac = 6^2 - 4(3)(4) = -12 < 0$
x-intercepts: none

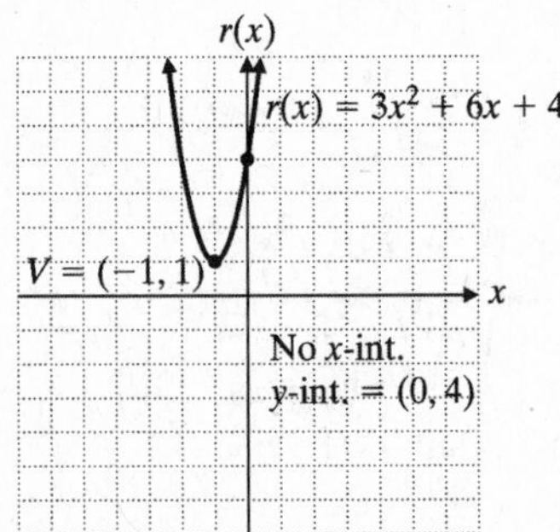

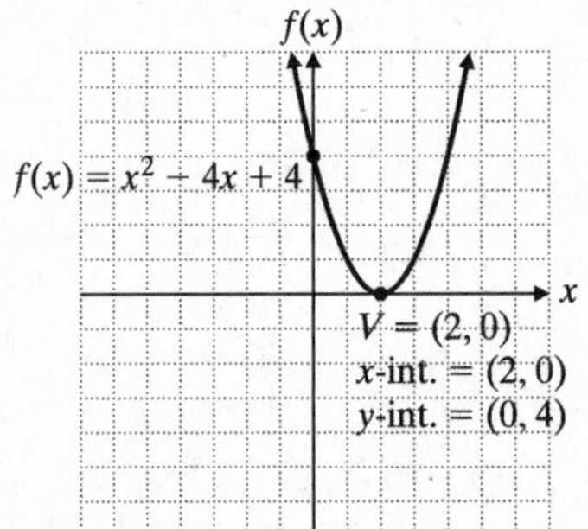

19. $f(x)=x^2-6x+5$

$\frac{-b}{2a}=\frac{-(-6)}{2(1)}=3$

$f(3)=3^2-6(3)+5=-4$

$V(3,-4)$

$f(0)=0^2-6(0)+5=5$

y-intercept: (0, 5)

$x^2-6x+5=0$

$(x-5)(x-1)=0$

$x=5, x=1$

x-intercepts: (5, 0), (1, 0)

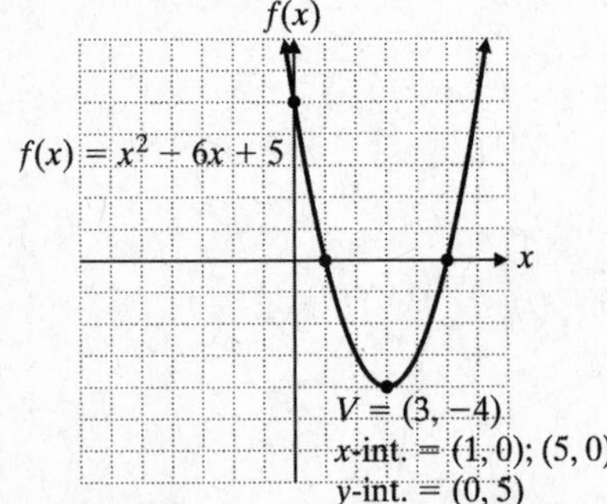

21. $f(x)=x^2-4x+4$

$\frac{-b}{2a}=\frac{-(-4)}{2(1)}=2$

$f(2)=2^2-4(2)+4=0$

$V(2, 0)$

$f(0)=0^2-4(0)+4=4$

y-intercept: (0, 4)

$x^2-4x+4=0$

$(x-2)(x-2)=0$

$x=2, x=2$

x-intercept: (2, 0)

23. $f(x)=x^2-4$

$\frac{-b}{2a}=\frac{-0}{2(1)}=0$

$f(0)=0^2-4=-4$

$V(0,-4)$

$f(0)=0^2-4=-4$

y-intercept: (0, −4)

$x^2-4=0$

$x^2=4$

$x=\pm 2$

x-intercepts: (±2, 0)

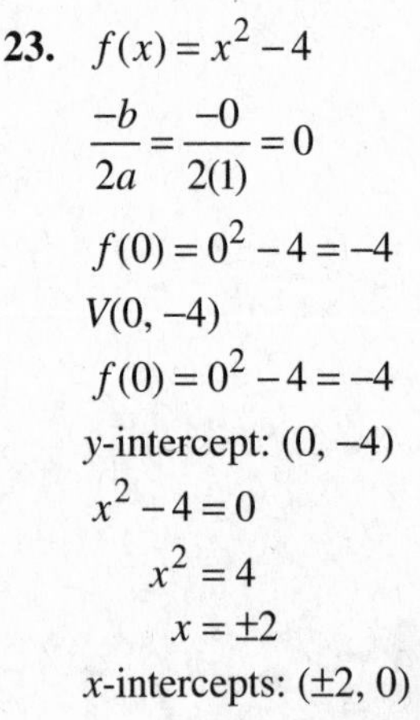

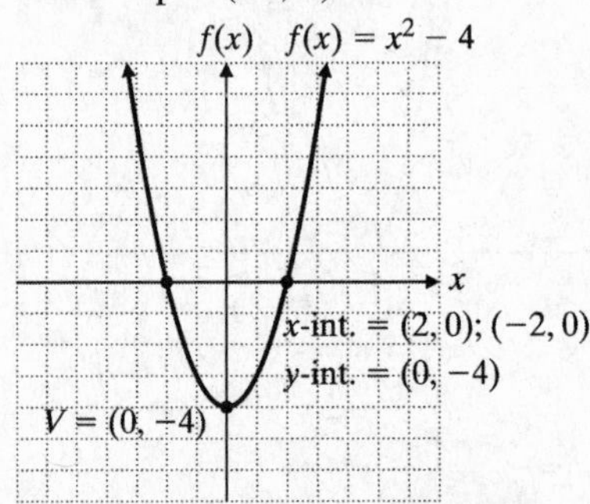

25. $N(x)=0.18x^2-3.18x+102.25$

$N(20)=0.18(20)^2-3.18(20)+102.25$
$=110.65$

$N(40)=0.18(40)^2-3.18(40)+102.25$
$=263.05$

$N(60)=0.18(60)^2-3.18(60)+102.25$
$=559.45$

$N(80)=0.18(80)^2-3.18(80)+102.25$
$=999.85$

$N(100)=0.18(100)^2-3.18(100)+102.25$
$=1584.25$

27. From the graph, $N(70) \approx 750.000$. There are approximately 750,000 people who scuba dive and have a mean income of \$70,000.

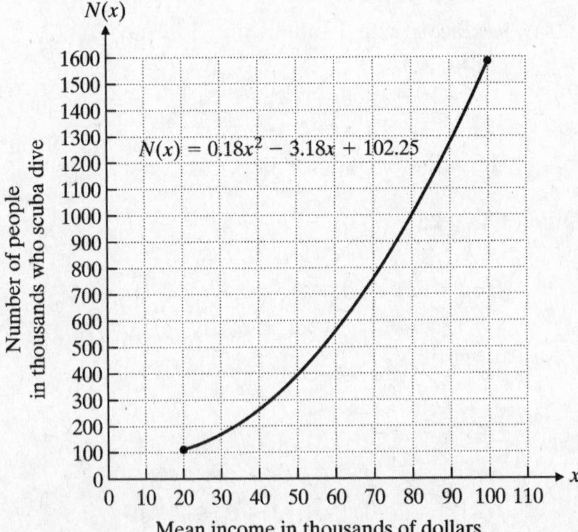

29. From the graph, $N(x) = 1000$ when $x \approx 80$. This means 1,000,000 people who scuba dive have a mean income of \$80,000.

31. $P(x) = -6x^2 + 312x - 3672$

$$P(16) = -6(16)^2 + 312(16) - 3672 = -216$$
$$P(20) = -6(20)^2 + 312(20) - 3672 = 168$$
$$P(24) = -6(24)^2 + 312(24) - 3672 = 360$$
$$P(30) = -6(30)^2 + 312(30) - 3672 = 288$$
$$P(35) = -6(35)^2 + 312(35) - 3672 = -102$$

33. $\dfrac{-b}{2a} = \dfrac{-312}{2(-6)} = 26$

$$P(26) = -6(26)^2 + 312(26) - 3672 = 384$$

26 tables per day will give a maximum profit of \$384 per day.

35. $P(x) = 0$

$$-6x^2 + 312x - 3672 = 0$$
$$x^2 - 52x + 612 = 0$$
$$(x-18)(x-34) = 0$$

$x = 18, x = 34$

18 or 34 tablets per day will give a daily profit of zero dollars.

37. $d(t) = -16t^2 + 32t + 40$

$$\frac{-b}{2a} = \frac{-32}{2(-16)} = 1$$
$$d(1) = -16(1^2) + 32(1) + 40 = 56$$
$$-16^2 + 32t + 40 = 0$$
$$t = \frac{-32 \pm \sqrt{32^2 - 4(-16)(40)}}{2(-16)}$$

$t \approx 2.9,\ t \approx -0.9$

The maximum height is 56 feet and the ball will reach the ground about 2.9 seconds after being thrown upward.

39. $y = 2.3x^2 - 5.4x - 1.6$

x-intercepts: $(-0.3, 0), (2.6, 0)$

41. From $(0, -10)$:

$$-10 = a \cdot 0^2 + b \cdot 0 + c$$
$$-10 = c$$

From $(3, 41)$:

$$41 = a \cdot 3^2 + b \cdot 3 - 10$$
$$41 = 9a + 3b - 10$$
$$51 = 9a + 3b \quad \textbf{(1)}$$

From $(-1, -15)$:

$$-15 = a(-1)^2 + b(-1) - 10$$
$$-15 = a - b - 10$$
$$-5 = a - b \quad \textbf{(2)}$$

Solve equation **(2)** for *a*.

$a = b - 5$

Substitute $b - 5$ for *a* in equation **(1)**.

$$51 = 9(b-5) + 3b$$
$$51 = 9b - 45 + 3b$$
$$96 = 12b$$
$$8 = b$$
$$a = b - 5 = 8 - 5 = 3$$

$a = 3,\ b = 8,\ c = -10$

Cumulative Review

43. $3x - y + 2z = 12$ **(1)**

$2x - 3y + z = 5$ **(2)**

$x + 3y + 8z = 22$ **(3)**

Eliminate *y* by multiplying equation **(1)** by 3 and adding the result to equation **(3)**.

$$\begin{array}{r} 9x-3y+6z=36 \\ x+3y+8z=22 \\ \hline 10x \quad +14z=58 \\ 5x+7z=29 \quad \textbf{(4)} \end{array}$$

Eliminate y by adding equation **(2)** and **(3)**.

$$\begin{array}{r} 2x-3y+z=5 \\ x+3y+8z=22 \\ \hline 3x \quad +9z=27 \\ x+3z=9 \quad \textbf{(5)} \end{array}$$

Multiply equation **(5)** by –5 and add the result to equation **(4)**.

$$\begin{array}{r} 5x+7z=29 \\ -5x-15z=-45 \\ \hline -8z=-16 \\ z=2 \end{array}$$

Replace z with 2 in equation **(5)**.

$$\begin{aligned} x+3(2)&=9 \\ x+6&=9 \\ x&=3 \end{aligned}$$

Replace x with 3 and z with 2 in equation **(1)**.

$$\begin{aligned} 3(3)-y+2(2)&=12 \\ 9-y+4&=12 \\ -y&=-1 \\ y&=1 \end{aligned}$$

The solution is (3, 1, 2).

44. $7x+3y-z=-2$ **(1)**
$x+5y+3z=2$ **(2)**
$x+2y+z=1$ **(3)**

Eliminate z by multiplying equation **(1)** by 3 and adding the result to equation **(2)**.

$$\begin{array}{r} 21x+9y-3z=-6 \\ x+5y+3z=2 \\ \hline 22x+14y \quad =-4 \\ 11x+7y=-2 \quad \textbf{(4)} \end{array}$$

Eliminate z by adding equations **(1)** and **(3)**.

$$\begin{array}{r} 7x+3y-z=-2 \\ x+2y+z=1 \\ \hline 8x+5y \quad =-1 \quad \textbf{(5)} \end{array}$$

Multiply equation **(4)** by 5, equation **(5)** by –7, and add the results.

$$\begin{array}{r} 55x+35y=-10 \\ -56x-35y=7 \\ \hline -x \quad =-3 \\ x=3 \end{array}$$

Replace x with 3 in equation **(4)**.

$$\begin{aligned} 11(3)+7y&=-2 \\ 33+7y&=-2 \\ 7y&=-35 \\ y&=-5 \end{aligned}$$

Replace x with 3 and y with –5 in equation **(3)**.

$$\begin{aligned} 3+2(-5)+z&=1 \\ 3-10+z&=1 \\ z&=8 \end{aligned}$$

The solution is (3, –5, 8).

Quick Quiz 8.5

1. $f(x)=-2x^2-4x+6$

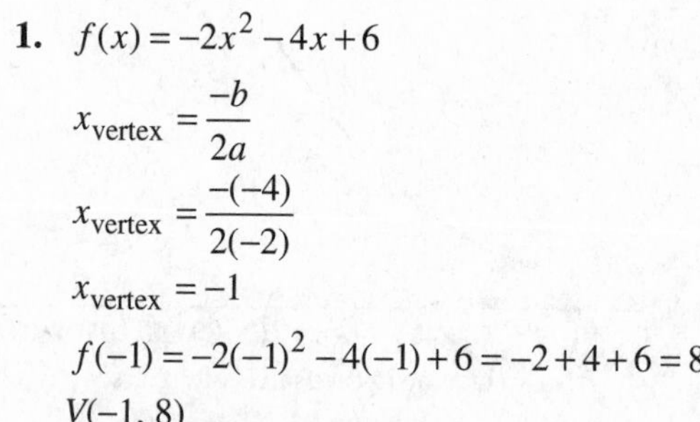

$$x_{\text{vertex}}=\frac{-b}{2a}$$

$$x_{\text{vertex}}=\frac{-(-4)}{2(-2)}$$

$$x_{\text{vertex}}=-1$$

$$f(-1)=-2(-1)^2-4(-1)+6=-2+4+6=8$$

$V(-1, 8)$

2. $f(0)=-2(0)^2-4(0)+6=6$

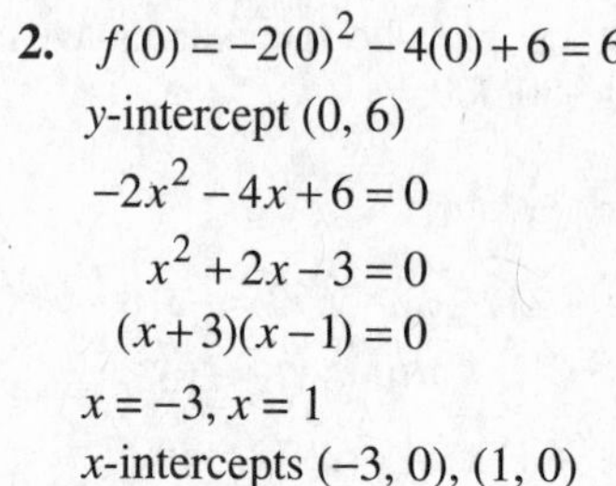

y-intercept (0, 6)

$$\begin{aligned} -2x^2-4x+6&=0 \\ x^2+2x-3&=0 \\ (x+3)(x-1)&=0 \end{aligned}$$

$x=-3, x=1$

x-intercepts (–3, 0), (1, 0)

3.

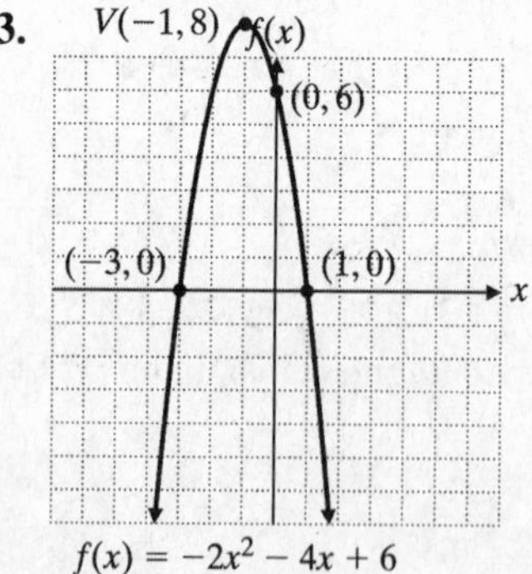

4. Answers may vary. Possible solution: The function is in standard form, $f(x)=ax^2+bx+c$ with $a = 4$, $b = -9$, and $c = -5$. The x-coordinate of the vertex is $x_{\text{vertex}}=\frac{-b}{2a}=\frac{-(-9)}{2(4)}=\frac{9}{8}$. The y-coordinate of the vertex is $f(x_{\text{vertex}})=f\left(\frac{9}{8}\right)$.

8.6 Exercises

1. The boundary points divide the number line into regions. All values of x in a given region produce results that are greater than zero, or else all the values of x in a given region produce results that are less than zero.

3. $x^2 + x - 12 < 0$
$x^2 + x - 12 = 0$
$(x+4)(x-3) = 0$
$x = -4, x = 3$
Boundary points: −4, 3
Test: $x^2 + x - 12$

Region	Test	Result
$x < -4$	−5	$8 > 0$
$-4 < x < 3$	0	$-12 < 0$
$x > 3$	4	$8 > 0$

$(x + 4)(x - 3) = 0$
$-4 < x < 3$

5. $x^2 \geq 4$
$x^2 - 4 \geq 0$
$x^2 - 4 = 0$
$(x+2)(x-2) = 0$
$x = -2, x = 2$
Boundary points: −2, 2
Test: $x^2 - 4$

Region	Test	Result
$x < -2$	−3	$5 > 0$
$-2 < x < 2$	0	$-4 < 0$
$x > 2$	3	$5 > 0$

$(x + 2)(x - 2) = 0$
$x \leq -2$ or $x \geq 2$

7. $2x^2 + x - 3 < 0$
$2x^2 + x - 3 = 0$
$(2x+3)(x-1) = 0$
$x = -\frac{3}{2}, x = 1$

Boundary points: $-\frac{3}{2}$, 1

Test: $2x^2 + x - 3$

Region	Test	Result
$x < -\frac{3}{2}$	−2	$3 > 0$
$-\frac{3}{2} < x < 1$	0	$-3 < 0$
$x > 1$	2	$7 > 0$

$(2x + 3)(x - 1) = 0$
$-\frac{3}{2} < x < 1$

9. $x^2 + x - 20 > 0$
$x^2 + x - 20 = 0$
$(x+5)(x-4) = 0$
$x = -5, x = 4$
Boundary points: −5, 4
Test: $x^2 + x - 20$

Region	Test	Result
$x < -5$	−6	$10 > 0$
$-5 < x < 4$	0	$-20 < 0$
$x > 4$	5	$10 > 0$

$x < -5$ or $x > 4$

11. $4x^2 \leq 11x + 3$
$4x^2 - 11x - 3 \leq 0$
$4x^2 - 11x - 3 = 0$
$(4x+1)(x-3) = 0$
$x = -\frac{1}{4}, x = 3$

Boundary points: $-\frac{1}{4}$, 3

Test: $4x^2 - 11x - 3$

Region	Test	Result
$x < -\frac{1}{4}$	-1	$12 > 0$
$-\frac{1}{4} < x < 3$	0	$-3 < 0$
$x > 3$	4	$17 > 0$

$-\frac{1}{4} \le x \le 3$

13. $6x^2 - 5x > 6$
$6x^2 - 5x - 6 > 0$
$6x^2 - 5x - 6 = 0$
$(2x-3)(3x+2) = 0$
$x = \frac{3}{2},\ x = -\frac{2}{3}$
Boundary points: $\frac{3}{2}, -\frac{2}{3}$
Test: $6x^2 - 5x - 6$

Region	Test	Result
$x < -\frac{2}{3}$	-1	$5 > 0$
$-\frac{2}{3} < x < \frac{3}{2}$	0	$-6 < 0$
$x > \frac{3}{2}$	2	$8 > 0$

$x < -\frac{2}{3}$ or $x > \frac{3}{2}$

15. $-2x + 30 \ge x(x+5)$
$-2x + 30 \ge x^2 + 5x$
$0 \ge x^2 + 7x - 30$
$x^2 + 7x - 30 \le 0$
$x^2 + 7x - 30 = 0$
$(x+10)(x-3) = 0$
$x = -10, x = 3$
Boundary points: $-10, 3$
Test: $x^2 + 7x - 30$

Region	Test	Result
$x < -10$	-12	$30 > 0$
$-10 < x < 3$	0	$-30 < 0$
$x > 3$	5	$30 > 0$

$-10 \le x \le 3$

17. $x^2 - 2x \ge -1$
$x^2 - 2x + 1 \ge 0$
$(x-1)^2 \ge 0$
Since the square of any real number must be 0 or positive, all real numbers satisfy this inequality.

19. $x^2 - 4x \le -4$
$x^2 - 4x + 4 \le 0$
$x^2 - 4x + 4 = 0$
$(x-2)(x-2) = 0$
$x = 2$
Boundary point: 2
Test: $x^2 - 4x + 4$

Region	Test	Result
$x < 2$	0	$4 > 0$
$x > 2$	3	$1 > 0$

$x = 2$

21. $x^2 - 2x > 4$
$x^2 - 2x - 4 > 0$
$x^2 - 2x - 4 = 0$
$a = 1, b = -2, c = -4$
$x = \frac{-(-2) \pm \sqrt{(-2)^2 - 4(1)(-4)}}{2(1)}$
$x = \frac{2 \pm \sqrt{20}}{2} = 1 \pm \sqrt{5}$
$x \approx 3.2,\ x \approx -1.2$
Approximate boundary points: 3.2, −1.2
Test: $x^2 - 2x - 4$

Region	Test	Result
$x < 1 - \sqrt{5}$	-2	$4 > 0$
$1 - \sqrt{5} < x < 1 + \sqrt{5}$	0	$-4 < 0$
$x > 1 + \sqrt{5}$	4	$4 > 0$

$x < 1 - \sqrt{5}$ or $x > 1 + \sqrt{5}$
Approximately $x < -1.2$ or $x > 3.2$

23. $x^2 - 4x < 3$

$x^2 - 4x - 3 < 0$

$x^2 - 4x - 3 = 0$

$a = 1, b = -4, c = -3$

$x = \frac{-(-4) \pm \sqrt{(-4)^2 - 4(1)(-3)}}{2(1)}$

$x = \frac{4 \pm \sqrt{28}}{2}$

$x = 2 \pm \sqrt{7}$

$x \approx -0.6, x \approx 4.6$

Approximate boundary points: −0.6, 4.6

Test: $x^2 - 4x - 3$

Region	Test	Result
$x < 2 - \sqrt{7}$	−1	$2 > 0$
$2 - \sqrt{7} < x < 2 + \sqrt{7}$	0	$-3 < 0$
$x > 2 + \sqrt{7}$	5	$2 > 0$

$2 - \sqrt{7} < x < 2 + \sqrt{7}$

Approximately $-0.6 < x < 4.6$

25. $2x^2 \ge x^2 - 4$

$x^2 \ge -4$

Since the square of any real number must be 0 or positive, all real numbers satisfy this inequality.

27. $5x^2 \le 4x^2 - 1$

$x^2 \le -1$

Since the square of any real number must be 0 or positive, no real number satisfies this inequality.

29. $s = -16t^2 + 640t$

$-16t^2 + 640t > 6000$

$0 > 16t^2 - 640t + 6000$

$0 > t^2 - 40t + 375$

$t^2 - 40t + 375 < 0$

$t^2 - 40t + 375 = 0$

$(t - 15)(t - 25) = 0$

$t = 15, t = 25$

Boundary points: 15, 25

Test: $t^2 - 40t + 375$

Region	Test	Result
$t < 15$	0	$375 > 0$
$15 < t < 25$	20	$-25 < 0$
$t > 25$	30	$75 > 0$

$15 < t < 25$

The height will be greater than 6000 feet for times greater than 15 seconds but less than 25 seconds.

31. a. Profit $= -10(x^2 - 200x + 1800)$

$-10(x^2 - 200x + 1800) > 0$

$x^2 - 200x + 1800 < 0$

$x^2 - 200x + 1800 = 0$

$a = 1, b = -200, c = 1800$

$x = \frac{-(-200) \pm \sqrt{(-200)^2 - 4(1)(1800)}}{2(1)}$

$x = \frac{200 \pm \sqrt{32,800}}{2}$

$x \approx 9.4, x \approx 190.6$

Approximate boundary points: 190.6, 9.4

Test: $x^2 - 200x + 1800$

Approximate Region	Test	Result
$x < 9.4$	0	$1800 > 0$
$9.4 < x < 190.6$	10	$-100 < 0$
$x > 190.6$	200	$1800 > 0$

Approximately $9.4 < x < 190.6$

The profit is greater than zero when more than 9.4 but fewer than 190.6 units are manufactured each day.

b. $-10(50^2 - 200(50) + 1800) = 57,000$

Daily profit is $57,000 when 50 units are manufactured.

c. $-10(60^2 - 200(60) + 1800) = 66,000$

Daily profit is $66,000 when 60 units are manufactured.

Cumulative Review

33. Let E5 and E6 represent her scores on the fifth and sixth tests.

$$\frac{0+81+92+80+E5+E6}{6} \geq 70$$
$$E5+E6+253 \geq 420$$
$$E5+E6 \geq 167$$

She must score a combined total of 167 points on the two remaining tests. Any two scores that total 167 will be sufficient to participate in synchronized swimming.

34. Let x = ounces of chocolate banana bites. Then $2x$ = ounces of Dutch coffee and $x - 1.6$ = ounces of Belgian chocolate.

$$x+2x+x-1.6=16$$
$$4x=17.6$$
$$x=4.4$$

$2x = 2(4.4) = 8.8$

$x - 1.6 = 4.4 - 1.6 = 2.8$

The assortment contains 4.4 ounces of chocolate banana bites, 8.8 ounces of Dutch coffee, and 2.8 ounces of Belgian chocolate.

35. 10(18) + 14(10) + 5(16) = 400

10(22) + 14(12) + 5(19) = 483

It would cost the family \$400 for the 2-hour cruise and \$483 for the 3-hour cruise.

36. There are 10 + 14 + 5 = 29 total family members, so 23 people plan to take a cruise. Let x = the number of adults, y = the number of children, and z = the number of seniors that plan to take the cruise.

$$x+y+z=23 \quad \textbf{(1)}$$
$$18x+10y+16z=314 \quad \textbf{(2)}$$
$$22x+12y+19z=380 \quad \textbf{(3)}$$

Multiply equation **(1)** by −10 and add the result to equation **(2)** to eliminate y.

$$-10x-10y-10z=-230$$
$$\underline{18x+10y+16z=314}$$
$$8x \quad +6z=84$$
$$4x+3z=42 \quad \textbf{(4)}$$

Multiply equation **(1)** by −12 and add the result to equation **(3)** to eliminate y.

$$-12x-12y-12z=-276$$
$$\underline{22x+12y+19z=380}$$
$$10x \quad +7z=104 \quad \textbf{(5)}$$

Multiply equation **(4)** by 5 and equation **(5)** by −2.

$$20x+15z=210$$
$$\underline{-20x-14z=-208}$$
$$z=2$$

Replace z with 2 in equation **(4)**.

$$4x+3(2)=42$$
$$4x+6=42$$
$$4x=36$$
$$x=9$$

Replace x with 9 and z with 2 in equation **(1)**.

$$9+y+2=23$$
$$y+11=23$$
$$y=12$$

9 adults, 12 children, and 2 seniors plan to take a cruise.

Quick Quiz 8.6

1. $x^2-7x+6>0$

$x^2-7x+6=0$

$(x-6)(x-1)=0$

$x = 6, x = 1$

Boundary points: 1, 6

Test: x^2-7x+6

Region	Test	Result
$x<1$	0	$6>0$
$1<x<6$	2	$-4<0$
$x>6$	7	$6>0$

$x<1$ or $x>6$

2. $6x^2-x-2<0$

$6x^2-x-2=0$

$(3x-2)(2x+1)=0$

$x=\frac{2}{3}, x=-\frac{1}{2}$

Boundary points: $-\frac{1}{2}, \frac{2}{3}$

Test: $6x^2-x-2$

Region	Test	Result
$x<-\frac{1}{2}$	−1	$5>0$
$-\frac{1}{2}<x<\frac{2}{3}$	0	$-2<0$
$x>\frac{2}{3}$	1	$3>0$

$-\frac{1}{2}<x<\frac{2}{3}$

3. $x^2 + 4x - 8 \geq 0$

$x^2 + 4x - 8 = 0$

$a = 1, b = 4, c = -8$

$$x = \frac{-4 \pm \sqrt{(4)^2 - 4(1)(-8)}}{2(1)}$$

$$x \approx \frac{-4 \pm \sqrt{48}}{2}$$

$x \approx -5.5, x \approx 1.5$

Approximate boundary points: -5.5, 1.5

Test: $x^2 + 4x - 8$

Approximate Region	Test	Result
$x < -5.5$	-10	$52 > 0$
$-5.5 < x < 1.5$	0	$-8 < 0$
$x > 1.5$	10	$132 > 0$

Approximately $x \leq -5.5$ or $x \geq 1.5$

4. Answers may vary. Possible solution:

The equation $x^2 + 2x + 8 = 0$ does not have any real solutions. Thus there are no boundary points. Also, the quadratic function $f(x) = x^2 + 2x + 8$ has no x-intercepts, so the graph lies entirely above or below the x-axis. This means that the inequality is either true for all real numbers, or has no real number solutions.

Putting Your Skills to Work

1. Since all four plans include more than 100 minutes, his best choices are Plan A or Plan B.

2. For Plan A, 350 minutes is 50 additional minutes.

$\$39.99 + \$0.20(50) = \$39.99 + \$10.00 = \$49.99$

Thus, Plan B is the best choice for 350 minutes, since it includes more than 350 minutes and costs only \$39.99.

3. For Plan A, 470 minutes is 170 additional minutes.

$$\begin{aligned}\$39.99 + \$0.20(170) &= \$39.99 + \$34.00 \\ &= \$73.99\end{aligned}$$

For Plan B, 470 minutes is 20 additional minutes.

$\$39.99 + \$0.45(20) = \$39.99 + \$9.00 = \$48.91$

Thus, Plan B is the best choice for 470 minutes, since it costs less than the plans that include more than 470 minutes.

4. For Plan A, 550 minutes is 250 additional minutes.

$$\begin{aligned}\$39.99 + \$0.20(250) &= \$39.99 + \$50.00 \\ &= \$89.99\end{aligned}$$

For Plan B, 550 minutes is 100 additional minutes.

$$\begin{aligned}\$39.99 + \$0.45(100) &= \$39.99 + \$45.00 \\ &= \$84.99\end{aligned}$$

Thus, Plan C is the best choice for 550 minutes, since it includes more than 550 minutes and costs only \$49.99.

5. 11 hours is 11(60) = 660 minutes. For Plan A, 11 hours is 360 additional minutes.

$$\begin{aligned}\$39.99 + \$0.20(360) &= \$39.99 + \$72.00 \\ &= \$111.99\end{aligned}$$

For Plan B, 11 hours is 210 additional minutes.

$$\begin{aligned}\$39.99 + \$0.45(210) &= \$39.99 + \$94.50 \\ &= \$134.49\end{aligned}$$

For Plan C, 11 hours is 60 additional minutes.

$\$49.99 + \$0.20(60) = \$49.99 + \$12.00 = \$61.99$

Thus, Plan D is the best choice for 11 hours, since it includes more than 660 minutes and costs only \$59.99.

6. 16 hours, 40 minutes is 16(60) + 40 = 1000 minutes.

For Plan A, 16 hours, 40 minutes is 700 additional minutes.

$$\begin{aligned}\$39.99 + \$0.20(700) &= \$39.99 + \$140.00 \\ &= \$179.99\end{aligned}$$

For Plan B, 16 hours, 40 minutes is 550 additional minutes.

$$\begin{aligned}\$39.99 + \$0.45(550) &= \$39.99 + \$247.50 \\ &= \$287.49\end{aligned}$$

For Plan C, 16 hours, 40 minutes is 400 additional minutes.

$$\begin{aligned}\$49.99 + \$0.20(400) &= \$49.99 + \$80.00 \\ &= \$129.99\end{aligned}$$

For Plan D, 16 hours, 40 minutes is 100 additional minutes.

$$\begin{aligned}\$59.99 + \$0.40(100) &= \$59.99 + \$40.00 \\ &= \$99.99\end{aligned}$$

Thus, Plan D is the best choice for 16 hours, 40 minutes of talking.

7. Combining Calling Plan C and Bundle D would cost \$49.99 + \$35.00 = \$84.99, which is within his budget.

8. Combining Calling Plan D and Bundle D would cost \$59.99 + \$35.00 = \$94.99, which is not within his budget.

9. For Plan C, 750 minutes is 150 additional minutes.
$\$49.99+\$0.20(150) = \$49.99+\$30.00$
$= \$79.99$
Combining this with Bundle D would cost $79.99 + $35.00 = $114.99.

10. Answers may vary.

Chapter 8 Review Problems

1. $6x^2 = 24$
$x^2 = 4$
$x = \pm\sqrt{4}$
$x = \pm 2$

2. $(x+8)^2 = 81$
$x+8 = \pm\sqrt{81}$
$x+8 = \pm 9$
$x = -8 \pm 9$
$x = 1,\ x = -17$

3. $x^2 + 8x + 13 = 0$
$x^2 + 8x = -13$
$x^2 + 8x + 16 = -13 + 16$
$(x+4)^2 = 3$
$x+4 = \pm\sqrt{3}$
$x = -4 \pm \sqrt{3}$

4. $4x^2 - 8x + 1 = 0$
$4x^2 - 8x = -1$
$x^2 - 2x = -\frac{1}{4}$
$x^2 - 2x + 1 = -\frac{1}{4} + 1$
$(x-1)^2 = \frac{3}{4}$
$x - 1 = \pm\frac{\sqrt{3}}{2}$
$x = 1 \pm \frac{\sqrt{3}}{2} = \frac{2 \pm \sqrt{3}}{2}$

5. $x^2 - 4x - 2 = 0$
$a = 1, b = -4, c = -2$
$x = \frac{-(-4) \pm \sqrt{(-4)^2 - 4(1)(-2)}}{2(1)}$
$x = \frac{4 \pm \sqrt{24}}{2}$
$x = \frac{4 \pm 2\sqrt{6}}{2} = \frac{2\left(2 \pm \sqrt{6}\right)}{2}$
$x = 2 \pm \sqrt{6}$

6. $3x^2 - 8x + 4 = 0$
$a = 3, b = -8, c = 4$
$x = \frac{-(-8) \pm \sqrt{(-8)^2 - 4(3)(4)}}{2(3)}$
$x = \frac{8 \pm \sqrt{16}}{6}$
$x = \frac{8 \pm 4}{6}$
$x = \frac{2}{3},\ x = 2$

7. $4x^2 - 12x + 9 = 0$
$(2x-3)^2 = 0$
$2x - 3 = 0$
$x = \frac{3}{2}$

8. $x^2 - 14 = 5x$
$x^2 - 5x - 14 = 0$
$(x-7)(x+2) = 0$
$x = 7, x = -2$

9. $6x^2 - 23x = 4x$
$6x^2 - 27x = 0$
$3x(2x - 9) = 0$
$x = 0,\ x = \frac{9}{2}$

10. $2x^2 = 5x - 1$
$2x^2 - 5x + 1 = 0$
$a = 2, b = -5, c = 1$
$x = \frac{-(-5) \pm \sqrt{(-5)^2 - 4(2)(1)}}{2(2)}$
$x = \frac{5 \pm \sqrt{17}}{4}$

11. $x^2-3x-23=5$
$x^2-3x-28=0$
$(x-7)(x+4)=0$
$x=7,\ x=-4$

12. $5x^2-10=0$
$5x^2=10$
$x^2=2$
$x=\pm\sqrt{2}$

13. $3x^2-2x=15x-10$
$3x^2-17x+10=0$
$(3x-2)(x-5)=0$
$x=\frac{2}{3},\ x=5$

14. $6x^2+12x-24=0$
$x^2+2x-4=0$
$a=1,\ b=2,\ c=-4$
$$x=\frac{-2\pm\sqrt{2^2-4(1)(-4)}}{2(1)}$$
$$x=\frac{-2\pm\sqrt{20}}{2}=\frac{-2\pm2\sqrt{5}}{2}=\frac{2\left(-1\pm\sqrt{5}\right)}{2}$$
$x=-1\pm\sqrt{5}$

15. $7x^2+24=5x^2$
$2x^2=-24$
$x^2=-12$
$x=\pm\sqrt{-12}$
$x=\pm2i\sqrt{3}$

16. $3x^2+5x+1=0$
$a=3,\ b=5,\ c=1$
$$x=\frac{-5\pm\sqrt{5^2-4(3)(1)}}{2(3)}$$
$$x=\frac{-5\pm\sqrt{13}}{6}$$

17. $2x(x-4)-4=-x$
$2x^2-8x-4+x=0$
$2x^2-7x-4=0$
$(2x+1)(x-4)=0$
$x=-\frac{1}{2},\ x=4$

18. $9x(x+2)+2=12x$
$9x^2+18x+2=12x$
$9x^2+6x+2=0$
$a=9,\ b=6,\ c=2$
$$x=\frac{-6\pm\sqrt{6^2-4(9)(2)}}{2(9)}$$
$$x=\frac{-6\pm\sqrt{-36}}{18}=\frac{-6\pm6i}{18}=\frac{6(-1\pm i)}{18}$$
$$x=\frac{-1\pm i}{3}$$

19. The LCD is x.
$$\frac{x-5}{x}+9x=1$$
$$\frac{x-5}{x}\cdot x+9x\cdot x=1\cdot x$$
$x-5+9x^2=x$
$9x^2=5$
$x^2=\frac{5}{9}$
$$x=\pm\sqrt{\frac{5}{9}}$$
$$x=\pm\frac{\sqrt{5}}{3}$$

20. $\frac{4}{5}x^2+x+\frac{1}{5}=0$
$4x^2+5x+1=0$
$(4x+1)(x+1)=0$
$x=-\frac{1}{4},\ x=-1$

21. The LCD is $6y$.
$$y+\frac{5}{3y}+\frac{17}{6}=0$$
$$y\cdot6y+\frac{5}{3y}\cdot6y+\frac{17}{6}\cdot6y=0\cdot6y$$
$6y^2+10+17y=0$
$6y^2+17y+10=0$
$(6y+5)(y+2)=0$
$y=-\frac{5}{6},\ y=-2$

22. The LCD is y^2.

$$\frac{19}{y}-\frac{15}{y^2}+10=0$$

$$\frac{19}{y}\cdot y^2-\frac{15}{y^2}\cdot y^2+10\cdot y^2=0\cdot y^2$$

$$19y-15+10y^2=0$$

$$10y^2+19y-15=0$$

$$(2y+5)(5y-3)=0$$

$$y=-\frac{5}{2},\ y=\frac{3}{5}$$

23. The LCD is y^2.

$$\frac{15}{y^2}-\frac{2}{y}=1$$

$$\frac{15}{y^2}\cdot y^2-\frac{2}{y}\cdot y^2=1\cdot y^2$$

$$15-2y=y^2$$

$$y^2+2y-15=0$$

$$(y+5)(y-3)=0$$

$$y=-5,\ y=3$$

24. The LCD is y.

$$y-18+\frac{81}{y}=0$$

$$y\cdot y-18\cdot y+\frac{81}{y}\cdot y=0\cdot y$$

$$y^2-18y+81=0$$

$$(y-9)^2=0$$

$$y-9=0$$

$$y=9$$

25. $(3y+2)(y-1)=7(-y+1)$

$$3y^2-y-2=-7y+7$$

$$3y^2+6y-9=0$$

$$y^2+2y-3=0$$

$$(y+3)(y-1)=0$$

$$y=-3,\ y=1$$

26. $y(y+1)+(y+2)^2=4$

$$y^2+y+y^2+4y+4=4$$

$$2y^2+5y=0$$

$$y(2y+5)=0$$

$$y=0,\ y=-\frac{5}{2}$$

27. The LCD is $(x + 3)(x + 1)$.

$$\frac{2x}{x+3}+\frac{3x-1}{x+1}=3$$
$$\frac{2x}{x+3}(x+3)(x+1)+\frac{3x-1}{x+1}(x+3)(x+1)=3(x+3)(x+1)$$
$$2x(x+1)+(3x-1)(x+3)=3(x+3)(x+1)$$
$$2x^2+2x+3x^2+8x-3=3x^2+12x+9$$
$$2x^2-2x-12=0$$
$$x^2-x-6=0$$
$$(x-3)(x+2)=0$$

$x=3, x=-2$

28. The LCD is $(2x + 5)(x + 4)$.

$$\frac{4x+1}{2x+5}+\frac{3x}{x+4}=2$$
$$\frac{4x+1}{2x+5}(2x+5)(x+4)+\frac{3x}{x+4}(2x+5)(x+4)=2(2x+5)(x+4)$$
$$(4x+1)(x+4)+3x(2x+5)=2(2x+5)(x+4)$$
$$4x^2+17x+4+6x^2+15x=4x^2+26x+40$$
$$6x^2+6x-36=0$$
$$x^2+x-6=0$$
$$(x+3)(x-2)=0$$

$x=-3, x=2$

29. $4x^2-5x-3=0$
$a=4, b=-5, c=-3$
$b^2-4ac=(-5)^2-4(4)(-3)=73$
Two irrational solutions

30. $2x^2-7x+6=0$
$a=2, b=-7, c=6$
$b^2-4ac=(-7)^2-4(2)(6)=1$
Two rational solutions

31. $3x^2-5x+6=0$
$a=3, b=-5, c=6$
$b^2-4ac=(-5)^2-4(3)(6)=-47$
Two nonreal complex solutions

32. $25x^2-20x+4=0$
$a=25, b=-20, c=4$
$b^2-4ac=(-20)^2-4(25)(4)=0$
One rational solution

33. $5, -5$

$x = 5 \qquad x = -5$
$x - 5 = 0 \qquad x + 5 = 0$
$(x-5)(x+5) = 0$
$x^2 - 25 = 0$

34. $3i, -3i$

$x = 3i \qquad x = -3i$
$x - 3i = 0 \qquad x + 3i = 0$
$(x-3i)(x+3i) = 0$
$x^2 - 9i^2 = 0$
$x^2 + 9 = 0$

35. $5\sqrt{2}, -5\sqrt{2}$

$x = 5\sqrt{2} \qquad x = -5\sqrt{2}$
$x - 5\sqrt{2} = 0 \qquad x + 5\sqrt{2} = 0$
$\left(x - 5\sqrt{2}\right)\left(x + 5\sqrt{2}\right) = 0$
$x^2 - 50 = 0$

36. $-\frac{1}{4}, -\frac{3}{2}$

$x = -\frac{1}{4} \qquad x = -\frac{3}{2}$
$4x = -1 \qquad 2x = -3$
$4x + 1 = 0 \qquad 2x + 3 = 0$
$(4x+1)(2x+3) = 0$
$8x^2 + 14x + 3 = 0$

37. $x^4 - 6x^2 + 8 = 0$

$y = x^2$
$y^2 - 6y + 8 = 0$
$(y-4)(y-2) = 0$
$y - 4 = 0 \qquad y - 2 = 0$
$y = 4 \qquad y = 2$
$x^2 = 4 \qquad x^2 = 2$
$x = \pm\sqrt{4} = \pm 2 \qquad x = \pm\sqrt{2}$

38. $2x^6 - 5x^3 - 3 = 0$

$y = x^3$
$2y^2 - 5y - 3 = 0$
$(2y+1)(y-3) = 0$
$2y + 1 = 0 \qquad y - 3 = 0$
$y = -\frac{1}{2} \qquad y = 3$
$x^3 = -\frac{1}{2} \qquad x^3 = 3$
$x = \sqrt[3]{-\frac{1}{2}} = -\frac{\sqrt[3]{4}}{2} \qquad x = \sqrt[3]{3}$

39. $x^{2/3} - 3 = 2x^{1/3}$

$x^{2/3} - 2x^{1/3} - 3 = 0$
$y = x^{1/3}$
$y^2 - 2y - 3 = 0$
$(y-3)(y+1) = 0$
$y = 3 \qquad y = -1$
$x^{1/3} = 3 \qquad x^{1/3} = -1$
$x = 27 \qquad x = -1$

40. $3x - 5x^{1/2} = 2$

$3x - 5x^{1/2} - 2 = 0$
$y = x^{1/2}$
$3y^2 - 5y - 2 = 0$
$(3y+1)(y-2) = 0$
$3y + 1 = 0 \qquad y - 2 = 0$
$y = -\frac{1}{3} \qquad y = 2$
$x^{1/2} = -\frac{1}{3}$ extraneous $\qquad x^{1/2} = 2$
$x = 4$

41. $(2x+1)^2 + 6(2x+1) + 8 = 0$

$y = 2x + 1$
$y^2 + 6y + 8 = 0$
$(y+4)(y+2) = 0$
$y + 4 = 0 \qquad y + 2 = 0$
$y = -4 \qquad y = -2$
$2x + 1 = -4 \qquad 2x + 1 = -2$
$2x = -5 \qquad 2x = -3$
$x = -\frac{5}{2} \qquad x = -\frac{3}{2}$

42. $1+4x^{-8}=5x^{-4}$

$4x^{-8}-5x^{-4}+1=0$

$y=x^{-4}$

$4y^2-5y+1=0$

$(y-1)(4y-1)=0$

$y-1=0 \qquad 4y-1=0$

$y=1 \qquad y=\frac{1}{4}$

$x^{-4}=1 \qquad x^{-4}=\frac{1}{4}$

$x^4=1 \qquad x^4=4$

$x=\pm\sqrt[4]{1}=\pm1 \qquad x=\pm\sqrt[4]{4}=\pm\sqrt{2}$

43. $3M=\frac{2A^2}{N}$

$A^2=\frac{3MN}{2}$

$A=\pm\sqrt{\frac{3MN}{2}}$

44. $3t^2+4b=t^2+6ay$

$2t^2=6ay-4b$

$t^2=3ay-2b$

$t=\pm\sqrt{3ay-2b}$

45. $yx^2-3x-7=0$

$a=y,\ b=-3,\ c=-7$

$x=\frac{-(-3)\pm\sqrt{(-3)^2-4(y)(-7)}}{2y}$

$x=\frac{3\pm\sqrt{9+28y}}{2y}$

46. $20d^2-xd-x^2=0$

$(4d-x)(5d+x)=0$

$4d-x=0 \qquad 5d+x=0$

$d=\frac{x}{4} \qquad d=-\frac{x}{5}$

47. $2y^2+4ay-3a=0$

$a=2,\ b=4a,\ c=-3a$

$y=\frac{-4a\pm\sqrt{(4a)^2-4(2)(-3)}}{2(2)}$

$y=\frac{-4a\pm\sqrt{16a^2+24}}{4}$

$y=\frac{-4a\pm\sqrt{4(4a^2+6)}}{4}$

$y=\frac{-2a\pm\sqrt{4a^2+6}}{2}$

48. $AB=3x^2+2y^2-4x$

$3x^2-4x+2y^2-AB=0$

$a=3,\ b=-4,\ c=2y^2-AB$

$x=\frac{-(-4)\pm\sqrt{(-4)^2-4(3)(2y^2-AB)}}{2(3)}$

$x=\frac{4\pm\sqrt{16-12(2y^2-AB)}}{6}$

$x=\frac{4\pm\sqrt{16-24y^2+12AB}}{6}$

$x=\frac{4\pm2\sqrt{4-6y^2+3AB}}{6}$

$x=\frac{2\pm\sqrt{4-6y^2+3AB}}{3}$

49. $c^2=a^2+b^2$

$c^2=\left(3\sqrt{2}\right)^2+2^2$

$c^2=18+4$

$c^2=22$

$c=\sqrt{22}$

50. $c^2=a^2+b^2$

$16^2=a^2+4^2$

$256=a^2+16$

$a^2=240$

$a=\sqrt{240}=4\sqrt{15}$

51. $c^2 = a^2 + b^2$
$6^2 = 5^2 + b^2$
$36 = 25 + b^2$
$b^2 = 11$
$b = \sqrt{11} \approx 3.3$
The car is approximately 3.3 miles from the observer.

52. Let b = the length of the base. Then $2b + 6$ = the length of the altitude.
$A = \frac{1}{2}ab$
$\frac{1}{2}b(2b+6) = 70$
$b(2b+6) = 140$
$2b^2 + 6b - 140 = 0$
$b^2 + 3b - 70 = 0$
$(b-7)(b+10) = 0$
$b - 7 = 0$ $\quad$ $b + 10 = 0$
$b = 7$ $\quad$ $b = -10$
Since the length of the base must be positive, $b = 7$.
$2b + 6 = 2(7) + 6 = 20$
The base is 7 cm and the altitude is 20 cm.

53. Let W = the width of the rectangle. Then $4W + 1$ = the length.
$(4W+1)W = 203$
$4W^2 + W - 203 = 0$
$(W-7)(4W+29) = 0$
$W - 7 = 0$ $\quad$ $4W + 29 = 0$
$W = 7$ $\quad$ $W = -\frac{29}{4}$
Since the width must be positive, $W = 7$.
$4W + 1 = 4(7) + 1 = 29$
The width is 7 m and the length is 29 m.

54. Let v = the faster speed. Then $v - 10$ = the slower speed.
$\text{Time} = \frac{\text{Distance}}{\text{Rate}}$
$5 = \frac{80}{v} + \frac{10}{v-10}$
$5v(v-10) = \frac{80}{v}v(v-10) + \frac{10}{v-10}v(v-10)$
$5v^2 - 50v = 80v - 800 + 10v$
$5v^2 - 140v + 800 = 0$
$v^2 - 28v + 160 = 0$
$(v-8)(v-20) = 0$
$v - 8 = 0$ $\quad$ $v - 20 = 0$
$v = 8$ $\quad$ $v = 20$
Since v must be greater than 10, $v = 20$.
The speed was 20 mph for 80 miles and 10 mph for 10 miles.

55. Let v = her speed with no rain. Then $x - 5$ = her speed in the rain.
$\text{Time} = \frac{\text{Distance}}{\text{Rate}}$
$\frac{200}{v} + \frac{90}{v-5} = 6$
$\frac{200}{v}v(v-5) + \frac{90}{v-5}v(v-5) = 6v(v-5)$
$200(v-5) + 90v = 6v(v-5)$
$200v - 1000 + 90v = 6v^2 - 30v$
$6v^2 - 320v + 1000 = 0$
$(v-50)(6v-20) = 0$
$v - 50 = 0$ $\quad$ $6v - 20 = 0$
$v = 50$ $\quad$ $v = \frac{10}{3}$
Since v must be greater than 5, $v = 50$.
$v - 5 = 45$
The speed before the rain was 50 mph and 45 mph in the rain.

56. The length of the garden plus the walkway is $10 + 2x$, and the width is $6 + 2x$. The area of brick is the combined area minus the area of the garden.
$(10+2x)(6+2x) - 10(6) = 100$
$60 + 32x + 4x^2 - 60 = 100$
$4x^2 + 32x - 100 = 0$
$x^2 + 8x - 25 = 0$
$x = \frac{-8 \pm \sqrt{8^2 - 4(1)(-25)}}{2}$
$x \approx -10.4,\ x \approx 2.4$
Since the width of the walkway must be positive, $x \approx 2.4$.
The walkway should be approximately 2.4 feet wide.

57. Let x = the width of walkway in meters. The length of the pool plus the walkway is $50 + 2x$, and the width $25 + 2x$. The area of nonslip surface is the combined area minus the area of the pool.

$76 = (50+2x)(25+2x) - 50(25)$
$76 = 1250 + 4x^2 - 1250$
$4x^2 + 150x - 76 = 0$
$(2x+76)(2x-1) = 0$
$2x + 76 = 0 \qquad 2x - 1 = 0$
$x = -\frac{76}{2} \qquad x = \frac{1}{2}$

Since the width of the walkway must be positive, $x = \frac{1}{2}$.

The width of the walkway is 0.5 meters.

58. $g(x) = -x^2 + 6x - 11$
$\frac{-b}{2a} = \frac{-6}{2(-1)} = 3$
$g(3) = -3^2 + 6(3) - 11 = -2$
$V(3, -2)$
$g(0) = -0^2 + 6(0) - 11 = -11$
y-intercept: $(0, -11)$
$b^2 - 4ac = 6^2 - 4(-1)(-11) = -8 < 0$
x-intercepts: none

59. $f(x) = x^2 + 10x + 25$
$\frac{-b}{2a} = \frac{-10}{2(1)} = -5$
$f(-5) = (-5)^2 + 10(-5) + 25 = 0$
$V(-5, 0)$
$f(0) = 0^2 + 10(0) + 25 = 25$
y-intercept: $(0, 25)$
$x^2 + 10x + 25 = 0$
$(x+5)^2 = 0$
$x = -5$
x-intercept: $(-5, 0)$

60. $f(x) = x^2 + 4x + 3$
$\frac{-b}{2a} = \frac{-4}{2(1)} = -2$
$f(-2) = (-2)^2 + 4(-2) + 3 = -1$
$V(-2, -1)$
$f(0) = 0^2 + 4(0) + 3 = 3$
y-intercept: $(0, 3)$
$x^2 + 4x + 3 = 0$
$(x+3)(x+1) = 0$
$x = -3, x = -1$
x-intercepts: $(-3, 0), (-1, 0)$

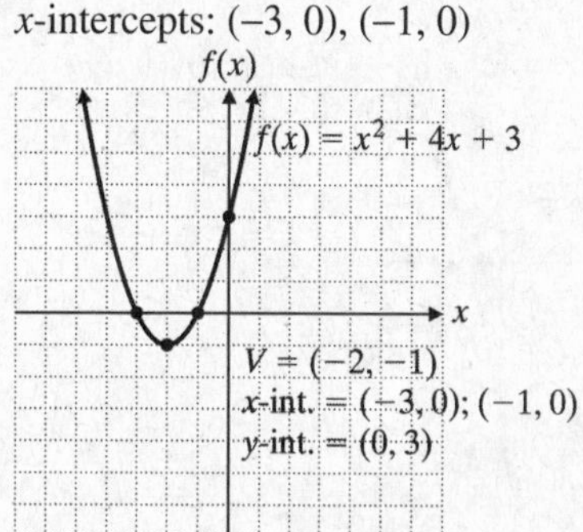

61. $f(x) = x^2 + 6x + 5$
$\frac{-b}{2a} = \frac{-6}{2(1)} = -3$
$f(-3) = (-3)^2 + 6(-3) + 5 = -4$
$V(-3, -4)$
$f(0) = 0^2 + 6(0) + 5 = 5$
y-intercept: $(0, 5)$
$x^2 + 6x + 5 = 0$
$(x+5)(x+1) = 0$
$x = -5, x = -1$
x-intercepts: $(-5, 0), (-1, 0)$

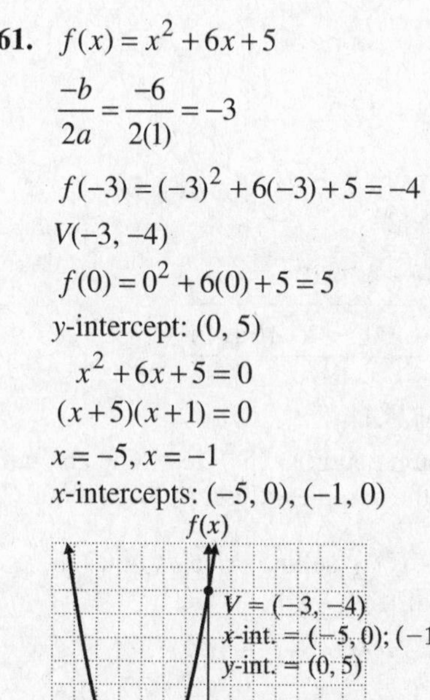
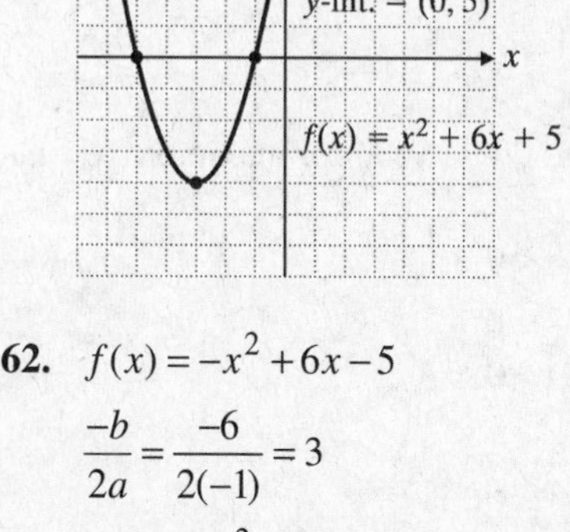

62. $f(x) = -x^2 + 6x - 5$
$\frac{-b}{2a} = \frac{-6}{2(-1)} = 3$
$f(3) = -3^2 + 6(3) - 5 = 4$
$V(3, 4)$
$f(0) = -0^2 + 6(0) - 5 = -5$
y-intercept: $(0, -5)$
$-x^2 + 6x - 5 = 0$
$x^2 - 6x + 5 = 0$
$(x-1)(x-5) = 0$
$x = 1, x = 5$
x-intercepts: $(1, 0), (5, 0)$

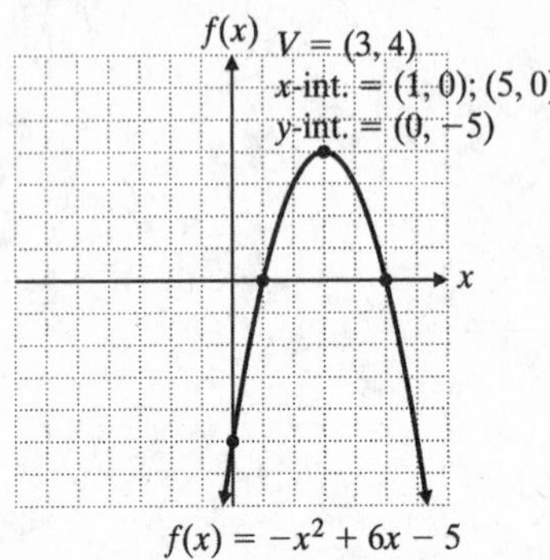

$f(x) = -x^2 + 6x - 5$

63. $h(t) = -16t^2 + 400t + 40$

$\frac{-b}{2a} = \frac{-400}{2(-16)} = 12.5$

$h(1.25) = -16(12.5)^2 + 400(12.5) + 40$
$h(12.5) = 2540$

$-16t^2 + 400t + 40 = 0$

$t = \frac{-400 \pm \sqrt{400^2 - 4(-16)(40)}}{2(-16)}$

$t \approx -0.1,\ t \approx 25.1$

The maximum height is 2540 feet. The amount of time for the complete flight is 25.1 seconds.

64. The revenue is the selling price, x, times the number sold, $1200 - x$.

$R(x) = x(1200 - x) = -x^2 + 1200x$

$\frac{-b}{2a} = \frac{-1200}{2(-1)} = 600$

The price should be \$600 for maximum revenue.

65. $x^2 + 7x - 18 < 0$
$x^2 + 7x - 18 = 0$
$(x+9)(x-2) = 0$
$x = -9, x = 2$
Boundary points: −9, 2
Test: $x^2 + 7x - 18$

Region	Test	Result
$x < -9$	−10	$12 > 0$
$-9 < x < 2$	0	$-18 < 0$
$x > 2$	3	$12 > 0$

$-9 < x < 2$

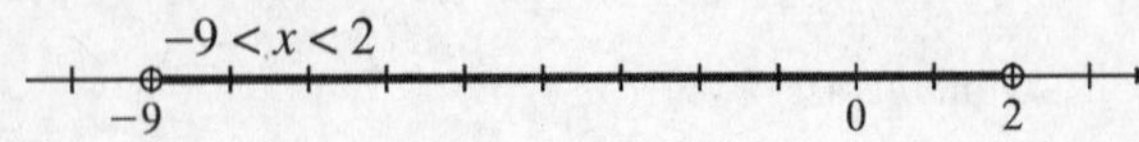

66. $x^2 + 4x - 21 < 0$
$x^2 + 4x - 21 = 0$
$(x+7)(x-3) = 0$
$x = -7, x = 3$
Boundary points: −7, 3
Test: $x^2 + 4x - 21$

Region	Test	Result
$x < -7$	−8	$11 > 0$
$-7 < x < 3$	0	$-21 < 0$
$x > 3$	4	$11 > 0$

$-7 < x < 3$

−7 0 3

67. $x^2 - 9x + 20 > 0$
$x^2 - 9x + 20 = 0$
$(x-5)(x-4) = 0$
$x = 5, x = 4$
Boundary points: 4, 5
Test: $x^2 - 9x + 20$

Region	Test	Result
$x < 4$	0	$20 > 0$
$4 < x < 5$	4.5	$-0.25 < 0$
$x > 5$	6	$2 > 0$

$x < 4$ or $x > 5$

4 5

68. $x^2 - 11x + 28 > 0$
$x^2 - 11x + 28 = 0$
$(x-7)(x-4) = 0$
$x = 7, x = 4$
Boundary points: 7, 4
Test: $x^2 - 11x + 28$

Region	Test	Result
$x < 4$	0	$28 > 0$
$4 < x < 7$	6	$-2 < 0$
$x > 7$	8	$4 > 0$

$x < 4$ or $x > 7$

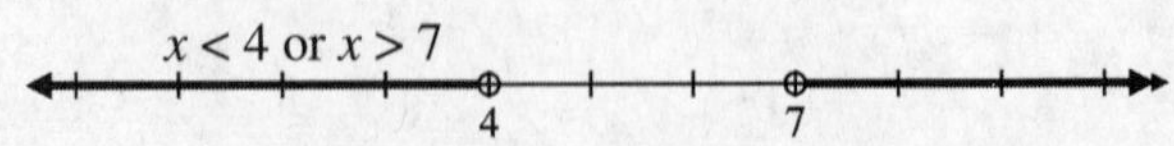

69. $2x^2 - x - 6 \le 0$

$2x^2 - x - 6 = 0$

$(x-2)(2x+3) = 0$

$x = 2, x = -\frac{3}{2}$

Boundary points: $2, -\frac{3}{2}$

Test: $2x^2 - x - 6$

Region	Test	Result
$x < -\frac{3}{2}$	-2	$4 > 0$
$-\frac{3}{2} < x < 2$	0	$-6 < 0$
$x > 2$	4	$22 > 0$

$-\frac{3}{2} \le x \le 2$

70. $3x^2 - 13x + 12 \le 0$

$3x^2 - 13x + 12 = 0$

$(3x-4)(x-3) = 0$

$x = \frac{4}{3}, x = 3$

Boundary points: $\frac{4}{3}, 3$

Test: $3x^2 - 13x + 12$

Region	Test	Result
$x < \frac{4}{3}$	0	$12 > 0$
$\frac{4}{3} < x < 3$	2	$-2 < 0$
$x > 3$	4	$8 > 0$

$\frac{4}{3} \le x \le 3$

71. $9x^2 - 4 > 0$

$9x^2 - 4 = 0$

$x^2 = \frac{4}{9}$

$x = \pm\sqrt{\frac{4}{9}} = \pm\frac{2}{3}$

Boundary points: $-\frac{2}{3}, \frac{2}{3}$

Test: $9x^2 - 4$

Region	Test	Result
$x < -\frac{2}{3}$	-2	$32 > 0$
$-\frac{2}{3} < x < \frac{2}{3}$	0	$-4 < 0$
$x > \frac{2}{3}$	2	$32 > 0$

$x < -\frac{2}{3}$ or $x > \frac{2}{3}$

72. $16x^2 - 25 > 0$

$16x^2 - 25 = 0$

$x^2 = \frac{25}{16}$

$x = \pm\sqrt{\frac{25}{16}} = \pm\frac{5}{4}$

Boundary points: $-\frac{5}{4}, \frac{5}{4}$

Test: $16x^2 - 25$

Region	Test	Result
$x < -\frac{5}{4}$	-2	$39 > 0$
$-\frac{5}{4} < x < \frac{5}{4}$	0	$-25 < 0$
$x > \frac{5}{4}$	2	$39 > 0$

$x < -\frac{5}{4}$ or $x > \frac{5}{4}$

73. $4x^2 - 8x \le 12 + 5x^2$

$0 \le x^2 + 8x + 12$

$x^2 + 8x + 12 \ge 0$

$x^2 + 8x + 12 = 0$

$(x+6)(x+2) = 0$

$x = -6, x = -2$

Boundary points: $-6, -2$

Test: $x^2 + 8x + 12$

Region	Test	Result
$x < -6$	-8	$12 > 0$
$-6 < x < -2$	-4	$-4 < 0$
$x > -2$	0	$12 > 0$

$x \le -6$ or $x \ge -2$

74. $x^2 - 9x > 4 - 7x$

$x^2 - 2x - 4 > 0$

$x^2 - 2x - 4 = 0$

$a = 1, b = -2, c = -4$

$$x = \frac{-(-2) \pm \sqrt{(-2)^2 - 4(1)(-4)}}{2(1)}$$

$$x = \frac{2 \pm \sqrt{20}}{2} = 1 \pm \sqrt{5}$$

$x \approx -1.2, x \approx 3.2$

Boundary points: $1-\sqrt{5}, 1+\sqrt{5}$

Test: $x^2 - 2x - 4$

Region	Test	Result
$x < 1-\sqrt{5}$	-2	$4 > 0$
$1-\sqrt{5} < x < 1+\sqrt{5}$	0	$-4 < 0$
$x > 1+\sqrt{5}$	4	$4 > 0$

$x < 1-\sqrt{5}$ or $x > 1+\sqrt{5}$

Approximately $x < -1.2$ or $x > 3.2$.

75. $x^2 + 13x > 16 + 7x$

$x^2 + 6x - 16 > 0$

$x^2 + 6x - 16 = 0$

$(x+8)(x-2) = 0$

$x = -8, x = 2$

Boundary points: -8, 2

Test: $x^2 + 6x - 16$

Region	Test	Result
$x < -8$	-10	$24 > 0$
$-8 < x < 2$	0	$-16 < 0$
$x > 2$	4	$24 > 0$

$x < -8$ or $x > 2$

76. $3x^2 - 12x > -11$

$3x^2 - 12x + 11 > 0$

$3x^2 - 12x + 11 = 0$

$a = 3, b = -12, c = 11$

$$x = \frac{-(-12) \pm \sqrt{(-12)^2 - 4(3)(11)}}{2(3)} = \frac{12 \pm \sqrt{12}}{6}$$

$$x = \frac{6 \pm \sqrt{3}}{3}$$

$x \approx 1.4, x \approx 2.6$

Approximate boundary points: 1.4, 2.6

Test: $3x^2 - 12x + 11$

Approximate Region	Test	Result
$x < 1.4$	0	$11 > 0$
$1.4 < x < 2.6$	2	$-1 < 0$
$x > 2.6$	3	$2 > 0$

$x < \frac{6-\sqrt{3}}{3}$ or $x > \frac{6+\sqrt{3}}{3}$

$x < 1.4$ $\quad$ $x > 2.6$

77. $4x^2 + 12x + 9 < 0$

$(2x+3)^2 < 0$

Since the square of any real number must be 0 or positive, there is no real solution.

78. $-2x^2 + 7x + 12 \le -3x^2 + x$

$x^2 + 6x + 12 \le 0$

$a = 1, b = 6, c = 12$

$b^2 - 4ac = 6^2 - 4(1)(12) = -12 < 0$

Since there are no real solutions to the equation, the inequality is either true for all real numbers or false for all real numbers.

Test $x^2 + 6x + 12$ at $x = 0$.

$0^2 + 6(0) + 12 = 12 > 0$

The inequality has no real solution.

79. $(x+4)(x-2)(3-x)>0$
$(x+4)(x-2)(3-x)=0$
$x=-4, x=2, x=3$
Boundary points: –4, 2, 3
Test: $(x+4)(x-2)(3-x)$

Region	Test	Result
$x<-4$	–5	$56>0$
$-4<x<2$	0	$-24<0$
$2<x<3$	2.5	1.625
$x>3$	4	$-16<0$

$x<-4$ or $2<x<3$

80. $(x+1)(x+4)(2-x)<0$
$(x+1)(x+4)(2-x)=0$
$x=-1, x=-4, x=2$
Boundary points: –4, –1, 2
Test: $(x+1)(x+4)(2-x)$

Region	Test	Result
$x<-4$	–5	$28>0$
$-4<x<-1$	–2	$-8<0$
$-1<x<2$	0	$8>0$
$x>2$	3	$-28<0$

$-4<x<-1$ or $x>2$

How Am I Doing? Chapter 8 Test

1. $8x^2+9x=0$
$x(8x+9)=0$
$x=0,\ x=-\dfrac{9}{8}$

2. $6x^2-3x=1$
$6x^2-3x-1=0$
$a=6, b=-3, c=-1$
$$x=\frac{-(-3)\pm\sqrt{(-3)^2-4(6)(-1)}}{2(6)}$$
$$x=\frac{3\pm\sqrt{33}}{12}$$

3. The LCD is $6x$.

$$\frac{3x}{2}-\frac{8}{3}=\frac{2}{3x}$$
$$\frac{3x}{2}\cdot 6x-\frac{8}{3}\cdot 6x=\frac{2}{3x}\cdot 6x$$
$$9x^2-16x=4$$
$$9x^2-16x-4=0$$
$$(9x+2)(x-2)=0$$
$$x=-\frac{2}{9},\ x=2$$

4.
$$x(x-3)-30=5(x-2)$$
$$x^2-3x-30=5x-10$$
$$x^2-8x-20=0$$
$$(x-10)(x+2)=0$$
$$x=10,\ x=-2$$

5.
$$7x^2-4=52$$
$$7x^2=56$$
$$x^2=8$$
$$x=\pm\sqrt{8}=\pm\sqrt{4\cdot 2}$$
$$x=\pm 2\sqrt{2}$$

6. The LCD is $(2x+1)(2x-1)=4x^2-1$.

$$\frac{2x}{2x+1}-\frac{6}{4x^2-1}=\frac{x+1}{2x-1}$$
$$\frac{2x}{2x+1}(2x+1)(2x-1)-\frac{6}{4x^2-1}(4x^2-1)=\frac{x+1}{2x-1}(2x+1)(2x-1)$$
$$2x(2x-1)-6=(x+1)(2x+1)$$
$$4x^2-2x-6=2x^2+3x+1$$
$$2x^2-5x-7=0$$
$$(2x-7)(x+1)=0$$
$$x=\frac{7}{2},\ x=-1$$

7. $2x^2 - 6x + 5 = 0$

$a = 2, b = -6, c = 5$

$$x = \frac{-(-6) \pm \sqrt{(-6)^2 - 4(2)(5)}}{2(2)}$$

$$x = \frac{6 \pm \sqrt{-4}}{4}$$

$$x = \frac{6 \pm 2i}{4} = \frac{2(3 \pm i)}{4}$$

$$x = \frac{3 \pm i}{2}$$

8. $2x(x-3) = -3$

$2x^2 - 6x = -3$

$2x^2 - 6x + 3 = 0$

$a = 2, b = -6, c = 3$

$$x = \frac{-(-6) \pm \sqrt{(-6)^2 - 4(2)(3)}}{2(2)}$$

$$x = \frac{6 \pm \sqrt{12}}{4}$$

$$x = \frac{6 \pm 2\sqrt{3}}{4} = \frac{2\left(3 \pm \sqrt{3}\right)}{4}$$

$$x = \frac{3 \pm \sqrt{3}}{2}$$

9. $x^4 - 11x^2 + 18 = 0$

$y = x^2$

$y^2 - 11y + 18 = 0$

$(y-9)(y-2) = 0$

$y - 9 = 0 \qquad y - 2 = 0$

$y = 9 \qquad y = 2$

$x^2 = 9 \qquad x^2 = 2$

$x = \pm 3 \qquad x = \pm\sqrt{2}$

10. $3x^{-2} - 11x^{-1} - 20 = 0$

$y = x^{-1}$

$3y^2 - 11y - 20 = 0$

$(y-5)(3y+4) = 0$

$y - 5 = 0 \qquad 3y + 4 = 0$

$y = 5 \qquad y = -\frac{4}{3}$

$x^{-1} = 5 \qquad x^{-1} = -\frac{4}{3}$

$x = \frac{1}{5} \qquad x = -\frac{3}{4}$

11. $x^{2/3} - 3x^{1/3} - 4 = 0$

$y = x^{1/3}$

$y^2 - 3y - 4 = 0$

$(y-4)(y+1) = 0$

$y - 4 = 0 \qquad y + 1 = 0$

$y = 4 \qquad y = -1$

$x^{1/3} = 4 \qquad x^{1/3} = -1$

$x = 64 \qquad x = -1$

12. $B = \frac{xyw}{z^2}$

$z^2 = \frac{xyw}{B}$

$z = \pm\sqrt{\frac{xyw}{B}}$

13. $5y^2 + 2by + 6w = 0$

$$y = \frac{-2b \pm \sqrt{(2b)^2 - 4(5)(6w)}}{2(5)}$$

$$y = \frac{-2b \pm \sqrt{4(b^2 - 30w)}}{10}$$

$$y = \frac{-2b \pm 2\sqrt{b^2 - 30w}}{10}$$

$$y = \frac{-b \pm \sqrt{b^2 - 30w}}{5}$$

14. Let x = the width of the rectangle. Then $3x + 1$ = the length.

$x(3x+1) = 80$

$3x^2 + x = 80$

$3x^2 + x - 80 = 0$

$(x-5)(3x+16) = 0$

$x = 5,\ x = -\frac{16}{3}$

Since the width must be positive, $x = 5$. The width is 5 miles and the length is $3(5) + 1 = 16$ miles.

15. $c^2 = a^2 + b^2$

$c^2 = 6^2 + \left(2\sqrt{3}\right)^2$

$c^2 = 36 + 12 = 48$

$c = \sqrt{48} = \sqrt{16 \cdot 3}$

$c = 4\sqrt{3}$

16. Let v = the speed for the first 6 miles. Then $v + 1$ is their speed for the other 3 miles.

$$\text{Time} = \frac{\text{Distance}}{\text{Rate}}$$

$$\frac{6}{v}+\frac{3}{v+1}=4$$
$$6(v+1)+3v=4v(v+1)$$
$$6v+6+3v=4v^2+4v$$
$$4v^2-5v-6=0$$
$$(v-2)(4v+3)=0$$
$$v=2,\ v=-\frac{3}{4}$$

Since the speed must be positive, $v = 2$.
The paddled 2 mph on the first part and 3 mph on the second part.

17. $f(x)=-x^2-6x-5$

$$\frac{-b}{2a}=\frac{-(-6)}{2(-1)}=-3$$

$f(-3)=-(-3)^2-6(-3)-5$
$f(-3)=4$
$V(-3, 4)$
$f(0)=-0^2-6(0)-5=-5$
y-intercept: $(0, -5)$
$-x^2-6x-5=0$
$x^2+6x+5=0$
$(x+1)(x+5)=0$
$x+1=0$ $\quad$ $x+5=0$
$x=-1$ $\quad$ $x=-5$
x-intercepts: $(-1, 0)$, $(-5, 0)$

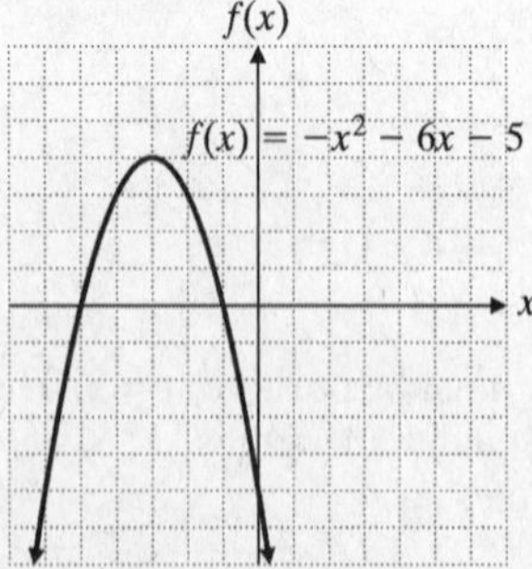

18. $2x^2+3x\ge 27$
$2x^2+3x-27\ge 0$
$2x^2+3x-27=0$
$(2x+9)(x-3)=0$
$x=-\frac{9}{2},\ x=3$

Boundary points: $-\frac{9}{2}$, 3

Test: $2x^2+3x-27$

Region	Test	Result
$x<-\frac{9}{2}$	-5	$8>0$
$-\frac{9}{2}<x<3$	0	$-27<0$
$x>3$	4	$17>0$

$x\le -\frac{9}{2}$ or $x\ge 3$

19. $x^2-5x-14<0$
$x^2-5x-14=0$
$(x+2)(x-7)=0$
$x=-2,\ x=7$
Boundary points: -2, 7
Test: $x^2-5x-14$

Region	Test	Result
$x<-2$	-3	$10>0$
$-2<x<7$	0	$-14<0$
$x>7$	8	$10>0$

$-2<x<7$

20. $x^2+3x-7>0$
$x^2+3x-7=0$
$a = 1, b = 3, c = -7$

$$x=\frac{-3\pm\sqrt{3^2-4(1)(-7)}}{2(1)}=\frac{-3\pm\sqrt{37}}{2}$$

$x\approx -4.5,\ x\approx 1.5$
Approximate boundary points: -4.5, 1.5
Test: x^2+3x-7

Approximate Region	Test	Result
$x<-4.5$	-5	$3>0$
$-4.5<x<1.5$	0	$-7<0$
$x>1.5$	2	$3>0$

Approximately $x < -4.5$ or $x > 1.5$.

Cumulative Test for Chapters 1–8

1. $(-3x^{-2}y^3)^4 = (-3)^4 x^{-2\cdot 4} y^{3\cdot 4}$
$= 81x^{-8}y^{12}$
$= \dfrac{81y^{12}}{x^8}$

2. $\frac{1}{2}a^3 - 2a^2 + 3a - \frac{1}{4}a^3 - 6a + a^2 = \frac{1}{4}a^3 - a^2 - 3a$

3.
$$\begin{aligned}\frac{1}{3}(x-3)+1 &= \frac{1}{2}x-2\\ 6\left[\frac{1}{3}(x-3)+1\right] &= 6\left(\frac{1}{2}x-2\right)\\ 2(x-3)+6 &= 3x-12\\ 2x-6+6 &= 3x-12\\ 2x &= 3x-12\\ -x &= -12\\ x &= 12\end{aligned}$$

4. $6x - 3y = -12$

x	y
0	4
−2	0

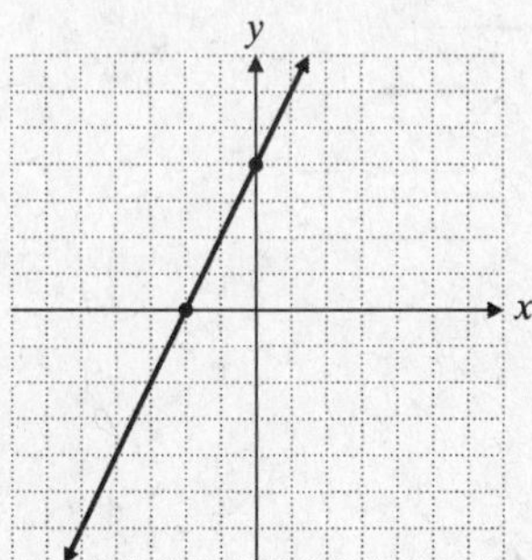

5.
$$\begin{aligned}2y+x &= 8\\ 2y &= -x+8\\ y &= -\frac{1}{2}x+4\\ m &= -\frac{1}{2}\end{aligned}$$

The slope of a parallel line is $m = -\frac{1}{2}$.
$$\begin{aligned}y - y_1 &= m(x-x_1)\\ y-(-1) &= -\frac{1}{2}(x-6)\\ y+1 &= -\frac{1}{2}x+3\\ 2y+2 &= -x+6\\ x+2y &= 4\end{aligned}$$

6. $3x+4y=-14$ **(1)**
$-x-3y=13$ **(2)**

Multiply equation **(2)** by 3 and add the result to equation **(1)** to eliminate *x*.
$$\begin{aligned}3x+4y &= -14\\ -3x-9y &= 39\\ \hline -5y &= 25\\ y &= -5\end{aligned}$$

Substitute −5 for *y* in equation **(2)**.
$$\begin{aligned}-x-3(-5) &= 13\\ -x+15 &= 13\\ -x &= -2\\ x &= 2\end{aligned}$$

7. $125x^3 - 27y^2 = (5x)^3 - (3y)^3$
$= (5x-3y)(25x^2+15xy+9y^2)$

8. $\sqrt{72x^3y^6} = \sqrt{36x^2y^6 \cdot 2x} = 6xy^3\sqrt{2x}$

9.
$$\begin{aligned}\left(5+\sqrt{3}\right)\left(\sqrt{6}-\sqrt{2}\right) &= 5\sqrt{6}-5\sqrt{2}+\sqrt{18}-\sqrt{6}\\ &= 5\sqrt{6}-5\sqrt{2}+\sqrt{9\cdot 2}-\sqrt{6}\\ &= 5\sqrt{6}-5\sqrt{2}+3\sqrt{2}-\sqrt{6}\\ &= 4\sqrt{6}-2\sqrt{2}\end{aligned}$$

10. $\dfrac{3x}{\sqrt{6}} = \dfrac{3x}{\sqrt{6}} \cdot \dfrac{\sqrt{6}}{\sqrt{6}} = \dfrac{3x\sqrt{6}}{6} = \dfrac{x\sqrt{6}}{2}$

11.
$$\begin{aligned}3x^2+12x &= 26x\\ 3x^2-14x &= 0\\ x(3x-14) &= 0\end{aligned}$$
$x = 0,\ x = \dfrac{14}{3}$

12. $12x^2 = 11x - 2$

$12x^2 - 11x + 2 = 0$

$(4x-1)(3x-2) = 0$

$4x - 1 = 0 \qquad 3x - 2 = 0$

$x = \frac{1}{4} \qquad x = \frac{2}{3}$

13. $44 = 3(2x-3)^2 + 8$

$3(2x-3)^2 = 36$

$(2x-3)^2 = 12$

$2x - 3 = \pm\sqrt{12}$

$2x - 3 = \pm 2\sqrt{3}$

$2x = 3 \pm 2\sqrt{3}$

$x = \frac{3 \pm 2\sqrt{3}}{2}$

14. The LCD is x^2.

$3 - \frac{4}{x} + \frac{5}{x^2} = 0$

$3x^2 - \frac{4}{x} \cdot x^2 + \frac{5}{x^2} \cdot x^2 = 0 \cdot x^2$

$3x^2 - 4x + 5 = 0$

$a = 3,\ b = -4,\ c = 5$

$x = \frac{-(-4) \pm \sqrt{(-4)^2 - 4(3)(5)}}{2(3)}$

$x = \frac{4 \pm \sqrt{-44}}{6}$

$x = \frac{4 \pm 2i\sqrt{11}}{6} = \frac{2\left(2 \pm i\sqrt{11}\right)}{6}$

$x = \frac{2 \pm i\sqrt{11}}{3}$

15. $\sqrt{6x+12} - 2 = x$

$\sqrt{6x+12} = x + 2$

$\left(\sqrt{6x+12}\right)^2 = (x+2)^2$

$6x + 12 = x^2 + 4x + 4$

$0 = x^2 - 2x - 8$

$0 = (x+2)(x-4)$

$x = -2,\ x = 4$

Check $x = -2$: $\sqrt{6(-2)+12} \stackrel{?}{=} -2 + 2$

$\sqrt{-12+12} \stackrel{?}{=} 0$

$0 = 0$ True

Check $x = 4$: $\sqrt{6(4)+12} \stackrel{?}{=} 4 + 2$

$\sqrt{24+12} \stackrel{?}{=} 6$

$\sqrt{36} \stackrel{?}{=} 6$

$6 = 6$ True

The solutions are −2 and 4.

16. $x^{2/3} + 9x^{1/3} + 18 = 0$

$y = x^{1/3}$

$y^2 + 9y + 18 = 0$

$(y+6)(y+3) = 0$

$y = -6,\ y = -3$

$x^{1/3} = -6 \qquad x^{1/3} = -3$

$x = (-6)^3 = -216 \qquad x = (-3)^3 = -27$

17. $2y^2 + 5wy - 7z = 0$

$a = 2,\ b = 5w,\ c = -7z$

$y = \frac{-5w \pm \sqrt{(5w)^2 - 4(2)(-7z)}}{2(2)}$

$y = \frac{-5w \pm \sqrt{25w^2 + 56z}}{4}$

18. $3y^2 + 16z^2 = 5w$

$3y^2 = 5w - 16z^2$

$y^2 = \frac{5w - 16z^2}{3}$

$y = \pm\sqrt{\frac{5w - 16z^2}{3}}$

$y = \pm\frac{\sqrt{15w - 48z^2}}{3}$

19. $c^2 = a^2 + b^2$

$\left(\sqrt{38}\right)^2 = 5^2 + b^2$

$38 = 25 + b^2$

$b^2 = 13$

$b = \sqrt{13}$

20. Let b = the length of the base. Then $3b + 3$ = the length of the altitude.

$A = \frac{1}{2}ab$

$\frac{1}{2}b(3b+3) = 45$

$3b^2 + 3b = 90$

$3b^2 + 3b - 90 = 0$

$b^2 + b - 30 = 0$

$(b+6)(b-5) = 0$

$b = -6,\ b = 5$

Since the length of the base must be positive, $b = 5$.

The base of the triangle is 5 meters and the altitude of the triangle is 3(5) + 3 = 18 meters.

21. $f(x) = -x^2 + 8x - 12$

$a = -1,\ b = 8,\ c = -12$

$x_{\text{vertex}} = \frac{-b}{2a}$

$x_{\text{vertex}} = \frac{-8}{2(-1)}$

$x_{\text{vertex}} = 4$

$f(x_{\text{vertex}}) = f(4) = -4^2 + 8(4) - 12 = 4$

$V(4, 4)$

$f(0) = -0^2 + 8(0) - 12 = -12$

y-intercept: (0, −12)

$-x^2 + 8x - 12 = 0$

$x^2 - 8x + 12 = 0$

$(x-6)(x-2) = 0$

$x - 6 = 0 \qquad x - 2 = 0$

$x = 6 \qquad x = 2$

x-intercepts: (6, 0), (2, 0)

22.

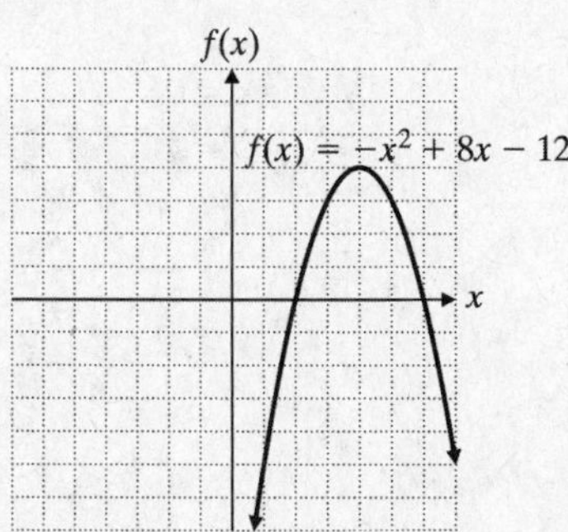

23. $6x^2 - x \le 2$

$6x^2 - x - 2 \le 0$

$6x^2 - x - 2 = 0$

$(3x-2)(2x+1) = 0$

$x = \frac{2}{3},\ x = -\frac{1}{2}$

Boundary points: $\frac{2}{3}, -\frac{1}{2}$

Test: $6x^2 - x - 2$

Region	Test	Result
$x < -\frac{1}{2}$	−1	$5 > 0$
$-\frac{1}{2} < x < \frac{2}{3}$	0	$-2 < 0$
$x > \frac{2}{3}$	1	$3 > 0$

$-\frac{1}{2} \le x \le \frac{2}{3}$

24. $x^2 > -3x + 18$

$x^2 + 3x - 18 > 0$

$x^2 + 3x - 18 = 0$

$(x+6)(x-3) = 0$

$x = -6,\ x = 3$

Boundary points: −6, 3

Test: $x^2 + 3x - 18$

Region	Test	Result
$x < -6$	−10	$52 > 0$
$-6 < x < 3$	0	$-18 < 0$
$x > 3$	6	$36 > 0$

$x < -6$ or $x > 3$

Chapter 9

9.1 Exercises

1. Subtract the values of the points and use the absolute value: $|-2-4| = |-6| = 6$.

3. The equation, $(x-1)^2+(y+2)^2=9=3^2$, is in standard form. Determine the values of h, k, and r to find the center and radius: $h = 1$, $k = -2$, and $r = 3$. The center is $(1, -2)$ and the radius is 3.

5.
$$d=\sqrt{(x_2-x_1)^2+(y_2-y_1)^2}$$
$$d=\sqrt{(1-2)^2+(6-4)^2}$$
$$d=\sqrt{1+4}=\sqrt{5}$$

7.
$$d=\sqrt{(x_2-x_1)^2+(y_2-y_1)^2}$$
$$d=\sqrt{(5-2)^2+(1-(-5))^2}$$
$$d=\sqrt{9+36}=\sqrt{45}=3\sqrt{5}$$

9.
$$d=\sqrt{(x_2-x_1)^2+(y_2-y_1)^2}$$
$$d=\sqrt{(4-(-2))^2+(-5-(-13))^2}$$
$$d=\sqrt{36+64}=\sqrt{100}=10$$

11.
$$d=\sqrt{(x_2-x_1)^2+(y_2-y_1)^2}$$
$$d=\sqrt{\left(\frac{5}{4}-\frac{1}{4}\right)^2+\left(-\frac{1}{3}-\left(-\frac{2}{3}\right)\right)^2}$$
$$d=\sqrt{1+\frac{1}{9}}=\sqrt{\frac{10}{9}}=\frac{\sqrt{10}}{\sqrt{9}}$$
$$d=\frac{\sqrt{10}}{3}$$

13.
$$d=\sqrt{(x_2-x_1)^2+(y_2-y_1)^2}$$
$$d=\sqrt{\left(\frac{7}{3}-\frac{1}{3}\right)^2+\left(\frac{1}{5}-\frac{3}{5}\right)^2}$$
$$d=\sqrt{4+\frac{4}{25}}=\sqrt{\frac{104}{25}}=\frac{\sqrt{4\cdot 26}}{\sqrt{25}}$$
$$d=\frac{2\sqrt{26}}{5}$$

15.
$$d=\sqrt{(x_2-x_1)^2+(y_2-y_1)^2}$$
$$d=\sqrt{(1.3-(-5.7))^2+(2.6-1.6)^2}$$
$$d=\sqrt{7^2+1^2}$$
$$d=\sqrt{50}$$
$$d=5\sqrt{2}$$

17.
$$d=\sqrt{(x_2-x_1)^2+(y_2-y_1)^2}$$
$$10=\sqrt{(1-7)^2+(y-2)^2}$$
$$100=y^2-4y+40$$
$$y^2-4y-60=0$$
$$(y-10)(y+6)=0$$
$$y-10=0 \qquad y+6=0$$
$$y=10 \qquad y=-6$$

19.
$$d=\sqrt{(x_2-x_1)^2+(y_2-y_1)^2}$$
$$2.5=\sqrt{(0-1.5)^2+(y-2)^2}$$
$$6.25=2.25+y^2-4y+4$$
$$y^2-4y=0$$
$$y(y-4)=0$$
$$y=0 \qquad y-4=0$$
$$y=4$$

21.
$$d=\sqrt{(x_2-x_1)^2+(y_2-y_1)^2}$$
$$\sqrt{5}=\sqrt{(x-4)^2+(5-7)^2}$$
$$\sqrt{5}=\sqrt{x^2-8x+16+4}$$
$$5=x^2-8x+20$$
$$x^2-8x+15=0$$
$$(x-3)(x-5)=0$$
$$x-3=0 \qquad x-5=0$$
$$x=3 \qquad x=5$$

23.
$$d=\sqrt{(x_2-x_1)^2+(y_2-y_1)^2}$$
$$4=\sqrt{(2-5)^2+(y-7)^2}$$
$$16=9+y^2-14y+49$$
$$y^2-14y+42=0$$
$$y=\frac{14\pm\sqrt{14^2-4(1)(42)}}{2(1)}$$
$$y=\frac{14\pm\sqrt{28}}{2}\approx\begin{cases}9.6\\4.4\end{cases}$$
(select the shortest distance)

$y\approx 4.4$ miles

The plane can be detected at 4.4 miles.

25.
$$(x-h)^2+(y-k)^2=4^2$$
$$(x-(-3))^2+(y-7)^2=6^2$$
$$(x+3)^2+(y-7)^2=36$$

27.
$$(x-h)^2+(y-k)^2=r^2$$
$$(x-(-2.4))^2+(y-0)^2=\left(\frac{3}{4}\right)^2$$
$$(x+2.4)^2+y^2=\frac{9}{16}$$

29. $(x-h)^2+(y-k)^2=r^2$
$$\left(x-\frac{3}{8}\right)^2+(y-0)^2=\sqrt{3}^2$$
$$\left(x-\frac{3}{8}\right)^2+y^2=3$$

31.
$$x^2+y^2=25$$
$$(x-0)^2+(y-0)^2=5^2$$
$C(0, 0)$, $r=5$

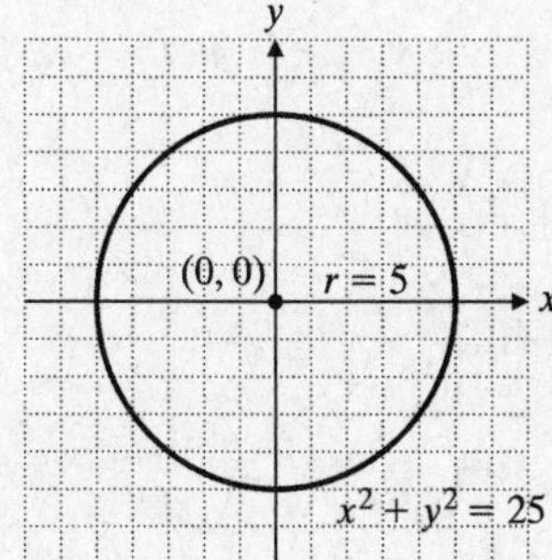

33. $(x-5)^2+(y-3)^2=16=4^2$

$C(5, 3)$, $r=4$

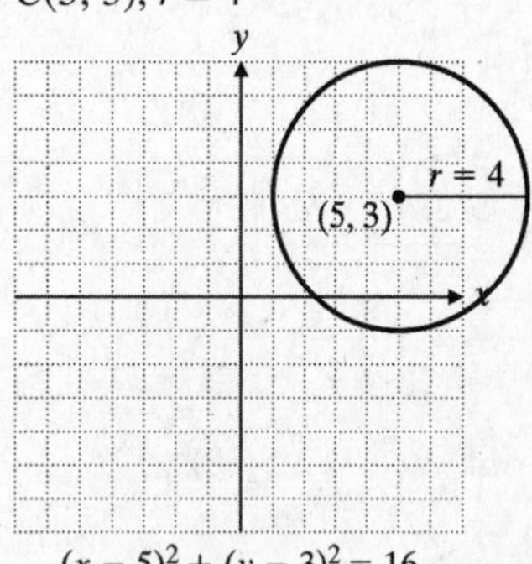

$(x-5)^2+(y-3)^2=16$

35.
$$(x+2)^2+(y-3)^2=25$$
$$[x-(-2)]^2+(y-3)^2=5^2$$
$C(-2, 3)$, $r=5$

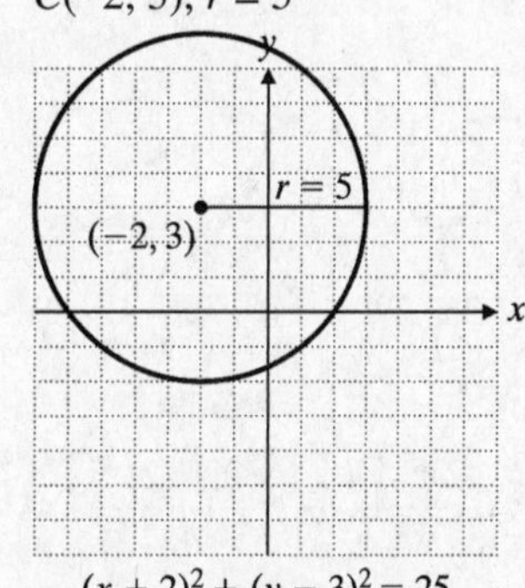

$(x+2)^2+(y-3)^2=25$

37.
$$x^2+y^2+8x-6y-24=0$$
$$x^2+8x+16+y^2-6y+9=24+16+9$$
$$(x+4)^2+(y-3)^2=49=7^2$$
$C(-4, 3)$, $r=7$

39.
$$x^2+y^2-10x+6y-2=0$$
$$x^2-10x+25+y^2+6y+9=2+25+9$$
$$(x-5)^2+(y+3)^2=36$$
$C(5, -3)$, $r=6$

41. $x^2+y^2+3x-2=0$
$$x^2+3x+y^2=2$$
$$x^2+3x+\frac{9}{4}+y^2=2+\frac{9}{4}=\frac{17}{4}$$
$$\left(x+\frac{3}{2}\right)^2+y^2=\left(\frac{\sqrt{17}}{2}\right)^2$$
$$C\left(-\frac{3}{2}, 0\right),\ r=\frac{\sqrt{17}}{2}$$

43. $C(44.8, 31.8),\ r = 25.3,\ r^2 = 640.09$

$$(x-44.8)^2 + (y-31.8)^2 = 640.09$$

45. $(x-5.32)^2 + (y+6.54)^2 = 47.28$

$$(y+6.54)^2 = 47.28 - (x-5.32)^2$$
$$y+6.54 = \pm\sqrt{47.28-(x-5.32)^2}$$
$$y_1 = -6.54 + \sqrt{47.28-(x-5.32)^2}$$
$$y_2 = -6.54 - \sqrt{47.28-(x-5.32)^2}$$

Cumulative Review

47. $4x^2 + 2x = 1$

$$4x^2 + 2x - 1 = 0$$
$$x = \frac{-2 \pm \sqrt{2^2 - 4(4)(-1)}}{2(4)}$$
$$x = \frac{-2 \pm \sqrt{20}}{8} = \frac{-2 \pm \sqrt{4(5)}}{8} = \frac{-2 \pm 2\sqrt{5}}{8}$$
$$x = \frac{-1 \pm \sqrt{5}}{4}$$

48. $5x^2 - 6x - 7 = 0$

$$x = \frac{-(-6) \pm \sqrt{(-6)^2 - 4(5)(-7)}}{2(5)}$$
$$x = \frac{6 \pm \sqrt{176}}{10} = \frac{6 \pm \sqrt{16(11)}}{10}$$
$$x = \frac{6 \pm 4\sqrt{11}}{10} = \frac{3 \pm 2\sqrt{11}}{5}$$

49. $V = Ah$

$$V = 20\text{ mi}^2(150\text{ ft})\left(\frac{5280^2\text{ ft}^2}{\text{mi}^2}\right)$$
$$V = 8.364 \times 10^{10}\text{ ft}^3$$

There was 8.364×10^{10} ft^3 of rock and sediments that settled in this region.

50. $d = rt$

$$t = \frac{d}{r} = \frac{15}{670}\text{ hr} \cdot \frac{3600\text{ sec}}{\text{hr}}$$
$$t \approx 81\text{ sec}$$

He had approximately 81 seconds the run.

Quick Quiz 9.1

1. $d = \sqrt{(x_2-x_1)^2 + (y_2-y_1)^2}$

$$d = \sqrt{(-2-3)^2 + (-6-(-4))^2}$$
$$d = \sqrt{25+4} = \sqrt{29}$$

2. $(x-h)^2 + (y-k)^2 = r^2$

$$(x-5)^2 + (y-(-6))^2 = 7^2$$
$$(x-5)^2 + (y+6)^2 = 49$$

3. $x^2 + 4x + y^2 - 6y + 4 = 0$

$$x^2 + 4x + 4 + y^2 - 6y + 9 = -4 + 4 + 9$$
$$(x+2)^2 + (y-3)^2 = 9$$

center $(-2, 3)$, $r = 3$

4. Answers may vary. Possible solution: Using the distance formula:

$$d = \sqrt{(x_2-x_1)^2 + (y_2-y_1)^2}$$

we fill in the known variables, and solve for the one unknown variable x.

9.2 Exercises

1. The graph of $y = x^2$ is symmetric about the y-axis. The graph of $x = y^2$ is symmetric about the x-axis.

3. Since $y = 2(x-3)^2 + 4$ is in standard form, $y = a(x-h)^2 + k$, the vertex is $(h, k) = (3, 4)$.

5. $y = -4x^2$

$V(0, 0)$

Let $x = 0$: $y = -4(0)^2$

$$y = 0$$

y-intercept = $(0, 0)$

x	-2	-1	0	1	2
y	-16	-4	0	-4	-16

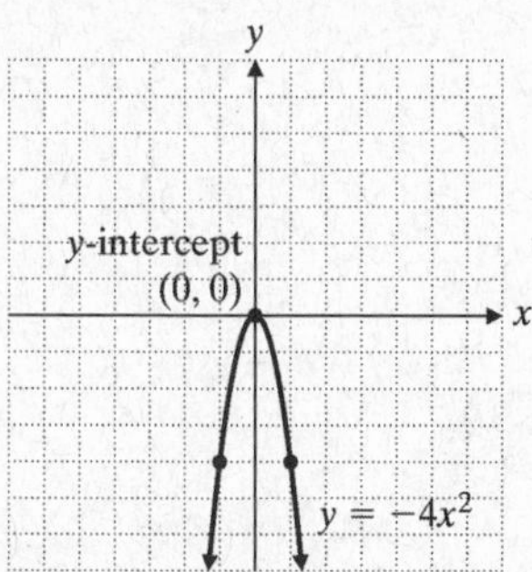

7. $y = x^2 - 6$

$V(0, -6)$

Let $x = 0$.

$y = 0^2 - 6 = -6$

y-intercept: $(0, -6)$

x	−3	−2	0	2	3
y	3	−2	−6	−2	6

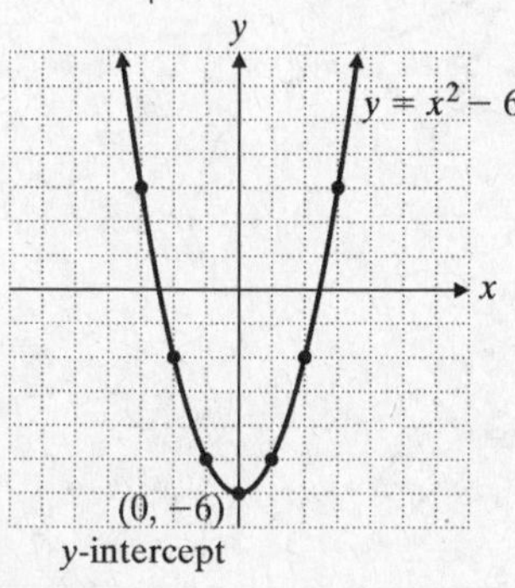

9. $y = \frac{1}{2}x^2 - 2$

$V(0, -2)$

Let $x = 0$.

$y = \frac{1}{2}(0)^2 - 2 = -2$

y-intercept: $(0, -2)$

x	−4	−2	0	2	4
y	6	0	−2	0	6

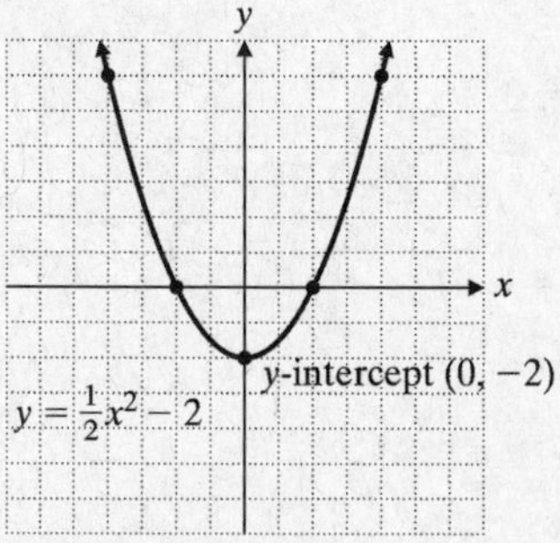

11. $y = (x-3)^2 - 2$

$V(3, -2)$

Let $x = 0$.

$y = (0-3)^2 - 2 = 7$

y-intercept: $(0, 7)$

x	1	2	3	4	5
y	2	−1	−2	−1	2

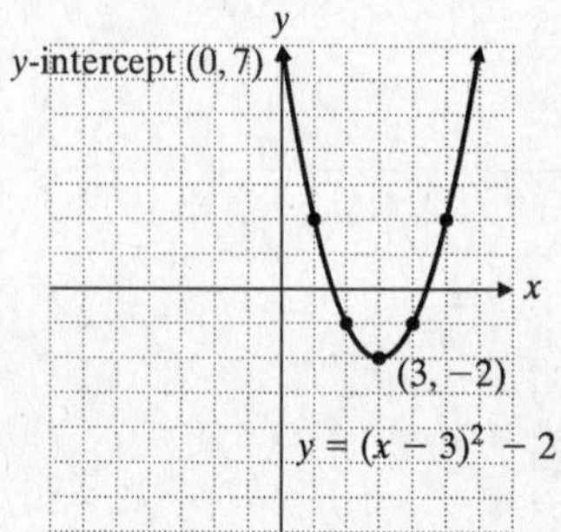

13. $y = 2(x-1)^2 + \frac{3}{2}$

$V\left(1, \frac{3}{2}\right)$

Let $x = 0$.

$y = 2(0-1)^2 + \frac{3}{2} = \frac{7}{2}$

y-intercept: $\left(0, \frac{7}{2}\right)$

x	−1	0	1	2	3
y	$\frac{19}{2}$	$\frac{7}{2}$	$\frac{3}{2}$	$\frac{7}{2}$	$\frac{19}{2}$

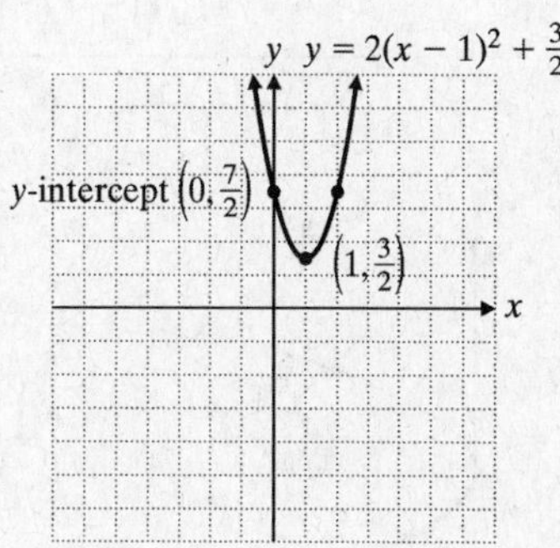

15. $y=-4\left(x+\frac{3}{2}\right)^2+5$

$V\left(-\frac{3}{2}, 5\right)$

Let $x = 0$.

$y=-4\left(0+\frac{3}{2}\right)^2+5$

$y=-4$

y-intercept: $(0, -4)$

x	-3	-2	$-\frac{3}{2}$	-1	0
y	-4	4	5	4	-4

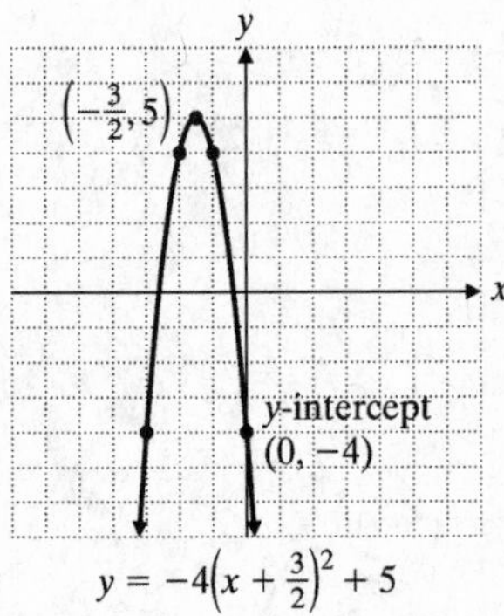

17. $x=\frac{1}{2}y^2$

$V(0, 0)$

Let $y = 0$.

$x=\frac{1}{2}(0)^2=0$

x-intercept: $(0, 0)$

x	0	2	2	4	4
y	0	2	-2	-8	8

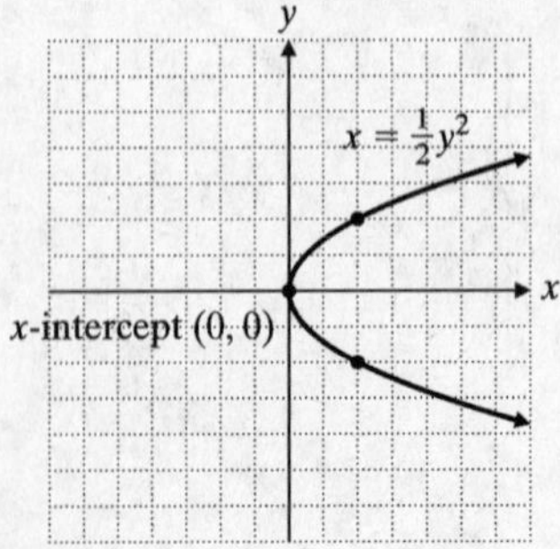

19. $x=\frac{1}{4}y^2-2$

$V(-2, 0)$

Let $y = 0$.

$x=\frac{1}{4}(0)^2-2=-2$

x-intercept: $(-2, 0)$

x	-2	-1	-1	2	2
y	0	-2	2	-4	4

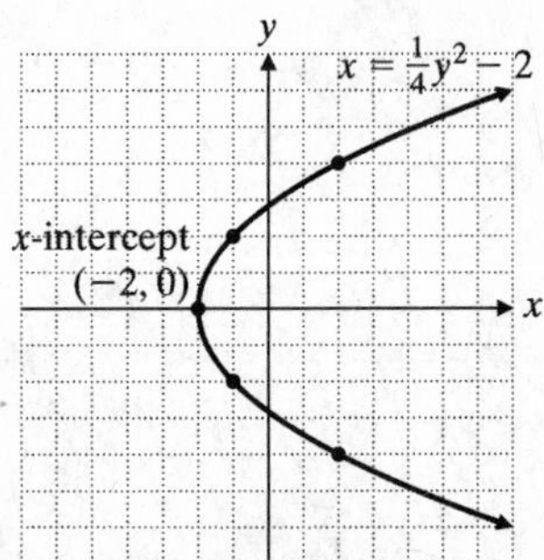

21. $x=-y^2+2$

$V(2, 0)$

Let $y = 0$.

$x=-0^2+2=2$

x-intercept: $(2, 0)$

x	2	-2	-2	-7	-7
y	0	-2	2	-3	3

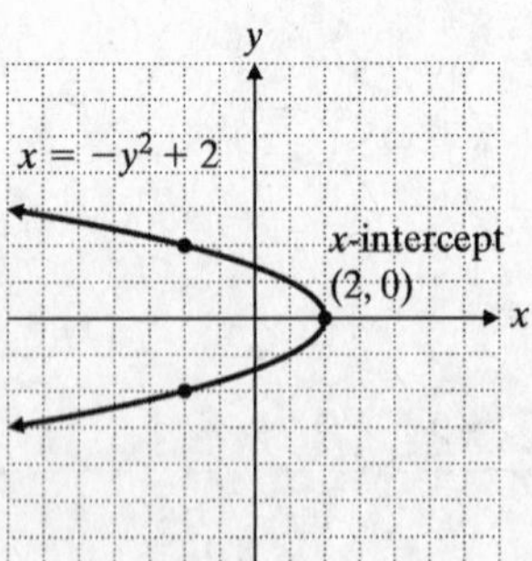

23. $x=(y-2)^2+3$

$V(3, 2)$

Let $y = 0$.

$x=(0-2)^2+3=7$

x-intercept: $(7, 0)$

x	3	4	4	7	7
y	2	1	3	4	0

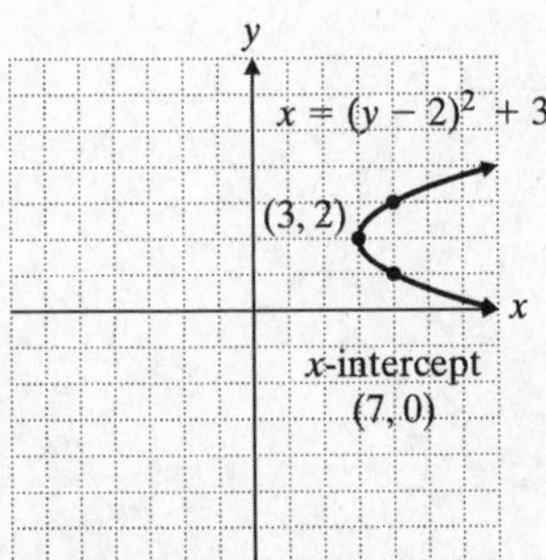

25. $x = -3(y+1) - 2$
$V(-2, -1)$
Let $y = 0$.
$x = -3(0+1) - 2$
$x = -5$
x-intercept: $(-5, 0)$

x	−2	−5	−5
y	−1	0	−2

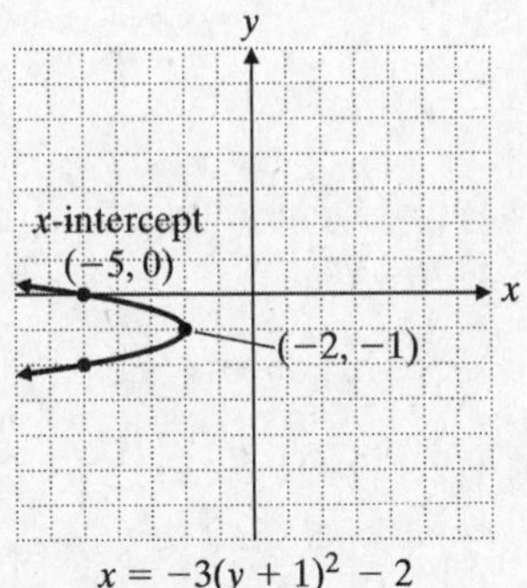

27. $y = x^2 - 4x - 1$
$y = x^2 - 4x + 4 - 4 - 1$
$y = (x-2)^2 - 5$

a. Vertical since the equation has the form $y = a(x-h)^2 + k$.

b. Opens upward since $a > 0$.

c. $V(2, -5)$

29. $y = -2x^2 + 8x - 1$
$y = -2(x^2 - 4x + 4) - 1 + 8$
$y = -2(x-2)^2 + 7$

a. Vertical since the equation has the form $y = a(x-h)^2 + k$.

b. Opens downward since $a < 0$.

c. vertex $(2, 7)$

31. $x = y^2 + 8y + 9$
$x = y^2 + 8y + 16 - 7$
$x = (y+4)^2 - 7$

a. Horizontal since the equation has the form $x = a(y-k)^2 + h$.

b. Opens right since $a > 0$.

c. $V(-7, -4)$

33. $y = ax^2$
$8 = a(16)^2$
$a = \frac{1}{32}$
$y = \frac{1}{32}x^2$

35. $a = \frac{1}{4p} = \frac{1}{32} \Rightarrow p = 8$
The distance from (0, 0) to the focus point is 8 inches.

37. $y = 2x^2 + 6.48x - 0.1312$
$y = 2(x^2 + 3.24x + 2.6244) - 0.1312 - 5.2488$
$y = 2(x+1.62)^2 - 5.38$
$V(-1.62, -5.38)$
y-intercept: $(0, -0.1312)$

$$\frac{-6.48 \pm \sqrt{6.48^2 - 4(2)(-0.1312)}}{2(2)}$$
$$= \begin{cases} 0.020121947 \\ -3.260121947 \end{cases}$$

x-intercepts: $(0.020121947, 0)$, $(-3.26012194, 0)$

39. $P = -2x^2 + 400x + 35{,}000$
$P = -2(x^2 - 200x + 10{,}000) + 35{,}000 + 20{,}000$
$P = -2(x - 100)^2 + 55{,}000$
Vertex (100, 55,000)
The maximum profit is \$55,000. The number of watches produced is 100.

41. $E = x(900 - x) = -x^2 + 900x$
$E = -(x^2 - 900x + 202{,}500) + 202{,}500$
$E = -(x - 450)^2 + 202{,}500$
Vertex (450, 202,500)
The maximum yield is 202,500 and the number of trees per acre is 450.

Cumulative Review

43. $d = rt$
$d = 90$
$t_1 = 1.5$
$t_2 = 2.5 - 1.5 = 1$
$r_2 = r_1 + 20$
$d = r_1t_1 + r_2t_2$
$90 = r_1(1.5) + (r_1 + 20)1$
$90 = 1.5r_1 + r_1 + 20$
$70 = 2.5r_1$
$r_1 = 28$ mph
His speed was 28 mph on the way to the warehouse.

44. $d = rt$
$d = 40t = \frac{40}{60}(56 - 15) = 27\frac{1}{3}$
Matthew lives $27\frac{1}{3}$ miles from work.

45. $8(1050)(0.88) = 7392$
Sir George can expect 7392 blooms if there is heavy rainfall this year.

46. $\frac{2900}{6(0.44)} \approx 1098$
There will be 1098 buds on each bush.

Quick Quiz 9.2

1. $y = -3(x + 2)^2 + 5$
vertex (–2, 5)
Let $x = 0$.
$y_{\text{int}} = -3(0 + 2)^2 + 5 = -7$
y-intercept (0, –7)

2. $x = (y - k)^2 + h$
vertex: (3, 2)
y-intercept (7, 0)
$x = (y - 2)^2 + 3$
Test for point (7, 0).
$7 = (0 - 2)^2 + 3$
$7 = 4 + 3$
$7 = 7$ True

3. $y = 2(x - 3)^2 - 6$
vertex (3, –6)

x	1	3	5
y	2	–6	2

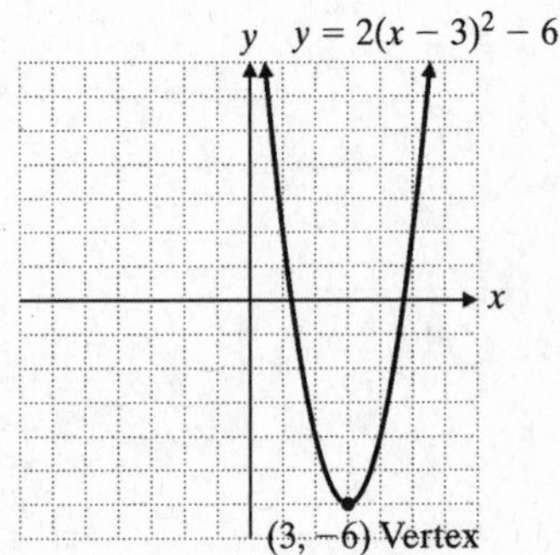

4. Answers may vary. Possible solution:
$y = ax^2$, vertical, opens up
$y = -ax^2$, vertical, opens down
$x = ay^2$, horizontal, opens right
$x = -ay^2$, horizontal, opens left

9.3 Exercises

1. $\frac{(x+2)^2}{4} + \frac{(y-3)}{9} = 1$ is in the form
$\frac{(x-h)^2}{a^2} + \frac{(y-k)^2}{b^2} = 1$ where (h, k) is the center of the ellipse. Therefore, the center of the ellipse is (–2, 3).

3. $\frac{x^2}{36}+\frac{y^2}{4}=1 \Rightarrow \frac{x^2}{6^2}+\frac{y^2}{2^2}=1$

$a=6, b=2$

Intercepts: $(\pm 6, 0), (0, \pm 2)$

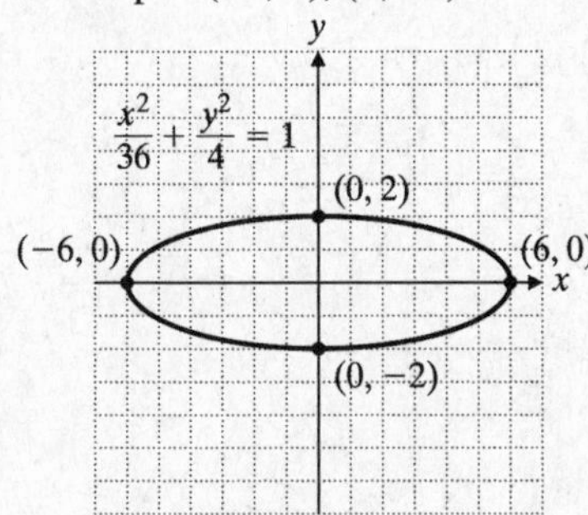

5. $\frac{x^2}{81}+\frac{y^2}{100}=1 \Rightarrow \frac{x^2}{9^2}+\frac{y^2}{10^2}=1$

$a=9, b=10$

Intercepts: $(\pm 9, 0), (0, \pm 10)$

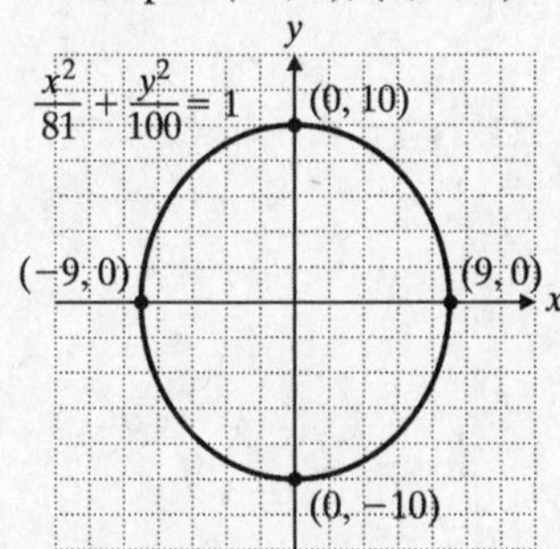

Scale: Each unit = 2

7. $4x^2+y^2-36=0$

$\frac{x^2}{9}+\frac{y^2}{36}=1 \Rightarrow \frac{x^2}{3^2}+\frac{y^2}{6^2}=1$

$a=3, b=6$

Intercepts: $(\pm 3, 0), (0, \pm 6)$

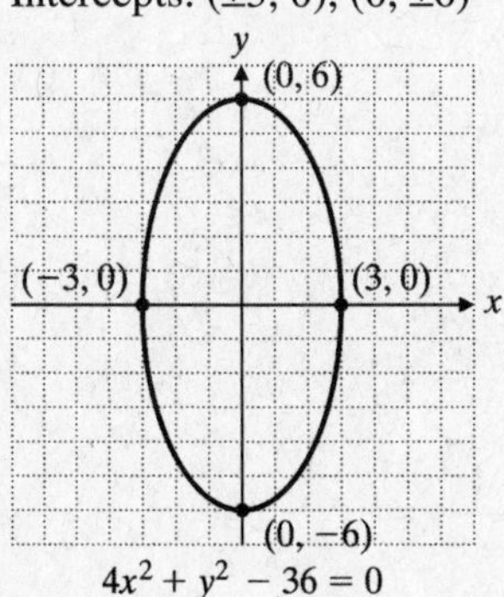

9. $x^2+9y^2=81$

$\frac{x^2}{81}+\frac{y^2}{9}=1$

$\frac{x^2}{9^2}+\frac{y^2}{3^2}=1$

$a=9, b=3$

Intercepts: $(\pm 9, 0), (0, \pm 3)$

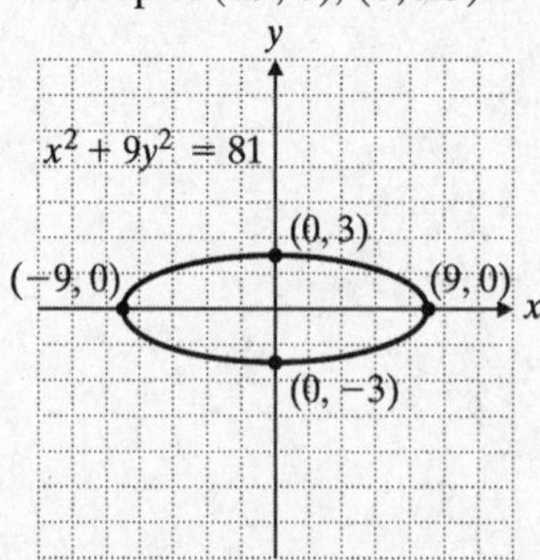

Scale: Each unit = 2

11. $x^2+12y^2=36$

$\frac{x^2}{36}+\frac{y^2}{3}=1$

$\frac{x^2}{6^2}+\frac{y^2}{\sqrt{3}^2}=1$

$a=6,\ b=\sqrt{3}$

Intercepts: $(\pm 6, 0), \left(0, \pm\sqrt{3}\right)$

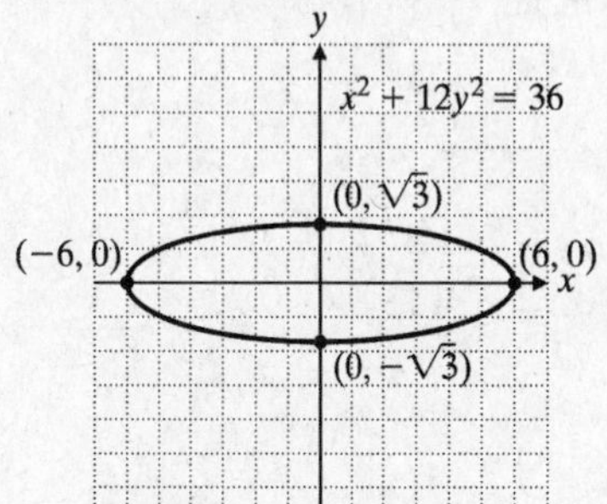

13. $\frac{x^2}{\frac{25}{4}}+\frac{y^2}{\frac{16}{9}}=1$

$\frac{x^2}{\left(\frac{5}{2}\right)^2}+\frac{y^2}{\left(\frac{4}{3}\right)^2}=1$

$a=\frac{5}{2}, b=\frac{4}{3}$

Intercepts: $\left(\pm\frac{5}{2}, 0\right), \left(0, \pm\frac{4}{3}\right)$

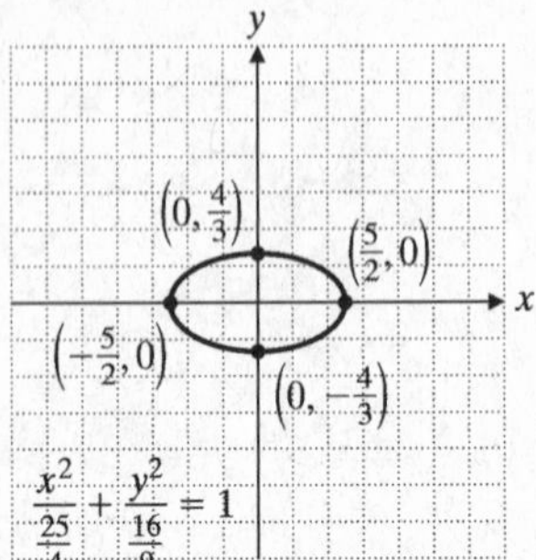

15. $121x^2 + 64y^2 = 7744$

$$\frac{x^2}{64} + \frac{y^2}{121} = 1$$

$$\frac{x^2}{8^2} + \frac{y^2}{11^2} = 1$$

$a = 8$, $b = 11$, $C(0, 0)$

Intercepts: $(0, \pm 11)$, $(\pm 8, 0)$

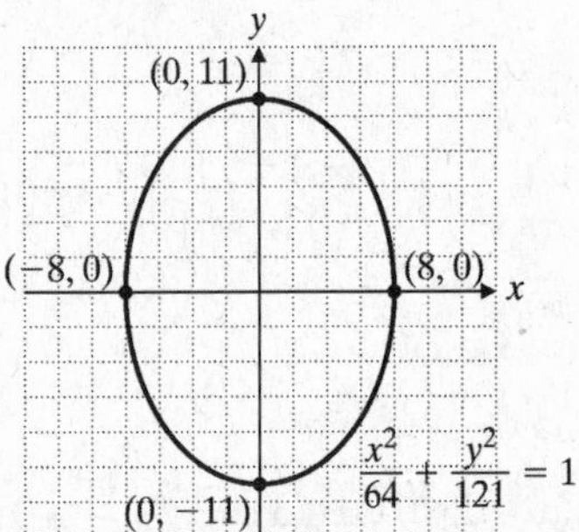

Scale: Each unit = 2

17. $\frac{(x-h)^2}{a^2} + \frac{(y-k)^2}{b^2}$

$C(0, 0) \Rightarrow (h, k) = (0, 0)$

x-int$(13, 0) \Rightarrow a = 13$

y-int$(0, -12) \Rightarrow b = 12$

$$\frac{x^2}{13^2} + \frac{y^2}{12^2} = 1$$

$$\frac{x^2}{169} + \frac{y^2}{144} = 1$$

19. $\frac{(x-h)^2}{a^2} + \frac{(y-k)^2}{b^2} = 1$

$C(0, 0) \Rightarrow (h, k) = (0, 0)$

x-int$(9, 0) \Rightarrow a = 9$

y-int$\left(0, 3\sqrt{2}\right) \Rightarrow b = 3\sqrt{2}$

$$\frac{x^2}{9^2} + \frac{y^2}{\left(3\sqrt{2}\right)^2} = 1$$

$$\frac{x^2}{81} + \frac{y^2}{18} = 1$$

21. $\frac{x^2}{5013} + \frac{y^2}{4970} = 1 \Rightarrow a^2 = 5013$

$a = \sqrt{5013}$

$d = 2a = 2\sqrt{5013} \approx 142$

The largest possible distance across the ellipse is 142 million miles.

23. $\frac{(x-h)^2}{a^2} + \frac{(y-k)^2}{b^2} = 1$

$$\frac{(x-5)^2}{9} + \frac{(y-2)^2}{1} = 1$$

$$\frac{(x-5)^2}{3^2} + \frac{(y-2)^2}{1^2} = 1$$

$a = 3$, $b = 1$, $C(5, 2)$

Vertices: (2, 2), (8, 2), (5, 1), (5, 3)

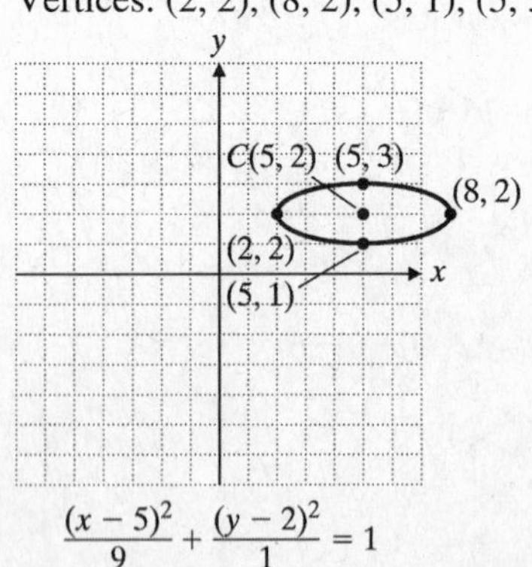

25. $\frac{(x-h)^2}{a^2} + \frac{(y-k)^2}{b^2} = 1$

$$\frac{x^2}{25} + \frac{(y-4)^2}{16} = 1$$

$$\frac{x^2}{5^2} + \frac{(y-4)^2}{4^2} = 1$$

$a = 5$, $b = 4$, $C(0, 4)$

Vertices: (5, 4), (0, 8), (–5, 4), (0, 0)

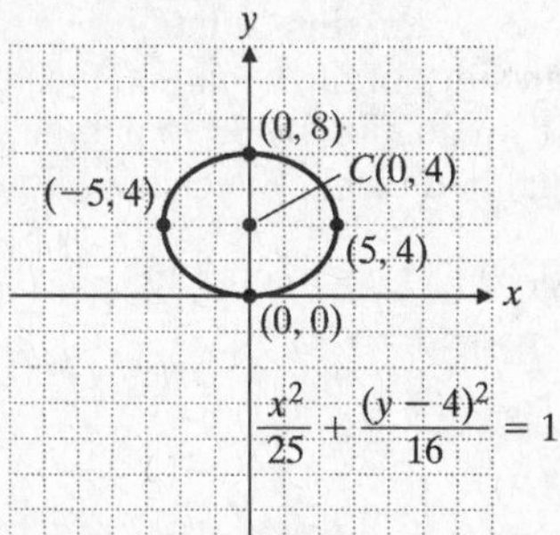

Scale: Each unit = 2

27. $\frac{(x-h)^2}{a^2}+\frac{(y-k)^2}{b^2}=1$

$\frac{(x+5)^2}{16}+\frac{(y+2)^2}{36}=1$

$\frac{(x+5)^2}{4^2}+\frac{(y+2)^2}{6^2}=1$

$a = 4$, $b = 6$, $C(-5, -2)$

Vertices: (−1, −2), (−5, 4), (−9, −2), (−5, −8)

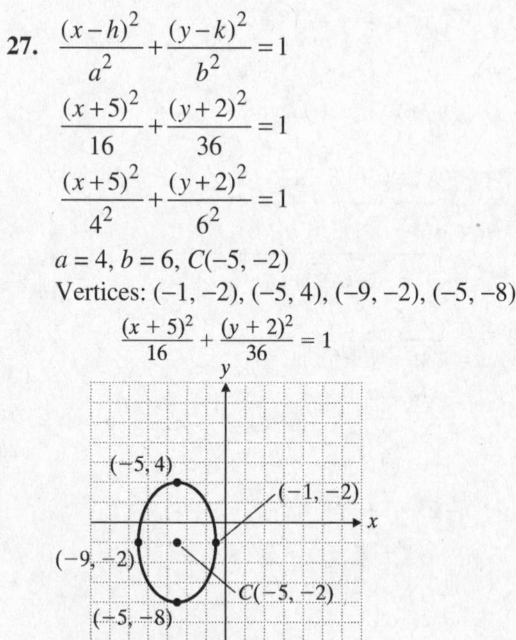

Scale: Each unit = 2

29. $\left(\frac{2+6}{2}, \frac{3+3}{2}\right)=(4, 3)$

$a = |6 - 4| = 2$

$b = |7 - 3| = 4$

$\frac{(x-4)^2}{2^2}+\frac{(y-3)^2}{4^2}=1$

$\frac{(x-4)^2}{4}+\frac{(y-3)^2}{16}=1$

31. $C\left(\frac{0+60}{2}, \frac{0+40}{2}\right)=C(30, 20)$

$2a = 60$

$a = 30$

$2b = 40$

$b = 20$

$\frac{(x-30)^2}{30^2}+\frac{(y-20)^2}{20^2}=1$

$\frac{(x-30)^2}{900}+\frac{(y-20)^2}{400}=1$

33. $\frac{(x-3.6)^2}{14.98}+\frac{(y-5.3)^2}{28.98}=1$

$(y-5.3)^2=28.98\left(1-\frac{12.96}{14.98}\right)$

$y=\pm\sqrt{28.98\left(1-\frac{12.967}{14.98}\right)}+5.3$

$y=\begin{cases}7.2768\\3.3232\end{cases}$

$\frac{(x-3.6)^2}{14.98}+\frac{(0-5.3)^2}{28.98}=1$

$(x-3.6)^2=14.98\left(1-\frac{28.09}{28.98}\right)$

$x=\pm\sqrt{14.98\left(1-\frac{28.09}{28.98}\right)}+3.6$

$x=\begin{cases}4.2783\\2.9217\end{cases}$

x-intercepts: (4.2783, 0), (2.9217, 0)

y-intercepts: (0, 7.2768), (0, 3.3232)

Cumulative Review

35. $\frac{5}{\sqrt{2x}-\sqrt{y}}=\frac{5}{\sqrt{2x}-\sqrt{y}}\cdot\frac{\sqrt{2x}+\sqrt{y}}{\sqrt{2x}+\sqrt{y}}$

$=\frac{5\left(\sqrt{2x}+\sqrt{y}\right)}{2x-y}$

36. $\left(2\sqrt{3}+4\sqrt{2}\right)\left(5\sqrt{6}-\sqrt{2}\right)$

$=10\sqrt{18}-2\sqrt{6}+20\sqrt{12}-8$

$=10\sqrt{9\cdot 2}-2\sqrt{6}+20\sqrt{4\cdot 3}-8$

$=30\sqrt{2}-2\sqrt{6}+40\sqrt{3}-8$

37. $4.5x = 102 \Rightarrow x = 22\frac{2}{3}$

It took $22\frac{2}{3}$ weeks to complete the framework.

Quick Quiz 9.3

1. $$\frac{x^2}{a^2}+\frac{y^2}{b^2}=1$$
$$\frac{x^2}{5^2}+\frac{y^2}{(-6)^2}=1$$
$$\frac{x^2}{25}+\frac{y^2}{36}=1$$

2. $$\frac{(x-h)^2}{a^2}+\frac{(y-k)^2}{b^2}=1$$
$$h=x_1+\frac{(x_2-x_1)}{2}=-7+\frac{1-(-7)}{2}=-3$$
$$k=y_1+\frac{(y_2-y_1)}{2}=-5+\frac{1-(-5)}{2}=-2$$
$$a=x_2-h=1-(-3)=4$$
$$b=y_2-k=1-(-2)=3$$
$$\frac{(x-(-3))^2}{4^2}+\frac{[y-(-2)]^2}{3^2}=1$$
$$\frac{(x+3)^2}{16}+\frac{(y+2)^2}{9}=1$$

3. $$\frac{(x-2)^2}{9}+\frac{(y+1)}{25}=1$$
$h=2$
$k=-1$
$a=3$
$b=5$
Center $(2, -1)$
Vertices: $(-1, -1)$, $(2, 4)$, $(5, -1)$, $(2, -6)$

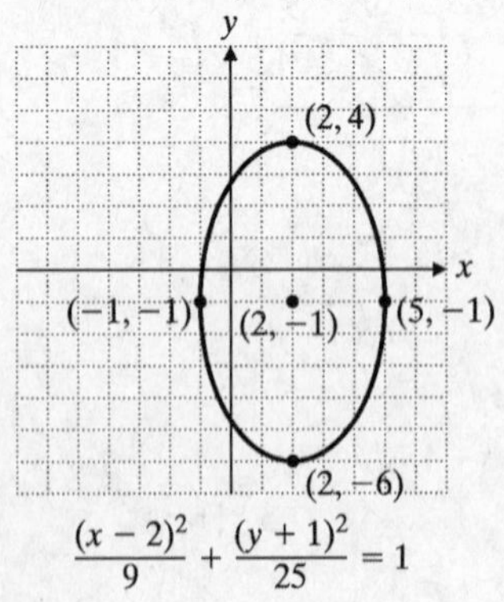

$$\frac{(x-2)^2}{9}+\frac{(y+1)^2}{25}=1$$

4. Answers may vary. Possible solution:
Add 36 to both sides of the equation, then divide both sides of the equation by 36 to put the equation in standard form. The center of the ellipse is at (0, 0) so y-intercepts are $\left(0, \pm\frac{b}{2}\right)$ and x-intercepts are $\left(\pm\frac{a}{2}, 0\right)$.

How Am I Doing? Sections 9.1–9.3

1. $$(x-h)^2+(y-k)^2=r^2$$
$$(x-8)^2+(y-(-2))^2=\sqrt{7}^2$$
$$(x-8)^2+(y+2)^2=7$$

2. $(x_1, y_1)=(-6, -2)$, $(x_2, y_2)=(-3, 4)$
$$d=\sqrt{(x_2-x_1)^2+(y_2-y_1)^2}$$
$$d=\sqrt{(-3-(-6))^2+(4-(-2))^2}$$
$$d=\sqrt{9+36}$$
$$d=\sqrt{45}=\sqrt{9(5)}$$
$$d=3\sqrt{5}$$

3. 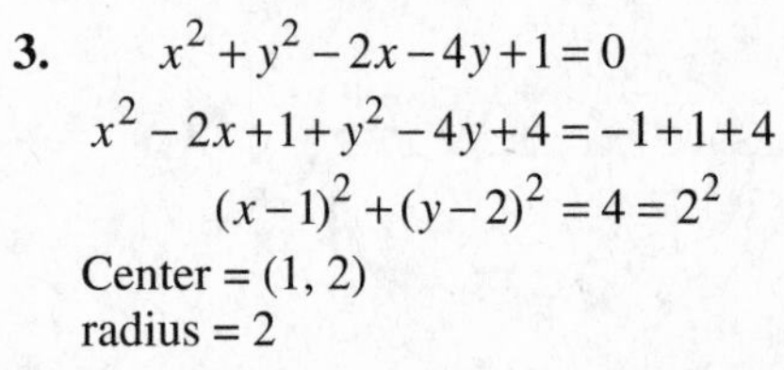
$$x^2+y^2-2x-4y+1=0$$
$$x^2-2x+1+y^2-4y+4=-1+1+4$$
$$(x-1)^2+(y-2)^2=4=2^2$$
Center = (1, 2)
radius = 2

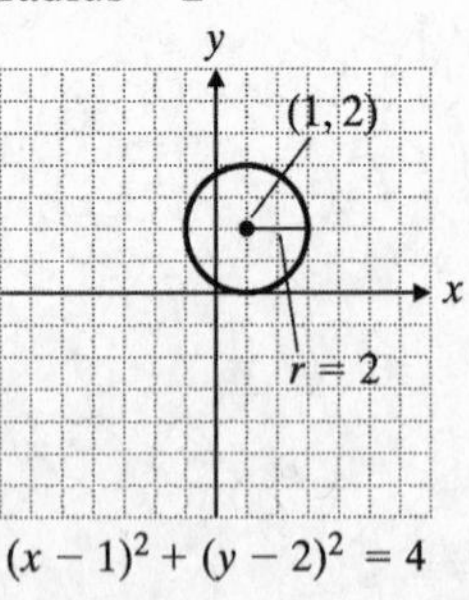

$(x-1)^2+(y-2)^2=4$

4. $y=a(x-h)^2+k$ has $x=h$ as its axis of symmetry. $y=4(x-3)^2+5$ has $x=3$ as its axis of symmetry.

5. $y=a(x-h)^2+k$ has $V(h, k)$. Therefore,
$y=\frac{1}{3}(x+4)^2+6=\frac{1}{3}(x-(-4))^2+6$ has $V(-4, 6)$.

6. $x = (y+1)^2 + 2$
$a > 0$, opens right; horizontal
$V(2, -1)$
Let $y = 0$.
$x = (0+1)^2 + 2 = 3$
$(3, 0)$

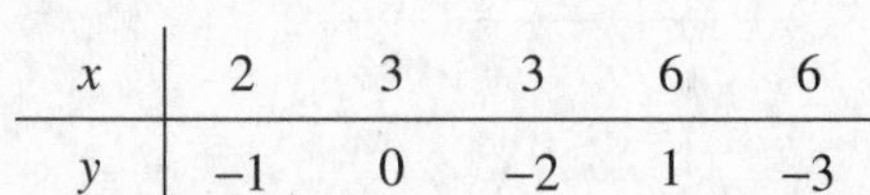

x	2	3	3	6	6
y	−1	0	−2	1	−3

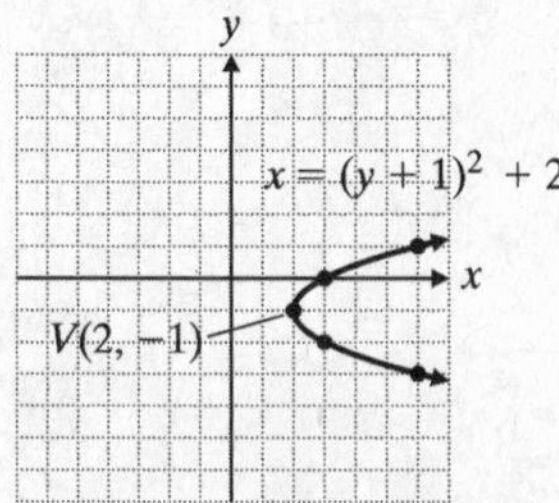

7. $x^2 = y - 4x - 1$
$y = x^2 + 4x + 1$
$y = x^2 + 4x + 4 - 3$
$y = (x+2)^2 - 3$
$a > 0$, opens up; vertical
$V(-2, -3)$
Let $x = 0$.
$y = (0+2)^2 - 3$
$y = 1$
$(0, 1)$

x	−4	0	−1	−3	−2
y	1	1	−2	−2	−3

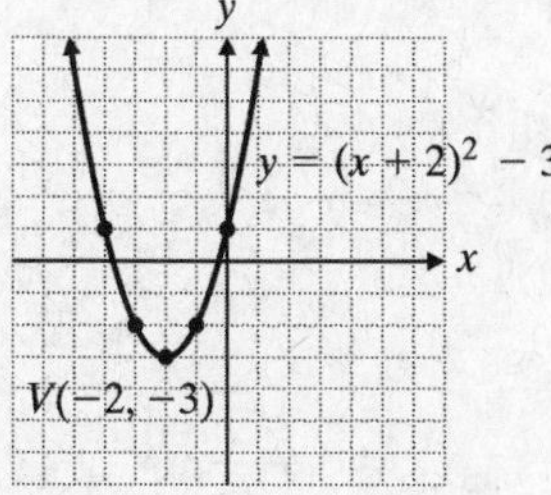

8. $C(0, 0) \Rightarrow h = 0, k = 0$
x-intercept: $(-10, 0) \Rightarrow a = 10$
y-intercept: $(0, 7) \Rightarrow b = 7$

$$\frac{(x-h)^2}{a^2} + \frac{(y-k)^2}{b^2} = 1$$

$$\frac{x^2}{10^2} + \frac{y^2}{7^2} = 1$$

$$\frac{x^2}{100} + \frac{y^2}{49} = 1$$

9. $4x^2 + y^2 - 36 = 0$

$$4x^2 + y^2 = 36$$

$$\frac{x^2}{9} + \frac{y^2}{36} = 1$$

$a = 3, b = 6$
Intercepts: $(0, \pm 6)$, $(\pm 3, 0)$

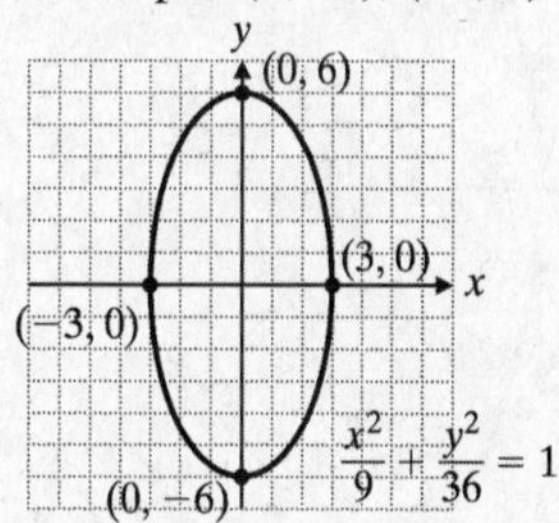

10. $\dfrac{(x+3)^2}{25} + \dfrac{(y-1)^2}{16} = 1$
$a = 5, b = 4$
Vertices: $(-8, 1)$, $(2, 1)$, $(-3, 5)$, $(-3, -3)$

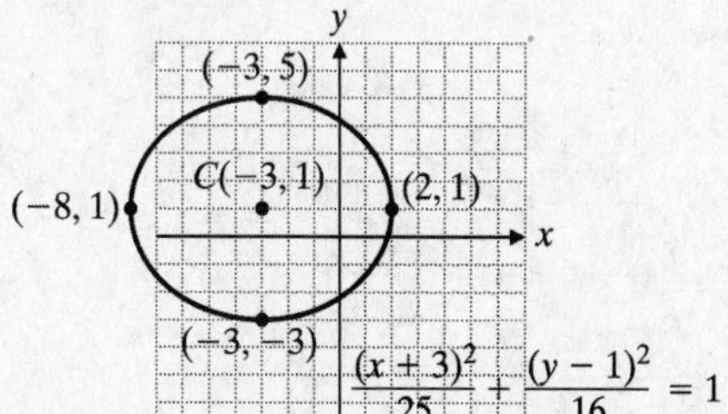

9.4 Exercises

1. The standard form of a horizontal hyperbola centered at the origin is $\dfrac{x^2}{a^2} - \dfrac{y^2}{b^2} = 1$ where a, b are real numbers and $a, b > 0$.

3. $\dfrac{x^2}{16}-\dfrac{y^2}{4}=1$ is a horizontal hyperbola centered at origin with vertices at (4, 0) and (−4, 0). Draw a fundamental rectangle with corners at (4, 2), (4, −2), (−4, 2), (−4, −2). Extend the diagonals through the rectangle as asymptotes of the hyperbola. Construct each branch of the hyperbola passing through the vertex and approaching the asymptotes.

5. $\dfrac{x^2}{4}-\dfrac{y^2}{25}=1 \Rightarrow \dfrac{x^2}{2^2}-\dfrac{y^2}{5^2}=1$

$a=2, b=5$

$V(\pm 2, 0),\ y_{\text{asymptote}}=\pm\dfrac{5}{2}x$

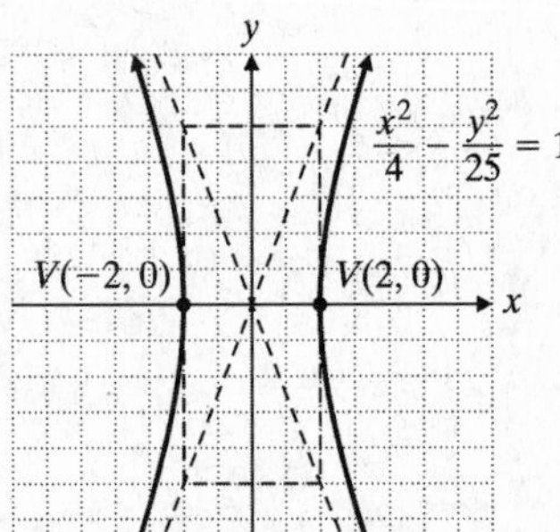

7. $\dfrac{y^2}{25}-\dfrac{x^2}{16}=1 \Rightarrow \dfrac{y^2}{5^2}-\dfrac{x^2}{4^2}=1$

$a=4, b=5$

$V(0, \pm 5)$

$y_{\text{asymptote}}=\pm\dfrac{5}{4}x$

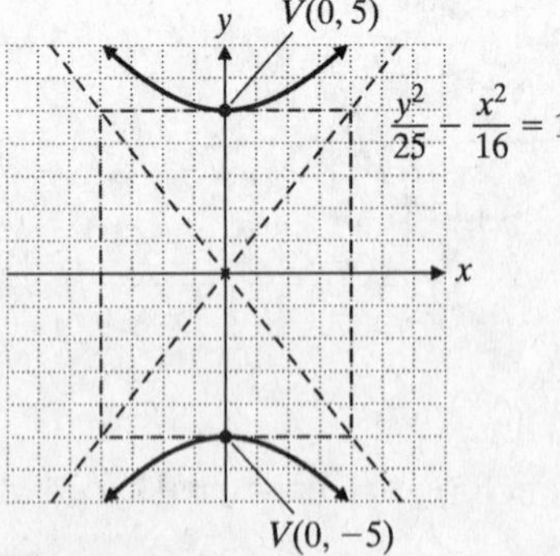

9. $4x^2-y^2=64$

$\dfrac{x^2}{16}-\dfrac{y^2}{64}=1$

$\dfrac{x^2}{4^2}-\dfrac{y^2}{8^2}=1$

$a=4, b=8$

$V(\pm 4, 0),\ y_{\text{asymptote}}=\pm\dfrac{8}{4}x$

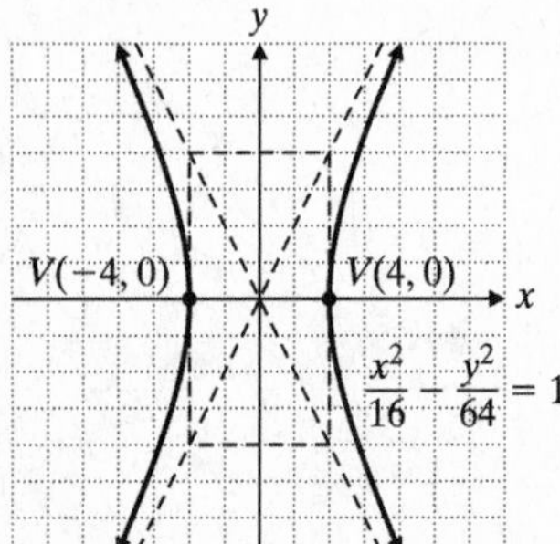

Scale: Each unit = 2

11. $8x^2-y^2=16$

$\dfrac{x^2}{2}-\dfrac{y^2}{16}=1$

$\dfrac{x^2}{\sqrt{2}^2}-\dfrac{y^2}{4^2}=1$

$a=\sqrt{2}, b=4$

$V\left(\pm\sqrt{2}, 0\right),\ y_{\text{asymptote}}=\pm\dfrac{4}{\sqrt{2}}x$

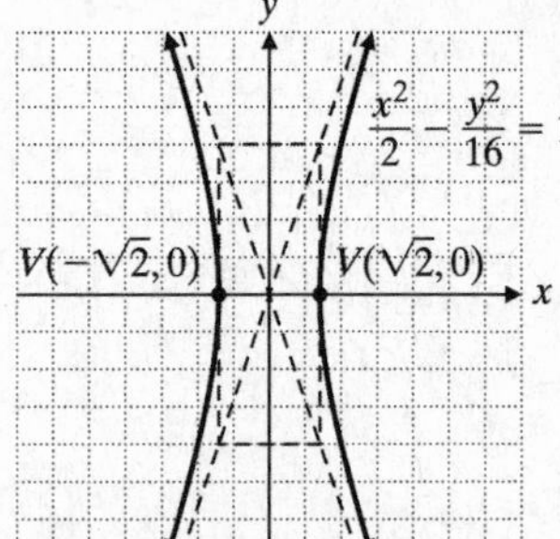

13. $4y^2-3x^2=48$

$\dfrac{y^2}{12}-\dfrac{x^2}{16}=1$

$\dfrac{y^2}{\left(2\sqrt{3}\right)^2}-\dfrac{x^2}{4^2}=1$

$b=2\sqrt{3}, a=4$

$V\left(0, \pm 2\sqrt{3}\right),\ y_{\text{asymptote}}=\pm\dfrac{2\sqrt{3}}{4}x=\pm\dfrac{\sqrt{3}}{2}x$

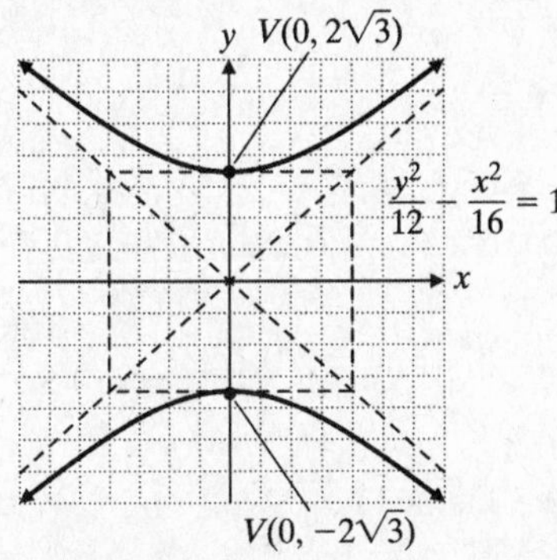

15. $V(\pm 3, 0) \Rightarrow a = 3,\ a^2 = 9$

$y_{\text{asymptote}} = \pm \frac{4}{3}x$

$b = 4,\ b^2 = 16$

$\frac{x^2}{9} - \frac{y^2}{16} = 1$

17. $V(0, \pm 11)$

$b = 11,\ b^2 = 121$

$y_{\text{asymptote}} = \pm \frac{11}{13}x$

$a = 13,\ a^2 = 169$

$\frac{y^2}{121} - \frac{x^2}{169} = 1$

19. $\frac{x^2}{a^2} - \frac{y^2}{b^2} = 1$, from graph, $a = 120$.

$y_{\text{asymptote}} = \frac{3}{1}x$

$\frac{b}{a} = \frac{3}{1}$

$\frac{b}{120} = \frac{3}{1}$

$b = 360$

$\frac{x^2}{120^2} - \frac{y^2}{360^2} = 1$

$\frac{x^2}{14,400} - \frac{y^2}{129,600} = 1,$

where x and y are measured in millions of miles.

21. $\frac{(x-1)^2}{4} - \frac{(y+2)^2}{9} = 1$

$\frac{(x-1)^2}{2^2} - \frac{(y+2)^2}{3^2} = 1$

$C(1, -2)$

$a = 2, b = 3$

$V(1 \pm 2, -2) = (3, -2), (-1, -2)$

$y_{\text{asymptote}} = \pm \frac{3}{2}(x-1) - 2$

$\frac{(x-1)^2}{4} - \frac{(y+2)^2}{9} = 1$

y
x
$C(1, -2)$
$V(-1, -2)$
$V(3, -2)$

23. $\frac{(y+2)^2}{36} - \frac{(x+1)^2}{81} = 1$

$\frac{(y+2)^2}{6^2} - \frac{(x+1)^2}{9^2} = 1$

$C(-1, -2)$

$a = 6, b = 9$

$V(-1, -2 \pm 6) = (-1, 4), (-1, -8)$

$y_{\text{asymptote}} = \pm \frac{6}{9}(x-(-1)) + (-2)$

$y_{\text{asymptote}} = \pm \frac{2}{3}(x+1) - 2$

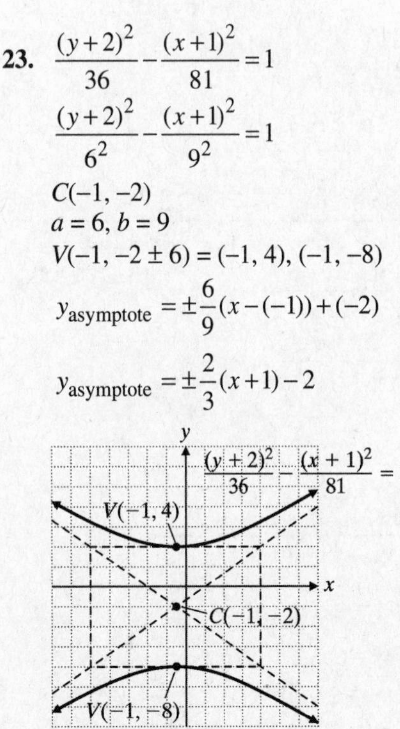

Scale: Each unit = 2

25. $\frac{(x+6)^2}{7} - \frac{y^2}{3} = 1$

$\frac{(x-(-6))^2}{\sqrt{7}^2} - \frac{(y-0)^2}{\sqrt{3}^2} = 1$

$C(-6, 0)$

$V\left(-6 \pm \sqrt{7}, 0\right)$

27. $C\left(4, \frac{0-14}{2}\right) = (4, -7)$

$y_{\text{asymptote}} = \frac{-7}{4}x = \frac{b}{a}x \Rightarrow a = 4,\ b = -7$

$\frac{(y-(-7))^2}{7^2} - \frac{(x-4)^2}{4^2} = 1$

$\frac{(y+7)^2}{49} - \frac{(x-4)^2}{16} = 1$

29. $8x^2 - y^2 = 16\big|_{x=3.5}$

$8(3.5)^2 - y^2 = 16$

$y^2 = 8(3.5)^2 - 16$

$y = \pm\sqrt{8(3.5)^2 - 16} = \pm 9.055385138$

Cumulative Review

31. $\frac{3}{x^2-5x+6} + \frac{2}{x^2-4}$

$= \frac{3(x+2)}{(x-3)(x-2)(x+2)} + \frac{2(x-3)}{(x-2)(x+2)(x-3)}$

$= \frac{3x+6+2x-6}{(x-3)(x-2)(x+2)}$

$= \frac{5x}{(x-3)(x-2)(x+2)}$

32. $\frac{2x}{5x^2+9x-2} - \frac{3}{5x-1}$

$= \frac{2x}{(5x-1)(x+2)} - \frac{3(x+2)}{(5x-1)(x+2)}$

$= \frac{2x-3(x+2)}{(5x-1)(x+2)}$

$= \frac{2x-3x-6}{(5x-1)(x+2)}$

$= \frac{-x-6}{(5x-1)(x+2)}$ or $-\frac{x+6}{(5x-1)(x+2)}$

33. x = amount grossed by top ten movies

$x = \frac{71.4}{0.616} = 115.9$

115.9 million dollars were grossed by top ten movies.

34. x = number of people on U.S. airlines in summer 2007

$x = 211.5 + 211.5(0.013) = 214.3$

214.3 million people flew during summer 2007.

Quick Quiz 9.4

1. $36y^2 - 9x^2 = 36$

$\frac{36y^2}{36} - \frac{9x^2}{36} = \frac{36}{36}$

$\frac{y^2}{1} - \frac{x^2}{4} = 1$

$b = \sqrt{1} = 1$

vertices (0, −1), (0, 1)

2. $\frac{x^2}{4^2} - \frac{y^2}{b} = 1$

$y = \frac{b}{a}x = \frac{5}{4}x,\ b = 5$

$\frac{x^2}{4^2} - \frac{y^2}{5^2} = 1$

$\frac{x^2}{16} - \frac{y^2}{25} = 1$

3.

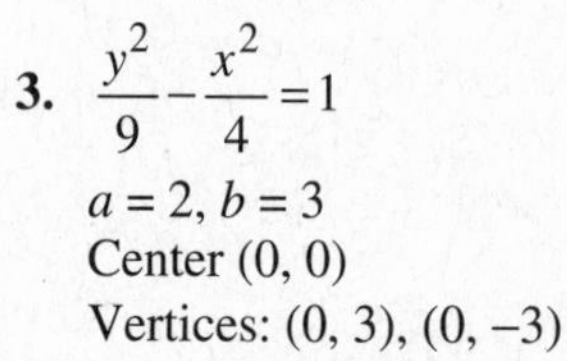

$\frac{y^2}{9} - \frac{x^2}{4} = 1$

$a = 2,\ b = 3$

Center (0, 0)

Vertices: (0, 3), (0, −3)

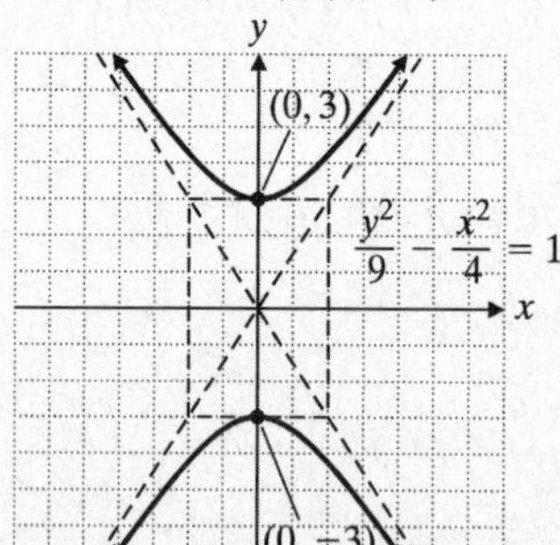

4. Answers may vary. Possible solution: Divide both sides of the equation by 196 in order to put the equation in standard form. This equation describes a horizontal hyperbola, the asymptote of which is $y = \frac{b}{a}x$. Substitution yields $y = \frac{7}{2}x$.

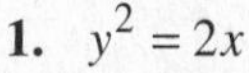

9.5 Exercises

1. $y^2 = 2x$

$y = -2x + 2$, substitute into first equation

$$(-2x+2)^2 = 2x$$
$$4x^2 - 8x + 4 = 2x$$
$$4x^2 - 10x + 4 = 0$$
$$2x^2 - 5x + 2 = 0$$
$$(2x-1)(x-2) = 0$$
$$x = \begin{cases} \frac{1}{2}, \ y = -2\left(\frac{1}{2}\right) + 2 = 1 \\ 2, \ y = -2(2) + 2 = -2 \end{cases}$$

$(2, -2)$, $\left(\frac{1}{2}, 1\right)$ is the solution.

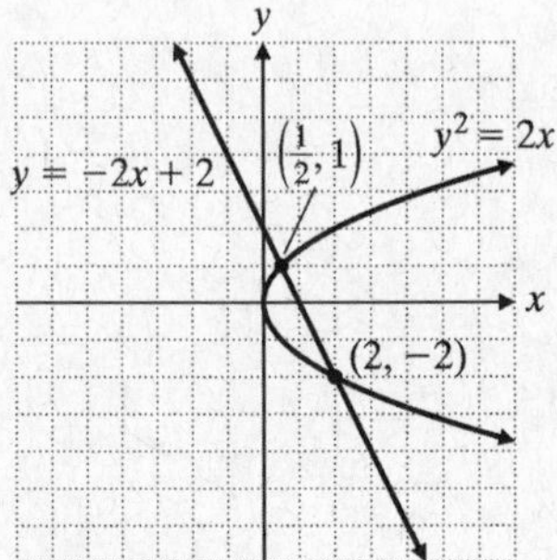

3. $x + 2y = 0$

$$x = -2y$$
$$x^2 + 4y^2 = 32$$
$$(-2y)^2 + 4y^2 = 32$$
$$4y^2 + 4y^2 = 8y^2 = 32$$
$$y^2 = 4$$
$$y = \begin{cases} 2, \ x = -2(2) = -4 \\ -2, \ x = -2(-2) = 4 \end{cases}$$

$(-4, 2)$, $(4, -2)$ is the solution.

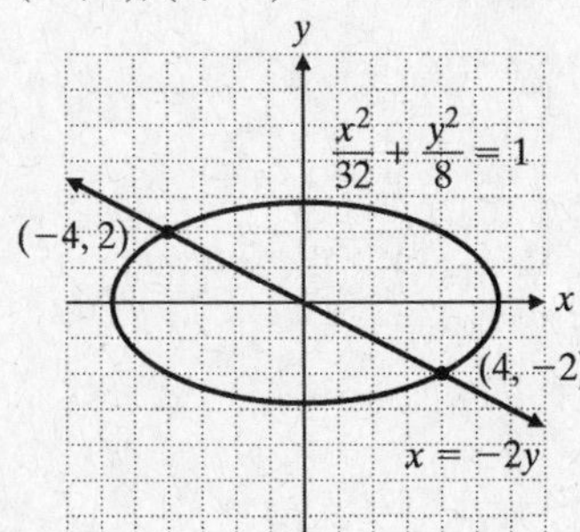

5. $\frac{x^2}{1} - \frac{y^2}{3} = 1$

$$3x^2 - y^2 = 3$$
$$x + y = 1$$
$$y = 1 - x$$
$$3x^2 - (1-x)^2 = 3$$
$$3x^2 - 1 + 2x - x^2 = 3$$
$$2x^2 + 2x - 4 = 0$$
$$x^2 + x - 2 = 0$$
$$(x+2)(x-1) = 0$$

$x + 2 = 0$　　　$x - 1 = 0$

$x = -2$　　　$x = 1$

$$y = 1 - x = \begin{cases} 1 - (-2) = 3 \\ 1 - 1 = 0 \end{cases}$$

$(-2, 3)$, $(1, 0)$ is the solution.

7. $x^2 + y^2 - 25 = 0$

$$3y = x + 5$$
$$x = 3y - 5$$
$$(3y-5)^2 + y^2 - 25 = 0$$
$$9y^2 - 30y + 25 + y^2 - 25 = 0$$
$$10y^2 - 30y = 0$$
$$10y(y-3) = 0$$

$10y = 0$　　　$y - 3 = 0$

$y = 0$　　　$y = 3$

$$x = 3y - 5 = \begin{cases} 3(0) - 5 = -5 \\ 3(3) - 5 = 4 \end{cases}$$

$(-5, 0)$, $(4, 3)$ is the solution.

9. $x^2 + 2y^2 = 4$

$$y = -x + 2$$
$$x^2 + 2(-x+2)^2 = 4$$
$$x^2 + 2x^2 - 8x + 8 = 4$$
$$3x^2 - 8x + 4 = 0$$
$$(3x-2)(x-2) = 0$$

$3x - 2 = 0$　　　$x - 2 = 2$

$x = \frac{2}{3}$　　　$x = 2$

$$y = -x + 2 = \begin{cases} -\frac{2}{3} + 2 = \frac{4}{3} \\ -2 + 2 = 0 \end{cases}$$

$\left(\frac{2}{3}, \frac{4}{3}\right)$, $(2, 0)$ is the solution.

11. $\frac{x^2}{4} - \frac{y^2}{4} = 1$

$x^2 - y^2 = 4$

$x + y - 4 = 0$

$x = 4 - y$

$(4-y)^2 - y^2 = 4$

$16 - 8y + y^2 - y^2 = 4$

$8y = 12$

$y = \frac{3}{2}$

$x = 4 - y = 4 - \frac{3}{2} = \frac{5}{2}$

$\left(\frac{5}{2}, \frac{3}{2}\right)$ is the solution.

13. $2x^2 - 5y^2 = -2 \xrightarrow{\times 2} 4x^2 - 10y^2 = -4$

$3x^2 + 2y^2 = 35 \xrightarrow{\times 5} 15x^2 + 10y^2 = 175$

$19x^2 = 171$

$x = \pm 3$

$y^2 = \frac{35 - 3x^2}{2}$

$y = \pm\sqrt{\frac{35-3x^2}{2}}\Bigg|_{x=\pm 3} = \pm\sqrt{\frac{35-3(\pm 3)^2}{2}}$

$y = \pm 2$

$(3, \pm 2,), (-3, \pm 2)$ is the solution.

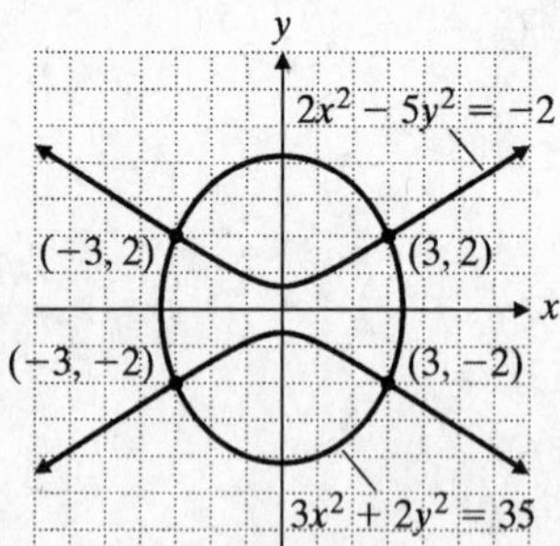

15. $x^2 + y^2 = 9$

$2x^2 - y^2 = 3$

$3x^2 = 12$

$x^2 = 4$

$x = \pm 2$

$(\pm 2)^2 + y^2 = 9$

$y^2 = 5$

$y = \pm\sqrt{5}$

$\left(2, \pm\sqrt{5}\right), \left(-2, \pm\sqrt{5}\right)$ is the solution.

17. $x^2 + 2y^2 = 8 \longrightarrow x^2 + 2y^2 = 8$

$x^2 - y^2 = 1 \xrightarrow{\times -1} -x^2 + y^2 = -1$

$3y^2 = 7$

$y^2 = \frac{7}{3} \Rightarrow y = \pm\frac{\sqrt{7}}{\sqrt{3}} \cdot \frac{\sqrt{3}}{\sqrt{3}}$

$y = \pm\frac{\sqrt{21}}{3}$

$x^2 = 1 + y^2 = 1 + \frac{7}{3} = \frac{10}{3}$

$x = \pm\frac{\sqrt{30}}{3}$

$\left(\frac{\sqrt{30}}{3}, \pm\frac{\sqrt{21}}{3}\right), \left(-\frac{\sqrt{30}}{3}, \pm\frac{\sqrt{21}}{3}\right)$ is the solution.

19. $x^2 + y^2 = 7 \longrightarrow x^2 + y^2 = 7$

$\frac{x^2}{3} - \frac{y^2}{9} = 1 \longrightarrow 3x^2 - y^2 = 9$

$4x^2 = 16$

$x^2 = 4$

$x = \pm 2$

$y^2 = 7 - x^2 = 7 - 4 = 3$

$y = \pm\sqrt{3}$

$\left(2, \pm\sqrt{3}\right), \left(-2, \pm\sqrt{3}\right)$ is the solution.

21. $2xy = 5$

$x = \frac{5}{2y}$

$x - 4y = 3$

$\frac{5}{2y} - 4y = 3$

$$8y^2+6y-5=0$$
$$(2y-1)(4y+5)=0$$
$$y=\frac{1}{2},\ y=-\frac{5}{4}$$
$$y=\frac{1}{2},\ x=\frac{5}{2\cdot\frac{1}{2}}=5$$
$$y=-\frac{5}{4},\ x=\frac{5}{2\cdot\frac{-5}{4}}=-2$$
$\left(5, \frac{1}{2}\right), \left(-2, -\frac{5}{4}\right)$ is the solution.

23. $xy=-6$
$2x+y=-4 \Rightarrow y=-4-2x$
$$x(-4-2x)=-6$$
$$2x^2+4x-6=0$$
$$x^2+2x-3=0$$
$$(x+3)(x-1)=0$$
$x=-3,\ x=1$
$x=-3,\ y=-4-2(-3)=2$
$x=1,\ y=-4-2(1)=-6$
$(-3, 2), (1, -6)$ is the solution.

25. $x+y=5$
$$x=5-y$$
$$x^2+y^2=4$$
$$(5-y)^2+y^2=4$$
$$25-10y+y^2+y^2=4$$
$$2y^2-10y+21=0$$
$$y=\frac{-(-10)\pm\sqrt{(-10)^2-4(2)(21)}}{2(2)}$$
$$y=\frac{10\pm\sqrt{-68}}{4}$$
No real solution, line does not intersect the circle.

27. $x^2+y^2=16,000,000$
$$y^2=16,000,000-x^2$$
$$25,000,000x^2-9,000,000y^2=2.25\times10^{14}$$
$$25x^2-9y^2=2.25\times10^8$$
$$25x^2-9(16,000,000-x^2)=2.25\times10^8$$
$$34x^2=369,000,000$$
$$x^2=10,852,941.18$$
$$x=3294.380242$$
$$y^2=16,000,000-10,852,941.18$$
$$y^2=5,147,058.824$$
$$y=2268.713032$$
The hyperbola intersects the circle when $(x, y) \approx (3290, 2270)$.

Cumulative Review

29. $(3x^3-8x^2-33x-10)\div(3x+1)$

$$\begin{array}{r|l} & x^2-3x-10 \\ 3x+1 & 3x^3-8x^2-33x-10 \\ & \underline{3x^3+x^2} \\ & -9x^2-33x \\ & \underline{-9x^2-3x} \\ & -30x-10 \\ & \underline{-30x-10} \\ & 0 \end{array}$$

$(3x^3-8x^2-33x-10)\div(3x+1)=x^2-3x-10$

30.
$$\frac{6x^4-24x^3-30x^2}{3x^3-21x^2+30x}=\frac{6x^2(x^2-4x-5)}{3x(x^2-7x+10)}=\frac{2x(x-5)(x+1)}{(x-5)(x-2)}=\frac{2x(x+1)}{x-2}$$

Quick Quiz 9.5

1. $2x-y=4$ (1)
$y^2-4x=0$ (2)
Solve (1) for *y*.
$2x-y=4$
$-y=4-2x$
$y=2x-4$
Substitute $2x-4$ for *y* in (2).
$$(2x-4)^2-4x=0$$
$$4x^2-16x+16-4x=0$$
$$4x^2-20x+16=0$$
$$(4x-16)(x-1)=0$$
$x=4,\ 1$
Solving (1) for *y* with $x=4$:
$2(4)-y=4$
$-y=-4$
$y=4$
$(4, 4)$
Solving (1) for *y* with $x=1$:

$$\begin{aligned} 2(1)-y&=4 \\ -y&=2 \\ y&=-2 \end{aligned}$$

(1, –2)

2. $y-x^2=-4$ (1)
$x^2+y^2=16$ (2)

Solve (1) for x^2.

$$\begin{aligned} y-x^2&=-4 \\ -x^2&=-4-y \\ x^2&=y+4 \end{aligned}$$

Substitute $y + 4$ for x^2 in (2).

$$\begin{aligned} y+4+y^2&=16 \\ y^2+y-12&=0 \\ (y+4)(y-3)&=0 \end{aligned}$$

$y = 3, -4$
Solving (1) for x with $y = 3$:

$$\begin{aligned} 3-x^2&=-4 \\ -x^2&=-4-3 \\ x^2&=7 \\ x&=\pm\sqrt{7} \end{aligned}$$

$\left(\sqrt{7}, 3\right), \left(-\sqrt{7}, 3\right)$

Solving (1) for x with $y = -4$:

$$\begin{aligned} -4-x^2&=-4 \\ -x^2&=0 \\ x&=0 \end{aligned}$$

(0, –4)

3. $(x+2)^2+(y-1)^2=9$ (1)
$x=2-y$ (2)

Substitute $2 - y$ for x in (1).

$$\begin{aligned} (2-y+2)^2+(y-1)^2&=9 \\ 2y^2-10y+8&=0 \\ y^2-5y+4&=0 \\ (y-4)(y-1)&=0 \end{aligned}$$

$y = 1, 4$
Solving (2) for x with $y = 1$:
$x = 2 - 1 = 1$
(1, 1)
Solving (2) for x with $y = 4$:
$x = 2 - 4 = -2$
(–2, 4)

4. Answers may vary. Possible solution:
Use the substitution method. Start by labeling the equations.

$y^2+2x^2=18$ (1)
$xy=4$ (2)

Because (2) is a linear equation, and because the y^2 term of (1) has a coefficient of 1, choose to solve (2) for the variable y and substitute the found value for y into (1).

$y=\frac{4}{x}$ (2)

$\left(\frac{4}{x}\right)^2+2x^2=18$ (2) and (1)

Solve the resulting equation, in terms of x only, for x.

$x=\pm1,\ \pm2\sqrt{2}$

Solve (2) for y, and substitute each of the four found values of x to find corresponding y values.

(1, 4), (–1, –4), $\left(2\sqrt{2}, \sqrt{2}\right)$, $\left(-2\sqrt{2}, -\sqrt{2}\right)$

Check for extraneous answers.

Putting Your Skills to Work

1. Store A: $\frac{1}{5}(400+400)=\frac{1}{5}(800)=160$
Total $= \$400+\$400-\$160=\$640$

Store B: $\frac{1}{5}(500)=100$
Total $= \$500+\$500-\$100=\$900$

Store A has the better buy.

2. a. about $30 · 0.10 = $3

b. about $30 · 0.20 = $6

c. about $30 · 0.30 = $9

3. a. about $60 · 0.20 = $12

b. about $60 · 0.25 = $15

c. about $60 · 0.30 = $18

4.
$$\begin{aligned} (52+36+2\times15)\times0.2 &\approx (50+40+30)\times0.2 \\ &=120\times0.2 \\ &=24 \end{aligned}$$

She would save about $24.

5. 1st savings: $95 × 0.10 = $9.50
New price: $95 − $9.50 = $85.50
2nd savings: $85.50 × 0.2 = $17.10
Total savings ≈ $10 + $17 = $27

6. A: (6 × 40) × 0.80 = $192
B: (4 × 30) + (2 × 30) × 0.9 = 120 + 54 = $174
Store B is offering the better buy.

7. Answers may vary.

Chapter 9 Review Problems

1. (0, −6) and (−3, 2)
$d=\sqrt{(0-(-3))^2+(-6-2)^2}=\sqrt{9+64}=\sqrt{73}$

2. (−7, 3) and (−2, −1)
$d=\sqrt{(-2-(-7))^2+(-1-3)^2}=\sqrt{25+16}=\sqrt{41}$

3. $(x-h)^2+(y-k)^2=r^2$
$(x-(-6))^2+(y-3)^2=\sqrt{15}^2$
$(x+6)^2+(y-3)^2=15$

4. $(x-h)^2+(y-k)^2=r^2$
$(x-0)^2+(y-(-7))^2=5^2$
$x^2+(y+7)^2=25$

5. $x^2+y^2+2x-6y+5=0$
$x^2+2x+1+y^2-6y+9=-5+1+9$
$(x+1)^2+(y-3)^2=5=\sqrt{5}^2$
$C(-1, 3),\ r=\sqrt{5}$

6. $x^2+y^2-10x+12y+52=0$
$x^2-10x+25+y^2+12y+36=-52+25+36$
$(x-5)^2+(y+6)^2=9=3^2$
$C(5, -6),\ r=3$

7. $x=\frac{1}{3}y^2$
Let $y = 0$.
$x=\frac{1}{3}(0)^2=0$

x	y	
0	0	$V(0, 0)$
3	3	y-intercept: (0, 0)
3	−3	

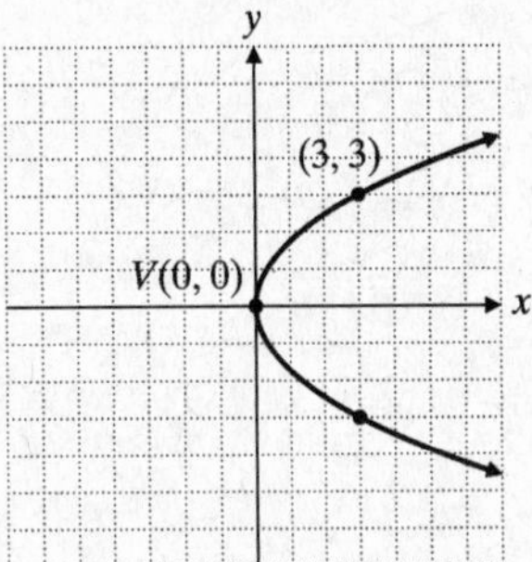

8. $x=\frac{1}{2}(y-2)^2+4$
Let $y = 0$.
$x=\frac{1}{2}(0-2)^2+4$
$x=6$

x	y	
4	2	$V(4, 2)$
6	0	y-intercept: (6, 0)
6	4	

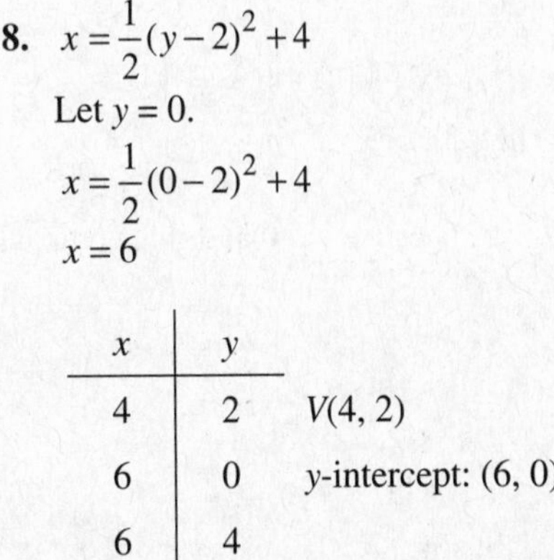

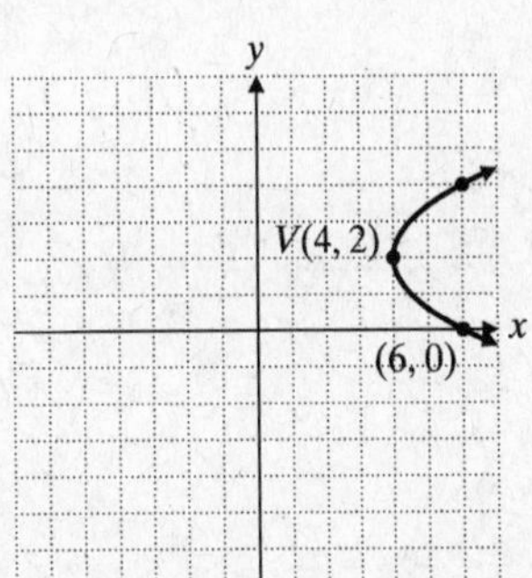

9. $y=-2(x+1)^2-3$
Let $x = 0$.
$y=-2(0+1)^2-3$
$y=-5$

x	y	
−2	−5	$V(-1, -3)$
−1	−3	y-intercept: (0, −5)
0	−5	

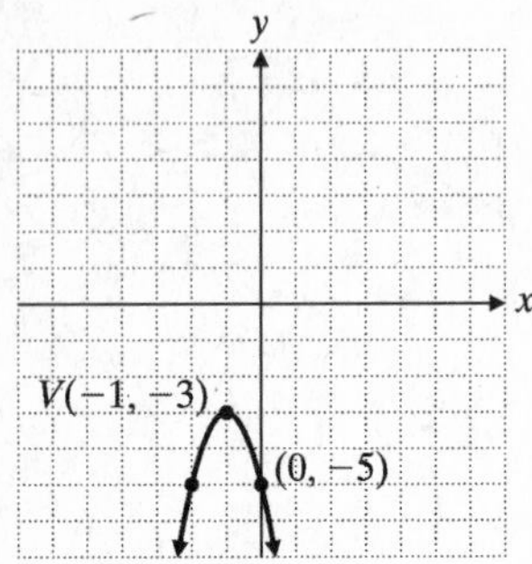

10. $x^2+6x=y-4$

$$y=x^2+6x+9-9+4$$
$$y=(x+3)^2-5$$

$V(-3, -5)$

Opens upward because $a>0$ and it is vertical.

11. $x+8y=y^2+10$

$$x=y^2-8y+16-6$$
$$x=(y-4)^2-6$$

$V(-6, 4)$

Opens to right since $a>0$ and it is horizontal.

12. $\frac{x^2}{4}+\frac{y^2}{1}=1 \Rightarrow \frac{x^2}{2^2}+\frac{y^2}{1^2}=1$

$a=2$, $b=1$, $C(0, 0)$

Vertices: (0, 1), (0, −1), (−2, 0), (2, 0)

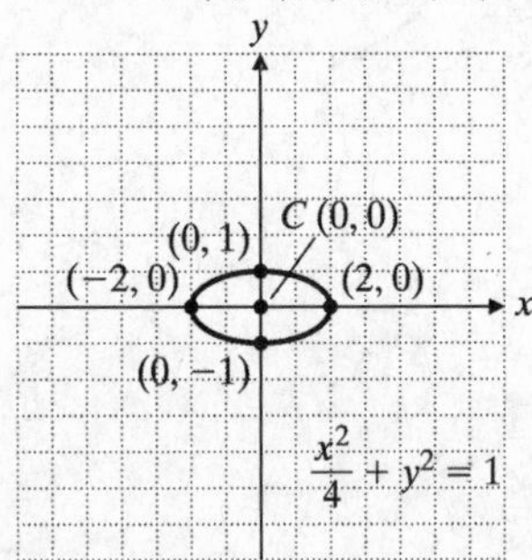

13. $16x^2+y^2-32=0$

$$\frac{x^2}{2}+\frac{y^2}{32}=1$$
$$\frac{x^2}{\sqrt{2}^2}+\frac{y^2}{\left(4\sqrt{2}\right)^2}=1$$

$a=\sqrt{2}$, $b=4\sqrt{2}$

$C(0, 0)$

Vertices: $\left(0, 4\sqrt{2}\right), \left(0, -4\sqrt{2}\right)$

$\left(-\sqrt{2}, 0\right)$ $\left(\sqrt{2}, 0\right)$

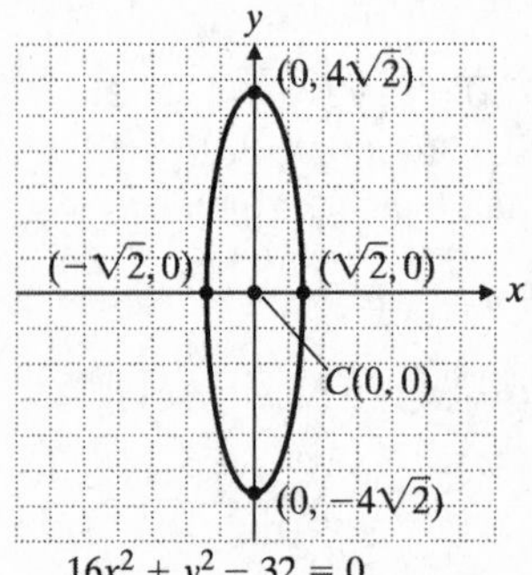

14. $\frac{(x+5)^2}{4}+\frac{(y+3)^2}{25}=1$

$$\frac{(x-(-5))^2}{2^2}+\frac{(y-(-3))^2}{5^2}=1$$

$C(-5, -3)$

$a=2$, $b=5$

Vertices: (−3, −3), (−7, −3), (−5, 2), (−5, −8)

15. $\frac{(x+1)^2}{9}+\frac{(y-2)^2}{16}=1$

$$\frac{(x-(-1))^2}{3^2}+\frac{(y-2)^2}{4^2}=1$$

$C(-1, 2)$

$a=3$, $b=4$

Vertices: (2, 2), (−4, 2), (−1, 6), (−1, −2)

16. $x^2-4y^2-16=0$

$$\frac{x^2}{16}-\frac{y^2}{4}=1$$
$$\frac{x^2}{4^2}-\frac{y^2}{2^2}=1$$

$C(0, 0)$

$a=4$, $b=2$

Vertices: (−4, 0), (4, 0)

$$y_{\text{asymptote}}=\pm\frac{2}{4}x=\pm\frac{1}{2}x$$

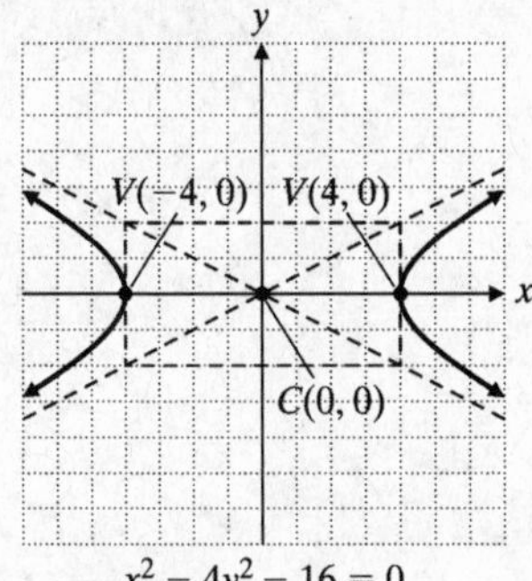

17. $3y^2 - x^2 = 27$

$\frac{y^2}{9} - \frac{x^2}{27} = 1$

$\frac{y^2}{3^2} - \frac{x^2}{\sqrt{27}^2} = 1$

$a = 3,\ b = \sqrt{27}$

$C(0, 0)$

Vertices: (0, 3), (0, −3)

$y_{\text{asymptote}} = \pm\frac{3}{\sqrt{27}}x = \pm\frac{\sqrt{3}}{3}x$

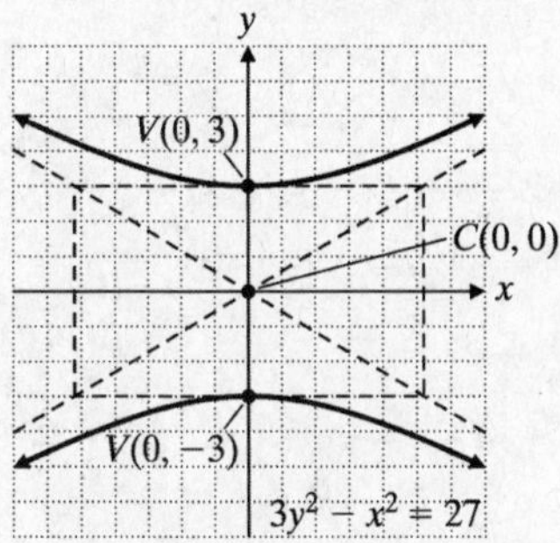

18. $\frac{(x-2)^2}{4} - \frac{(y+3)^2}{25} = 1$

$\frac{(x-2)^2}{2^2} - \frac{(y-(-3))^2}{5^2} = 1$

$C(2, -3),\ a = 2,\ b = 5$

Vertices: (0, −3), (4, −3)

19. $9(y-2)^2 - (x+5)^2 - 9 = 0$

$\frac{(y-2)^2}{1^2} - \frac{(x-(-5))^2}{3^2} = 0$

$C(-5, 2),\ a = 3,\ b = 1$

Vertices: (−5, 3), (−5, 1)

20. $x^2 + y = 9 \longrightarrow x^2 + y = 9$

$y - x = 3 \xrightarrow{\times -1} x - y = -3$

$x^2 + x = 6$

$(x+3)(x-2) = 0$

$x + 3 = 0 \qquad x - 2 = 0$

$x = -3 \qquad x = 2$

$y = x + 3 = \begin{cases} -3+3=0 \\ 2+3=5 \end{cases}$

(−3, 0), (2, 5) is the solution.

21. $x^2 + y^2 = 4$

$x + y = 2$

$y = 2 - x$

$x^2 + (2-x)^2 = 4$

$x^2 + x^2 - 4x + 4 = 4$

$x^2 - 2x = 0$

$x(x-2) = 0$

$x = 0, x = 2$

$x = 0, y = 2 - 0 = 2$

$x = 2, y = 2 - 2 = 0$

(0, 2), (2, 0) is the solution.

22. $2x^2 + y^2 = 17$

$y^2 = 17 - 2x^2$

$x^2 + 2y^2 = 22$

$x^2 + 2(17 - 2x^2) = 22$

$x^2 + 34 - 4x^2 = 22$

$3x^2 = 12$

$x^2 = 4$

$x = \pm 2$

$y = \pm\sqrt{17 - 2x^2}$

$y = \pm\sqrt{17 - 2(4)}$

$y = \pm 3$

(2, ±3), (−2, ±3) is the solution.

23. $xy = -2$

$y = \frac{-2}{x}$

$x^2 + y^2 = 5$

$x^2 + \left(\frac{-2}{x}\right)^2 = 5$

$x^4 + 4 = 5x^2$

$x^4 - 5x^2 + 4 = 0$

$(x^2 - 4)(x^2 - 1) = 0$

$x^2-4=0 \qquad x^2-1=0$
$x^2=4 \qquad x^2=1$
$x=\pm 2 \qquad x=\pm 1$

$$y=\frac{-2}{x}=\begin{cases}\frac{-2}{2}=-1\\ \frac{-2}{-2}=1\\ \frac{-2}{1}=-2\\ \frac{-2}{-1}=2\end{cases}$$

(2, −1), (−2, 1), (1, −2), (−1, 2) is the solution.

24. $3x^2-4y^2=12$
$7x^2-y^2=8$
$y^2=7x^2-8$
$3x^2-4(7x^2-8)=12$
$3x^2-28x^2+32=12$
$25x^2=20$
$x^2=\frac{20}{25}$
$y^2=7x^2-8=7\cdot\frac{20}{25}-8=-\frac{12}{5}$
$y^2>0\Rightarrow$ no real solution, hyperbolas do not intersect.

25. $y=x^2+1$
$x^2=y-1$
$x^2+y^2-8y+7=0$
$y-1+y^2-8y+7=0$
$y^2-7y+6=0$
$(y-1)(y-6)=0$
$y-1=0 \qquad y-6=0$
$y=1 \qquad y=6$

$$x^2=y-1=\begin{cases}1-1=0\\ 6-1=5\end{cases}$$

$$x=\begin{cases}0\\ \pm\sqrt{5}\end{cases}$$

(0, 1), $(\sqrt{5}, 6)$, $(-\sqrt{5}, 6)$ is the solution.

26. $2x^2+y^2=18$
$xy=4$
$y=\frac{4}{x}$
$2x^2+\left(\frac{4}{x}\right)^2=18$
$2x^4+16=18x^2$
$x^4-9x^2+8=0$
$(x^2-8)(x^2-1)=0$
$x^2=8 \qquad x^2=1$
$x=\pm 2\sqrt{2} \qquad x=\pm 1$

$$y=\frac{4}{x}=\begin{cases}\frac{4}{2\sqrt{2}}=\sqrt{2}\\ \frac{4}{-2\sqrt{2}}=-\sqrt{2}\\ \frac{4}{1}=4\\ \frac{4}{-1}=-4\end{cases}$$

$(2\sqrt{2}, \sqrt{2})$, $(-2\sqrt{2}, -\sqrt{2})$, (1, 4), (−1, −4) is the solution.

27. $y^2-2x^2=2 \xrightarrow{\times -2} -2y^2+4x^2=-4$
$2y^2-3x^2=5 \longrightarrow \underline{2y^2-3x^2=5}$
$x^2=1,\ x=\pm 1$

$y^2=2x^2+2=2(1)+2=4$
$y=\pm 2$
(1, ±2), (−1, ±2) is the solution.

28. $y^2=2x$
$y=\frac{1}{2}x+1$
$x=2y-2$
$y^2=2(2y-2)=4y-4$
$y^2-4y+4=0$
$(y-2)^2=0$
$y=2$
$x=2y-2=2(2)-2=4-2=2$
$x=2$
(2, 2) is the solution.

29. $y^2 = \frac{1}{2}x$

$y = x - 1$

$x = y + 1$

$$y^2 = \frac{1}{2}(y+1)$$

$2y^2 - y - 1 = 0$

$(2y+1)(y-1) = 0$

$2y + 1 = 0 \qquad y - 1 = 0$

$y = -\frac{1}{2} \qquad y = 1$

$$x = y + 1 = \begin{cases} -\frac{1}{2} + 1 = \frac{1}{2} \\ 1 + 1 = 2 \end{cases}$$

$\left(\frac{1}{2}, -\frac{1}{2}\right)$, (2, 1) is the solution.

30. $y^2 = 4px$

$y^2 = 4(2)x = 8x$

$y^2 = 8x$

$2.5^2 = 8x$

$x = 0.78125$

The searchlight should be 0.78 feet deep.

31. $y^2 = 4px$

$5^2 = 4p(4)$

$p = \frac{25}{16} = 1.5625$

The receiver should be placed 1.56 feet from the center of the dish.

How Am I Doing? Chapter 9 Test

1. (−6, −8) and (−2, 5)

$d = \sqrt{(-2-(-6))^2 + (5-(-8))^2}$

$d = \sqrt{16 + 169}$

$d = \sqrt{185}$

2. $y^2 - 6y - x + 13 = 0$

$x - 13 + 9 = y^2 - 6y + 9$

$x = (y-3)^2 + 4$

Parabola: $V(4, 3)$

Let $y = 0$.

$x = (0-3)^2 + 4$

$x = 13$

x-int: (13, 0)

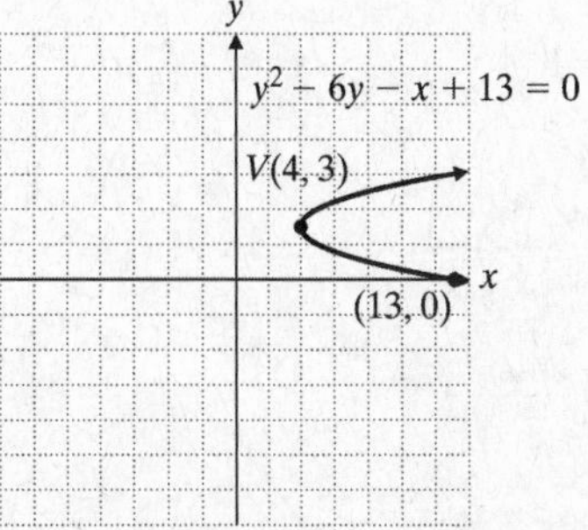

Scale: Each unit = 2

3. $x^2 + y^2 + 6x - 4y + 9 = 0$

$x^2 + 6x + 9 + y^2 - 4y + 4 = -9 + 9 + 4$

$(x+3)^2 + (y-2)^2 = 4 = 2^2$

Circle: $C(-3, 2)$, $r = 2$

x-int: (−3, 0)

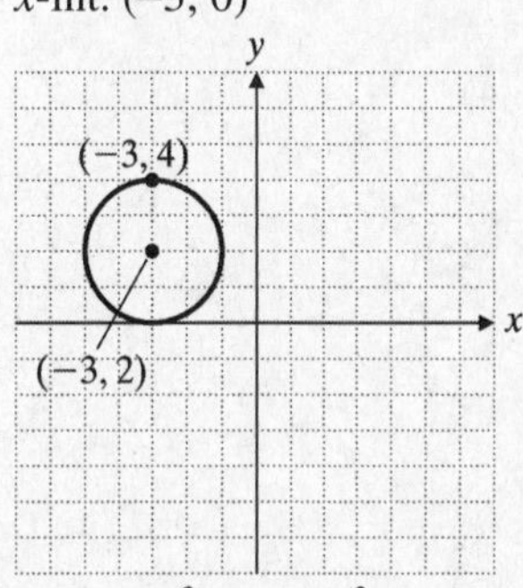

$(x + 3)^2 + (y - 2)^2 = 4$

4. $\frac{x^2}{25} + \frac{y^2}{1} = 1$

$\frac{x^2}{5^2} + \frac{y^2}{1^2} = 1$

Ellipse: $C(0, 0)$

$a = 5$, $b = 1$

Vertices: (5, 0), (−5, 0), (0, 1), (0, −1)

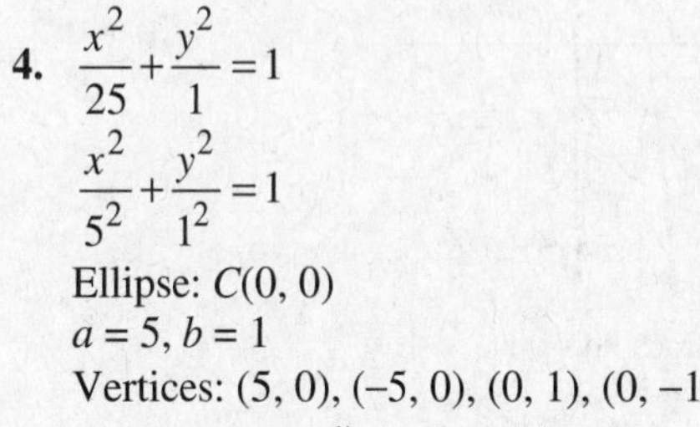

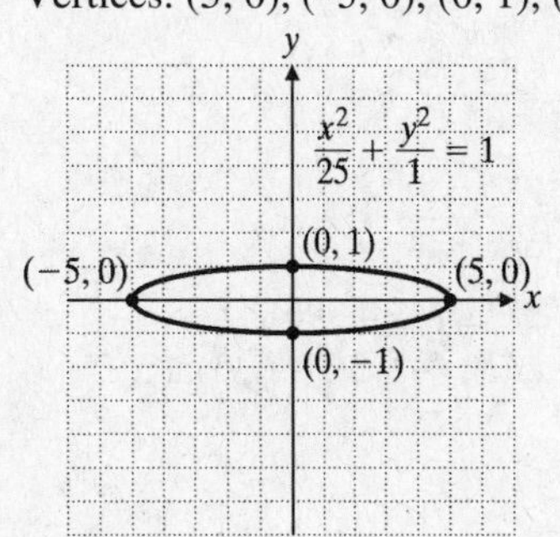

5. $\frac{x^2}{10}-\frac{y^2}{9}=1$

Hyperbola
Center: $C(0, 0)$
Vertices: $V\left(\pm\sqrt{10}, 0\right)$

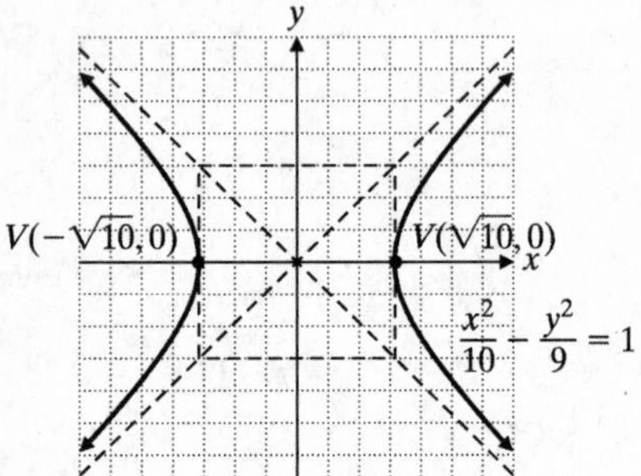

6. $y=-2(x+3)^2+4$

Parabola: $V(-3, 4)$
Let $x = 0$.

$y=-2(0+3)^2+4$
$y=-18+4$
$y=-14$

y-intercept: $(0, -14)$

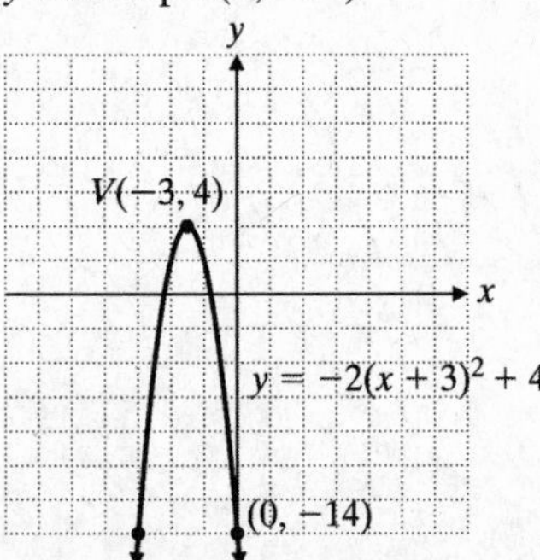

Scale: Each unit = 2

7. $\frac{(x+2)^2}{16}+\frac{(y-5)^2}{4}=1$

$\frac{(x-(-2))^2}{4^2}+\frac{(y-5)^2}{2^2}=1$

Ellipse: $C(-2, 5)$, $a = 4$, $b = 2$
Vertices: $(-2, 7)$, $(-2, 3)$, $(-6, 5)$, $(2, 5)$

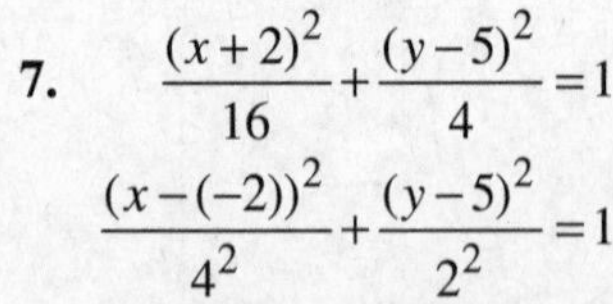

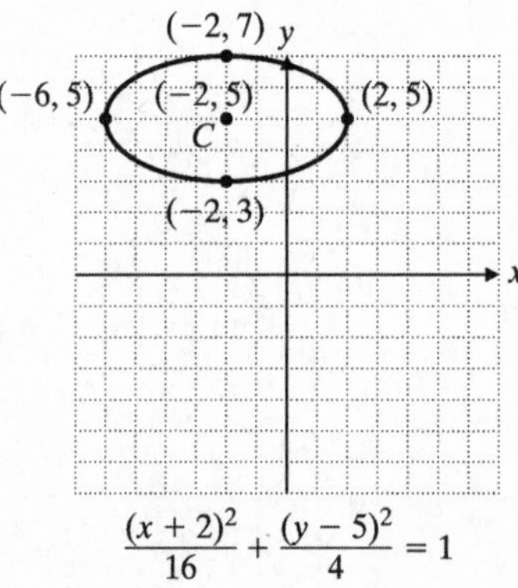

8. $7y^2-7x^2=28$

$\frac{7y^2}{28}-\frac{7x^2}{28}=\frac{28}{28}$

$\frac{y^2}{4}-\frac{x^2}{4}=1$

$\frac{y^2}{2^2}-\frac{x^2}{2^2}=1$

Hyperbola
$C(0, 0)$
Vertices: $(0, 2)$, $(0, -2)$, $a = 2$, $b = 2$

$y_{\text{asymptote}}=\pm\frac{2}{2}x=\pm x$

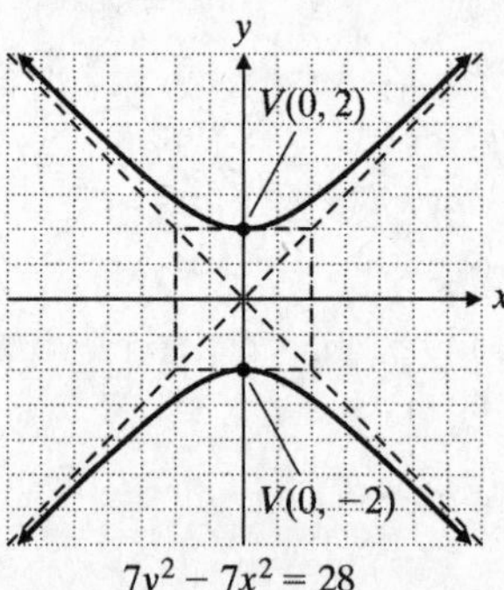

9. $(x-h)^2+(y-k)^2=r^2$

$(x-3)^2+(y-(-5))^2=\sqrt{8}^2$

$(x-3)^2+(y+5)^2=8$

10. $C(h, k) = C(0, 0)$
$h = 0$, $k = 0$
$(3, 0) \Rightarrow a = 3$
$(0, 5) \Rightarrow b = 5$

$$\frac{(x-h)^2}{a^2}+\frac{(y-k)^2}{b^2}=1$$
$$\frac{(x-0)^2}{3^2}+\frac{(y-0)^2}{5^2}=1$$
$$\frac{x^2}{9}+\frac{y^2}{25}=1$$

11. $x=(y-k)^2+h,\ (h,k)=(-7,3)$
$x=(y-3)^2+(-7)$
$x=(y-3)^2-7$
Check: (2, 0)
$2 \stackrel{?}{=} (0-3)^2-7$
$2 \stackrel{?}{=} 9-7$
$2=2$

12. $C(h,k)=C(0,0)\Rightarrow h=0,\ k=0$
$V(\pm 3,0)\Rightarrow a=3$
$y_{\text{asymptote}}=\pm\frac{b}{a}x=\frac{5}{3}x\Rightarrow b=5$
$$\frac{(x-h)^2}{a^2}-\frac{(y-k)^2}{b^2}=1$$
$$\frac{(x-0)^2}{3^2}-\frac{(y-0)^2}{5^2}=1$$
$$\frac{x^2}{9}-\frac{y^2}{25}=1$$

13. $-2x+y=5$
$y=2x+5$
$x^2+y^2-25=0$
$x^2+(2x+5)^2-25=0$
$x^2+4x^2+20x+25-25=0$
$5x^2+20x=0$
$5x(x+4)=0$
$x=0,\ x=-4$
$y=2x+5=2(0)+5=5$
$y=2(-4)+5=-3$
(0, 5), (–4, –3) is the solution.

14. $x^2+y^2=9$
$y=x-3$
$x^2+(x-3)^2=9$
$x^2+x^2-6x+9=9$
$x^2-3x=0$
$x(x-3)=0$
$x=0,\ x=3$
$y=x-3=0-3=-3$
$y=x-3=3-3=0$
(0, –3), (3, 0) is the solution.

15. $4x^2+y^2-4=0$
$y^2=4-4x^2$
$9x^2-4y^2-9=0$
$9x^2-4(4-4x^2)-9=0$
$9x^2-16+16x^2-9=0$
$25x^2=25$
$x^2=1$
$x=\pm 1$
$y^2=4-4x^2=4-4(1)=0$
$y=0$
(1, 0), (–1, 0) is the solution.

16. $x^2+2y^2=15$
$x^2-\ y^2=6$ subtract
$3y^2=9\Rightarrow y=\pm\sqrt{3}$
$x^2=6+y^2=6+3=9\Rightarrow x=\pm 3$
$\left(3,\pm\sqrt{3}\right), \left(-3,\pm\sqrt{3}\right)$ is the solution.

Cumulative Test for Chapters 1–9

1. 5(–3) = –3(5) illustrates the commutative property of multiplication.

2. $2\{x-3[x-2(x+1)]\}=2\{x-3[x-2x-2]\}$
$=2\{x-3[-x-2]\}$
$=2\{x+3x+6\}$
$=2\{4x+6\}$
$=8x+12$

3. $3(4-6)^3+\sqrt{25}=3(-2)^3+\sqrt{25}$
$=3(-8)+5$
$=-24+5$
$=-19$

4. $A=3bt+prt$
$prt=A-3bt$
$p=\frac{A-3bt}{rt}$

5. $4x^3-16x=4x(x^2-4)=4x(x+2)(x-2)$

6. $\frac{3x}{x-2}+\frac{5}{x-1}=\frac{3x(x-1)+5(x-2)}{(x-2)(x-1)}$
$=\frac{3x^2-3x+5x-10}{(x-2)(x-1)}$
$=\frac{3x^2+2x-10}{(x-2)(x-1)}$

7. $\frac{3}{2x+3}=\frac{1}{2x-3}+\frac{2}{4x^2-9}$
$\frac{3}{(2x+3)}=\frac{1}{2x-3}+\frac{2}{(2x+3)(2x-3)}$
$(2x+3)(2x-3)\left(\frac{3}{2x+3}\right)=(2x+3)(2x-3)\left(\frac{1}{2x-3}\right)+(2x+3)(2x-3)\left(\frac{2}{(2x+3)(2x-3)}\right)$
$3(2x-3)=2x+3+2$
$6x-9=2x+5$
$4x=14$
$x=\frac{7}{2}$

8. $3x-2y-9z=9$
$x-y+z=8$
$2x+3y-z=-2$
Switch the first and second equations.
$x-y+z=8$
$3x-2y-9z=9$
$2x+3y-z=-2$
Multiply the first equation by –3 and add to the second equation and multiply the first equation by –2 and add to the third equation.
$x-y+z=8$
$y-12z=-15$
$5y-3z=-18$
Multiply the second equation by –5 and add to the third equation.
$x-y+z=8$
$y-12z=-15$
$57z=57$
$z=1$
$y=12z-15=12(1)-15$
$y=-3$
$x-y+z=8$
$x-(-3)+1=8$
$x=4$
(4, –3, 1) is the solution.

9. $\left(\sqrt{2}+\sqrt{3}\right)\left(2\sqrt{6}-\sqrt{3}\right)=2\sqrt{12}-\sqrt{6}+2\sqrt{18}-3$
$=2\sqrt{4\cdot 3}+2\sqrt{9\cdot 2}-\sqrt{6}-3$
$=4\sqrt{3}+6\sqrt{2}-\sqrt{6}-3$

10. $\sqrt{12x^2} + 2x\sqrt{27} - \sqrt{18x}$
$= \sqrt{4x^2 \cdot 3} + 2x\sqrt{9 \cdot 3} - \sqrt{9 \cdot 2x}$
$= 2x\sqrt{3} + 6x\sqrt{3} - 3\sqrt{2x}$
$= 8x\sqrt{3} - 3\sqrt{2x}$

11. $x + 4(x+2) > 7x + 8$
$x + 4x + 8 > 7x + 8$
$-2x > 0$
$x < 0$

12. $\frac{6(x-4)}{5} \ge \frac{3(x+2)}{4}$
$24(x-4) \ge 15(x+2)$
$24x - 96 \ge 15x + 30$
$9x \ge 126$
$x \ge 14$

13. $d = \sqrt{(-3-6)^2 + (-4-(-1))^2}$
$d = \sqrt{81+9}$
$d = \sqrt{90}$
$d = \sqrt{9 \cdot 10}$
$d = 3\sqrt{10}$

14. $y = -\frac{1}{2}(x+2)^2 - 3$, Parabola: $V(-2, -3)$
Opens down since $a < 0$.
Let $x = 0$.
$y = -\frac{1}{2}(0+2)^2 - 3$
$y = -5$
y-intercept: $(0, -5)$

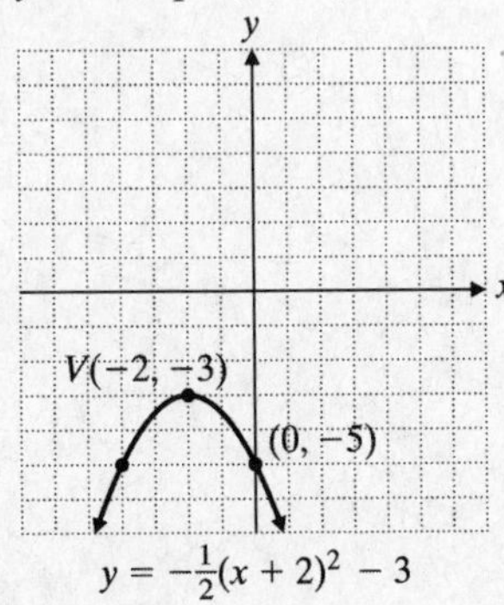

15. $25x^2 + 25y^2 = 125$
$x^2 + y^2 = 5 = \sqrt{5}^2$
Circle: $C(0, 0)$, $r = \sqrt{5}$

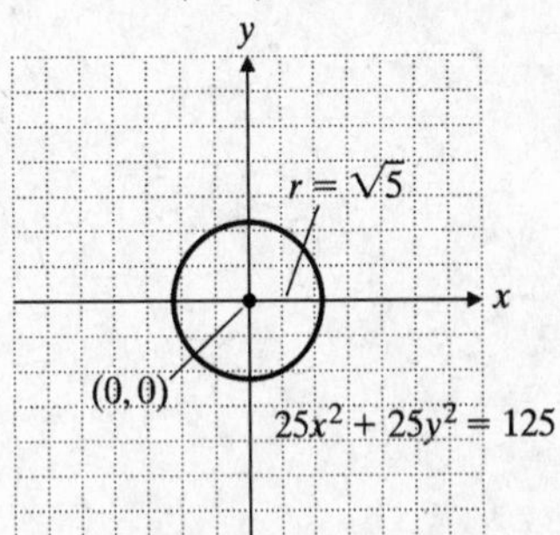

16. $16x^2 - 4y^2 = 64$
$\frac{x^2}{4} - \frac{y^2}{16} = 1$
$\frac{x^2}{2^2} - \frac{y^2}{4^2} = 1$
Hyperbola: $C(0, 0)$
$a = 2, b = 4$
Vertices: $(-2, 0)$, $(2, 0)$
$y_{\text{asymptote}} = \pm\frac{4}{2}x = \pm 2x$

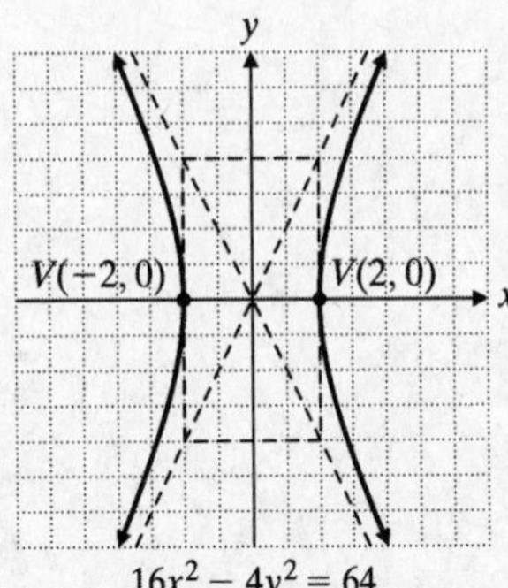

17. $\frac{(x-2)^2}{25} + \frac{(y-3)^2}{16} = 1$
$\frac{(x-2)^2}{5^2} + \frac{(y-3)^2}{4^2} = 1$
Ellipse: $C(2, 3)$
$a = 5, b = 4$
Vertices: $(2, 7)$, $(-3, 3)$, $(2, -1)$, $(7, 3)$

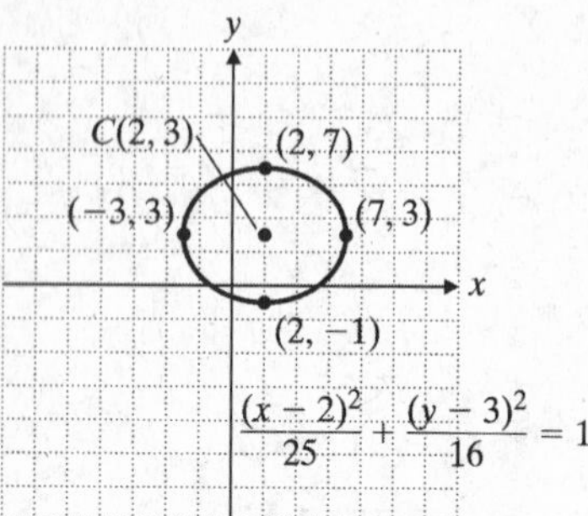

Scale: Each unit = 2

18. $y = 2x^2$

$y = 2x + 4$

$2x^2 = 2x + 4$

$x^2 - x - 2 = 0$

$(x-2)(x+1) = 0$

$x - 2 = 0 \qquad x + 1 = 0$

$x = 2 \qquad x = -1$

$y = 2x^2 = 2(2)^2 = 8$

$y = 2(-1)^2 = 2$

(2, 8), (−1, 2) is the solution.

19. $y = x^2 + 1$

$x^2 = y - 1$

$4y^2 = 4 - x^2 = 4 - (y-1)$

$4y^2 = 4 - y + 1$

$4y^2 + y - 5 = 0$

$(y-1)(4y+5) = 0$

$y = 1,\ y = -\frac{5}{4}$

$$y = \pm\sqrt{y-1} = \begin{cases} \pm\sqrt{1-1} = 0 \\ \pm\sqrt{1-\frac{5}{4}} = \pm\sqrt{-\frac{1}{4}}, \text{ not a real number} \end{cases}$$

(0, 1) is the solution.

20. $x^2 + y^2 = 25$

$x - 2y = -5$

$x = 2y - 5$

$(2y-5)^2 + y^2 = 25$

$4y^2 - 20y + 25 + y^2 = 25$

$5y^2 - 20y = 0$

$5y(y-4) = 0$

$5y = 0 \qquad y - 4 = 0$

$y = 0 \qquad y = 4$

$x = 2y - 5 = 2(0) - 5 = -5$

$x = 2y - 5 = 2(4) - 5 = 3$

(3, 4), (−5, 0) is the solution.

21. $xy = -15$

$y = -\frac{15}{x}$

$4x + 3y = 3$

$4x + 3\left(-\frac{15}{x}\right) = 3$

$4x^2 - 45 = 3x$

$4x^2 - 3x - 45 = 0$

$(x+3)(4x-15) = 0$

$x + 3 = 0 \qquad 4x - 15 = 0$

$x = -3 \qquad x = \frac{15}{4}$

$y = -\frac{15}{x} = -\frac{15}{-3} = 5 \qquad y = -\frac{15}{x} = -\frac{15}{\frac{15}{4}} = -4$

$(-3, 5), \left(\frac{15}{4}, -4\right)$ is the solution.

Chapter 10

10.1 Exercises

1. $f(x)=3x-5$

$$f\left(-\frac{2}{3}\right)=3\left(-\frac{2}{3}\right)-5$$
$$f\left(-\frac{2}{3}\right)=-2-5$$
$$f\left(-\frac{2}{3}\right)=-7$$

3. $f(x) = 3x - 5$

$$f(a-4)=3(a-4)-5$$
$$f(a-4)=3a-12-5$$
$$f(a-4)=3a-17$$

5. $g(x)=\frac{1}{2}x-3$

$$g(4)+g(a)=\frac{1}{2}(4)-3+\frac{1}{2}a-3$$
$$=2-3+\frac{1}{2}a-3$$
$$=\frac{1}{2}a-4$$

7. $g(x)=\frac{1}{2}x-3$

$$g(4a)-g(a)=\frac{1}{2}(4a)-3-\left[\frac{1}{2}a-3\right]$$
$$g(4a)-g(a)=2a-3-\frac{1}{2}a+3$$
$$g(4a)-g(a)=\frac{3}{2}a$$

9. $g(x)=\frac{1}{2}x-3$

$$g(2a-4)=\frac{1}{2}(2a-4)-3$$
$$g(2a-4)=a-2-3$$
$$g(2a-4)=a-5$$

11. $g(x)=\frac{1}{2}x-3$

$$g(a^2)-g\left(\frac{2}{5}\right)=\left(\frac{1}{2}a^2-3\right)-\left(\frac{1}{2}\cdot\frac{2}{5}-3\right)$$
$$=\frac{1}{2}a^2-3-\frac{1}{5}+3$$
$$=\frac{1}{2}a^2-\frac{1}{5}$$

13. $p(x)=3x^2+4x-2$

$$p(-2)=3(-2)^2+4(-2)-2$$
$$p(-2)=3(4)+4(-2)-2$$
$$p(-2)=12-8-2$$
$$p(-2)=2$$

15. $p(x)=3x^2+4x-2$

$$p\left(\frac{1}{2}\right)=3\left(\frac{1}{2}\right)^2+4\left(\frac{1}{2}\right)-2$$
$$p\left(\frac{1}{2}\right)=\frac{3}{4}+2-2$$
$$p\left(\frac{1}{2}\right)=\frac{3}{4}$$

17. $p(x)=3x^2+4x-2$

$$p(a+1)=3(a+1)^2+4(a+1)-2$$
$$p(a+1)=3a^2+6a+3+4a+4-2$$
$$p(a+1)=3a^2+10a+5$$

19. $p(x)=3x^2+4x-2$

$$p\left(-\frac{2a}{3}\right)=3\left(-\frac{2a}{3}\right)^2+4\left(-\frac{2a}{3}\right)-2$$
$$p\left(-\frac{2a}{3}\right)=\frac{4a^2}{3}-\frac{8a}{3}-2$$

21. $h(x)=\sqrt{x+5}$

$$h(4)=\sqrt{4+5}=\sqrt{9}=3$$

23. $h(x)=\sqrt{x+5}$

$$h(7)=\sqrt{7+5}=\sqrt{12}=\sqrt{4\cdot 3}$$
$$h(7)=2\sqrt{3}$$

25. $h(x)=\sqrt{x+5}$

$$h(a^2-1)=\sqrt{a^2-1+5}=\sqrt{a^2+4}$$

27. $h(x)=\sqrt{x+5}$

$$h(-2b)=\sqrt{-2b+5}$$

29. $h(x)=\sqrt{x+5}$

$$\begin{aligned}h(4a-1)&=\sqrt{4a-1+5}\\&=\sqrt{4a+4}\\&=\sqrt{4(a+1)}\\&=2\sqrt{a+1}\end{aligned}$$

31. $h(x)=\sqrt{x+5}$

$$h(b^2+b)=\sqrt{b^2+b+5}$$

33. $r(x)=\dfrac{7}{x-3}$

$$r(7)=\frac{7}{7-3}$$

$$r(7)=\frac{7}{4}$$

35. $r(x)=\dfrac{7}{x-3}$

$$r(3.5)=\frac{7}{3.5-3}=\frac{7}{0.5}=14$$

37. $r(x)=\dfrac{7}{x-3}$

$$r(a^2)=\frac{7}{a^2-3}$$

39. $r(x)=\dfrac{7}{x-3}$

$$r(a+2)=\frac{7}{a+2-3}$$

$$r(a+2)=\frac{7}{a-1}$$

41. $r(x)=\dfrac{7}{x-3}$

$$r\left(\frac{1}{2}\right)+r(8)=\frac{7}{\frac{1}{2}-3}+\frac{7}{8-3}=-\frac{14}{5}+\frac{7}{5}=-\frac{7}{5}$$

43. $f(x) = 2x - 3$

$$\begin{aligned}\frac{f(x+h)-f(x)}{h}&=\frac{2(x+h)-3-(2x-3)}{h}\\&=\frac{2x+2h-3-2x+3}{h}\\&=\frac{2h}{h}\\&=2\end{aligned}$$

45. $f(x)=x^2-x$

$$\begin{aligned}\frac{f(x+h)-f(x)}{h}&=\frac{(x+h)^2-(x+h)-(x^2-x)}{h}\\&=\frac{x^2+2xh+h^2-x-h-x^2+x}{h}\\&=\frac{2xh+h^2-h}{h}\\&=\frac{h(2x+h-1)}{h}\\&=2x+h-1\end{aligned}$$

47. $P=2.5w^2$

a. $P(w)=2.5w^2$

b. $P(20)=2.5(20)^2=1000$ kilowatts

c. $P=2.5(20+e)^2=2.5(400+40e+e^2)$

$P(e)=2.5e^2+100e+1000$

d. $P(2)=2.5(2)^2+100(2)+1000$

$=1210$ kilowatts

49. The function values associated with $p(x) - 13$ would be the function values of $p(x)$ decreased by 13.

$p(3) - 13 \approx 39 - 13 = 26$

51. $f(x)=3x^2-4.6x+1.23$

$f(0.026a)=3(0.026a)^2-4.6(0.026a)+1.23$

$f(0.026a)=0.002a^2-0.120a+1.23$

Cumulative Review

53. $\dfrac{7}{6}+\dfrac{5}{x}=\dfrac{3}{2x}$

$$\begin{aligned}7x+30&=9\\7x&=-21\\x&=-3\end{aligned}$$

54.
$$\frac{1}{6}-\frac{2}{3x+6}=\frac{1}{2x+4}$$
$$6(x+2)\cdot\frac{1}{6}-6(x+2)\cdot\frac{2}{3(x+2)}=6(x+2)\cdot\frac{1}{2(x+2)}$$
$$x+2-4=3$$
$$x=5$$

55.
$$\frac{V_{\text{Earth}}}{V_{\text{Mercury}}}=\frac{\frac{4}{3}\pi\left(\frac{7927}{2}\right)^3}{\frac{4}{3}\pi\left(\frac{3031}{2}\right)^3}$$
$$V_{\text{Earth}}=17.88828747...(V_{\text{Mercury}})$$
The volume of the Earth is approximately 17.9 times greater than the volume of Mercury.

56.
$$\frac{V_{\text{Jupiter}}}{V_{\text{Uranus}}}=\frac{\frac{4}{3}\pi(43,348)^3}{\frac{4}{3}\pi(14,584)^3}$$
$$V_{\text{Jupiter}}=26.25894219...(V_{\text{Uranus}})$$
Jupiter's volume is approximately 26.3 times greater than the volume of Uranus.

Quick Quiz 10.1

1. $f(x)=\frac{3}{5}x-4$
$$f(a)=\frac{3}{5}a-\frac{20}{5}$$
$$f(-3)=\frac{3}{5}(-3)-\frac{20}{5}=-\frac{9}{5}-\frac{20}{5}=-\frac{29}{5}$$
$$f(a)-f(-3)=\frac{3a}{5}-\frac{20}{5}+\frac{29}{5}=\frac{3a}{5}+\frac{9}{5}$$

2. $g(x)=2x^2-3x+4$
$$g\left(\frac{2}{3}a\right)=2\left(\frac{2}{3}a\right)^2-3\left(\frac{2}{3}a\right)+4$$
$$=2\left(\frac{4}{9}a^2\right)-\frac{6}{3}a+4$$
$$=\frac{8}{9}a^2-2a+4$$

3. $h(x)=\frac{3}{x+4}$
$$h(a-6)=\frac{3}{a-6+4}=\frac{3}{a-2}$$

4. Answers may vary. Possible solution:
For the function $k(x) = \sqrt{3x+1}$ evaluated at $k(2a - 1)$, substitute $2a - 1$ for x in the function and solve:
$k(2a-1) = \sqrt{3(2a-1)+1} = \sqrt{6a-2}$

10.2 Exercises

1. No, $f(x + 2)$ means to substitute $x + 2$ for x in the function $f(x)$. $f(x) + f(2)$ means to evaluate $f(x)$ and $f(2)$ and then add the two results. One example is $f(x) = 2x + 1$
$f(x + 2) = 2(x + 2) + 1 = 2x + 5$
$f(x) + f(2) = 2x + 1 + 2(2) + 1 = 2x + 6$

3. To obtain the graph of $f(x) + k$ for $k > 0$, shift the graph of $f(x)$ up k units.

5. Graph fails vertical line test and does not represent a function.

7. Graph passes vertical line test and does represent a function.

9. Graph passes vertical line test and does represent a function.

11. Graph fails vertical line test and does not represent a function.

13. Graph passes vertical line test and does represent a function.

For Exercises 15, 17, and 19:

x	$f(x) = x^2$
−2	4
−1	1
0	0
1	1
2	4

15. $f(x) = x^2$, $h(x) = x^2 - 3$
Shift $f(x)$ down 3 units.

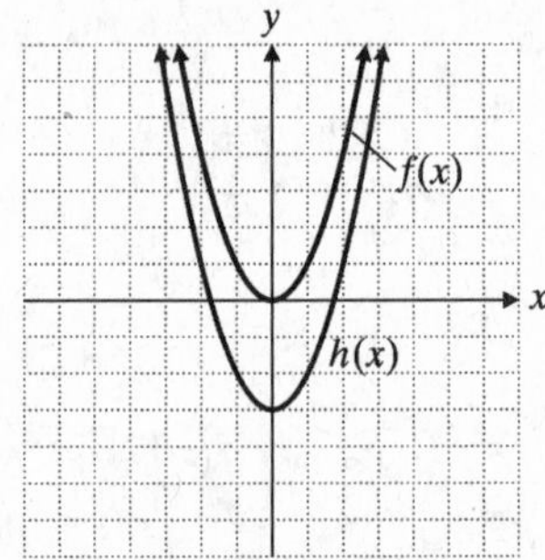

17. $f(x) = x^2$, $p(x) = (x+1)^2$
Shift $f(x)$ left 1 unit.

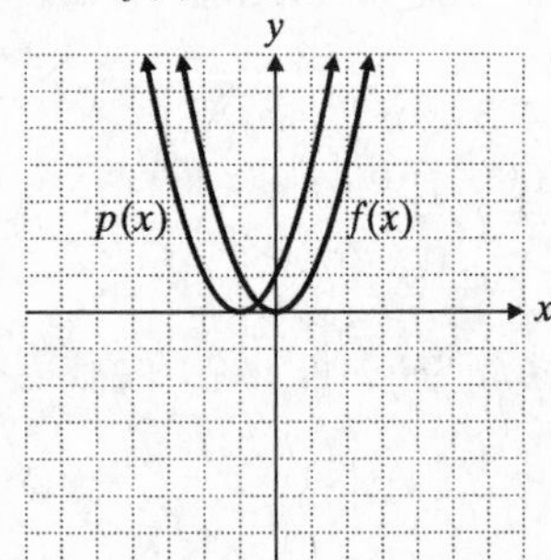

19. $f(x) = x^2$, $g(x) = (x-2)^2 + 1$
Shift $f(x)$ right 2 units and up 1 unit.

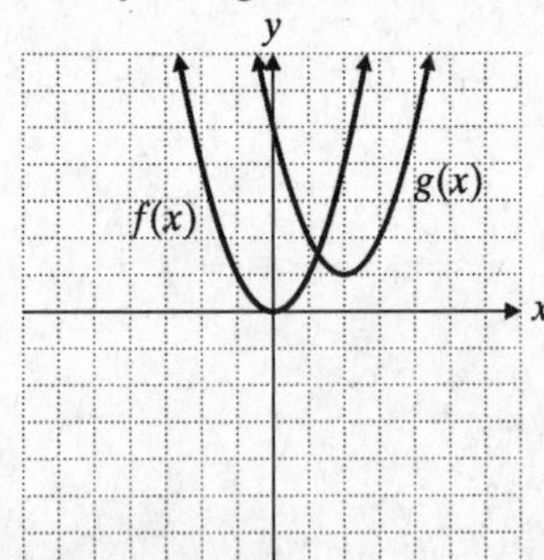

21. $f(x) = x^3$, $r(x) = x^3 - 1$
Shift $f(x)$ down 1 unit.

x	$f(x) = x^3$
−2	−8
−1	−1
0	0
1	1
2	8

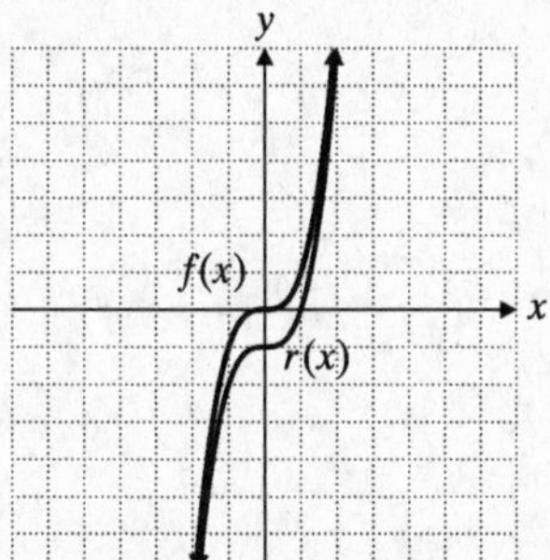

For Exercises 23 and 25:

x	$f(x) = \|x\|$
−2	2
−1	1
0	0
1	1
2	2

23. $f(x) = |x|$, $s(x) = |x + 4|$
Shift $f(x)$ left 4 units.

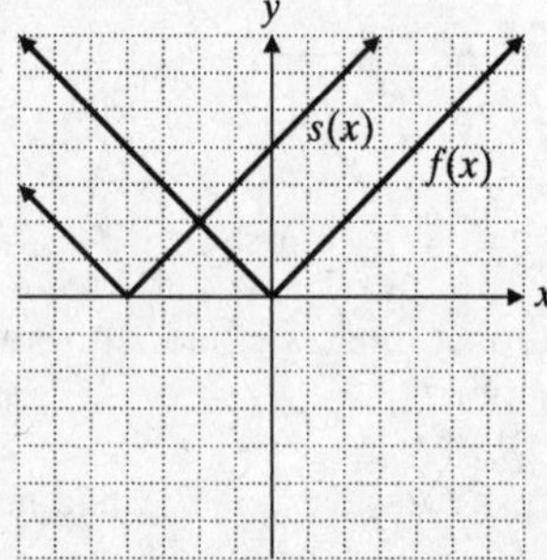

25. $f(x) = |x|$, $t(x) = |x - 3| - 4$
Shift right 3 and down 4 units.

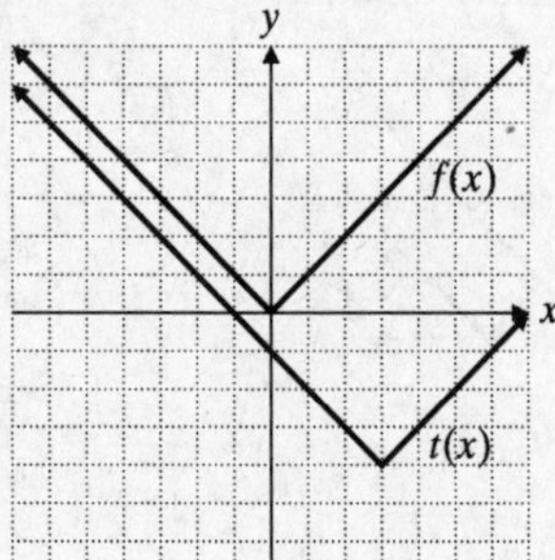

27. $f(x) = \frac{3}{x}$, $g(x) = \frac{3}{x} - 2$
Shift $f(x)$ down 2 units.

x	$f(x) = \frac{3}{x}$
−3	−1
−1	−3
0	undefined
1	3
3	1

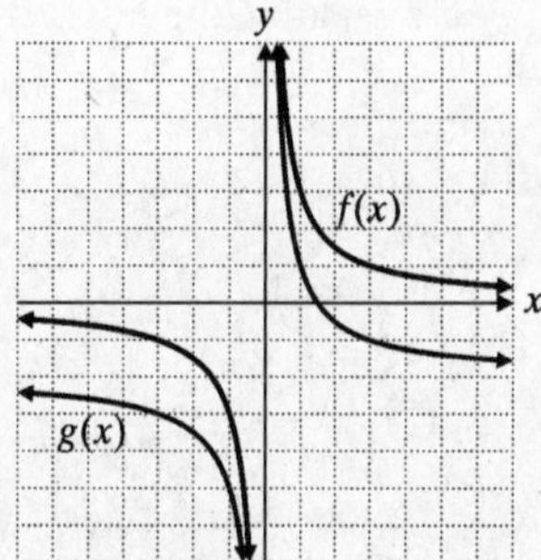

Cumulative Review

29. $$\begin{aligned}\sqrt{12} + 3\sqrt{50} - 4\sqrt{27} &= \sqrt{4 \cdot 3} + 3\sqrt{25 \cdot 2} - 4\sqrt{9 \cdot 3} \\ &= 2\sqrt{3} + 3 \cdot 5\sqrt{2} - 4 \cdot 3\sqrt{3} \\ &= 2\sqrt{3} + 15\sqrt{2} - 12\sqrt{3} \\ &= 15\sqrt{2} - 10\sqrt{3}\end{aligned}$$

30. $$\begin{aligned}\left(\sqrt{5x} + \sqrt{2}\right)^2 &= \sqrt{5x}^2 + 2\sqrt{5x}\sqrt{2} + \sqrt{2}^2 \\ &= 5x + 2\sqrt{10x} + 2\end{aligned}$$

31. $$\frac{\sqrt{5} - 2}{\sqrt{5} + 1} = \frac{\sqrt{5} - 2}{\sqrt{5} + 1} \cdot \frac{\sqrt{5} - 1}{\sqrt{5} - 1} = \frac{5 - 3\sqrt{5} + 2}{5 - 1} = \frac{7 - 3\sqrt{5}}{4}$$

Quick Quiz 10.2

1. The graph of $h(x)$ is 3 units to the right of the graph of $f(x)$.

2. The graph of $g(x)$ is 4 units below the graph of $f(x)$.

3. $f(x) = |x|$, $k(x) = |x - 1| - 4$
Shift $f(x)$ to the right 1 unit and down 4 units.

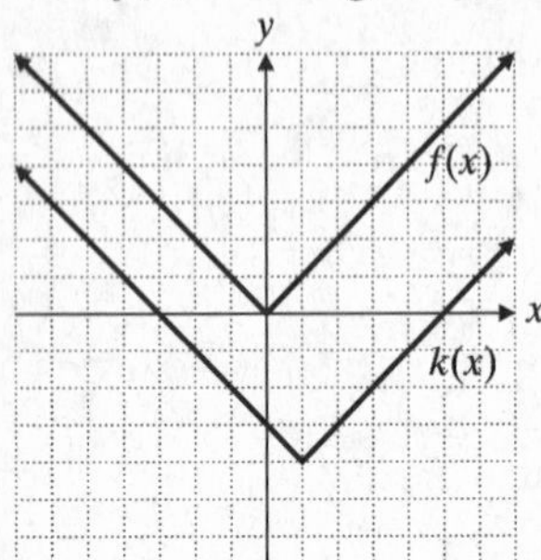

4. Answers may vary. Possible solution:
If a vertical line passes through more than one point of the graph of a relation, the relation is not a function.

How Am I Doing Sections 10.1–10.2

1. $f(x) = 2x - 6$
$f(-3) = 2(-3) - 6 = -6 - 6 = -12$

2. $f(x) = 2x - 6$
$f(a) = 2a - 6$

3. $f(x) = 2x - 6$
$f(2a) = 2(2a) - 6 = 4a - 6$

4. $f(x) = 2x - 6$
$f(a + 2) = 2(a + 2) - 6 = 2a + 4 - 6 = 2a - 2$

5. $f(x) = 5x^2 + 2x - 3$
$f(-2) = 5(-2)^2 + 2(-2) - 3 = 20 - 4 - 3 = 13$

6. $f(x) = 5x^2 + 2x - 3$
$f(a) = 5a^2 + 2a - 3$

7. $f(x) = 5x^2 + 2x - 3$
$$\begin{aligned} f(a+1) &= 5(a+1)^2 + 2(a+1) - 3 \\ &= 5(a^2 + 2a + 1) + 2a + 2 - 3 \\ &= 5a^2 + 10a + 5 + 2a - 1 \\ &= 5a^2 + 12a + 4 \end{aligned}$$

8. $f(x) = 5x^2 + 2x - 3$
$f(3a) = 5(3a)^2 + 2(3a) - 3 = 45a^2 + 6a - 3$

9. $f(x) = \dfrac{3x}{x+2}$
$$\begin{aligned} f(a) + f(a-2) &= \frac{3a}{a+2} + \frac{3(a-2)}{a-2+2} \\ &= \frac{3a^2 + 3(a+2)(a-2)}{a(a+2)} \\ &= \frac{3a^2 + 3a^2 - 12}{a(a+2)} \\ &= \frac{6a^2 - 12}{a(a+2)} \\ &= \frac{6(a^2 - 2)}{a(a+2)} \end{aligned}$$

10. $f(x) = \dfrac{3x}{x+2}$
$$\begin{aligned} f(3a) - f(3) &= \frac{3(3a)}{3a+2} - \frac{3(3)}{3+2} \\ &= \frac{9a}{3a+2} - \frac{9}{5} \\ &= \frac{45a - 9(3a+2)}{5(3a+2)} \\ &= \frac{45a - 27a - 18}{5(3a+2)} \\ &= \frac{18(a-1)}{5(3a+2)} \end{aligned}$$

11. Graph passes vertical line test and therefore represents a function.

12. Graph does not pass vertical line test and hence does not represent a function.

13. $f(x) = |x|$, $s(x) = |x - 3|$
Shift $f(x)$ to the right 3 units.

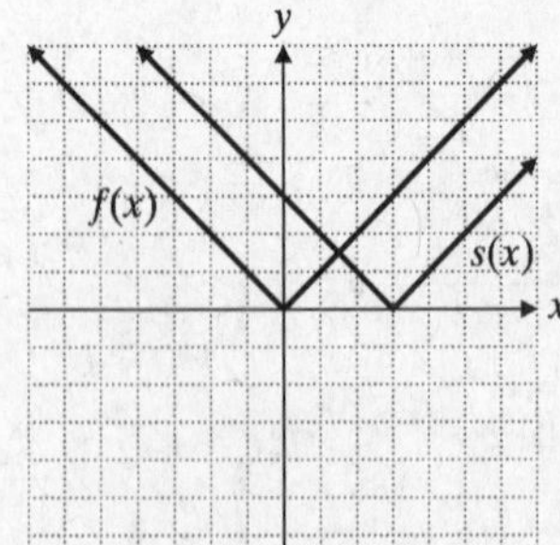

14. $f(x)=x^2,\ h(x)=(x+2)^2+3$

Shift $f(x)$ to the left 2 units and up 3 units.

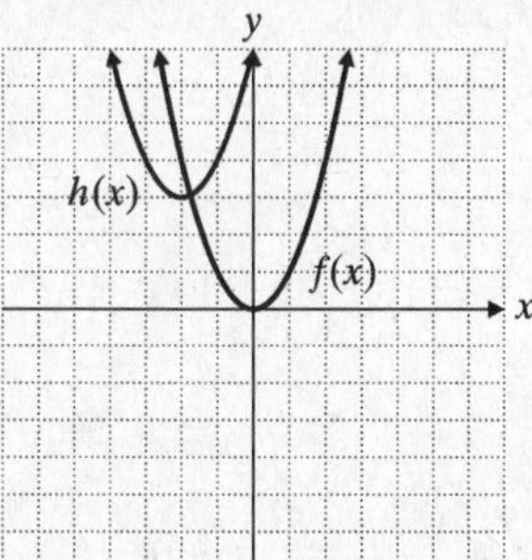

15. $f(x)=\dfrac{4}{x+2}$

x	$f(x)=\frac{4}{x+2}$
−5	$-\frac{4}{3}$
−4	−2
−3	−4
−2	undefined
0	2
1	$\frac{4}{3}$
2	1

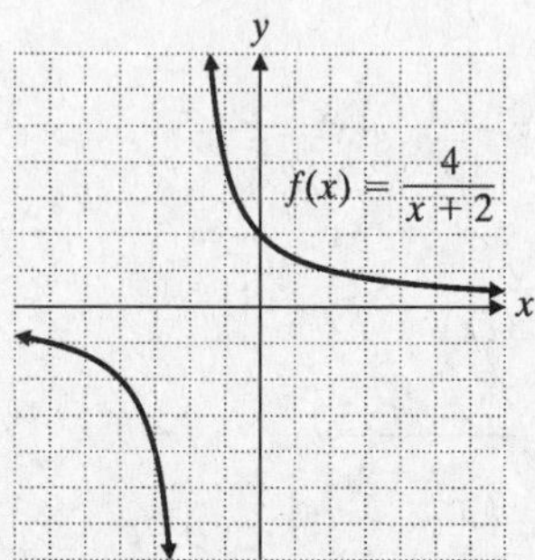

16. The graph of $g(x)=\dfrac{4}{x+2}-2$ is the graph of $f(x)=\dfrac{4}{x+2}$ shifted down 2 units.

10.3 Exercises

1. $f(x) = -2x + 3,\ g(x) = 2 + 4x$

a. $$\begin{aligned}(f+g)(x)&=f(x)+g(x)\\&=-2x+3+2+4x\\&=2x+5\end{aligned}$$

b. $$\begin{aligned}(f-g)(x)&=f(x)-g(x)\\&=-2x+3-(2+4x)\\&=-2x+3-2-4x\\&=-6x+1\end{aligned}$$

c. $(f+g)(2) = 2(2) + 5 = 9$

d. $(f-g)(-1) = -6(-1) + 1 = 7$

3. $f(x)=3x^2-x,\ g(x)=5x+2$

a. $$\begin{aligned}(f+g)(x)&=f(x)+g(x)\\&=3x^2-x+5x+2\\&=3x^2+4x+2\end{aligned}$$

b. $$\begin{aligned}(f-g)(x)&=f(x)-g(x)\\&=3x^2-x-(5x+2)\\&=3x^2-6x-2\end{aligned}$$

c. $$\begin{aligned}(f+g)(2)&=3(2)^2+4(2)+2\\&=12+8+2\\&=22\end{aligned}$$

d. $$\begin{aligned}(f-g)(-1)&=3(-1)^2-6(-1)-2\\&=3+6-2\\&=7\end{aligned}$$

5. $f(x)=x^3-\dfrac{1}{2}x^2+x,\ g(x)=x^2-\dfrac{x}{4}-5$

a. $$\begin{aligned}(f+g)(x)&=f(x)+g(x)\\&=x^3-\frac{1}{2}x^2+x+x^2-\frac{x}{4}-5\\&=x^3+\frac{1}{2}x^2+\frac{3x}{4}-5\end{aligned}$$

b. $$\begin{aligned}(f-g)(x)&=f(x)-g(x)\\&=x^3-\frac{1}{2}x^2+x-\left(x^2-\frac{x}{4}-5\right)\\&=x^3-\frac{1}{2}x^2+x-x^2+\frac{x}{4}+5\\&=x^3-\frac{3}{2}x^2+\frac{5x}{4}+5\end{aligned}$$

c. $(f+g)(2)=(2)^3+\frac{1}{2}(2)^2+\frac{3(2)}{4}-5=\frac{13}{2}$

d. $(f-g)(-1)=(-1)^3-\frac{3}{2}(-1)^2+\frac{5(-1)}{4}+5=\frac{5}{4}$

7. $f(x)=-5\sqrt{x+6},\ g(x)=8\sqrt{x+6}$

a. $$\begin{aligned}(f+g)(x)&=f(x)+g(x)\\&=-5\sqrt{x+6}+8\sqrt{x+6}\\&=3\sqrt{x+6}\end{aligned}$$

b. $$\begin{aligned}(f-g)(x)&=f(x)-g(x)\\&=-5\sqrt{x+6}-8\sqrt{x+6}\\&=-13\sqrt{x+6}\end{aligned}$$

c. $(f+g)(2)=3\sqrt{2+6}=3\sqrt{8}=6\sqrt{2}$

d. $(f-g)(-1)=-13\sqrt{-1+6}=-13\sqrt{5}$

9. $f(x)=2x-3,\ g(x)=-2x^2-3x+1$

a. $$\begin{aligned}(fg)(x)&=f(x)g(x)\\&=(2x-3)(-2x^2-3x+1)\\&=-4x^3-6x^2+2x+6x^2+9x-3\\&=-4x^3+11x-3\end{aligned}$$

b. $$\begin{aligned}(fg)(-3)&=-4(-3)^3+11(-3)-3\\&=108-33-3\\&=72\end{aligned}$$

11. $f(x)=\frac{2}{x^2},\ g(x)=x^2-x$

a. $$\begin{aligned}(fg)(x)&=f(x)g(x)\\&=\frac{2}{x^2}(x^2-x)\\&=2-\frac{2}{x}\\&=\frac{2x-2}{x}\\&=\frac{2(x-1)}{x}\end{aligned}$$

b. $(fg)(-3)=\frac{2(-3-1)}{-3}=\frac{8}{3}$

13. $f(x)=\sqrt{-2x+1},\ g(x)=-3x$

a. $$\begin{aligned}(fg)(x)&=f(x)g(x)\\&=\sqrt{-2x+1}(-3x)\\&=-3x\sqrt{-2x+1}\end{aligned}$$

b. $(fg)(-3)=-3(-3)\sqrt{-2(-3)+1}=9\sqrt{7}$

15. $f(x)=x-6,\ g(x)=3x$

a. $\left(\frac{f}{g}\right)(x)=\frac{f(x)}{g(x)}=\frac{x-6}{3x},\ x\neq 0$

b. $\left(\frac{f}{g}\right)(2)=\frac{2-6}{3(2)}=\frac{-4}{6}=-\frac{2}{3}$

17. $f(x)=x^2-1,\ g(x)=x-1$

a. $$\begin{aligned}\left(\frac{f}{g}\right)(x)&=\frac{f(x)}{g(x)}\\&=\frac{x^2-1}{x-1}\\&=\frac{(x-1)(x+1)}{x-1}\\&=x+1,\ x\neq 1\end{aligned}$$

b. $\left(\frac{f}{g}\right)(2)=2+1=3$

19. $f(x)=x^2+10x+25,\ g(x)=x+5$

a. $$\begin{aligned}\left(\frac{f}{g}\right)(x)&=\frac{f(x)}{g(x)}\\&=\frac{x^2+10x+25}{x+5}\\&=\frac{(x+5)(x+5)}{(x+5)}\\&=x+5,\ x\neq -5\end{aligned}$$

b. $\left(\frac{f}{g}\right)(2)=2+5=7$

21. $f(x)=4x-1,\ g(x)=4x^2+7x-2$

a. $\left(\frac{f}{g}\right)(x)=\frac{f(x)}{g(x)}=\frac{4x-1}{4x^2+7x-2}$
$=\frac{(4x-1)}{(4x-1)(x+2)}$
$=\frac{1}{x+2},\ x\neq -2,\ \frac{1}{4}$

b. $\left(\frac{f}{g}\right)(2)=\frac{1}{2+2}=\frac{1}{4}$

23. $f(x)=3x+2,\ g(x)=x^2-2x$
$(f-g)(x)=f(x)-g(x)=3x+2-(x^2-2x)$
$=3x+2-x^2+2x$
$=-x^2+5x+2$

25. $g(x)=x^2-2x,\ h(x)=\frac{x-2}{3}$
$\left(\frac{g}{h}\right)(x)=\frac{g(x)}{h(x)}=\frac{x^2-2x}{\frac{x-2}{3}}=\frac{3x(x-2)}{x-2}=3x,$
$x\neq 2$

27. $g(x)=x^2-2x,\ f(x)=3x+2$
$(fg)(x)=f(x)g(x)$
$=(3x+2)(x^2-2x)$
$=3x^3-6x^2+2x^2-4x$
$=3x^3-4x^2-4x$
$(fg)(-1)=3(-1)^3-4(-1)^2-4(-1)$
$=-3-4+4$
$=-3$

29. $g(x)=x^2-2x,\ f(x)=3x+2$
$\left(\frac{g}{f}\right)(-1)=\frac{g(-1)}{f(-1)}=\frac{(-1)^2-2(-1)}{3(-1)+2}=\frac{3}{-1}=-3$

31. $f(x)=2-3x,\ g(x)=2x+5$
$f[g(x)]=f[2x+5]$
$=2-3(2x+5)$
$=2-6x-15$
$=-6x-13$

33. $f(x)=2x^2+5,\ g(x)=x-1$
$f[g(x)]=f[x-1]$
$=2(x-1)^2+5$
$=2(x^2-2x+1)+5$
$=2x^2-4x+7$

35. $f(x)=8-5x,\ g(x)=x^2+3$
$f[g(x)]=f[x^2+3[$
$=8-5(x^2+3)$
$=8-5x^2-15$
$=-5x^2-7$

37. $f(x)=\frac{7}{2x-3},\ g(x)=x+2$
$f[g(x)]=f(x+2)$
$=\frac{7}{2(x+2)-3}$
$=\frac{7}{2x+1},\ x\neq -\frac{1}{2}$

39. $f(x)=|x+3|,\ g(x)=2x-1$
$f[g(x)]=f[2x-1]$
$=|2x-1+3|$
$=|2x+2|$

41. $f(x)=x^2+2,\ g(x)=3x+5$
$f[g(x)]=f(3x+5)$
$=(3x+5)^2+2$
$=9x^2+30x+25+2$
$=9x^2+30x+27$

43. $f(x)=x^2+2,\ g(x)=3x+5$
$g[f(x)]=g[x^2+2]$
$=3(x^2+2)+5$
$=3x^2+6+5$
$=3x^2+11$

45. From Exercise 43, $g[f(x)]=3x^2+11.$
$g[f(0)]=3(0)^2+11=11$

47. $p(x)=\sqrt{x-1},\ f(x)=x^2+2$
$$\begin{aligned}(p\circ f)(x)&=p[f(x)]\\&=p[x^2+2]\\&=\sqrt{x^2+2-1}\\&=\sqrt{x^2+1}\end{aligned}$$

49. $g(x)=3x+5;\ h(x)=\frac{1}{x}$
$$\begin{aligned}(g\circ h)\left(\sqrt{2}\right)&=g\left[h\left(\sqrt{2}\right)\right]\\&=g\left[\frac{1}{\sqrt{2}}\right]\\&=3\cdot\frac{1}{\sqrt{2}}+5\\&=3\cdot\frac{1}{\sqrt{2}}\cdot\frac{\sqrt{2}}{\sqrt{2}}+5\\&=\frac{3\sqrt{2}}{2}+5\end{aligned}$$

51. $p(x)=\sqrt{x-1},\ f(x)=x^2+2$
$$\begin{aligned}(p\circ f)(-5)&=p[f(-5)]\\&=p[(-5)^2+2]\\&=p(27)\\&=\sqrt{27-1}\\&=\sqrt{26}\end{aligned}$$

53.
$$\begin{aligned}K[C(F)]&=K\left[\frac{5F-160}{9}\right]\\&=\frac{5F-160}{9}+273\\&=\frac{5F-160+9(273)}{9}\\&=\frac{5F+2297}{9}\end{aligned}$$

55. $r(t)=3t,\ a(r)=3.14r^2$
$$a[r(t)]=a[3t]=3.14(3t)^2=28.26t^2$$
$$a[r(20)]=28.26(20)^2=11{,}304\text{ ft}^2$$

Cumulative Review

56. $36x^2-12x+1=(6x)^2-2(6x)(1)+1^2=(6x-1)^2$

57. $25x^4-1=(5x^2)^2-1^2=(5x^2-1)(5x^2+1)$

58.
$$\begin{aligned}x^4-10x^2+9&=(x^2-9)(x^2-1)\\&=(x+3)(x-3)(x+1)(x-1)\end{aligned}$$

59. $3x^2-7x+2=(3x-1)(x-2)$

Quick Quiz 10.3

1. $f(x)=2x^2-4x-8$
$g(x)=-3x^2+5x-2$
$$\begin{aligned}(f-g)(x)&=f(x)-g(x)\\&=(2x^2-4x-8)-(-3x^2+5x-2)\\&=2x^2-4x-8+3x^2-5x+2\\&=5x^2-9x-6\end{aligned}$$

2. $f(x)=x^2-3$
$g(x)=\frac{x-4}{2}$
$$\begin{aligned}f[g(x)]&=f\left(\frac{x-4}{2}\right)\\&=\left(\frac{x-4}{2}\right)^2-3\\&=\frac{x^2-8x+16}{4}-3\\&=\frac{x^2}{4}-\frac{8x}{4}+\frac{16}{4}-3\\&=\frac{x^2}{4}-2x+1\end{aligned}$$

3. $f(x)=x-7$
$g(x)=2x-5$
$$\left(\frac{g}{f}\right)(2)=\frac{g(2)}{f(2)}=\frac{2(2)-5}{2-7}=\frac{-1}{-5}=\frac{1}{5}$$

4. Answers may vary. Possible solution: Evaluate both functions for −4, then subtract the results of $g(-4)$ from the results of $f(-4)$.

10.4 Exercises

1. A one-to-one function is a function in which no ordered pairs have the same second coordinate.

3. The graphs of a function f and its inverse f^{-1} are symmetric about the line $y=x$.

5. The graph of a horizontal line is the graph of a function because it passes the vertical line test. A horizontal line is not the graph of a one-to-one function because it fails the horizontal line test.

7. $B = \{(0, 1), (1, 0), (10, 0)\}$ is not one-to-one since two ordered pairs, (1, 0) and (10, 0), have the same second coordinate.

9. $F = \left\{\left(\frac{3}{2}, 2\right), \left(3, -\frac{4}{5}\right), \left(-\frac{2}{3}, -2\right), \left(-3, \frac{4}{5}\right)\right\}$

is a one-to-one function since no two ordered pairs have the same second coordinate.

11. $E = \{(1, 2.8), (3, 6), (-1, -2.8), (2.8, 1)\}$
is a one-to-one function since no two ordered pairs have the same second coordinate.

13. Graph of function passes the horizontal line test and therefore, function is one-to-one.

15. Graph of function fails the horizontal line test and therefore, function is not one-to-one.

17. Graph of function fails the horizontal line test and therefore, function is not one-to-one.

19. $J = \{(8, 2), (1, 1), (0, 0), (-8, -2)\}$

$J^{-1} = \{(2, 8), (1, 1), (0, 0), (-2, -8)\}$

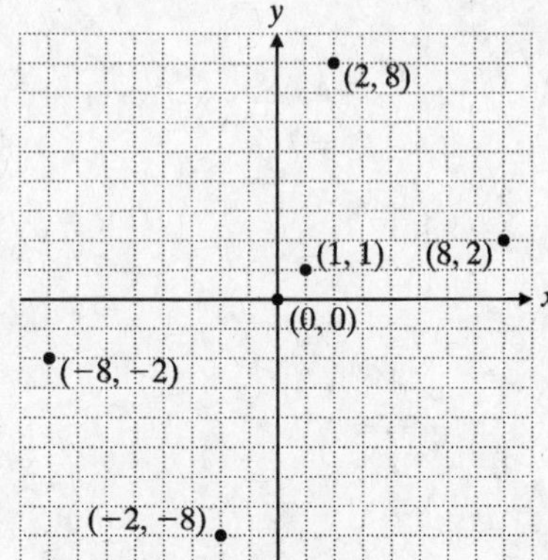

21. $f(x) = 4x - 5, f(x) \to y$

$y = 4x - 5,\ x \leftrightarrow y$

$x = 4y - 5$

$4y = x + 5$

$y = \frac{x+5}{4},\ y \to f^{-1}(x)$

$f^{-1}(x) = \frac{x+5}{4}$

23. $f(x) = x^3 + 7,\ f(x) \to y$

$y = x^3 + 7,\ x \leftrightarrow y$

$x = y^3 + 7$

$y^3 = x - 7$

$y = \sqrt[3]{x-7},\ y \to f^{-1}(x)$

$f^{-1}(x) = \sqrt[3]{x-7}$

25. $f(x) = -\frac{4}{x},\ f(x) \to y$

$y = -\frac{4}{x},\ x \leftrightarrow y$

$x = -\frac{4}{y}$

$y = -\frac{4}{x},\ y \to f^{-1}(-x)$

$f^{-1}(x) = -\frac{4}{x}$

27. $f(x) = \frac{4}{x-5},\ f(x) \to y$

$y = \frac{4}{x-5},\ x \leftrightarrow y$

$x = \frac{4}{y-5}$

$y - 5 = \frac{4}{x}$

$y = \frac{4}{x} + 5,\ y \to f^{-1}(x)$

$f^{-1}(x) = \frac{4}{x} + 5$ or $f^{-1}(x) = \frac{4+5x}{x}$

29. $g(x) = 2x + 5,\ g(x) \to y$

$y = 2x + 5,\ x \leftrightarrow y$

$x = 2y + 5$

$2y = x - 5$

$y = \frac{x-5}{2}$

$g^{-1}(x) = \frac{x-5}{2}$

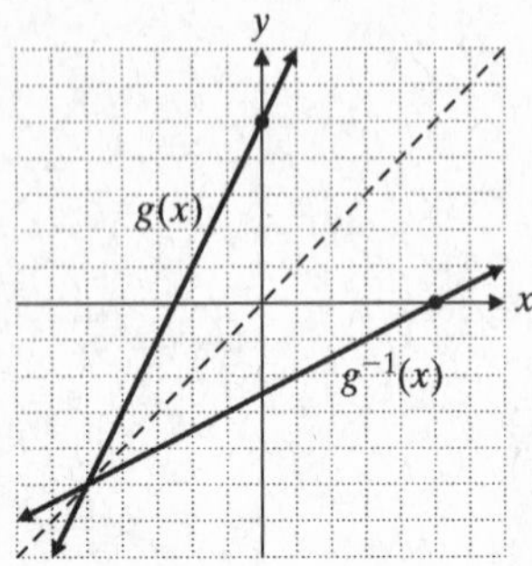

31. $h(x)=\frac{1}{2}x-2,\ h(x)\to y$

$$y=\frac{1}{2}x-2,\ x\leftrightarrow y$$
$$x=\frac{1}{2}y-2$$
$$2x=y-4$$
$$y=2x+4$$
$$h^{-1}(x)=2x+4$$

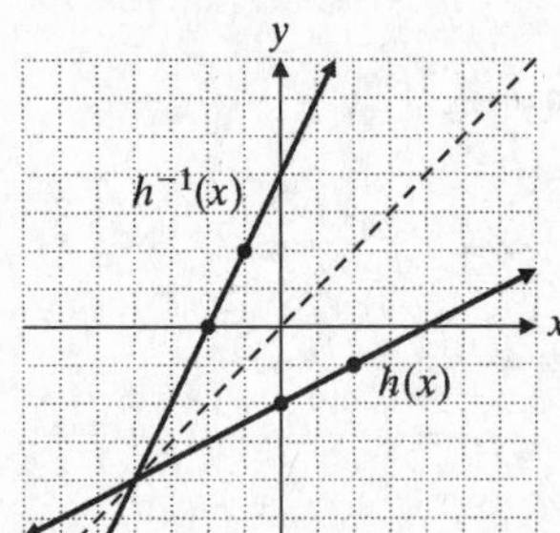

33. $r(x)=-3x-1,\ r(x)\to y$

$$y=-3x-1,\ x\leftrightarrow y$$
$$x=-3y-1$$
$$3y=-x-1$$
$$y=-\frac{x+1}{3},\ r^{-1}(x)$$
$$r^{-1}(x)=-\frac{x+1}{3}$$

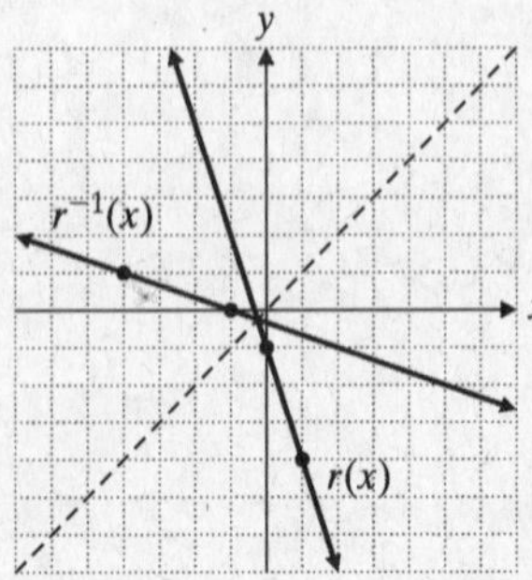

35. No; $f(x)=2x^2+3$ is a vertical parabola and fails the horizontal line test; it is not one-to-one and therefore, does not have an inverse.

37. $f(x)=2x+\frac{3}{2},\ f^{-1}(x)=\frac{1}{2}x-\frac{3}{4}$

$$f[f^{-1}(x)]=f\left(\frac{1}{2}x-\frac{3}{4}\right)$$
$$=2\left(\frac{1}{2}x-\frac{3}{4}\right)+\frac{3}{2}$$
$$=x-\frac{3}{2}+\frac{3}{2}$$
$$=x$$
$$f^{-1}[f(x)]=f^{-1}\left(2x+\frac{3}{2}\right)$$
$$=\frac{1}{2}\left(2x+\frac{3}{2}\right)-\frac{3}{4}$$
$$=x+\frac{3}{4}-\frac{3}{4}$$
$$=x$$

Cumulative Review

39.
$$x=\sqrt{15-2x}$$
$$x^2=15-2x$$
$$x^2+2x-15=0$$
$$(x+5)(x-3)=0$$
$$x+5=0 \quad\text{or}\quad x-3=0$$
$$x=-5 \qquad x=3$$

$x=-5$ does not check.
The solution is $x=3$.

40.
$$x^{2/3}+7x^{1/3}+12=0$$
$$(x^{1/3}+4)(x^{1/3}+3)=0$$
$$x^{1/3}+4=0 \qquad x^{1/3}+3=0$$
$$x^{1/3}=-4 \qquad x^{1/3}=-3$$
$$x=-64 \qquad x=-27$$

41. x = number of people working in forest related jobs in Canada in 2005

$$x=\frac{17{,}300{,}000}{15}=1{,}153{,}333$$

1,153,333 people worked in a job related to forests.

42. x = percent decrease in Ukrainian population between 2007 and 2050

$$x = 100\% \cdot \frac{46{,}800{,}000 - 33{,}400{,}000}{46{,}800{,}000}$$

$x = 28.6$

The expected percent of decrease is 28.6%.

Quick Quiz 10.4

1. $A = \{(3, -4), (2, -6), (5, 6), (-3, 4)\}$

a. Yes since no two ordered pairs have the same first coordinate.

b. Yes since no two ordered pairs have the same second coordinate.

c. $A^{-1} = \{(-4, 3), (-6, 2), (6, 5), (4, -3)\}$

2. $f(x) = 5 - 2x$

$y = 5 - 2x$

$x = 5 - 2y$

$x - 5 = -2y$

$y = \frac{-x+5}{2}$

$f^{-1}(x) = \frac{-x+5}{2}$

3. $f(x) = 2 - x^3$

$y = 2 - x^3$

$x = 2 - y^3$

$x - 2 = -y^3$

$2 - x = y^3$

$y = \sqrt[3]{2-x}$

$f^{-1}(x) = \sqrt[3]{2-x}$

4. Answers may vary. Possible solution:

To find the inverse of the function $f(x) = \frac{x-5}{3}$, substitute y for $f(x)$.

$y = \frac{x-5}{3}$

Interchange x and y.

$x = \frac{y-5}{3}$

Solve for y in terms of x.

$x = \frac{y-5}{3}$

$3x = y - 5$

$y = 3x + 5$

Replace y with $f^{-1}(x)$.

$f^{-1}(x) = 3x + 5$

Putting Your Skills to Work

1. $72° - 68° = 4°$

$4 \times 0.02 \times 205 = 16.4$

They could expect to save $16.40 per month.

2. $84° - 78° = 6°$

$6 \times 0.02 \times 205 = 24.6$

They could expect to save $24.60 per month.

3. $\frac{5°}{1\text{ mo}} \times \frac{12\text{ mo}}{1} \times \frac{\$205}{1\text{ mo}} \times \frac{0.02}{1°} = \$246$

They could expect to save $246 per year.

4. Answers will vary.

5. Answers will vary.

6. $2.95 × 100 = $295

$4.45 × 100 = $445

$445 – $295 = $150

The difference in cost is $150.

7. $4.45 × 100 × 5 = $2225

He can expect to pay $2225.

8. 4° lower: $\frac{4}{72} = 0.08$ or 8% savings

8% of 2225 = 0.08 × 2225 = 178

He will save $178.

9. 8° lower: $\frac{8}{72} = 0.16$ or 16% savings

16% of 2225 = 0.16 × 2225 = 356

He will save $356.

10. $445 = what percent of $2225

$\frac{445}{2225} = 20\%$; 10° lower

He needs to set it at 72° – 10° = 62°.

Chapter 10 Review Problems

1. $f(x)=\frac{1}{2}x+3$

$f(a-1)=\frac{1}{2}(a-1)+3=\frac{1}{2}a-\frac{1}{2}+\frac{6}{2}=\frac{1}{2}a+\frac{5}{2}$

2. $f(x)=\frac{1}{2}x+3$

$f(a+2)=\frac{1}{2}(a+2)+3=\frac{1}{2}a+1+3=\frac{1}{2}a+4$

3. $f(x)=\frac{1}{2}x+3$

$$\begin{aligned}f(a-1)-f(a)&=\frac{1}{2}(a-1)+3-\left(\frac{1}{2}a+3\right)\\&=\frac{1}{2}a-\frac{1}{2}+3-\frac{1}{2}a-3\\&=-\frac{1}{2}\end{aligned}$$

4. $f(x)=\frac{1}{2}x+3$

$$\begin{aligned}f(a+2)-f(a)&=\frac{1}{2}(a+2)+3-\left(\frac{1}{2}a+3\right)\\&=\frac{1}{2}a+1+3-\frac{1}{2}a-3\\&=1\end{aligned}$$

5. $f(x)=\frac{1}{2}x+3$

$$\begin{aligned}f(b^2-3)&=\frac{1}{2}(b^2-3)+3\\&=\frac{1}{2}b^2-\frac{3}{2}+3\\&=\frac{1}{2}b^2+\frac{3}{2}\end{aligned}$$

6. $f(x)=\frac{1}{2}x+3$

$$\begin{aligned}f(4b^2+1)&=\frac{1}{2}(4b^2+1)+3\\&=2b^2+\frac{1}{2}+3\\&=2b^2+\frac{7}{2}\end{aligned}$$

7. $p(x)=-2x^2+3x-1$

$p(-3)=-2(-3)^2+3(-3)-1=-18-9-1=-28$

8. $p(x)=-2x^2+3x-1$

$p(-1)=-2(-1)^2+3(-1)-1=-2-3-1=-6$

9. $p(x)=-2x^2+3x-1$

$$\begin{aligned}&p(2a)+p(-2)\\&=-2(2a)^2+3(2a)-1+[-2(-2)^2+3(-2)-1]\\&=-8a^2+6a-1+(-8-6-1)\\&=-8a^2+6a-16\end{aligned}$$

10. $p(x)=-2x^2+3x-1$

$$\begin{aligned}&p(-3a)+p(1)\\&=-2(-3a)^2+3(-3a)-1+[-2(1)^2+3(1)-1]\\&=-18a^2-9a-1+(-2+3-1)\\&=-18a^2-9a-1\end{aligned}$$

11. $p(x)=-2x^2+3x-1$

$$\begin{aligned}p(a+2)&=-2(a+2)^2+3(a+2)-1\\&=-2(a^2+4a+4)+3(a+2)-1\\&=-2a^2-8a-8+3a+6-1\\&=-2a^2-5a-3\end{aligned}$$

12. $p(x)=-2x^2+3x-1$

$$\begin{aligned}p(a-3)&=-2(a-3)^2+3(a-3)-1\\&=-2(a^2-6a+9)+3a-9-1\\&=-2a^2+12a-18+3a-10\\&=-2a^2+15a-28\end{aligned}$$

13. $h(x)=|2x-1|$
$h(0)=|2(0)-1|=|-1|=1$

14. $h(x)=|2x-1|$
$h(-5)=|2(-5)-1|=|-11|=11$

15. $h(x)=|2x-1|$

$h\left(\frac{1}{4}a\right)=\left|2\left(\frac{1}{4}a\right)-1\right|=\left|\frac{1}{2}a-1\right|$

16. $h(x)=|2x-1|$

$h\left(\frac{3}{2}a\right)=\left|2\left(\frac{3}{2}a\right)-1\right|=|3a-1|$

17. $h(x)=|2x-1|$

$h(a^2+a)=\left|2(a^2+a)-1\right|=\left|2a^2+2a-1\right|$

18. $h(x) = |2x - 1|$

$$h(2a^2 - 3a) = \left|2(2a^2 - 3a) - 1\right|$$
$$h(2a^2 - 3a) = \left|4a^2 - 6a - 1\right|$$

19. $r(x) = \dfrac{3x}{x+4},\ x \neq -4$

$$r(5) = \frac{3(5)}{5+4} = \frac{15}{9} = \frac{5}{3}$$

20. $r(x) = \dfrac{3x}{x+4},\ x \neq -4$

$$r(-6) = \frac{3(-6)}{-6+4} = \frac{-18}{-2} = 9$$

21. $r(x) = \dfrac{3x}{x+4},\ x \neq -4$

$$r(2a-5) = \frac{3(2a-5)}{2a-5+4} = \frac{6a-15}{2a-1}$$

22. $r(x) = \dfrac{3x}{x+4},\ x \neq -4$

$$r(1-a) = \frac{3(1-a)}{1-a+4} = \frac{3-3a}{5-a}$$

23. $r(x) = \dfrac{3x}{x+4},\ x \neq -4$

$$\begin{aligned} r(3) + r(a) &= \frac{3(3)}{3+4} + \frac{3(a)}{a+4} \\ &= \frac{9}{7} + \frac{3a}{a+4} \\ &= \frac{9(a+4) + 7(3a)}{7(a+4)} \\ &= \frac{9a + 36 + 21a}{7a+28} \\ &= \frac{30a+36}{7a+28} \end{aligned}$$

24. $r(x) = \dfrac{3x}{x+4},\ x \neq -4$

$$\begin{aligned} r(a) + r(-2) &= \frac{3a}{a+4} + \frac{3(-2)}{-2+4} \\ &= \frac{3a}{a+4} + \frac{-6}{2} \\ &= \frac{3a}{a+4} - 3 \\ &= \frac{3a - 3(a+4)}{a+4} \\ &= \frac{3a - 3a - 12}{a+4} \\ &= -\frac{12}{a+4} \end{aligned}$$

25. $f(x) = 7x - 4$

$$\begin{aligned} \frac{f(x+h) - f(x)}{h} &= \frac{7(x+h) - 4 - (7x-4)}{h} \\ &= \frac{7x + 7h - 4 - 7x + 4}{h} \\ &= \frac{7h}{h} \\ &= 7 \end{aligned}$$

26. $f(x) = 6x - 5$

$$\begin{aligned} \frac{f(x+h) - f(x)}{h} &= \frac{6(x+h) - 5 - (6x-5)}{h} \\ &= \frac{6x + 6h - 5 - 6x + 5}{h} \\ &= \frac{6h}{h} \\ &= 6 \end{aligned}$$

27. $f(x) = 2x^2 - 5x$

$$\begin{aligned} &\frac{f(x+h) - f(x)}{h} \\ &= \frac{2(x+h)^2 - 5(x+h) - (2x^2 - 5x)}{h} \\ &= \frac{2x^2 + 4xh + 2h^2 - 5x - 5h - 2x^2 + 5x}{h} \\ &= \frac{4xh + 2h^2 - 5h}{h} \\ &= 4x + 2h - 5 \end{aligned}$$

28. $f(x)=2x-3x^2$

$\frac{f(x+h)-f(x)}{h}$

$=\frac{2(x+h)-3(x+h)^2-(2x-3x^2)}{h}$

$=\frac{2x+2h-3x^2-6xh-3h^2-2x+3x^2}{h}$

$=\frac{2h-6xh-3h^2}{h}$

$=2-6x-3h$

$=-6x-3h+2$

29. a. Yes, the graph passes the vertical line test and therefore represents a function.

b. Yes, the graph passes the horizontal line test and therefore represents a one-to-one function.

30. a. No, the graph fails the vertical line test and therefore does not represent a function.

b. No, unless the graph represents a function first it cannot represent a one-to-one function.

31. a. Yes, the graph passes the vertical line test and therefore represents a function.

b. No, the graph fails the horizontal line test and therefore does not represent a one-to-one function.

32. a. Yes, the graph passes the vertical line test and therefore represents a function.

b. No, the graph fails the horizontal line test and therefore does not represent a one-to-one function.

33. a. No, the graph fails the vertical line test and therefore does not represent a function.

b. No, unless the graph represents a function first it cannot represent a one-to-one function.

34. a. Yes, the graph passes the vertical line test and therefore represents a function.

b. Yes, the graph passes the horizontal line test and therefore represents a one-to-one function.

35. $f(x)=x^2$

$g(x)=(x+2)^2+4$ is $f(x)$ shifted left 2 units and up 4 units.

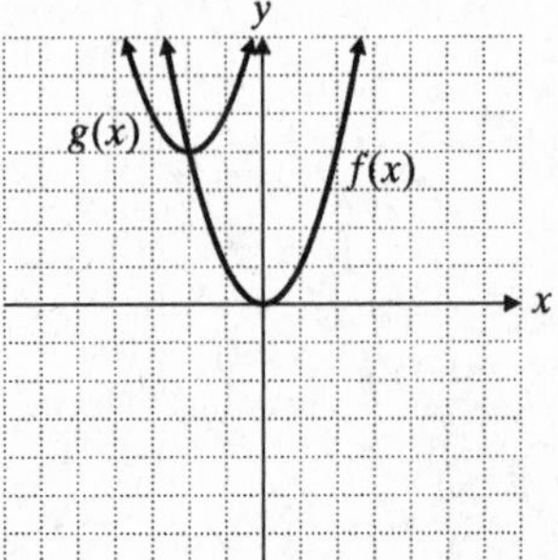

36. $f(x)=|x|$
$g(x)=|x+3|$ is $f(x)$ shifted left 3 units.

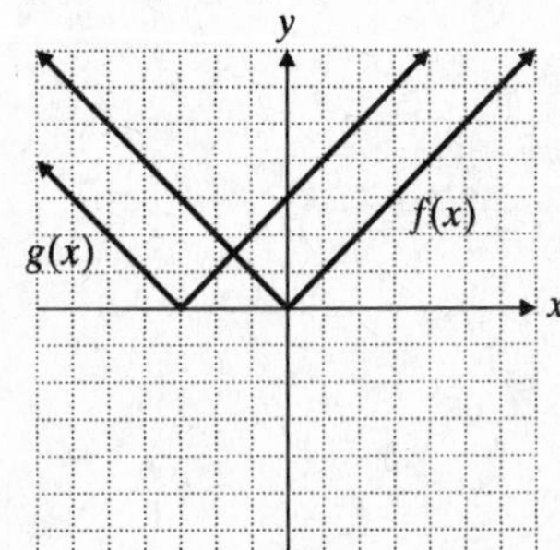

37. $f(x)=|x|$
$g(x)=|x-4|$ is $f(x)$ shifted right 4 units.

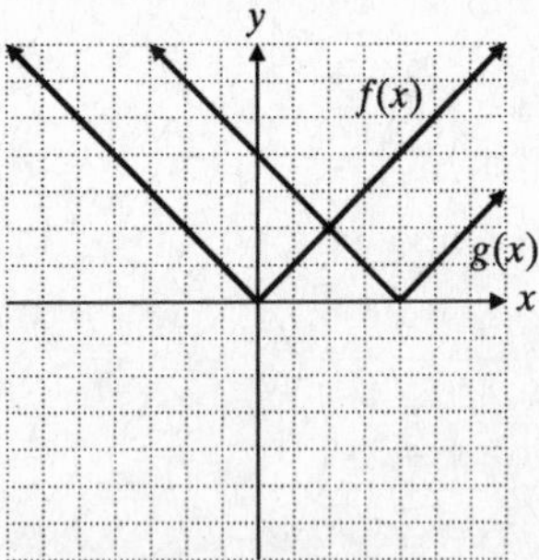

38. $f(x)=|x|$
$h(x)=|x|+3$ is $f(x)$ shifted up 3 units.

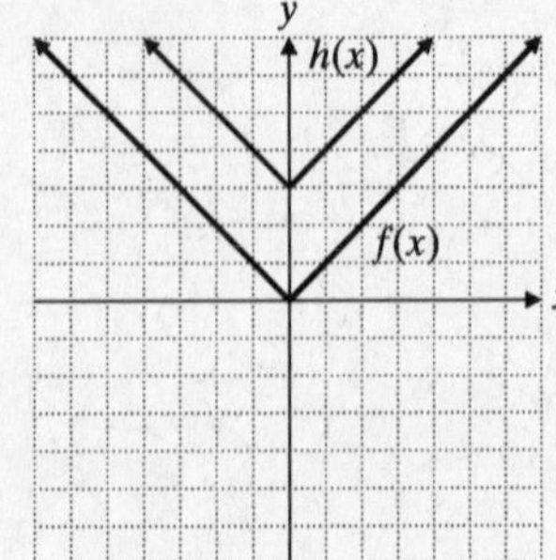

39. $f(x) = |x|$
$h(x) = |x| - 2$ is $f(x)$ shifted down 2 units.

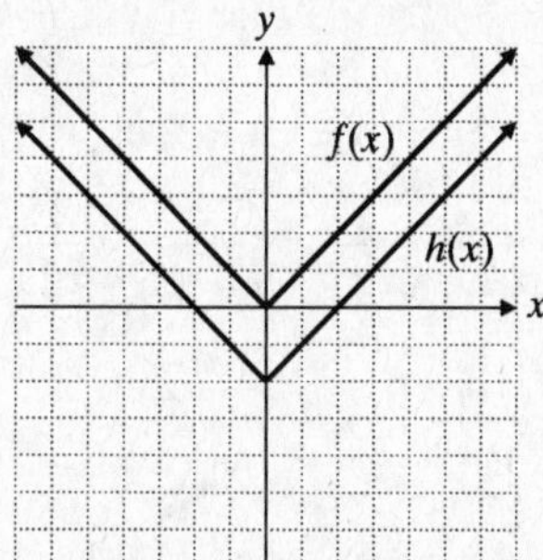

40. $f(x) = x^3$

$r(x) = (x+3)^3 + 1$ is $f(x)$ shifted left 3 units and up 1 unit.

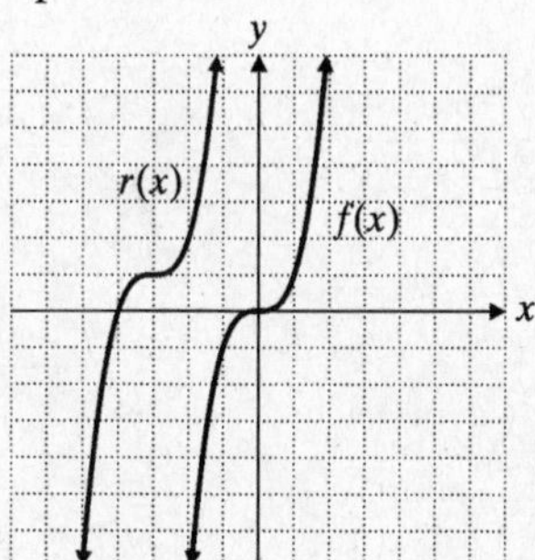

41. $f(x) = x^3$

$r(x) = (x-1)^3 + 5$ is $f(x)$ shifted right 1 unit and up 5 units.

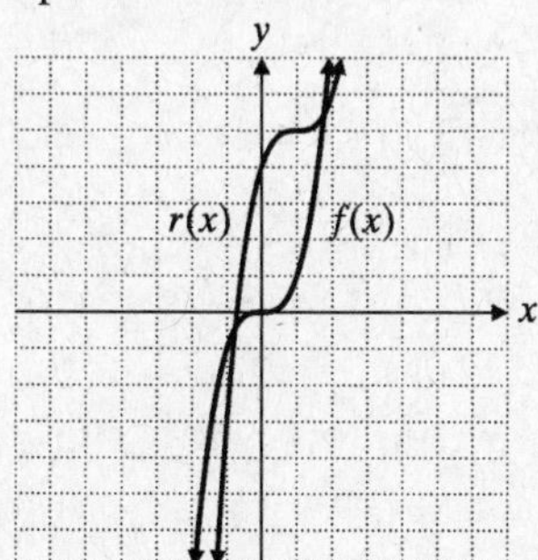

42. $f(x) = \frac{2}{x},\ x \neq 0$

$r(x) = \frac{2}{x+3} - 2,\ x \neq -3$ is $f(x)$ shifted left 3 units and down 2 units.

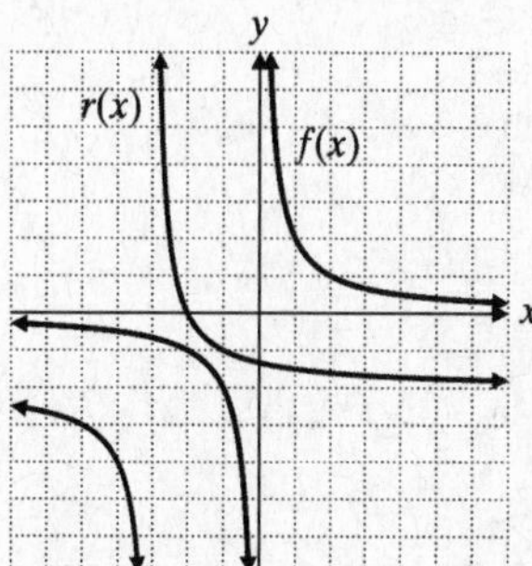

43. $f(x) = \frac{4}{x},\ x \neq 0$

$r(x) = \frac{4}{x+2},\ x \neq -2$ is $f(x)$ shifted left 2 units.

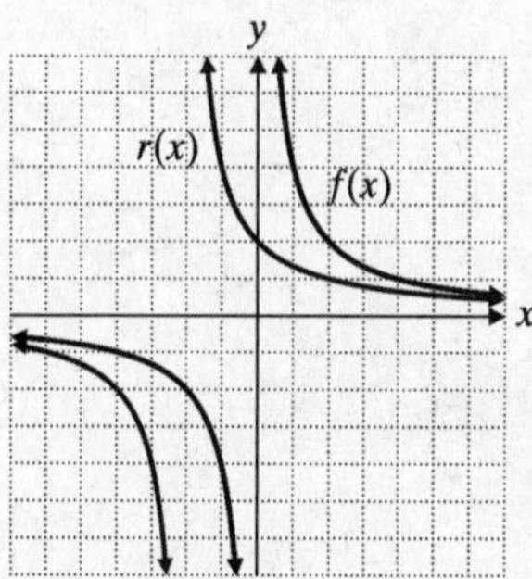

In Exercises 44–63,

$f(x) = 3x + 5;\ g(x) = \frac{2}{x},\ x \neq 0$

$s(x) = \sqrt{x-2},\ x \geq 2;\ h(x) = \frac{x+1}{x-4},\ x \neq 4$

$p(x) = 2x^2 - 3x + 4;\ t(x) = -\frac{1}{2}x - 3$

44.
$$\begin{aligned}(f+t)(x) &= f(x) + t(x)\\ &= 3x + 5 + \left(-\frac{1}{2}x - 3\right)\\ &= \frac{5}{2}x + 2\end{aligned}$$

45.
$$\begin{aligned}(f+p)(x) &= f(x) + p(x)\\ &= 3x + 5 + 2x^2 - 3x + 4\\ &= 2x^2 + 9\end{aligned}$$

46.
$$\begin{aligned}(p-f)(x) &= p(x) - f(x)\\ &= 2x^2 - 3x + 4 - (3x+5)\\ &= 2x^2 - 3x + 4 - 3x - 5\\ &= 2x^2 - 6x - 1\end{aligned}$$

47. $(t-f)(x) = t(x) - f(x)$
$= -\frac{1}{2}x - 3 - (3x+5)$
$= -\frac{7}{2}x - 8$

48. From Exercise 46, $(p-f)(x) = 2x^2 - 6x - 1$.
$(p-f)(2) = 2(2)^2 - 6(2) - 1$
$= 2(4) - 12 - 1$
$= 8 - 12 - 1$
$= -5$

49. From Exercise 47, $(t-f)(x) = -\frac{7}{2}x - 8$.
$(t-f)(-3) = -\frac{7}{2}(-3) - 8 = \frac{21}{2} - \frac{16}{2} = \frac{5}{2}$

50. $(fg)(x) = f(x)g(x) = (3x+5)\left(\frac{2}{x}\right) = \frac{6x+10}{x}$,
$x \neq 0$

51. $(tp)(x) = t(x)p(x)$
$= \left(-\frac{1}{2}x - 3\right)(2x^2 - 3x + 4)$
$= -x^3 + \frac{3x^2}{2} - 2x - 6x^2 + 9x - 12$
$= -x^3 - \frac{9}{2}x^2 + 7x - 12$

52. $\left(\frac{g}{h}\right)(x) = \frac{g(x)}{h(x)} = \frac{\frac{2}{x}}{\frac{x+1}{x-4}} = \frac{2}{x} \cdot \frac{x-4}{x+1} = \frac{2x-8}{x^2+x}$,
$x \neq 0, 4, -1$

53. $\left(\frac{g}{f}\right)(x) = \frac{g(x)}{f(x)}$
$= \frac{\frac{2}{x}}{3x+5}$
$= \frac{2}{x} \cdot \frac{1}{3x+5}$
$= \frac{2}{3x^2+5x}, \; x \neq 0, -\frac{5}{3}$

54. From Exercise 52, $\left(\frac{g}{h}\right)(x) = \frac{2x-8}{x^2+x}$,
$x \neq -1, 0, 4$.
$\left(\frac{g}{h}\right)(-2) = \frac{2(-2)-8}{(-2)^2+(-2)} = -6$

55. From Exercise 53, $\left(\frac{g}{f}\right)(x) = \frac{2}{3x^2+5x}$,
$x \neq 0, -\frac{5}{3}$
$\left(\frac{g}{f}\right)(-3) = \frac{2}{3(-3)^2+5(-3)} = \frac{2}{27-15} = \frac{2}{12} = \frac{1}{6}$

56. $p[f(x)] = p[3x+5]$
$= 2(3x+5)^2 - 3(3x+5) + 4$
$= 18x^2 + 60x + 50 - 9x - 15 + 4$
$= 18x^2 + 51x + 39$

57. $t[s(x)] = t\left(\sqrt{x-2}\right) = -\frac{1}{2}\sqrt{x-2} - 3, \; x \geq 2$

58. $s[p(x)] = s[2x^2 - 3x + 4]$
$= \sqrt{2x^2 - 3x + 4 - 2}$
$= \sqrt{2x^2 - 3x + 2}$

59. $s[t(x)] = s\left[-\frac{1}{2}x - 3\right]$
$= \sqrt{-\frac{1}{2}x - 3 - 2}$
$= \sqrt{-\frac{1}{2}x - 5}, \; x \leq -10$

60. From Exercise 58, $s[p(x)] = \sqrt{2x^2 - 3x + 2}$.
$s[p(2)] = \sqrt{2(2)^2 - 3(2) + 2} = 2$

61. From Exercise 59, $s[t(x)] = \sqrt{-\frac{1}{2}x - 5}, \; x \leq -10$.
$s[t(-18)] = \sqrt{-\frac{1}{2}(-18) - 5} = \sqrt{9-5} = \sqrt{4} = 2$

62. $f[g(x)] = f\left[\frac{2}{x}\right], \; x \neq 0$
$= 3\left(\frac{2}{x}\right) + 5$
$= \frac{6}{x} + 5$
$= \frac{6+5x}{x}$
$g[f(x)] = g[3x+5] = \frac{2}{3x+5}$
$f[g(x)] \neq g[f(x)]$

63. $p[g(x)] = p\left[\frac{2}{x}\right],\ x \neq 0$

$= 2\left(\frac{2}{x}\right)^2 - 3\left(\frac{2}{x}\right) + 4$

$= \frac{8 - 6x + 4x^2}{x^2}$

$g[p(x)] = g[2x^2 - 3x + 4] = \frac{2}{2x^2 - 3x + 4}$

$p[g(x)] \neq g[p(x)]$

64. $B = \{(3, 7), (7, 3), (0, 8), (0, -8)\}$

a. $D = \{0, 3, 7\}$

b. $R = \{-8, 3, 7, 8\}$

c. No, the set does not define a function since two of the ordered pairs have the same first coordinate.

d. No, since the set does not defined a function it cannot define a one-to-one function.

65. $A = \{(100, 10), (200, 20), (300, 30), (400, 10)\}$

a. $D = \{100, 200, 300, 400\}$

b. $R = \{10, 20, 30\}$

c. Yes, the set defines a function since no two of the ordered pairs have the same first coordinate.

d. No, the set does not define a one-to-one function since two of the ordered pairs have the same second coordinate.

66. $D = \left\{\left(\frac{1}{2}, 2\right), \left(\frac{1}{4}, 4\right), \left(-\frac{1}{3}, -3\right), \left(4, \frac{1}{4}\right)\right\}$

a. $D = \left\{\frac{1}{2}, \frac{1}{4}, -\frac{1}{3}, 4\right\}$

b. $R = \left\{2, 4, -3, \frac{1}{4}\right\}$

c. Yes, the set defines a function since no two of the ordered pairs have the same first coordinate.

d. Yes, the set defines a one-to-one function since it is a function and no two ordered pairs have the same second coordinate.

67. $C = \{(12, 6), (0, 6), (0, -1), (-6, -12)\}$

a. $D = \{12, 0, -6\}$

b. $R = \{-1, -12, 6\}$

c. No, the set does not define a function since two of the ordered pairs have the same first coordinate.

d. No, since the set does not define a function it cannot define a one-to-one function.

68. $F = \{(3, 7), (2, 1), (0, -3), (1, 1)\}$

a. $D = \{0, 1, 2, 3\}$

b. $R = \{-3, 1, 7\}$

c. Yes, the set defines a function since no two of the ordered pairs have the same first coordinate.

d. No, the set does not define a one-to-one function since two of the ordered pairs have the same second coordinate.

69. $E = \{(0, 1), (1, 2), (2, 9), (-1, -2)\}$

a. $D = \{-1, 0, 1, 2\}$

b. $R = \{-2, 1, 2, 9\}$

c. Yes, the set defines a function since no two of the ordered pairs have the same first coordinate.

d. Yes, the set defines a one-to-one function since it is a function and no two ordered pairs have the same second coordinate.

70. $A = \left\{\left(3, \frac{1}{3}\right), \left(-2, -\frac{1}{2}\right), \left(-4, -\frac{1}{4}\right), \left(5, \frac{1}{5}\right)\right\}$

$A^{-1} = \left\{\left(\frac{1}{3}, 3\right), \left(-\frac{1}{2}, -2\right), \left(-\frac{1}{4}, -4\right), \left(\frac{1}{5}, 5\right)\right\}$

71. $B = \{(1, 10), (3, 7), (12, 15), (10, 1)\}$

$B^{-1} = \{(10, 1), (7, 3), (15, 12), (1, 10)\}$

72. $f(x) = -\frac{3}{4}x + 2,\ f(x) \to y$

$y = -\frac{3}{4}x + 2,\ x \leftrightarrow y$

$x = -\frac{3}{4}y + 2$

$-4x = 3y - 8$

$-4x + 8 = 3y$

$y = -\frac{4}{3}x + \frac{8}{3},\ y \to f^{-1}(x)$

$f^{-1}(x) = -\frac{4}{3}x + \frac{8}{3}$

73. $g(x) = -8 - 4x,\ g(x) \to y$

$y = -8 - 4x,\ x \leftrightarrow y$

$x = -8 - 4y$

$4y = -x - 8$

$y = -\frac{1}{4}x - 2,\ y \to g^{-1}(x)$

$g^{-1}(x) = -\frac{1}{4}x - 2$

74. $h(x) = \frac{6}{x+5},\ h(x) \to y$

$y = \frac{6}{x+5},\ x \leftrightarrow y$

$x = \frac{6}{y+5}$

$y + 5 = \frac{6}{x}$

$y = \frac{6}{x} - 5$ or $y = \frac{6-5x}{x},\ y \to h^{-1}(x)$

$h^{-1}(x) = \frac{6}{x} - 5$ or $\frac{6-5x}{x}$

75. $j(x) = \frac{-7}{2-x},\ j(x) \to y$

$y = \frac{-7}{2-x},\ x \leftrightarrow y$

$x = \frac{-7}{2-y}$

$2 - y = \frac{-7}{x}$

$-y = -\frac{7}{x} - 2$

$y = \frac{7}{x} + 2$ or $y = \frac{7+2x}{x},\ y = j^{-1}(x)$

$j^{-1}(x) = \frac{7}{x} + 2$ or $\frac{2x+7}{x}$

76. $p(x) = \sqrt[3]{x+1},\ p(x) \to y$

$y = \sqrt[3]{x+1},\ x \leftrightarrow y$

$x = \sqrt[3]{y+1}$

$x^3 = y + 1$

$y = x^3 - 1,\ y \to p^{-1}(x)$

$p^{-1}(x) = x^3 - 1$

77. $r(x) = x^3 + 2,\ r(x) \to y$

$y = x^3 + 2,\ x \leftrightarrow y$

$x = y^3 + 2$

$x - 2 = y^3$

$y = \sqrt[3]{x-2},\ y \to r^{-1}(x)$

$r^{-1}(x) = \sqrt[3]{x-2}$

78. $f(x) = \frac{-x-2}{3},\ f(x) \to y$

$y = \frac{-x-2}{3},\ x \leftrightarrow y$

$x = \frac{-y-2}{3}$

$3x = -y - 2$

$y = -3x - 2,\ y \to f^{-1}(x)$

$f^{-1}(x) = -3x - 2$

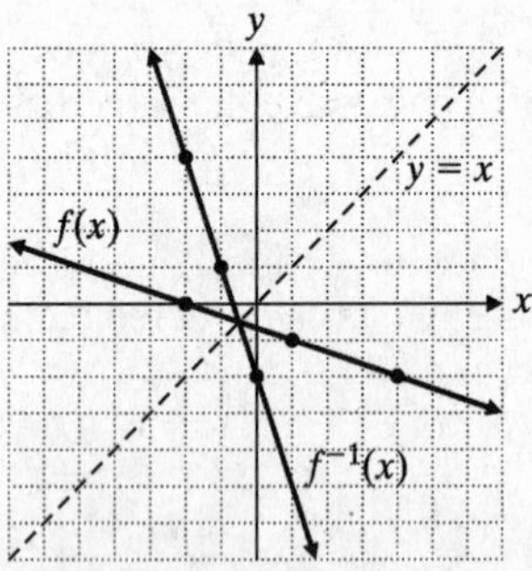

79. $f(x) = -\frac{3}{4}x + 1,\ f(x) \to y$

$y = -\frac{3}{4}x + 1,\ x \leftrightarrow y$

$x = -\frac{3}{4}y + 1$

$4x = -3y + 4$

$3y = -4x + 4$

$y = -\frac{4}{3}x + \frac{4}{3},\ y \to f^{-1}(x)$

$f^{-1}(x) = -\frac{4}{3}x + \frac{4}{3}$

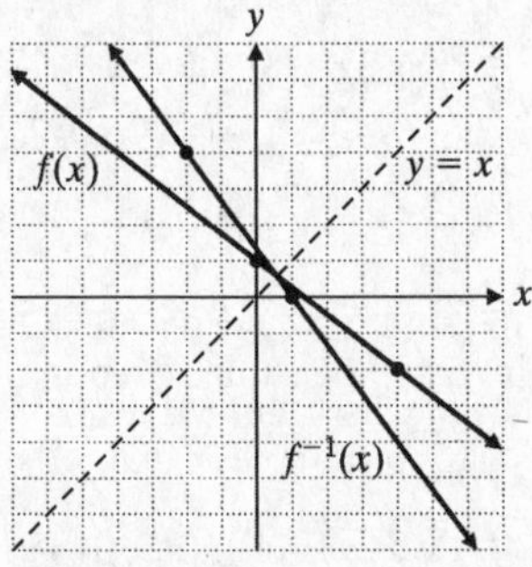

How Am I Doing? Chapter 10 Test

1. $f(x)=\frac{3}{4}x-2$

$f(-8)=\frac{3}{4}(-8)-2=-6-2=-8$

2. $f(x)=\frac{3}{4}x-2$

$f(3a)=\frac{3}{4}(3a)-2=\frac{9}{4}a-2$

3. $f(x)=\frac{3}{4}x-2$

$$\begin{aligned}f(a)-f(2)&=\frac{3}{4}a-2-\left(\frac{3}{4}(2)-2\right)\\&=\frac{3}{4}a-2-\frac{3}{2}+2\\&=\frac{3}{4}a-\frac{3}{2}\end{aligned}$$

4. $f(x)=3x^2-2x+4$

$f(-6)=3(-6)^2-2(-6)+4=108+12+4=124$

5. $f(x)=3x^2-2x+4$

$$\begin{aligned}f(a+1)&=3(a+1)^2-2(a+1)+4\\&=3a^2+6a+3-2a-2+4\\&=3a^2+4a+5\end{aligned}$$

6. $f(x)=3x^2-2x+4$

$$\begin{aligned}f(a)+f(1)&=3a^2-2a+4+3(1)^2-2(1)+4\\&=3a^2-2a+4+3-2+4\\&=3a^2-2a+9\end{aligned}$$

7. $f(x)=3x^2-2x+4$

$$\begin{aligned}f(-2a)-2&=3(-2a)^2-2(-2a)+4-2\\&=3(4a^2)+4a+2\\&=12a^2+4a+2\end{aligned}$$

8. **a.** Graph passes vertical line test and therefore represents a function.

b. Graph fails horizontal line test and does not represent a one-to-one function.

9. **a.** Graph passes vertical line test and therefore represents a function.

b. Graph passes horizontal line test and therefore represents a one-to-one function.

10. $f(x)=x^2$

$g(x)=(x-1)^2+3$ is $f(x)$ shifted right 1 unit and up 3 units.

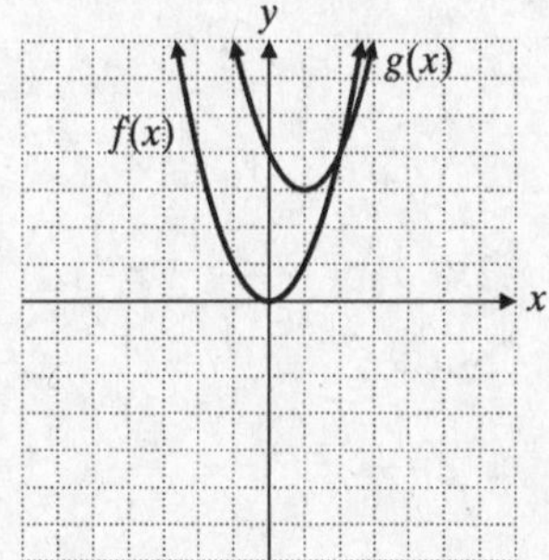

11. $f(x) = |x|$
$g(x) = |x + 1| + 2$ is $f(x)$ shifted left 1 unit and up 2 units.

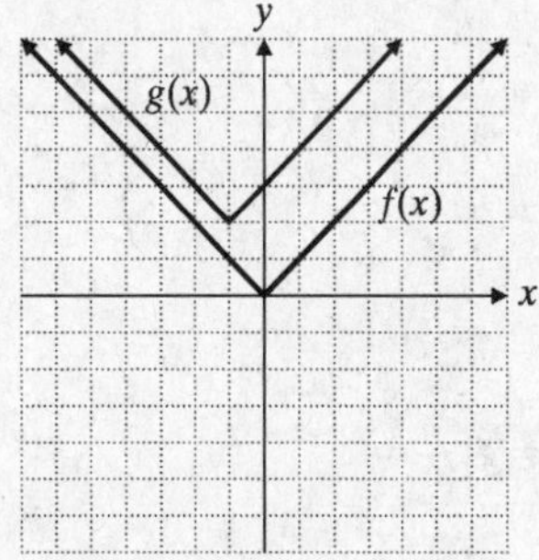

12. $f(x)=3x^2-x-6,\ g(x)=-2x^2+5x+7$

a.
$$\begin{aligned}(f+g)(x)&=f(x)+g(x)\\&=3x^2-x-6+(-2x^2+5x+7)\\&=3x^2-x-6-2x^2+5x+7\\&=x^2+4x+1\end{aligned}$$

b. $(f-g)(x) = f(x) - g(x)$
$$= 3x^2 - x - 6 - (-2x^2 + 5x + 7)$$
$$= 3x^2 - x - 6 + 2x^2 - 5x - 7$$
$$= 5x^2 - 6x - 13$$

c. From b, $(f-g)(x) = 5x^2 - 6x - 13.$
$$(f-g)(-2) = 5(-2)^2 - 6(-2) - 13$$
$$= 5(4) + 12 - 13$$
$$= 19$$

13. $f(x) = \dfrac{3}{x},\ x \neq 0;\ g(x) = 2x - 1$

a. $(fg)(x) = f(x)g(x) = \dfrac{3}{x}(2x-1) = \dfrac{6x-3}{x},$

$x \neq 0$

b. $\left(\dfrac{f}{g}\right)(x) = \dfrac{f(x)}{g(x)} = \dfrac{\frac{3}{x}}{2x-1} = \dfrac{3}{2x^2 - x},$

$x \neq 0, \dfrac{1}{2}$

c. $f[g(x)] = f[2x-1] = \dfrac{3}{2x-1},\ x \neq \dfrac{1}{2}$

14. $f(x) = \dfrac{1}{2}x - 3,\ g(x) = 4x + 5$

a. $(f \circ g)(x) = f[g(x)]$
$$= f[4x+5]$$
$$= \frac{1}{2}(4x+5) - 3$$
$$= 2x + \frac{5}{2} - \frac{6}{2}$$
$$= 2x - \frac{1}{2}$$

b. $(g \circ f)(x) = g[f(x)]$
$$= g\left[\frac{1}{2}x - 3\right]$$
$$= 4\left(\frac{1}{2}x - 3\right) + 5$$
$$= 2x - 12 + 5$$
$$= 2x - 7$$

c. $(f \circ g)\left(\dfrac{1}{4}\right) = 2 \cdot \dfrac{1}{4} - \dfrac{1}{2} = 0$

15. $B = \{(1, 8), (8, 1), (9, 10), (-10, 9)\}$

a. Yes, the function is one-to-one since no two ordered pairs have the same second coordinate.

b. $B^{-1} = \{(8, 1), (1, 8), (10, 9), (9, -10)\}$

16. $A = \{(1, 5), (2, 1), (4, -7), (0, 7), (-1, 5)\}$

a. The function is not one-to-one since two ordered pairs have the same second coordinate.

b. A has no inverse.

17. $f(x) = \sqrt[3]{2x-1},\ f(x) \to y$
$$y = \sqrt[3]{2x-1},\ x \leftrightarrow y$$
$$x = \sqrt[3]{2y-1}$$
$$x^3 = 2y - 1$$
$$2y = x^3 + 1$$
$$y = \frac{x^3+1}{2},\ y \to f^{-1}(x)$$
$$f^{-1}(x) = \frac{x^3+1}{2}$$

18. $f(x) = -3x + 2, f(x) \to y$

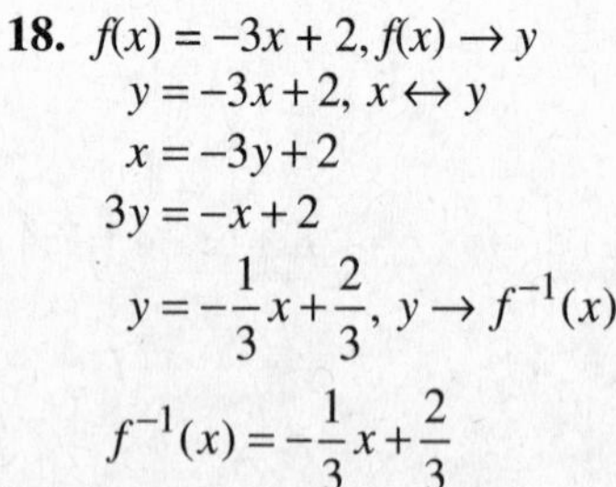

$$y = -3x + 2,\ x \leftrightarrow y$$
$$x = -3y + 2$$
$$3y = -x + 2$$
$$y = -\frac{1}{3}x + \frac{2}{3},\ y \to f^{-1}(x)$$
$$f^{-1}(x) = -\frac{1}{3}x + \frac{2}{3}$$

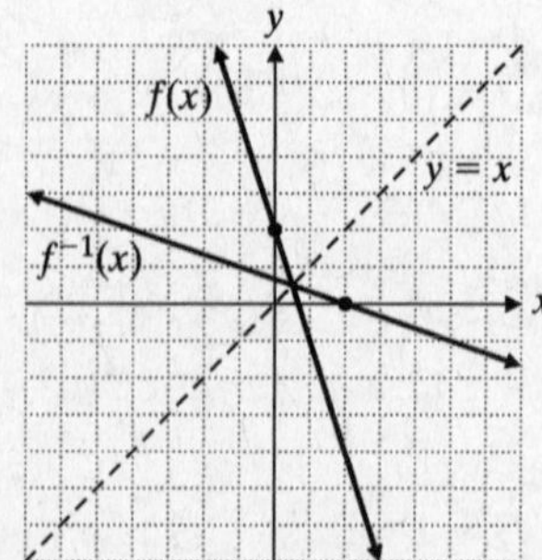

19. $f(x)=\frac{3}{7}x+\frac{1}{2},\ f^{-1}(x)=\frac{14x-7}{6}$

$$\begin{aligned}f^{-1}[f(x)]&=f^{-1}\left[\frac{3}{7}x+\frac{1}{2}\right]\\&=\frac{14\left[\frac{3}{7}x+\frac{1}{2}\right]-7}{6}\\&=\frac{6x+7-7}{6}\\&=\frac{6x}{6}\\&=x\end{aligned}$$

20. $f(x)=7-8x$

$$\begin{aligned}\frac{f(x+h)-f(x)}{h}&=\frac{7-8(x+h)-(7-8x)}{h}\\&=\frac{7-8x-8h-7+8x}{h}\\&=\frac{-8h}{h}\\&=-8\end{aligned}$$

Cumulative Test for Chapters 1–10

1. $$\begin{aligned}3x\{2y-3[x+2(x+2y)]\}&=3x\{2y-3[x+2x+4y]\}\\&=3x\{2y-3[3x+4y]\}\\&=3x\{2y-9x-12y\}\\&=3x\{-10y-9x\}\\&=-30xy-27x^2\end{aligned}$$

2. $$\begin{aligned}3y-2xy-x^2&=3(3)-2(-1)(3)-(-1)^2\\&=9+6-1\\&=14\end{aligned}$$

3. $$\begin{aligned}\frac{1}{2}(x-2)&=\frac{1}{3}(x+10)-2x\\3(x-2)&=2(x+10)-12x\\3x-6&=2x+20-12x\\13x&=26\\x&=2\end{aligned}$$

4. $$\begin{aligned}16x^4-1&=(4x^2)^2-1^2\\&=(4x^2+1)(4x^2-1)\\&=(4x^2+1)[(2x)^2-1^2]\\&=(4x^2+1)(2x+1)(2x-1)\end{aligned}$$

5. $$\begin{aligned}(x-1)(2x^2+x-5)&=2x^3+x^2-5x-2x^2-x+5\\&=2x^3-x^2-6x+5\end{aligned}$$

6.
$$\frac{3x}{x^2-4}=\frac{2}{x+2}+\frac{4}{2-x}$$
$$\frac{3x}{(x+2)(x-2)}=\frac{2}{x+2}-\frac{4}{x-2}$$
$$(x+2)(x-2)\frac{3x}{(x+2)(x-2)}=(x+2)(x-2)\frac{2}{x+2}-(x+2)(x-2)\frac{4}{x-2}$$
$$3x=2(x-2)-4(x+2)$$
$$3x=2x-4-4x-8$$
$$5x=-12$$
$$x=-\frac{12}{5}$$

7.
$$y-y_1=m(x-x_1)$$
$$y-(-1)=-3(x-2)$$
$$y+1=-3x+6$$
$$y=-3x+5$$

8. $3x+2y=5 \Rightarrow y=\frac{5-3x}{2}$
$$7x+5y=11$$
$$7x+5\cdot\frac{5-3x}{2}=11$$
$$14x+25-15x=22$$
$$-x=-3$$
$$x=3$$
$$y=\frac{5-3x}{2}=\frac{5-3(3)}{2}=-2$$
(3, –2) is the solution.

9. $\sqrt{18x^5y^6z^3}=\sqrt{9x^4y^6z^2\cdot 2xz}=3x^2y^3z\sqrt{2xz}$

10.
$$\left(\sqrt{2}+\sqrt{3}\right)\left(2\sqrt{2}-4\sqrt{3}\right)=2(2)-4\sqrt{6}+2\sqrt{6}-4(3)$$
$$=4-2\sqrt{6}-12$$
$$=-2\sqrt{6}-8$$

11. (6, –1) and (–3, –4)
$$d=\sqrt{(x_2-x_1)^2+(y_2-y_1)^2}$$
$$d=\sqrt{(-3-6)^2+(-4-(-1))^2}=\sqrt{81+9}=\sqrt{90}$$
$$d=3\sqrt{10}$$

12. $12x^2-11x+2=(4x-1)(3x-2)$

13.
$$x^3-5x^2-14x=x(x^2-5x-14)$$
$$=x(x-7)(x+2)$$

14. $(x-h)^2+(y-k)^2=r^2$

$(x-0)^2+(y-(-5))^2=\left(2\sqrt{2}\right)^2$

$x^2+(y+5)^2=8$

15. $f(x)=3x^2-2x+1$

a. $f(-2)=3(-2)^2-2(-2)+1=12+4+1=17$

b. $f(a-2)=3(a-2)^2-2(a-2)+1$
$=3a^2-12a+12-2a+4+1$
$=3a^2-14a+17$

c. From part a we know $f(-2)=17$.
$f(a)+f(-2)=3a^2-2a+1+17$
$=3a^2-2a+18$

16. $f(x)=x^3$, $g(x)=(x+2)^3+4$ is $f(x)$ shifted left 2 units and up 4 units.

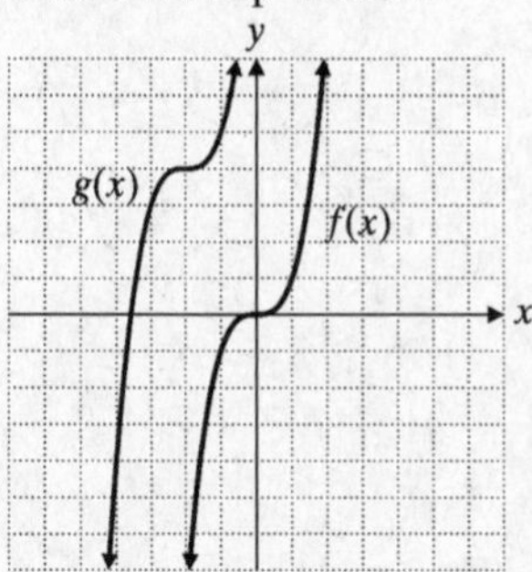

17. $f(x)=2x^2-5x-6,\ g(x)=5x+3$

a. $(fg)(x)$
$=f(x)g(x)$
$=(2x^2-5x-6)(5x+3)$
$=10x^3+6x^2-25x^2-15x-30x-18$
$=10x^3-19x^2-45x-18$

b. $\left(\dfrac{f}{g}\right)(x)=\dfrac{f(x)}{g(x)}=\dfrac{2x^2-5x-6}{5x+3},\ x\neq-\dfrac{3}{5}$

c. $f[g(x)]=f[5x+3]$
$=2(5x+3)^2-5(5x+3)-6$
$=50x^2+60x+18-25x-15-6$
$=50x^2+35x-3$

18. $A=\{(3, 6), (1, 8), (2, 7), (4, 4)\}$

a. Yes, A is a function; no two ordered pairs have the same first coordinate.

b. Yes, A is a one-to-one function since no two ordered pairs have the same second coordinate.

c. $A^{-1}=\{(6, 3), (8, 1), (7, 2), (4, 4)\}$

19. $f(x)=\sqrt[3]{7x-3},\ f(x)\to y$

$y=\sqrt[3]{7x-3},\ x\leftrightarrow y$

$x=\sqrt[3]{7y-3}$

$x^3=7y-3$

$7y=x^3+3$

$y=\dfrac{x^3+3}{7},\ y\to f^{-1}(x)$

$f^{-1}(x)=\dfrac{x^3+3}{7}$

20. $f(x)=5x^3-3x^2-6$

a. $f(5)=5(5)^3-3(5)^2-6=625-75-6=544$

b. $f(-3)=5(-3)^3-3(-3)^2-6$
$=-135-27-6$
$=-168$

c. $f(2a)=5(2a)^3-3(2a)^2-6$
$=5(8a^3)-3(4a^2)-6$
$=40a^3-12a^2-6$

21. a. $f(x)=-\dfrac{2}{3}x+2,\ f(x)\to y$

$y=-\dfrac{2}{3}x+2,\ x\leftrightarrow y$

$x=-\dfrac{2}{3}y+2$

$\dfrac{2}{3}y=-x+2$

$y=-\dfrac{3}{2}x+3,\ y\to f^{-1}(x)$

$f^{-1}(x)=-\dfrac{3}{2}x+3$

b.

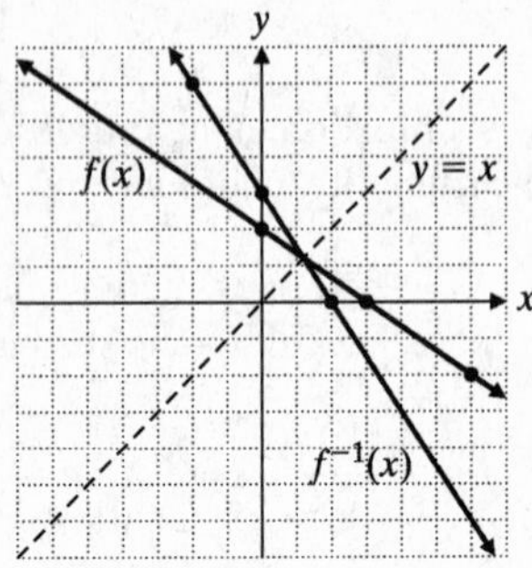

22. From Exercise 21,

$$\begin{aligned} f[f^{-1}(x)] &= f\left[-\frac{3}{2}x+3\right] \\ &= -\frac{2}{3}\left[-\frac{3}{2}x+3\right]+2 \\ &= x-2+2 \\ &= x \end{aligned}$$

Chapter 11

11.1 Exercises

1. An exponential function is a function of the form $f(x) = b^x$, where $b > 0$, $b \neq 1$, and x is a real number.

3. $f(x) = 3^x$

x	$y = f(x) = 3^x$
−2	$\frac{1}{9}$
−1	$\frac{1}{3}$
0	1
1	3
2	9

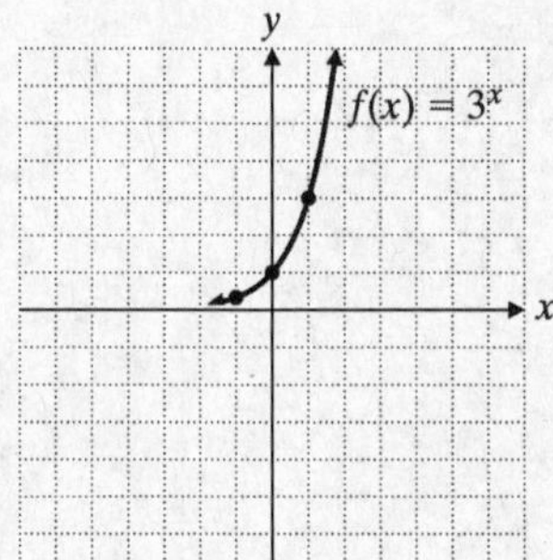

5. $f(x) = 2^{-x}$

x	$y = f(x) = 2^{-x}$
−2	4
−1	2
0	1
1	$\frac{1}{2}$
2	$\frac{1}{4}$

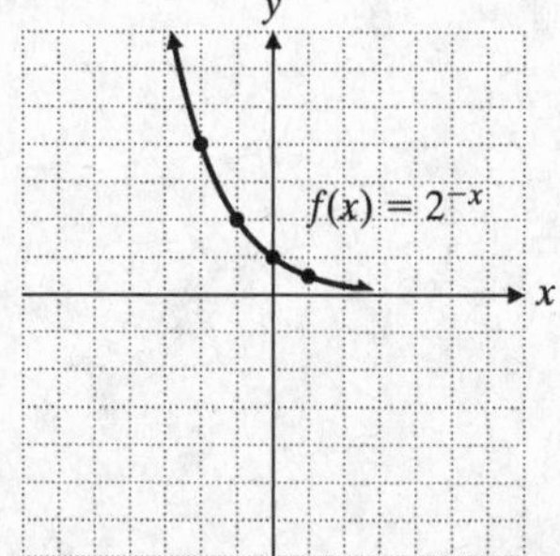

7. $f(x) = 3^{-x}$

x	$y = f(x) = 3^{-x}$
−2	9
−1	3
0	1
1	$\frac{1}{3}$
2	$\frac{1}{9}$

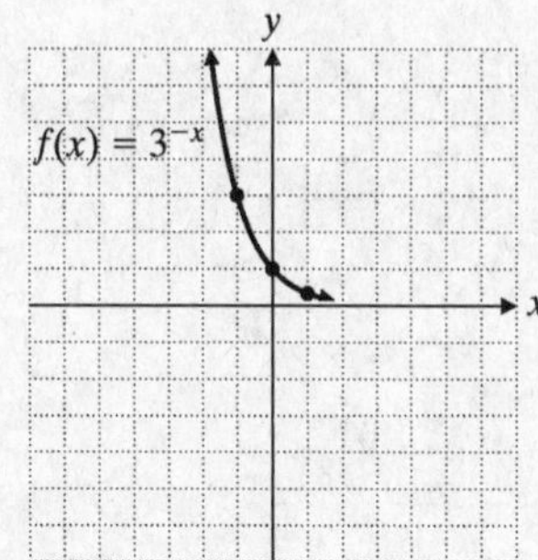

9. $f(x) = 2^{x+3}$

x	$y = f(x) = 2^{x+3}$
−3	1
−2	2
−1	4

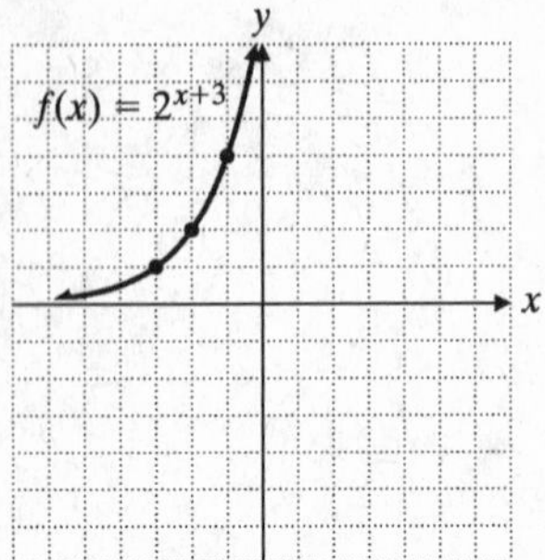

11. $f(x) = 3^{x-3}$

x	$y = f(x) = 3^{x-3}$
2	$\frac{1}{3}$
3	1
4	3

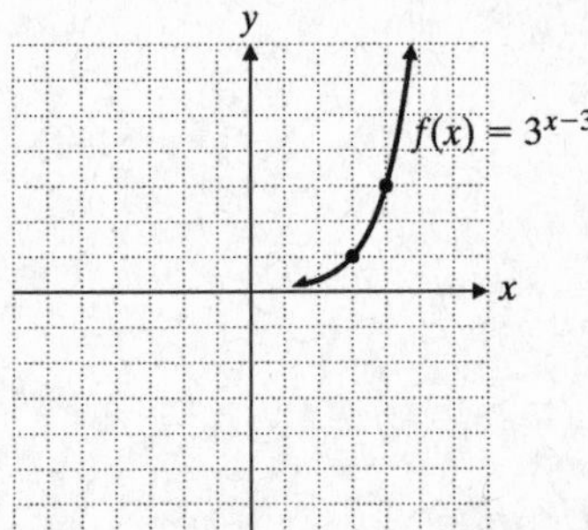

13. $f(x) = 2^x + 2$

x	$y = f(x) = 2^x + 2$
−2	$\frac{9}{4}$
−1	$\frac{5}{2}$
0	3
1	4
2	6

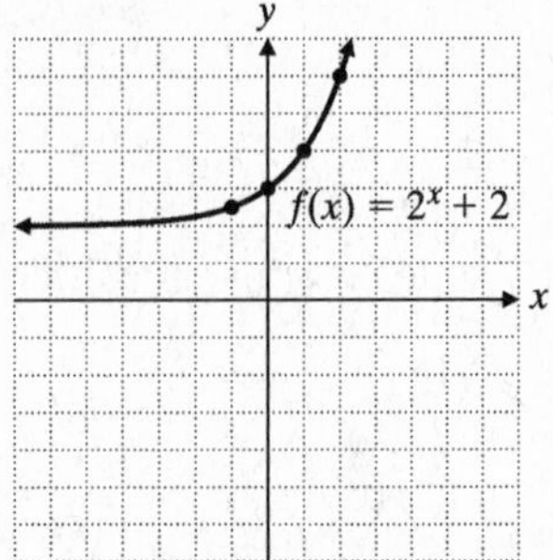

15. $f(x) = e^{x-1}$

x	$y = e^{x-1}$
−2	0.05
−1	0.14
0	0.37
1	1
2	2.7

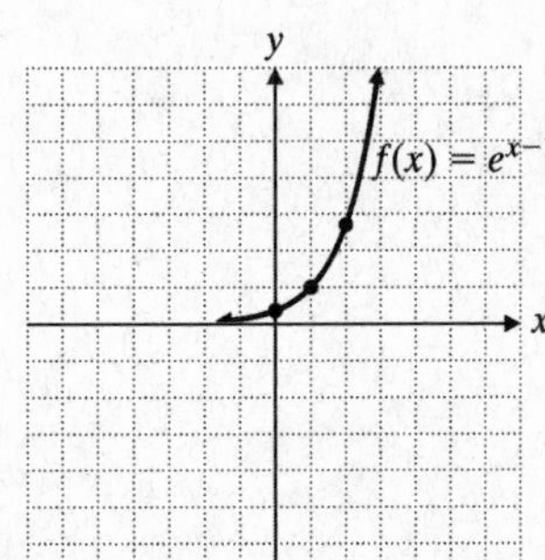

17. $f(x) = 2e^x$

x	$y = f(x) = 2e^x$
−2	0.27
−1	0.74
0	2
1	5.44
2	14.8

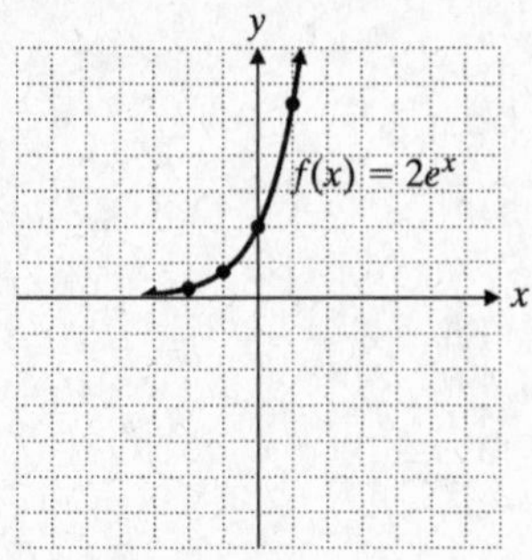

19. $f(x) = e^{1-x}$

x	$y = f(x) = e^{1-x}$
−2	20.1
−1	7.39
0	2.72
1	0.37
2	0.14

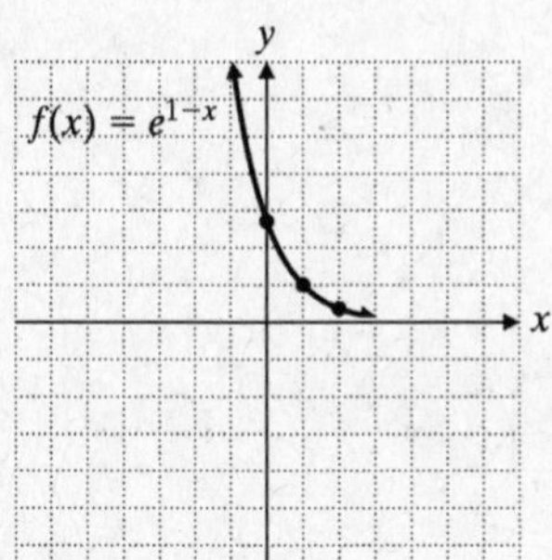

21. $2^x = 4$
$2^x = 2^2$
$x = 2$

23. $2^x = 1$
$2^x = 2^0$
$x = 0$

25. $2^x = \frac{1}{2}$
$2^x = \frac{1}{2^1} = 2^{-1}$
$x = -1$

27. $3^x = 81$
$3^x = 3^4$
$x = 4$

29. $3^x = 1$
$3^x = 3^0$
$x = 0$

31. $3^{-x} = \frac{1}{9} = \frac{1}{3^2}$
$3^{-x} = 3^{-2}$
$-x = -2$
$x = 2$

33. $4^x = 256$
$4^x = 4^4$
$x = 4$

35. $5^{x+1} = 125$
$5^{x+1} = 5^3$
$x + 1 = 3$
$x = 2$

37. $8^{3x-1} = 64 = 8^2$
$3x - 1 = 2$
$3x = 3$
$x = 1$

39. $A = P\left(1 + \frac{r}{n}\right)^{nt}$

$A = 2000\left(1 + \frac{0.063}{1}\right)^{1(3)} = 2402.314094$

Alicia will have \$2402.31 after 3 years.

41. $A = P\left(1 + \frac{r}{n}\right)^{nt}$

$A = 3000\left(1 + \frac{0.032}{4}\right)^{4(6)} = \$3632.34$

$A = 3000\left(1 + \frac{0.032}{12}\right)^{12(6)} = \$3634.08$

She will have \$3632.34 if it is compounded quarterly and \$3634.08 if it is compounded monthly.

43. $B(t) = 4000(2^t)$
$B(3) = 4000(2^3) = 32{,}000$
$B(9) = 4000(2^9) = 2{,}048{,}000$

At the end of 3 hours there will be 32,000 bacteria in the culture and at the end of 9 hours there will be 2,048,000 bacteria in the culture.

45. $S(f) = (1-0.18)^{f/4} = 0.82^{f/4}$

$S(20) = 0.82^{20/4} = 37\%$ at 20 ft

$S(48) = 0.82^{48/4} = 9\% < 10\%$

Yes, spotlights will be needed.

47. $A = Ce^{-0.0004297t}$

$A = 6e^{-0.0004297(1000)} \approx 3.91$

There will be approximately 3.91 mg of radium in the container after 1000 years.

49. $P = 14.7e^{-0.21d}$

$P = 14.7e^{-0.21(10)} \approx 1.80$

The pressure is approximately 1.80 lb/in.^{2} on a jet flying 10 miles above sea level.

51. $N = 0.00472e^{0.11596t}$

t = number of years since 1900

$N_{(90)} = 0.00472e^{0.11596(90)} = 160.8$ million

$N_{(105)} = 0.00472e^{0.11596(105)} = 915.9$ million

$\frac{N_{(105)} - N_{(85)}}{N_{(85)}} \times 100\% = 917\%$

The number of shares traded in 1990 was about 160.8 million and was about 915.9 million in 2005. This was an increase of about 470%.

53. From the graph the world's population reached three billion people sometime in 1955. (Answers may vary.)

55. $6.07(1.017)^{10} \approx 7.2$

The world's population would be about 7.2 billion people in 2010.

Cumulative Review

57. $5-2(3-x) = 2(2x+5)+1$

$5-6+2x = 4x+10+1$

$2x = -12$

$x = -6$

58. $\frac{7}{12} + \frac{3}{4}x + \frac{5}{4} = -\frac{1}{6}x$

$\frac{3}{4}x + \frac{1}{6}x = -\frac{7}{12} - \frac{5}{4}$

$\frac{9}{12}x + \frac{2}{12}x = -\frac{7}{12} - \frac{15}{12}$

$\frac{11}{12}x = -\frac{11}{6}$

$6x = -12$

$x = -2$

Quick Quiz 11.1

1. $f(x) = 2^{x+4}$

x	$f(x)$
−1	8
−2	4
−3	2
−4	1

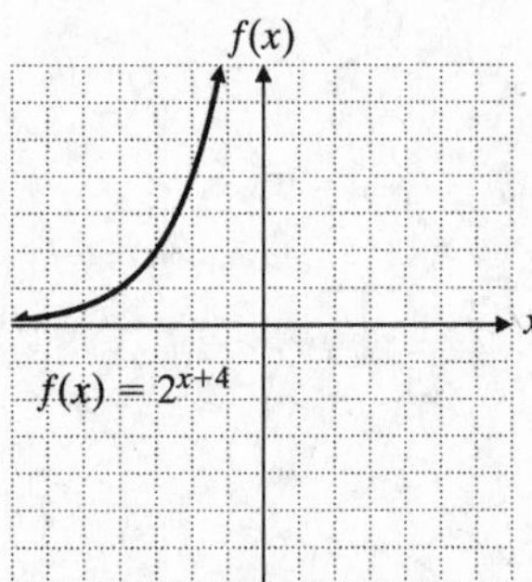

2. $f(x) = 3^x - 4$

$f(0) = 3^0 - 4 = 1 - 4 = -3$

$f(-1) = 3^{-1} - 4 = \frac{1}{3} - \frac{12}{3} = -\frac{11}{3}$

3. $3^{x+2} = 81$

$3^{x+2} = 3^4$

$x + 2 = 4$

$x = 2$

4. Answers may vary. Possible solution:
To solve the following equation for x:

$4^{-x} = \frac{1}{64}$, remember that $64 = 4^{-3}$, and replace the right side of the equation with 4^{-3}.

$4^{-x} = 4^{-3}$
This yields the same base on both sides of the equation, and allows the use of the property of exponential functions to simplify, which states that the exponents may be isolated and compared.
$-x = -3$ or $x = 3$

11.2 Exercises

1. A logarithm is an exponent.

3. In the equation $y = \log_b x$, the domain (the permitted values of x) is $\underline{x > 0}$.

5. $49 = 7^2 \Leftrightarrow \log_7 49 = 2$

7. $36 = 6^2 \Leftrightarrow \log_6 36 = 2$

9. $0.001 = 10^{-3} \Leftrightarrow \log_{10} 0.001 = -3$

11. $\frac{1}{32} = 2^{-5} \Leftrightarrow \log_2 \frac{1}{32} = -5$

13. $y = e^5 \Leftrightarrow \log_e y = 5$

15. $2 = \log_3 9 \Leftrightarrow 3^2 = 9$

17. $0 = \log_{17} 1 \Leftrightarrow 17^0 = 1$

19. $\frac{1}{2} = \log_{16} 4 \Leftrightarrow 16^{1/2} = 4$

21. $-2 = \log_{10}(0.01) \Leftrightarrow 10^{-2} = 0.01$

23. $-4 = \log_3\left(\frac{1}{81}\right) \Leftrightarrow 3^{-4} = \frac{1}{81}$

25. $-\frac{3}{2} = \log_e x \Leftrightarrow e^{-3/2} = x$

27. $\log_2 x = 4$
$2^4 = x$
$x = 16$

29. $\log_{10} x = -3$
$10^{-3} = x$
$x = \frac{1}{1000}$

31. $\log_4 64 = y$
$4^y = 64 = 4^3$
$y = 3$

33. $\log_8\left(\frac{1}{64}\right) = y$
$8^y = \frac{1}{64} = \frac{1}{8^2} = 8^{-2}$
$y = -2$

35. $\log_a 121 = 2$
$a^2 = 121 = 11^2$
$a = 11$

37. $\log_a 1000 = 3$
$a^3 = 1000 = 10^3$
$a = 10$

39. $\log_{25} 5 = w$
$25^w = 5$
$(5^2)^w = 5^{2w} = 5^1$
$2w = 1$
$w = \frac{1}{2}$

41. $\log_3\left(\frac{1}{3}\right) = w$
$3^w = \frac{1}{3} = \frac{1}{3^1} = 3^{-1}$
$w = -1$

43. $\log_{15} w = 0$
$15^0 = w$
$w = 1$

45. $\log_w 3 = \frac{1}{2}$
$w^{1/2} = 3$
$w = 9$

47. $\log_{10}(0.001) = x$
$10^x = 0.001 = 10^{-3}$
$x = -3$

49. $\log_2 128 = x$
$2^x = 128 = 2^7$
$x = 7$

51. $\log_{23} 1 = x$

$23^x = 1 = 23^0$

$x = 0$

53. $\log_6 \sqrt{6} = x$

$6^x = \sqrt{6} = 6^{1/2}$

$x = \frac{1}{2}$

55. $\log_{57} 1 = x$

$57^x = 1 = 57^0$

$x = 0$

57. $\log_3 x = y \Leftrightarrow 3^y = x$

$x = 3^y$	y
1	0
3	1
9	2

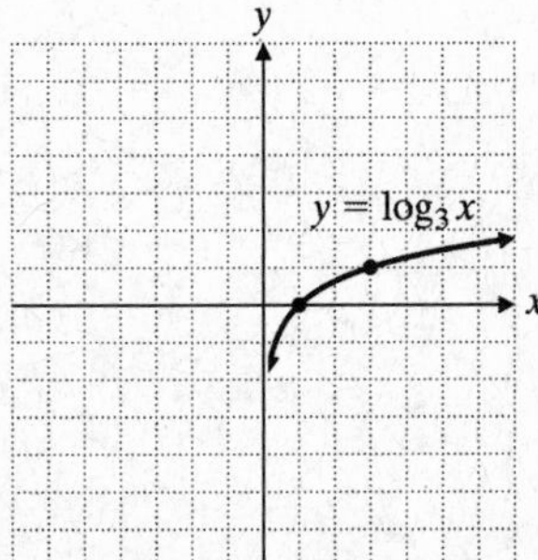

59. $\log_{1/4} x = y \Leftrightarrow \left(\frac{1}{4}\right)^y = x$

$y = \left(\frac{1}{4}\right)^y$	y
1	0
4	−1
16	−2

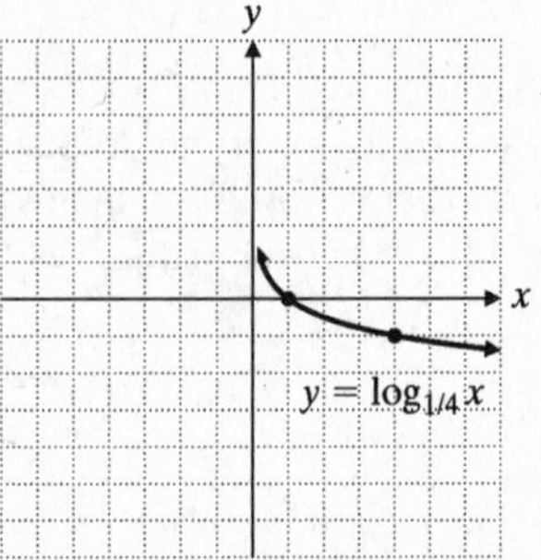

61. $\log_{10} x = y \Leftrightarrow 10^y = x$

$x = 10^y$	y
$\frac{1}{100}$	−2
$\frac{1}{10}$	−1
1	0
10	1
100	2

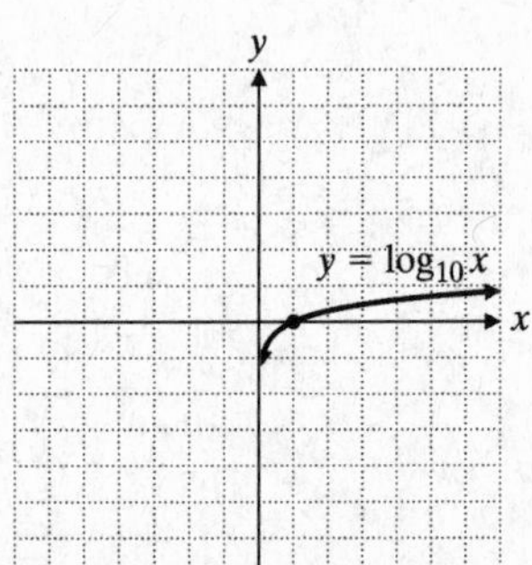

63. $f(x) = \log_3 x,\ f(x) \to y$

$y = \log_3 x \Leftrightarrow 3^y = x$

$x = 3^y$	y
$\frac{1}{9}$	−2
$\frac{1}{3}$	−1
1	0
3	1
9	2

$f^{-1}(x) = 3^x,\ f^{-1}(x) \to y$

$y = 3^x$

x	$y=3^x$
-2	$\frac{1}{9}$
-1	$\frac{1}{3}$
0	1
1	3
2	9

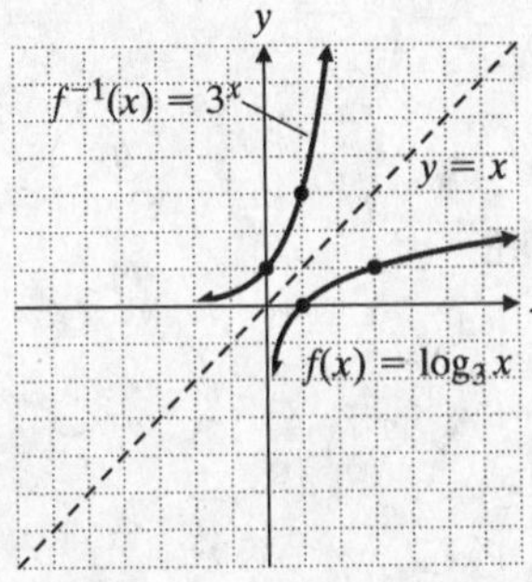

65. $\text{pH}=-\log_{10}[\text{H}^+]$
$\text{pH}=-\log_{10}[10^{-2}]$
$-\text{pH}=\log_{10}[10^{-2}]$
$10^{-\text{pH}}=10^{-2}$
$-\text{pH}=-2$
$\text{pH}=2$

67. $\text{pH}=-\log_{10}[\text{H}^+]$
$8=-\log_{10}[\text{H}^+]$
$-8=\log_{10}[\text{H}^+]$
$10^{-8}=\text{H}^+$

69. $\text{pH}=-\log_{10}[\text{H}^+]$
$\text{pH}=-\log_{10}[1.103\times10^{-3}]$
$\text{pH}=-\log_{10}[0.001103]$
$\text{pH}=-(-2.9574244588)$
$\text{pH}=2.957$

71. $N=1200+(2500)(\log_{10} d)$
$\log_{10} d=\frac{N-1200}{2500}$
$\log_{10} 10,000=\frac{N-1200}{2500}$
$10^{\frac{N-1200}{2500}}=10,000=10^4$
$\frac{N-1200}{2500}=4$
$N-1200=10,000$
$N=11,200$
11,200 sets of software sold.

73. $N=1200+(2500)(\log_{10} d)$
$\log_{10} d=\frac{18,700-1200}{2500}$
$\log_{10} d=7$
$d=10^7$
$d=10,000,000$
$10,000,000 should be spent on advertising.

Cumulative Review

75. $y=-\frac{2}{3}x+4$ has $m=-\frac{2}{3}\Rightarrow m_\perp=\frac{3}{2}$
$y-1=\frac{3}{2}(x-(-4))=\frac{3}{2}(x+4)$
$y-1=\frac{3}{2}x+6$
$y=\frac{3}{2}x+7$

76. $m=\frac{y_2-y_1}{x_2-x_1}=\frac{3-2}{-6-(-1)}=-\frac{1}{5}$

77. a. $A(t)=9000(2)^t$
$A(2)=9000(2)^2=36,000$
36,000 cells will grow in the first 2 hours.

b. $A(t)=9000(2)^t$
$A(12)=9000(2)^{12}=36,864,00$
36,864,000 cells will grow in the first 12 hours.

78. a. $C(t)=P(1.04)^t$
$C(5)=4400(1.04)^5=5353.27$
In 5 years, they will charge $5353.27.

b. $C(t) = P(1.04)^t$
$C(10) = 16,500(1.04)^{10} \doteq 24,424.03$
In 10 years, they will charge $24,424.03.

Quick Quiz 11.2

1. $\log_3 81 = x$
$81 = 3^x$
$3^4 = 3^x$
$4 = x$

2. $\log_2 x = 6$
$x = 2^6$
$x = 64$

3. $\log_{27} 3 = w$
$3 = 27^w$
$3^1 = (3^3)^w$
$3^1 = 3^{3w}$
$1 = 3w$
$w = \frac{1}{3}$

4. Answers may vary. Possible solution:
To solve the following equation: $-\frac{1}{2} = \log_e x$
Write an equivalent exponential equation:
$x = e^{-1/2}$

11.3 Exercises

1. $\log_3 AB = \log_3 A + \log_3 B$

3. $\log_5(7 \cdot 11) = \log_5 7 + \log_5 11$

5. $\log_b 9f = \log_b 9 + \log_b f$

7. $\log_9\left(\frac{2}{7}\right) = \log_9 2 - \log_9 7$

9. $\log_b\left(\frac{H}{10}\right) = \log_b H - \log_b 10$

11. $\log_a\left(\frac{E}{F}\right) = \log_a E - \log_a F$

13. $\log_8 a^7 = 7\log_8 a$

15. $\log_b A^{-2} = -2\log_b A$

17. $\log_5 \sqrt{w} = \log_5 w^{1/2} = \frac{1}{2}\log_5 w$

19. $\log_8 x^2 y = \log_8 x^2 + \log_8 y = 2\log_8 x + \log_8 y$

21. $\log_{11}\left(\frac{6M}{N}\right) = \log_{11}(6M) - \log_{11} N$
$= \log_{11} 6 + \log_{11} M - \log_{11} N$

23. $\log_2\left(\frac{5xy^4}{\sqrt{z}}\right)$
$= \log_2(5xy^4) - \log_2 \sqrt{z}$
$= \log_2 5 + \log_2 x + \log_2 y^4 - \log_2 z^{1/2}$
$= \log_2 5 + \log_2 x + 4\log_2 y - \frac{1}{2}\log_2 z$

25. $\log_a \sqrt[3]{\frac{x^4}{y}} = \log_a\left(\frac{x^4}{y}\right)^{1/3}$
$= \frac{1}{3}\log_a\left(\frac{x^4}{y}\right)$
$= \frac{1}{3}[\log_a x^4 - \log_b y]$
$= \frac{1}{3}[4\log_a x - \log_b y]$
$= \frac{4}{3}\log_a x - \frac{1}{3}\log_a y$

27. $\log_4 13 + \log_4 y + \log_4 3 = \log_4(13 \cdot y \cdot 3)$
$= \log_4(39y)$

29. $5\log_3 x - \log_3 7 = \log_3 x^5 - \log_3 7 = \log_3 \frac{x^5}{7}$

31. $2\log_b 7 + 3\log_b y - \frac{1}{2}\log_b z$
$= \log_b 7^2 + \log_b y^3 - \log_b z^{1/2}$
$= \log_b 49 + \log_b y^3 - \log_b \sqrt{z}$
$= \log_b \frac{49y^3}{\sqrt{z}}$

33. $\log_3 3 = 1$

35. $\log_e e = 1$

37. $\log_9 1 = 0$

39. $3\log_7 7 + 4\log_7 1 = 3(1) + 4(0) = 3$

41. $\log_8 x = \log_8 7$
$x = 7$

43. $\log_5(2x+7) = \log_5(29)$
$2x + 7 = 29$
$2x = 22$
$x = 11$

45. $\log_3 1 = x$
$x = 0$

47. $\log_7 7 = x$
$x = 1$

49. $\log_{10} x + \log_{10} 25 = 2$
$\log_{10}(25x) = 2$
$10^2 = 25x$
$25x = 100$
$x = 4$

51. $\log_2 7 = \log_2 x - \log_2 3$
$\log_2 7 = \log_2 \frac{x}{3}$
$\frac{x}{3} = 7$
$x = 21$

53. $3\log_5 x = \log_5 8$
$\log_5 x^3 = \log_5 8$
$x^3 = 8$
$x^3 = 2^3$
$x = 2$

55. $\log_e x = \log_e 5 + 1$
$\log_e \frac{x}{5} = 1$
$e^1 = \frac{x}{5}$
$x = 5e$

57. $\log_6(5x+21) - \log_6(x+3) = 1$
$\log_6 \frac{5x+21}{x+3} = 1$
$\frac{5x+21}{x+3} = 6$
$5x + 21 = 6x + 18$
$x = 3$

59. $5^{\log_5 4} + 3^{\log_3 2} = 4 + 2 = 6$

Cumulative Review

61. $V = \pi r^2 h = \pi(2)^2 5 \approx 62.83$

The volume is approximately $62.83\ \text{m}^3$.

62. $A = \pi r^2 = \pi(4)^2 \approx 50.27$

The area is approximately $50.3\ \text{m}^2$.

63. $5x + 3y = 9 \xrightarrow{\times 2} 10x + 6y = 18$
$7x - 2y = 25 \xrightarrow{\times 3} 21x - 6y = 75$
$31x \quad = 93$

$x = 3$
$5(3) + 3y = 9$
$3y = -6$
$y = -2$

(3, –2) is the solution.

64. Write the system as
$x + 2y + 2z = 1$
$2x - y + z = 3$
$4x + y + 2z = 0$

Add –2 times first equation to second and –4 times first to third.
$x + 2y + 2z = 1$
$-5y - 3z = 1$
$-7y - 6z = -4$

Add $-\frac{7}{5}$ times second equation to third.

$x + 2y + 2z = 1$
$-5y - 3z = 1$
$-\frac{9}{5}z = -\frac{27}{5}$
$z = 3$
$-5y - 3(3) = 1$
$-5y = 10$
$y = -2$

$x + 2(-2) + 2(3) = 1$
$x = -1$
$(x, y, z) = (-1, -2, 3)$ is the solution.

65. $\dfrac{5.9\times10^9 - 3.5\times10^9}{3.5\times10^9}\times100\% \approx 68.6\%$

$5.9\times10^9(1.686) \approx 9.9$

China will emit about 9.9×10^9 metric tons of carbon dioxide in 2017.

66. $\dfrac{1.20\times10^9 - 1.21\times10^9}{1.21\times10^9}\times100\% \approx 0.83\%$

$1.20\times10^9(1-0.0083) \approx 1.19$

Japan will emit approximately 1.19×10^8 metric tons of carbon dioxide in 2017.

Quick Quiz 11.3

1. $\log_5\left(\dfrac{\sqrt[3]{x}}{y^4}\right) = \log_5 x^{1/3} - \log_5 y^4$

$= \dfrac{1}{3}\log_5 x - 4\log_5 4$

2. $3\log_6 x + \log_6 y - \log_6 5$

$= \log_6 x^3 + \log_6 y - \log_6 5$

$= \log_6\left(\dfrac{x^3 y}{5}\right)$

3. $\dfrac{1}{2}\log_4 x = \log_4 25$

$\log_4 x^{1/2} = \log_4 25$

$x^{1/2} = 25$

$x = (25)^2$

$x = 625$

4. Answers may vary. Possible solution:
To simplify the following:
$\log_{10}(0.001)$
start by setting the expression equal to x.
$\log_{10}(0.001) = x$
What is really being asked is what ten must be raised to, to equal to 0.001.
$0.001 = 10^x = 10^{-3}$
Simplified, this expression $= 10^{-3}$.

How Am I Doing? Chapter 11.1–11.3

1. $f(x) = 2^{-x}$

x	$y = f(x) = 2^{-x}$
-2	4
-1	2
0	1
1	$\frac{1}{2}$
2	$\frac{1}{4}$

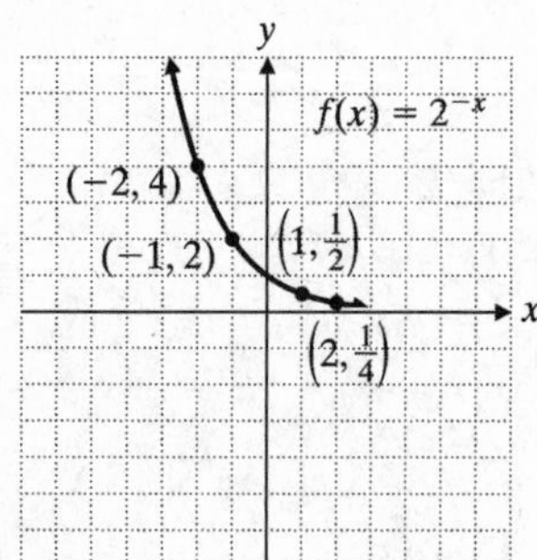

2. $3^{2x-1} = 27$

$3^{2x-1} = 3^3$

$2x - 1 = 3$

$2x = 4$

$x = 2$

3. $2^x = \dfrac{1}{32} = \dfrac{1}{2^5} = 2^{-5}$

$x = -5$

4. $125 = 5^{3x+4}$

$5^3 = 5^{3x+4}$

$3x + 4 = 3$

$3x = -1$

$x = -\dfrac{1}{3}$

5. $A = P(1+r)^t$

$A = 10{,}000(1+0.12)^4 = 15{,}735.1936$

In 4 years Nancy will have \$15,735.19.

6. $\frac{1}{49} = 7^{-2}$

$\log_7\left(\frac{1}{49}\right) = -2$

7. $-3 = \log_{10} 0.001$

$10^{-3} = 0.001$

8. $\log_5 x = 3$

$x = 5^3 = 125$

9. $\log_x 81 = -2$

$x^{-2} = 81$

$\frac{1}{x^2} = 81$

$x^2 = \frac{1}{81}$

$x = \pm\frac{1}{9}$

Pick + since $x > 0$.

$x = \frac{1}{9}$

10. Let $N = \log_{10}(10{,}000)$.

$10^N = 10{,}000$

$10^N = 10^4$

$N = 4$

11. $\log_5\left(\frac{x^2y^5}{z^3}\right) = \log_5(x^2y^5) - \log_5 z^3$

$= \log_5 x^2 + \log_5 y^5 - \log_5 z^3$

$= 2\log_5 x + 5\log_5 y - 3\log_5 z$

12. $\frac{1}{2}\log_4 x - 3\log_4 w = \log_4 x^{1/2} - \log_4 w^3$

$= \log_4 \sqrt{x} - \log_4 w^3$

$= \log_4\left(\frac{\sqrt{x}}{w^3}\right)$

13. $\log_3 x + \log_3 2 = 4$

$\log_3(2x) = 4$

$2x = 3^4$

$2x = 81$

$x = \frac{81}{2}$

14. $\log_7 x = \log_7 8$

$x = 8$

15. $\log_9 1 = x$

$9^x = 1 = 9^0$

$x = 0$

16. $\log_3 2x = 2$

$3^2 = 2x$

$x = \frac{9}{2}$

17. $1 = \log_4 3x$

$4^1 = 3x$

$x = \frac{4}{3}$

18. $\log_e x + \log_e 3 = 1$

$\log_e 3x = 1$

$e^1 = 3x$

$x = \frac{e}{3}$

11.4 Exercises

1. Error; you cannot take the logarithm of a negative number.

3. $\log 12.3 \approx 1.089905111$

5. $\log 25.6 \approx 1.408239965$

7. $\log 8 \approx 0.903089987$

9. $\log 125{,}000 \approx 5.096910013$

11. $\log 0.0123 \approx -1.910094889$

13. $\log x = 2.016$

$x = 10^{2.016} \approx 103.7528416$

15. $\log x = -0.3562$

$x = 10^{-0.3562} \approx 0.4403520272$

17. $\log x = 3.9304$

$x = 10^{3.9304} \approx 8519.223264$

19. $\log x = 6.4683$

$x = 10^{6.4683} \approx 2{,}939{,}679.609$

21. $\log x = -3.3893$
$x = 10^{-3.3893} \approx 0.000408037$

23. $\log x = -1.5672$
$x = 10^{-1.5672} \approx 0.027089438$

25. $\text{antilog}(7.6215) \approx 41{,}831{,}168.87$

27. $\text{antilog}(-1.0826) \approx 0.0826799109$

29. $\ln 5.62 \approx 1.726331664$

31. $\ln 1.53 \approx 0.4252677354$

33. $\ln 136{,}000 \approx 11.82041016$

35. $\ln 0.00579 \approx -5.151622987$

37. $\ln x = 0.95$
$x = e^{0.95} \approx 2.585709659$

39. $\ln x = 2.4$
$x = e^{2.4} \approx 11.02317638$

41. $\ln x = -0.05$
$x = e^{-0.05} \approx 0.951229425$

43. $\ln x = -2.7$
$x = e^{-2.7} \approx 0.0672055127$

45. $\text{antilog}_e(6.1582) \approx 472.5766708$

47. $\text{antilog}_e(-2.1298) \approx 0.1188610637$

49. $\log_3 9.2 = \dfrac{\log 9.2}{\log 3} \approx 2.020006063$

51. $\log_7(7.35) = \dfrac{\log 7.35}{\log 7} \approx 1.025073184$

53. $\log_6 0.127 = \dfrac{\log 0.127}{\log 6} \approx -1.151699337$

55. $\log_{15} 12 = \dfrac{\log 12}{\log 15} \approx 0.9175999207$

57. $\log_4 0.07733 = \dfrac{\ln 0.07733}{\ln 4} \approx -1.846414$

59. $\log_{21} 436 = \dfrac{\ln 436}{\ln 21} \approx 1.996254706$

61. $\ln 1537 \approx 7.337587744$

63. $\text{antilog}_e(-1.874) \approx 0.1535083985$

65. $\log x = 8.5634$
$x = 10^{8.5634} \approx 3.65931672 \times 10^8$

67. $\log_4 x = 0.8645$
$x = 4^{0.8645}$
$x \approx 3.314979618$

69. $y = \log_6 x$

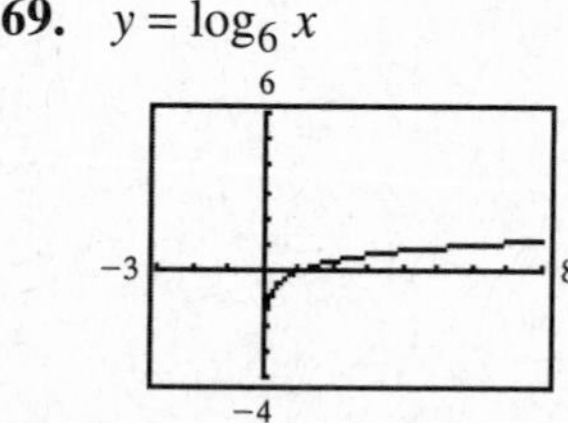

71. $y = \log_{0.4} x$

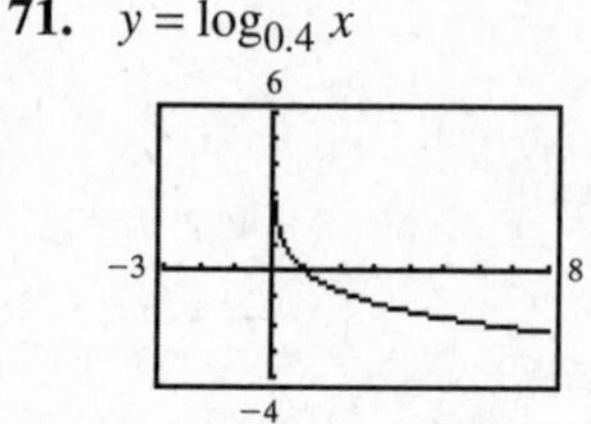

73. $N = 33.62 + 1.0249 \ln x$
$N(5) = 33.62 + 1.0249 \ln 5 \approx 35.27$
$N(15) = 33.62 + 1.0249 \ln 15 \approx 36.40$
1995 median age was approximately 35.27.
2005 median age was approximately 36.40.
$\dfrac{36.27 - 35.27}{35.27} \cdot 100\% = 3.2\%$
The median age increased by 3.2% between 1995 and 2005.

75. $R = \log x$
$R = \log 56{,}000$
$R \approx 4.75$

77. $R = \log x$
$6.6 = \log x$
$x = 10^{6.6} \approx 3{,}9891{,}071.706$
The shock wave is about 3,981,000 times greater than the smallest detectable shock wave.

Cumulative Review

79. $3x^2 - 11x - 5 = 0$

$$x = \frac{-(-11) \pm \sqrt{(-11)^2 - 4(3)(-5)}}{2(3)}$$

$$x = \frac{11 \pm \sqrt{181}}{6}$$

80. $2y^2 + 4y - 3 = 0$

$$y = \frac{-4 \pm \sqrt{4^2 - 4(2)(-3)}}{2(2)} = \frac{-4 \pm \sqrt{40}}{4}$$

$$y = \frac{-4 \pm \sqrt{4 \cdot 10}}{4} = \frac{-4 \pm 2\sqrt{10}}{4}$$

$$y = \frac{-2 \pm \sqrt{10}}{2}$$

81. Let x, y, z, w, s be the distance between adjacent exits, then:

$x + y + z + w + x = 36$

$x + y = 12 \rightarrow$ first eqn

$y + z = 15$

$z + w = 12 \rightarrow$ first eqn

$w + s = 15$

$12 + 12 + s = 36$, $s = 12$

$w + s = w + 12 = 15$, $w = 3$

$z + w = z + 3 = 12$, $z = 9$

$y + z = y + 9 = 15$, $y = 6$

$x + y = x + 6 = 12$, $x = 6$

distance between Exit 1 and Exit 2: $x = 6$ miles

distance between Exit 1 and Exit 3:

$x + y = 6 + 6 = 12$ miles

82. Let x, y, z, w, s be the distances between adjacent exits, then:

$x + y + z + w + s = 36$

$x + y = 12 \rightarrow$ first eqn

$y + z = 15$

$z + w = 12 \rightarrow$ first eqn

$w + s = 15$

$12 + 12 + s = 36$

$s = 12$

$w + s = w + 12 = 15$

$w = 3$

$z + w = z + 3 = 12$

$z = 9$

$y + z = y + 9 = 15$

$y = 6$

$x + y = x + 6 = 12$

$x = 6$

Distance between Exit 1 and Exit 4:

$x + y + z = 6 + 6 + 9 = 21$ miles

Distance between Exit 1 and Exit 5:

$x + y + z + w = 6 + 6 + 9 + 3 = 24$ miles

Quick Quiz 11.4

1. $\log 9.36 \approx 0.9713$

2.
$$\log x = 0.2253$$
$$\log_{10} x = 0.2253$$
$$x = 10^{0.2253}$$
$$x \approx 1.68$$

3. $\log_5 8.26 = \dfrac{\ln 8.26}{\ln 5} \approx 1.3119$

4. Answers may vary. Possible solution:
In order to solve for x in the following:
$\ln x = 1.7821$
take the antilog of both sides:
$x = e^{1.7821}$

11.5 Exercises

1.
$$\log_7\left(\frac{2}{3}x + 3\right) + \log_7 3 = 2$$
$$\log_7 3\left(\frac{2}{3}x + 3\right) = 2$$
$$\log_7(2x + 9) = 2$$
$$2x + 9 = 7^2$$
$$2x + 9 = 49$$
$$2x = 40$$
$$x = 20$$

Check: $\log_7\left(\frac{2}{3} \cdot 20 + 3\right) + \log_7 3 \stackrel{?}{=} 2$

$$\log_7 \frac{49}{3} + \log_7 3 \stackrel{?}{=} 2$$
$$\log_7 49 \stackrel{?}{=} 2$$
$$2 = 2$$

3.
$$\log_6(x + 3) + \log_6 4 = 2$$
$$\log_6(4x + 12) = 2$$
$$4x + 12 = 6^2$$
$$4x + 12 = 36$$
$$4x = 24$$
$$x = 6$$

Check: $\log_6(6+3)+\log_6 4 \stackrel{?}{=} 2$
$\log_6(9\cdot 4) \stackrel{?}{=} 2$
$\log_6 36 \stackrel{?}{=} 2$
$2=2$

5. $\log_2\left(x+\frac{4}{3}\right)=5-\log_2 6$
$\log_2\left(x+\frac{4}{3}\right)+\log_2 6=5$
$\log_2(6x+8)=5$
$6x+8=2^5$
$6x=24$
$x=4$

Check: $\log_2\left(4+\frac{4}{3}\right) \stackrel{?}{=} 5-\log_2 6$
$\log_2\left(\frac{16}{3}\right) \stackrel{?}{=} 5-\log_2(3\cdot 2)$
$\log_2 16-\log_2 \stackrel{?}{=} 5-\log_2 3-\log_2 2$
$4-\log_2 3 \stackrel{?}{=} 5-\log_2 3-1$
$4-\log_2 3=4-\log_2 3$

7. $\log(30x+40)=2+\log(x-1)$
$\log(30x+40)-\log(x-1)=2$
$\log\left(\frac{30x+40}{x-1}\right)=2$
$\frac{30x+40}{x-1}=10^2$
$30x+40=100x-100$
$70x=140$
$x=2$

Check: $\log(30(2)+40) \stackrel{?}{=} 2+\log(2-1)$
$\log 100 \stackrel{?}{=} 2+\log(1)$
$2 \stackrel{?}{=} 2+\log(1)=2+0$
$2=2$

9. $2+\log_6(x-1)=\log_6(12x)$
$\log_6(x-1)-\log_6(12x)=-2$
$\log_6\left(\frac{x-1}{12x}\right)=-2$
$\frac{x-1}{12x}=6^{-2}$
$36x-36=12x$
$24x=36$
$x=\frac{3}{2}$

Check: $2+\log_6\left(\frac{3}{2}-1\right) \stackrel{?}{=} \log_6\left(12\cdot\frac{3}{2}\right)$
$2+\log_6\frac{1}{2} \stackrel{?}{=} \log_6\frac{36}{2}$
$2+\log_6 1-\log_2 6 \stackrel{?}{=} \log_6\frac{36}{2}$
$2+0-\log_6 2 \stackrel{?}{=} \log_6 36-\log_6 2$
$2-\log_6 2=2-\log_6 2$

11. $\log(75x+50)-\log x=2$
$\log\left(\frac{75x+50}{x}\right)=2$
$\frac{75x+50}{x}=10^2$
$100x=75x+50$
$25x=50$
$x=\frac{50}{25}=2$

Check: $\log(75(2)+50)-\log(2) \stackrel{?}{=} 2$
$\log(200)-\log(2) \stackrel{?}{=} 2$
$\log\left(\frac{200}{2}\right) \stackrel{?}{=} 2$
$\log 100 \stackrel{?}{=} 2$
$2=2$

13. $\log_3(x+6)+\log_3 x=3$
$\log_3(x(x+6))=3$
$x(x+6)=3^3$
$x^2+6x-27=0$
$(x+9)(x-3)=0$
$x=-9$ gives $\log_3$(negative)
$x=3$ is the solution.

Check: $\log_3(3+6)+\log_3 3 \stackrel{?}{=} 3$
$\log_3 9+\log_3 3 \stackrel{?}{=} 3$
$2+1 \stackrel{?}{=} 3$
$3=3$

15. $1+\log(x-2)=\log(6x)$
$\log(6x)-\log(x-2)=1$
$\log\frac{6x}{x-2}=1$
$\frac{6x}{x-2}=10^1$
$6x=10x-20$
$4x=20$
$x=5$

Check: $1+\log(5-2) \stackrel{?}{=} \log(6\cdot 5) = \log 30$
$1+\log 3 \stackrel{?}{=} \log(10\cdot 3)$
$1+\log 3 \stackrel{?}{=} \log 10+\log 3$
$1+\log 3 = 1+\log 3$

17. $\log_2(x+5)-2 = \log_2 x$
$\log_2(x+5)-\log_2 x = 2$
$\log_2\dfrac{x+5}{x} = 2$
$\dfrac{x+5}{x} = 2^2 = 4$
$x+5 = 4x$
$3x = 5$
$x = \dfrac{5}{3}$

Check: $\log_2\left(\dfrac{5}{3}+5\right)-2 \stackrel{?}{=} \log_2\dfrac{5}{3}$
$\log_2\dfrac{20}{3}-2 \stackrel{?}{=} \log_2\dfrac{5}{3}$
$\log_2\left(4\cdot\dfrac{5}{3}\right)-2 \stackrel{?}{=} \log_2\dfrac{5}{3}$
$\log_2 4+\log_2\dfrac{5}{3}-2 \stackrel{?}{=} \log_2\dfrac{5}{3}$
$2+\log_2\dfrac{5}{3}-2 \stackrel{?}{=} \log_2\dfrac{5}{3}$
$\log_2\dfrac{5}{3} = \log_2\dfrac{5}{3}$

19. $2\log_7 x = \log_7(x+4)+\log_7 2$
$\log_7 x^2-\log_7(x+4) = \log_7 2$
$\log_7\dfrac{x^2}{(x+4)} = \log_7 2$
$\dfrac{x^2}{(x+4)} = 2$
$x^2-2x-8 = 0$
$(x-4)(x+2) = 0$
$x = 4, x = -2$, reject, gives $\log_7$ (negative)
Check: $2\log_7 4 \stackrel{?}{=} \log_7(4+4)+\log_7 2$
$\log_7 4^2 \stackrel{?}{=} \log_7(8)+\log_7 2$
$\log_7 16 \stackrel{?}{=} \log_7(8\cdot 2)$
$\log_7 16 = \log_7 16$

21. $\ln 10-\ln x = \ln(x-3)$
$\ln\dfrac{10}{x} = \ln(x-3)$
$\dfrac{10}{x} = x-3$
$x^2-3x-10 = 0$
$(x-5)(x+2) = 0$
$x = 5, x = -2$, reject, gives ln(negative)
Check: $\ln 10-\ln 5 \stackrel{?}{=} \ln(5-3)$
$\ln\dfrac{10}{5} \stackrel{?}{=} \ln 2$
$\ln 2 = \ln 2$

23. $7^{x+3} = 12$
$\log 7^{x+3} = \log 12$
$(x+3)\log 7 = \log 12$
$x+3 = \dfrac{\log 12}{\log 7}$
$x = \dfrac{\log 12}{\log 7}-3$
$x = \dfrac{\log 12-3\log 7}{\log 7}$

25. $2^{3x+4} = 17$
$\log 2^{3x+4} = \log 17$
$(3x+4)\log 2 = \log 17$
$3x+4 = \dfrac{\log 17}{\log 2}$
$3x = \dfrac{\log 17}{\log 2}-4\cdot\dfrac{\log 2}{\log 2}$
$x = \dfrac{\log 17}{3\log 2}-4\cdot\dfrac{\log 2}{3\log 2}$
$x = \dfrac{\log 17-4\log 2}{3\log 2}$

27. $8^{2x-1} = 90$
$\log 8^{2x-1} = \log 90$
$(2x-1)\log 8 = \log 90$
$2x-1 = \dfrac{\log 90}{\log 8}$
$2x = 1+\dfrac{\log 90}{\log 8}$
$x = \dfrac{1}{2}\left(1+\dfrac{\log 90}{\log 8}\right)$
$x \approx 1.582$

29.
$$5^x = 4^{x+1}$$
$$\log 5^x = \log 4^{x+1}$$
$$x\log 5 = (x+1)\log 4$$
$$x\log 5 = x\log 4 + \log 4$$
$$x(\log 5 - \log 4) = \log 4$$
$$x = \frac{\log 4}{\log 5 - \log 4}$$
$$x \approx 6.213$$

31.
$$e^{x-2} = 28$$
$$\ln e^{x-2} = \ln 28$$
$$(x-2)\ln e = \ln 28$$
$$x = 2 + \ln 28$$
$$x \approx 5.332$$

33.
$$88 = e^{2x+1}$$
$$\ln 88 = \ln e^{2x+1}$$
$$\ln 88 = (2x+1)\ln e$$
$$\ln 88 = (2x+1)(1)$$
$$\ln 88 = 2x+1$$
$$2x = \ln 88 - 1$$
$$x = \frac{\ln 88 - 1}{2}$$
$$x \approx 1.739$$

35.
$$A = P(1+r)^t$$
$$5000 = 1500(1+0.08)^t$$
$$1.08^t = \frac{10}{3}$$
$$\ln 1.08^t = \ln\frac{10}{3}$$
$$t = \frac{\ln\frac{10}{3}}{\ln 1.08} = 15.64392564...$$
It will take approximately 16 years.

37.
$$A = P(1+r)^t$$
$$3P = P(1+0.06)^t$$
$$1.06^t = 3$$
$$\ln 1.06^t = \ln 3$$
$$t\ln 1.06 = \ln 3$$
$$t = \frac{\ln 3}{\ln 1.06}$$
$$t = 18.85417668...$$
It will take approximately 19 years.

39.
$$A = P(1+r)^t$$
$$6500 = 5000(1+r)^6$$
$$1.3 = (1+r)^6$$
$$\ln 1.3 = \ln(1+r)^6 = 6\ln(1+r)$$
$$\ln(1+r) = \frac{\ln 1.3}{6}$$
$$e^{\ln(1+r)} = e^{\frac{\ln 1.3}{6}}$$
$$1+r = e^{\frac{\ln 1.3}{6}}$$
$$r = e^{\frac{\ln 1.3}{6}} - 1$$
$$r = 0.0446975079...$$
The rate is approximately 4.5%.

41.
$$A = A_0 e^{rt}$$
$$12 = 7e^{0.02t}$$
$$\frac{12}{7} = e^{0.02t}$$
$$\ln\frac{12}{7} = \ln e^{0.02t}$$
$$\ln\frac{12}{7} = 0.02t$$
$$t = \frac{\ln\frac{12}{7}}{0.02} = 26.94982504...$$
$$t \approx 27$$
It will take approximately 27 years.

43.
$$A = A_0 e^{rt}$$
$$2A_0 = A_0 e^{0.02t}$$
$$2 = e^{0.02t}$$
$$\ln 2 = \ln e^{0.02t}$$
$$\ln 2 = 0.02t$$
$$t = \frac{\ln 2}{0.02} = 34.65735903...$$
$$t \approx 35$$
It will take approximately 35 years.

45.
$$N = 20{,}800(1.264)^x$$
$$N = 20{,}800(1.264)^{13}$$
$$N = 437{,}295.79$$
In 2003 there will be approximately 437,000 employees.

47. $N = 20,800(1.264)^x$

$274,000 = 20,800(1.264)^x$

$1.264^x = \frac{274,000}{20,800}$

$\ln 1.264^x = \ln \frac{274,000}{20,800}$

$x = \frac{\ln \frac{274,000}{20,800}}{\ln 1.264} = 11.00461354...$

The number of employees will reach 274,000 sometime in 2001.

49. $A = A_0 e^{rt}$

$120,000 = 80,000e^{0.015t}$

$e^{0.015t} = 1.5$

$\ln e^{0.015t} = \ln 1.5$

$0.015t = \ln 1.5$

$t = \frac{\ln 1.5}{0.015}$

$t = 27.03111721...$

$t \approx 27$

It will take approximately 27 years.

51. $A = A_0 e^{rt}$

$1800 = 200e^{0.04t}$

$e^{0.04t} = 9$

$0.04t = \ln 9$

$t = \frac{\ln 9}{0.04} = 54.93061443...$

It will take approximately 55 hours.

53. $A = A_0 e^{rt}$

$A = 24,500e^{0.05(13)} = 46930.75031...$

By the end of 2010 approximately 46,931 people will be infected.

55. $R = \log\left(\frac{I}{I_0}\right)$

$7.1 = \log\left(\frac{I_{SF}}{I_0}\right) = \log I_{SF} - \log I_0$

$8.2 = \log\left(\frac{I_{KI}}{I_0}\right) = \log I_{KI} - \log I_0$

Subtracting the two equations gives

$-1.1 = \log I_{SF} - \log I_{KI} = \log \frac{I_{SF}}{I_{KI}}$

$10^{\log \frac{I_{SF}}{I_{KI}}} = 10^{-1.1}$

$\frac{I_{SF}}{I_{KI}} = 10^{-1.1}$

$I_{KI} = 10^{1.1} I_{KI} \approx 12.6 I_{SF}$

The Kurile island earthquake was about 12.6 times as intense as the San Francisco earthquake.

57. $R = \log\left(\frac{I}{I_0}\right)$

$8.3 = \log\left(\frac{I_S}{I_0}\right) = \log I_S - \log I_0$

$6.8 = \log\left(\frac{I_J}{I_0}\right) = \log I_J - \log I_0$

Subtracting the two equations gives

$1.5 = \log I_S - \log I_J = \log \frac{I_S}{I_J}$

$10^{\log \frac{I_S}{I_J}} = 10^{1.5}$

$\frac{I_S}{I_J} = 10^{1.5}$

$I_S = 10^{1.5} I_J \approx 31.6 I_J$

The intensity of the San Francisco earthquake was approximately 31.6 times greater than the intensity of the Japanese earthquake.

Cumulative Review

59. $\left(\sqrt{3}+2\sqrt{2}\right)\left(\sqrt{6}-\sqrt{2}\right) = \sqrt{18}-\sqrt{6}+2\sqrt{12}-4$

$= 3\sqrt{2}-\sqrt{6}+4\sqrt{3}-4$

60. $\sqrt{98x^3y^2} = \sqrt{49x^2y^2 \cdot 2x} = 7xy\sqrt{2x}$

61. Total students: 288 given
8 years olds: 2 given
9 year olds: 0 given
12 year olds: 54 given
13 year olds: 44 + 54 = 98
14–15 year olds: 91 given
Let y = number of 10 year olds.
Let $y + 17$ = number of 11 year olds.
How many of each category from 9–15?

$286 = 54+98+91+y+y+17$

$26 = 2y$

$y = 13$

$y + 17 = 30$

9 year olds: 0 students
10 year olds: 13 students
11 year olds: 30 students
12 year olds: 54 students
13 year olds: 98 students
14–15 year olds: 91 students

62. $\frac{2.85\times10^9 \text{ pounds}}{16 \text{ kilometers}} \cdot \frac{1 \text{ kilometer}}{0.62 \text{ miles}} \cdot \frac{1 \text{ U.S. dollar}}{0.51 \text{ pounds}}$
$= 0.563\times10^9$ dollars/mi
$= \frac{\$563,000,000}{\text{mile}}$
The extension cost approximately $563,000,000 per mile.

Quick Quiz 11.5

1.
$$\begin{aligned} 2-\log_6 x &= \log_6(x+5) \\ 2 &= \log_6 x+\log_6(x+5) \\ 2 &= \log_6[x(x+5)] \\ 2 &= \log_6(x^2+5x) \\ 6^2 &= x^2+5x \end{aligned}$$
$x^2+5-36=0$
$(x+9)(x-4)=0$
$x+9=0$ or $x-4=0$
$x=-9, 4$
−9 is extraneous. 4 is the only solution.

2.
$$\begin{aligned} \log_3(x-5)+\log_3(x+1) &= \log_3 7 \\ \log_3[(x-5)(x+1)] &= \log_3 7 \\ x^2+x-5x-5 &= 7 \\ x^2-4x-12 &= 0 \\ (x-6)(x+2) &= 0 \end{aligned}$$
$x-6=0$ or $x+2=0$
$x=6, -2$
−2 is extraneous. 6 is the only solution.

3.
$$\begin{aligned} 6^{x+2} &= 9 \\ \log 6^{x+2} &= \log 9 \\ (x+2)\log 6 &= \log 9 \\ x+2 &= \frac{\log 9}{\log 6} \\ x &= \frac{\log 9}{\log 6}-2 \\ x &\approx -0.774 \end{aligned}$$

4. Answers may vary. Possible solution:
To solve $26 = 52e^{3x}$, start by dividing both sides by 52.
$\frac{1}{2} = e^{3x}$
Then take the natural log of both sides.
$\ln\left(\frac{1}{2}\right) = 3x$
Isolate the x.
$\frac{\ln\left(\frac{1}{2}\right)}{3} = x$
Solve with a calculator.
$x \approx -0.231$

Putting Your Skills to Work

1. $4\times\frac{2}{3} = \frac{4}{1}\times\frac{2}{3} = \frac{8}{3}$
Lucy prepares $\frac{8}{3}$ cups of rice each week.

2. $\frac{21 \text{ ounces}}{1 \text{ cup}}\times\frac{8}{3}$ cups = 56 ounces
Lucy's family consumes 56 ounces of rice each week.

3. $\frac{1 \text{ lb}}{16 \text{ oz}}\times 56 \text{ oz} = \frac{7}{2}$ or $3\frac{1}{2}$ lb
Lucy's family eats $3\frac{1}{2}$ pounds each week.

4. a. $15.73 - 9.25 = 6.48$
$\frac{6.48}{25} \approx 0.26$
It increased by $0.26/lb.

b. $\frac{9.25}{25} = 0.37$; $\frac{0.26}{0.37} \approx 0.70$
It is a 70% increase.

5. $\frac{15.73}{25} = 0.6292$; $0.6292(3.5) \approx 2.20$
The cost in today's prices is $2.20 per week.

6. Answers will vary.

7. Answers will vary.

Chapter 11 Review Problems

1. $f(x) = 4^{3+x}$

x	$y = f(x) = 4^{3+x}$
–5	$\frac{1}{16}$
–4	$\frac{1}{4}$
–3	1
–2	4
–1	16

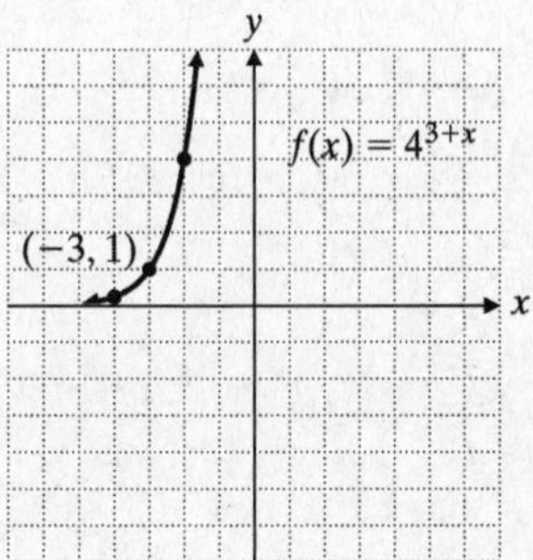

2. $f(x) = e^{x-3}$

x	$y = f(x) = e^{x-3}$
1	0.14
2	0.37
3	1
4	2.72
5	7.39

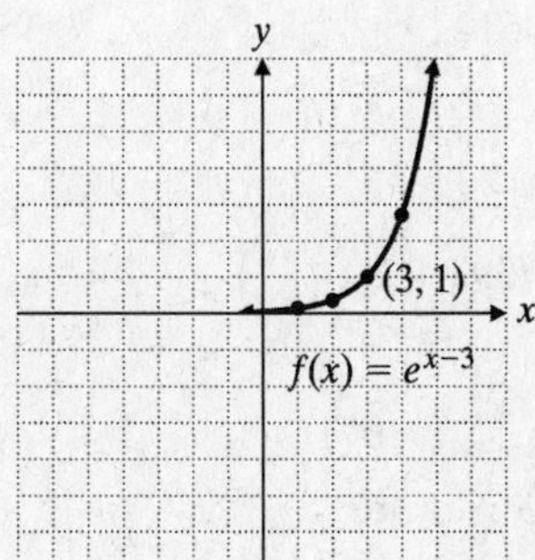

3. $3^{3x+1} = 81 = 3^4$
$3x + 1 = 4$
$3x = 3$
$x = 1$

4. $-2 = \log_{10}(0.01) \Leftrightarrow 10^{-2} = 0.01$

5. $8 = 4^{3/2} \Leftrightarrow \log_4 8 = \frac{3}{2}$

6. $\log_w 16 = 4$
$w^4 = 16 = 2^4$
$w = 2$

7. $\log_3 x = -2$
$3^{-2} = x$
$x = \frac{1}{9}$

8. $\log_8 x = 0$
$8^0 = x$
$x = 1$

9. $\log_7 w = -1$
$7^{-1} = w$
$w = \frac{1}{7}$

10. $\log_w 64 = 3$
$w^3 = 64 = 4^3$
$w = 4$

11. $\log_{10} w = -1$
$10^{-1} = w$
$w = 0.1$ or $\frac{1}{10}$

12. $\log_{10} 1000 = x$
$10^x = 1000 = 10^3$
$x = 3$

13. $\log_2 64 = x$
$2^x = 64 = 2^6$
$x = 6$

14. $\log_2 \frac{1}{4} = x$

$2^x = \frac{1}{4} = 2^{-2}$

$x = -2$

15. $\log_3 243 = x$

$3^x = 243 = 3^5$

$x = 5$

16. $\log_3 x = y \Leftrightarrow 3^y = x$

$x = 3^y$	y
$\frac{1}{9}$	-2
$\frac{1}{3}$	-1
1	0
3	1
9	2

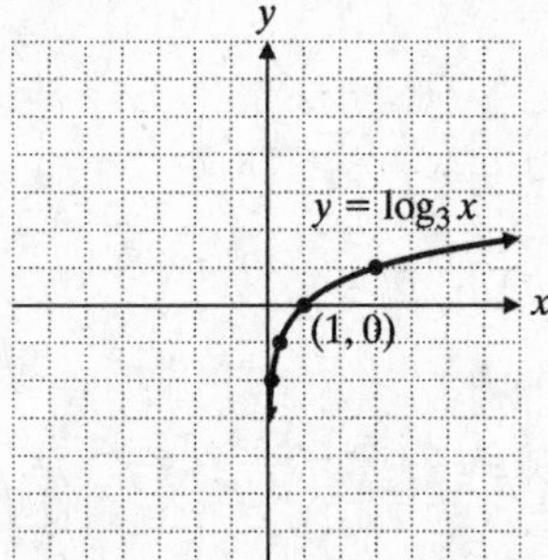

17. $\log_2\left(\frac{5x}{\sqrt{w}}\right) = \log_2(5x) - \log_2\sqrt{w}$

$= \log_2(5x) - \log_2 w^{1/2}$

$= \log_2 5 + \log_2 x - \frac{1}{2}\log_2 w$

18. $\log_2 x^3\sqrt{y} = \log_2 x^3 + \log_2\sqrt{y}$

$= 3\log_2 x + \log_2 y^{1/2}$

$= 3\log_2 x + \frac{1}{2}\log_2 y$

19. $\log_3 x + \log_3 w^{1/2} - \log_3 2$

$= \log_3 x + \log_3\sqrt{w} - \log_3 2$

$= \log_3\left(x\sqrt{w}\right) - \log_3 2$

$= \log_3 \frac{x\sqrt{w}}{2}$

20. $4\log_8 w - \frac{1}{3}\log_8 z = \log_8 w^4 - \log_8 z^{1/3}$

$= \log_8 w^4 - \log_8 \sqrt[3]{z}$

$= \log_8 \frac{w^4}{\sqrt[3]{z}}$

21. $\log_e e^6 = 6\log_e e = 6(1) = 6$

22. $\log_5 100 - \log_5 x = \log_5 4$

$\log_5 \frac{100}{x} = \log_5 4$

$\frac{100}{x} = 4$

$x = 25$

23. $\log_8 x + \log_8 3 = \log_8 75$

$\log_8(3x) = \log_8 75$

$3x = 75$

$x = 25$

24. $\log 23.8 = 1.376576957$

25. $\log 0.0817 = -1.087777943$

26. $\ln 3.92 = 1.366091654$

27. $\ln 803 = 6.688354714$

28. $\log n = 1.1367$

$n = 10^{1.1367}$

$n = 13.69935122$

29. $\ln n = 1.7$

$n = e^{1.7}$

$n = 5.473947392$

30. $\log_8 2.81 = \frac{\ln 2.81}{\ln 8}$

$\log_8 2.81 = 0.4968567101$

31. $\log_{11}\left(\frac{4}{3}x+7\right)+\log_{11}3=2$

$$\log_{11}\left[3\left(\frac{4}{3}x+7\right)\right]=2$$
$$\log_{11}(4x+21)=2$$
$$11^2=4x+21$$
$$4x+21=121$$
$$4x=100$$
$$x=25$$

Check: $\log_{11}\left(\frac{4}{3}\cdot 25+7\right)+\log_{11}3 \stackrel{?}{=} 2$

$$\log_{11}\left(\left(\frac{4}{3}\cdot 25+7\right)\cdot 3\right) \stackrel{?}{=} 2$$
$$\log_{11}(100+21) \stackrel{?}{=} 2$$
$$\log_{11}(121) \stackrel{?}{=} 2$$
$$2=2$$

32. $\log_8(x-3)=\log_8 6x-1$

$$\log_8(x-3)-\log_8 6x=-1$$
$$\log_8\frac{x-3}{6x}=-1$$
$$\frac{x-3}{6x}=8^{-1}=\frac{1}{8}$$
$$8x-24=6x$$
$$2x=24$$
$$x=12$$

Check: $\log_8(12-3) \stackrel{?}{=} \log_8(6\cdot 12)-1$

$$\log_8(9) \stackrel{?}{=} \log_8(72)-\log_8 8$$
$$\log_8(9) \stackrel{?}{=} \log_8\left(\frac{72}{8}\right)$$
$$\log_8(9)=\log_8(9)$$

33. $\log_5(x+1)-\log_5 8=\log_5 x$

$$\log_5(x+1)-\log_5 x=\log_5 8$$
$$\log_5\frac{x+1}{x}=\log_5 8$$
$$\frac{x+1}{x}=8$$
$$x+1=8x$$
$$7x=1$$
$$x=\frac{1}{7}$$

Check: $\log_5\left(\frac{1}{7}+1\right)-\log_5 8 \stackrel{?}{=} \log_5\left(\frac{1}{7}\right)$

$$\log_5\left(\frac{8}{7}\right)-\log_5 8 \stackrel{?}{=} \log_5\left(\frac{1}{7}\right)$$
$$\log_5\left(\frac{1}{7}\cdot 8\right)-\log_5 8 \stackrel{?}{=} \log_5\left(\frac{1}{7}\right)$$
$$\log_5\left(\frac{1}{7}\right)+\log_5 8-\log_5 8 \stackrel{?}{=} \log_5\left(\frac{1}{7}\right)$$
$$\log_5\left(\frac{1}{7}\right)=\log_5\left(\frac{1}{7}\right)$$

34. $\log_{12}(x+2)+\log_{12}3=1$

$$\log_{12}(3x+6)=\log_{12}12$$
$$3x+6=12$$
$$3x=6$$
$$x=2$$

Check: $\log_{12}(2+2)+\log_{12}3 \stackrel{?}{=} 1$

$$\log_{12}(4)+\log_{12}3 \stackrel{?}{=} 1$$
$$\log_{12}(12) \stackrel{?}{=} 1$$
$$1=1$$

35. $\log_2(x-2)+\log_2(x+5)=3$

$$\log_2((x-2)(x+5))=3$$
$$(x-2)(x+5)=2^3=8$$
$$x^2+3x-10=8$$
$$x^2+3x-18=0$$
$$(x+6)(x-3)=0$$

$x=3$, $x=-6$, reject -6 since it gives $\log_2$ (negative).

$x=3$ is the solution.

Check: $\log_2(3-2)+\log_2(3+5) \stackrel{?}{=} 3$

$$\log_2(1)+\log_2(8) \stackrel{?}{=} 3$$
$$0+\log_2(2^3) \stackrel{?}{=} 3$$
$$3\log_2 2 \stackrel{?}{=} 3$$
$$3=3$$

36. $\log_5(x+1)+\log_5(x-3)=1$

$$\log_5((x+1)(x-3))=1$$
$$(x+1)(x-3)=5^1=5$$
$$x^2-2x-3=5$$
$$x^2-2x-8=0$$
$$(x-4)(x+2)=0$$

$x=4$, $x=-2$, reject -2 since it gives $\log_5$ (negative).

Check: $\log_5(4+1)+\log_5(4-3) \stackrel{?}{=} 1$
$\log_5(5)+\log_5(1) \stackrel{?}{=} 1$
$1+0 \stackrel{?}{=} 1$
$1=1$

37. $\log(2t+3)+\log(4t-1)=2\log 3$
$\log((2t+3)(4t-1))=\log 3^2=\log 9$
$(2t+3)(4t-1)=9$
$8t^2+10t-3=9$
$8t^2+10t-12=0$
$4t^2+5t-6=0$
$(4t-3)(t+2)=0$

$t=\frac{3}{4}$, $t=-2$, reject -2 since it gives log(negative).

$t=\frac{3}{4}$ is the solution.

Check: $\log\left(2\cdot\frac{3}{4}+3\right)+\log\left(4\cdot\frac{3}{4}-1\right) \stackrel{?}{=} 2\log 3$
$\log\left(\frac{9}{2}\right)+\log(2) \stackrel{?}{=} 2\log 3$
$\log\left(\frac{9}{2}\cdot 2\right) \stackrel{?}{=} 2\log 3$
$\log(9) \stackrel{?}{=} 2\log 3$
$\log(3^2) \stackrel{?}{=} 2\log 3$
$2\log 3=2\log 3$

38. $\log(2t+4)-\log(3t+1)=\log 6$
$\log\left(\frac{2t+4}{3t+1}\right)=\log 6$
$\frac{2t+4}{3t+1}=6$
$2t+4=18t+6$
$16t=-2$
$t=-\frac{1}{8}$

Check: $\log\left(2\cdot\frac{-1}{8}+4\right)-\log\left(3\cdot\frac{-1}{8}+1\right) \stackrel{?}{=} \log 6$
$\log\left(\frac{15}{4}\right)-\log\left(\frac{5}{8}\right) \stackrel{?}{=} \log 6$
$\log\left(\frac{\frac{15}{4}}{\frac{5}{8}}\right) \stackrel{?}{=} \log 6$
$\log\left(\frac{\frac{15}{4}}{\frac{5}{8}}\right) \stackrel{?}{=} \log 6$
$\log 6=\log 6$

39. $3^x=14$
$\log 3^x=\log 14$
$x\log 3=\log 14$
$x=\frac{\log 14}{\log 3}$

40. $5^{x+3}=130$
$\log 5^{x+3}=\log 130$
$(x+3)\log 5=\log 130$
$x+3=\frac{\log 130}{\log 5}$
$x=\frac{\log 130}{\log 5}-3$
$x=\frac{\log 30-3\log 5}{\log 5}$

41. $e^{2x-1}=100$
$\ln(e^{2x-1})=\ln 100$
$(2x-1)\ln e=\ln 100$
$2x-1=\ln 100$
$2x=\ln 100+1$
$x=\frac{\ln 100+1}{2}$

42. $e^{2x}=30.6$
$\ln e^{2x}=30.6$
$2x\ln e=\ln 30.6$
$2x=\ln 30.6$
$x=\frac{\ln 30.6}{2}$

43. $2^{3x+1}=5^x$
$\ln 2^{3x+1}=\ln 5^x$
$(3x+1)\ln 2=x\ln 5$
$3x\ln 2+\ln 2=x\ln 5$
$x(3\ln 2-\ln 5)=-\ln 2$
$x=\frac{\ln 2}{\ln 5-3\ln 2}$
$x\approx -1.4748$

44. $3^{x+1}=7$
$\ln 3^{x+1}=\ln 7$
$(x+1)\ln 3=\ln 7$
$x=-1+\frac{\ln 7}{\ln 3}$
$x\approx 0.7712$

45.
$$e^{3x-4} = 20$$
$$\ln e^{3x-4} = \ln 20$$
$$(3x-4)\ln e = \ln 20$$
$$3x-4 = \ln 20$$
$$x = \frac{\ln 20 + 4}{3}$$
$$x \approx 2.3319$$

46.
$$1.03^x = 20$$
$$\ln 1.03^x = \ln 20$$
$$x \ln 1.03 = \ln 20$$
$$x = \frac{\ln 20}{\ln 1.03}$$
$$x \approx 101.3482$$

47.
$$A = P(1+r)^t$$
$$2P = P(1+0.08)^t$$
$$2 = 1.08^t$$
$$\ln 2 = \ln 1.08^t$$
$$\ln 2 = t \ln 1.08$$
$$t = \frac{\ln 2}{\ln 1.08} \approx 9$$
It will take about 9 years to double money in the account.

48. $A = P(1+r)^t$
$$A = 5000(1+0.06)^4 = 6312.38$$
He would have \$6312.38 in the account after 4 years.

49.
$$A = P(1+r)^t$$
$$20,000 = 12,000(1+0.07)^t$$
$$\frac{5}{3} = 1.07^t$$
$$\ln 1.07^t = \ln \frac{5}{3}$$
$$t \ln 1.07 = \ln \frac{5}{3}$$
$$t = \frac{\ln \frac{5}{3}}{\ln 1.07}$$
$$t = 7.550041795...$$
$$t \approx 8 \text{ years}$$
It will take approximately 8 years.

50. $A = P(1+r)^t$
$$A_{\text{Robert}} + 500 = A_{\text{Brother}}$$
$$3500(1+0.05)^t + 500 = 3500(1+0.06)^t$$
$$7(1.06)^t - 7(1.05)^t = 1$$
$$(1.06)^t - (1.06)^t = \frac{1}{7}$$
Solve with a graphing calculator. Graph $y_1 = 1.06^x - 1.05^x$ and $y_2 = \frac{1}{7}$ and use the *intersect* feature.
It will take approximately 9 years for Robert's amount to be \$500 less than his brother's amount.

51.
$$A = A_0 e^{rt}$$
$$16 = 7e^{0.02t}$$
$$\ln \frac{16}{7} = \ln e^{0.02t}$$
$$\ln \frac{16}{7} = 0.02t \ln e$$
$$t = \frac{\ln \frac{16}{7}}{0.02} = 41.33392866...$$
$$t \approx 41$$
It will take approximately 41 years.

52.
$$A = A_0 e^{rt}$$
$$10 = 6e^{0.02t}$$
$$\ln 1.\overline{6} = \ln e^{0.02t}$$
$$\ln 1.\overline{6} = 0.02t \ln e$$
$$t = \frac{\ln 1.\overline{6}}{0.02} = 25.54128119...$$
$$t \approx 26$$
It will take approximately 26 years.

53.
$$A = A_0 e^{rt}$$
$$2600 = 2000e^{0.03t}$$
$$\ln 1.3 = \ln e^{0.03t}$$
$$\ln 1.3 = 0.03t$$
$$t = \frac{\ln 1.3}{0.03} = 8.745475482...$$
$$t \approx 9$$
It will take approximately 9 years.

54.
$$A = A_0e^{rt}$$
$$95,000 = 40,000e^{0.08t}$$
$$2.375 = e^{0.08t}$$
$$\ln 2.375 = \ln e^{0.08t}$$
$$0.08t = \ln 2.375$$
$$t = \frac{\ln 2.375}{0.08} = 10.81246797...$$
$$t \approx 11$$
It will take approximately 11 years.

55. $W = p_0V_0 \ln\left(\frac{V_1}{V_0}\right)$

a. $W = 40(15)\ln\left(\frac{24}{15}\right)$
$W \approx 282$
The work done is approximately 282 lb.

b. $100 = p_0(8)\ln\left(\frac{40}{8}\right)$
$$p_0 = \frac{100}{(8)\ln\left(\frac{40}{8}\right)} \approx 7.77$$
The pressure is approximately 7.77 lb/in.^3.

56. $M = \log\left(\frac{I}{I_0}\right)$
$$8.4 = \log\left(\frac{I_A}{I_0}\right) = \log I_A - \log I_0$$
$$-\left(6.7 = \log\left(\frac{I_T}{I_0}\right) = \log I_T - \log I_0\right)$$
$$1.7 = \log I_A - \log I_T$$
$$\log\frac{I_A}{I_T} = 1.7$$
$$10^{\log\frac{I_A}{I_T}} = 10^{1.7}$$
$$\frac{I_A}{I_T} = 50.11872336...$$
$$I_A = 50.11872336...I_T$$
$$I_A \approx 50.1I_T$$
The Alaska earthquake was about 50.1 times more intense than the Turkey earthquake.

How Am I Doing? Chapter 11 Test

1. $f(x) = 3^{x+1}$

x	$y = f(x) = 3^{x+1}$
−1	1
0	3

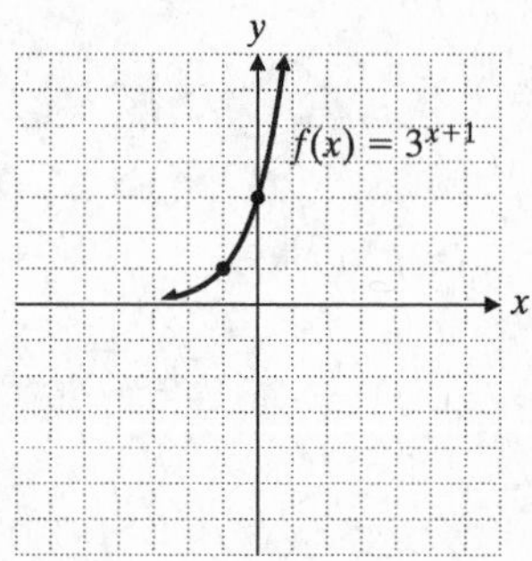

2. $f(x) = \log_2 x$

x	$y = f(x) = \log_2 x$
$\frac{1}{2}$	−1
1	0
2	2
4	2

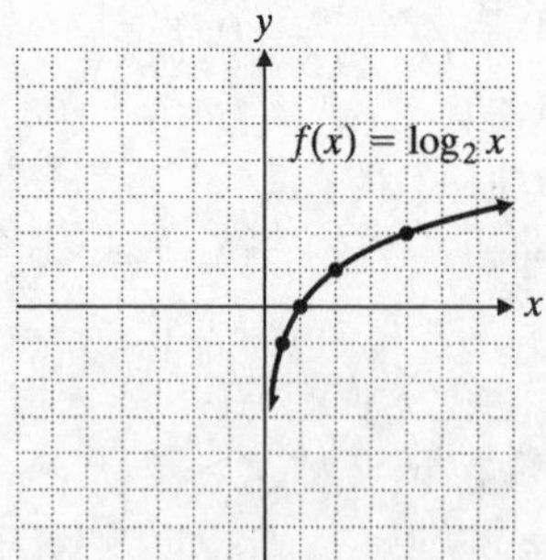

3. $4^{x+3} = 64 = 4^3$
$$x + 3 = 3$$
$$x = 0$$

4. $\log_w 125 = 3$
$$w^3 = 125 = 5^3$$
$$w = 5$$

5. $\log_8 x = -2 \Leftrightarrow x = 8^{-2} = \dfrac{1}{64}$

6. $$\begin{aligned} 2\log_7 x + \log_7 y - \log_7 4 &= \log_7 x^2 + \log_7 \frac{y}{4} \\ &= \log_7 \frac{x^2 y}{4} \end{aligned}$$

7. $\ln 5.99 = 1.7901$

8. $\log 23.6 = 1.3729$

9. $\log_3 1.62 = \dfrac{\log 1.62}{\log 3} = 0.4391$

10. $$\begin{aligned} \log x &= 3.7284 \\ x &= 10^{3.7284} \\ x &= 5350.569382 \end{aligned}$$

11. $\ln x = 0.14 \Leftrightarrow x = e^{0.14} = 1.150273799$

12. $$\begin{aligned} \log_8(x+3) - \log_8 2x &= \log_8 4 \\ \log_8\left(\frac{x+3}{2x}\right) &= \log_8 4 \\ \frac{x+3}{2x} &= 4 \\ x+3 &= 8x \\ 7x &= 3 \\ x &= \frac{3}{7} \end{aligned}$$

Check: $$\begin{aligned} \log_8\left(\frac{3}{7}+3\right) - \log_8\left(2\cdot\frac{3}{7}\right) &\stackrel{?}{=} \log_8 4 \\ \log_8\left(\frac{24}{7}\right) - \log_8\left(\frac{6}{7}\right) &\stackrel{?}{=} \log_8 4 \\ \log_8\left(\frac{\frac{24}{7}}{\frac{6}{7}}\right) &\stackrel{?}{=} \log_8 4 \\ \log_8 4 &= \log_8 4 \end{aligned}$$

13. $$\begin{aligned} \log_8 2x + \log_8 6 &= 2 \\ \log_8((2x)(6)) &= 2 \\ 12x &= 8^2 \\ 12x &= 64 \\ x &= \frac{16}{3} \end{aligned}$$

Check: $$\begin{aligned} \log_8 2\cdot\frac{16}{3} + \log_8 6 &\stackrel{?}{=} 2 \\ \log_8\left(2\cdot\frac{16}{3}\cdot 6\right) &\stackrel{?}{=} 2 \\ \log_8 64 &\stackrel{?}{=} 2 \\ \log_8 8^2 &\stackrel{?}{=} 2 \\ 2 &= 2 \end{aligned}$$

14. $$\begin{aligned} e^{5x-3} &= 57 \\ \ln e^{5x-3} &= \ln 57 \\ (5x-3)\ln e &= \ln 57 \\ 5x &= \ln 57 + 3 \\ x &= \frac{\ln 57 + 3}{5} \end{aligned}$$

15. $$\begin{aligned} 5^{3x+6} &= 17 \\ \ln 5^{3x+6} &= \ln 17 \\ (3x+6)\ln 5 &= \ln 17 \\ 3x+6 &= \frac{\ln 17}{\ln 5} \\ x &= \frac{-6 + \frac{\ln 17}{\ln 5}}{3} \approx -1.4132 \end{aligned}$$

16. $A = P(1+r)^t$

$A = 2000(1+0.08)^5 = 2938.656154...$

Henry will have \$2938.66.

17. $$\begin{aligned} A &= P(1+r)^t \\ 2P &= P(1+0.05)^t \\ 2 &= (1.05)^t \\ \ln 2 &= \ln 1.05^t \\ \ln 2 &= t\ln 1.05 \\ t &= \frac{\ln 2}{\ln 1.05} = 14.20669908... \approx 14 \end{aligned}$$

It will take about 14 years to double her money.

Practice Final Examination

1. $$\begin{aligned} (4-3)^2 + \sqrt{9} \div (-3) + 4 &= 1^2 + 3 \div (-3) + 4 \\ &= 1 + (-1) + 4 \\ &= 4 \end{aligned}$$

2. $36,250,000 = 3.625\times 10^7$

3. $3a + 6b - a + 5ab + 3a^2 + b = 2a + 7b + 5ab + 3a^2$

4. $3[2x-5(x+y)] = 3[2x-5x-5y]$
$= 3[-3x-5y]$
$= -9x-15y$

5. $F = \frac{9}{5}C + 32$
$F = \frac{9}{5}(-35) + 32$
$F = -31$

6. $\frac{1}{3}y - 4 = \frac{1}{2}y + 1$
$\frac{1}{6}y = -5$
$y = -30$

7. $A = \frac{1}{2}a(b+c)$
$2A = ab + ac$
$ab = 2A - ac$
$b = \frac{2A - ac}{a}$

8. $\left|\frac{2}{3}x - 4\right| = 2$
$\frac{2}{3}x - 4 = 2$ or $\frac{2}{3}x - 4 = -2$
$2x - 12 = 6$ $\quad$ $2x - 12 = -6$
$2x = 18$ $\quad$ $2x = 6$
$x = 9$ $\quad$ $x = 3$

9. $2x - 3 < x - 2(3x - 2)$
$2x - 3 < x - 6x + 4$
$7x < 7$
$x < 1$

10. $P = 2L + 2W = 1760$
$L + W = 880$
$2W - 200 + W = 880$
$3W = 1080$
$W = 360$
$L = 2W - 200 = 520$
The width is 360 meters and the length is 520 meters.

11. Let x = amount invested at 14%.
$4000 - x$ = amount invested at 12%.
$0.14x + 0.12(4000 - x) = 508$
$0.14x + 480 - 0.12x = 508$
$0.02x = 28$
$x = 1400$
$4000 - x = 2600$
\$1400 was invested at 14% and \$2600 at 12%.

12. $x + 5 \le -4$ or $2 - 7x \le 16$
$x \le -9$ $\quad$ $-7x \le 14$
$x \ge -2$

13. $|2x - 5| < 10$
$-10 < 2x - 5 < 10$
$-5 < 2x < 15$
$-\frac{5}{2} < x < \frac{15}{2}$

14. $7x - 2y = -14$
Let $x = 0$.
$7(0) - 2y = -14$
$y = 7$
Let $y = 0$.
$7x - 2(0) = -14$
$x = -2$

x	y
0	7
−2	0

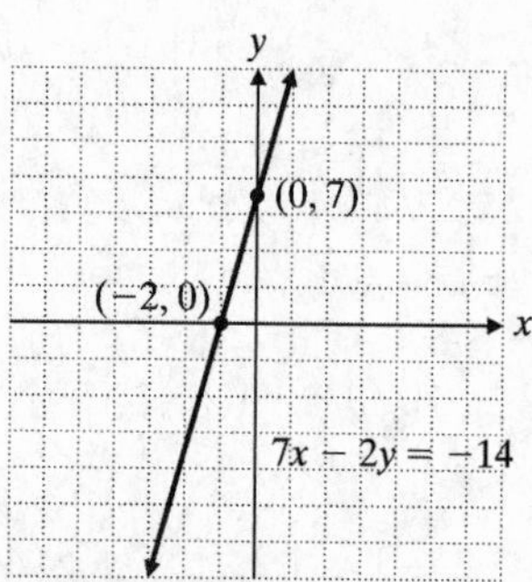

Scale: Each unit = 2

15. $3x - 4y \le 6$
Graph $3x - 4y = 6$ with a solid line.
Test point: (0, 0)
$3(0) - 4(0) \le 6$
$0 \le 6$, True
Shade the region containing (0, 0).

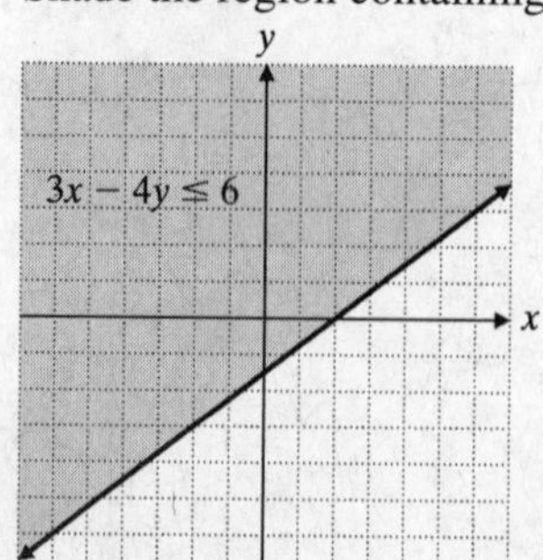

16. $m = \frac{y_2 - y_1}{x_2 - x_1} = \frac{-3-5}{-2-1} = \frac{8}{3}$

17. $3x + 2y = 8$

$y = -\frac{3}{2}x + 4,\ m = -\frac{3}{2}$

$m_{\|} = -\frac{3}{2},\quad y - 4 = -\frac{3}{2}(x - (-1))$

$y - 4 = -\frac{3}{2}(x+1)$

$2y - 8 = -3x - 3$

$3x + 2y = 5$

18. $f(x) = 3x^2 - 4x - 3$

$f(3) = 3(3^2) - 4(3) - 3$

$f(3) = 27 - 12 - 3$

$f(3) = 12$

19. $f(x) = 3x^2 - 4x - 3$

$f(-2) = 3((-2)^2) - 4(-2) - 3$

$f(-2) = 12 + 8 - 3$

$f(-2) = 17$

20. $f(x) = |2x - 4|$

x	$y = f(x) = \lvert 2x-4 \rvert$
1	2
2	0
3	2

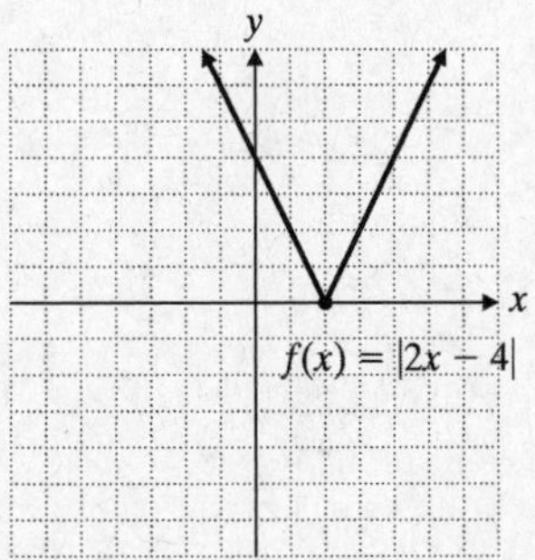

21. $2x + y = 15 \Rightarrow y = 15 - 2x$

$3x - 2y = 5$

$3x - 2(15 - 2x) = 5$

$3x - 30 + 4x = 5$

$7x = 35$

$x = 5$

$y = 15 - 2(5) = 5$

(5, 5) is the solution.

22. $4x - 3y = 12 \xrightarrow{\times 4} 16x - 12y = 48$

$3x - 4y = 2 \xrightarrow{\times -3} -9x + 12y = -6$

$7x = 42$

$x = 6$

$4(6) - 3y = 12$

$-3y = -12$

$y = 4$

(6, 4) is the solution.

23. Solve the system.

$2x + 3y - z = 16$

$x - y + 3z = -9$

$5x + 2y - z = 15$

Multiply first equation by 3 and add to the second equation.

$6x + 9y - 3z = 48$

$x - y + 3z = -9$

$7x + 8y = 39$

Multiply third equation by 3 and add to the second equation.

$x - y + 3z = -9$

$15x + 6y - 3z = 45$

$16x + 5y = 36$

Now solve the system.

$7x + 8y = 39 \xrightarrow{\times 5} 35x + 40y = 195$

$16x + 5y = 36 \xrightarrow{\times -8} -128x - 40y = -288$

$-93x = -93$

$x = 1$

$16x + 5y = 36$

$16(1) + 5y = 36$

$5y = 20$

$y = 4$

$x - y + 3z = -9$

$1 - 4 + 3z = -9$

$z = -2$

(1, 4, –2) is the solution.

24. $3x - 2y = 7 \xrightarrow{\times 3} 9x - 6y = 21$

$-9x + 6y = 2 \longrightarrow -9x + 6y = 2$

$0 = 23$

Inconsistent system. No solution.

25. A = number of adult tickets
$15 - A$ = number of children tickets
$8A + 3(15 - A) = 100$
$8A + 45 - 3A = 100$
$5a = 55$
$A = 11$ adult tickets
$15 - A = 4$ children tickets
He sold 11 adult and 4 children tickets.

26. $3y \geq 8x - 12$
Test point: (0, 0)
$3(0) \geq 8(0) - 12$
$0 \geq -12$, true

$2x + 3y \leq -6$
Test point: (0, 0)
$2(0) + 3(0) \leq -6$
$0 \leq -6$, false

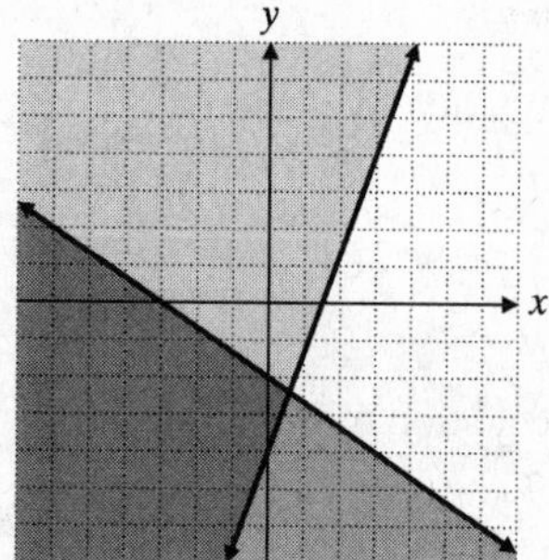

27. $(3x - 2)(2x^2 - 4x + 3)$
$= 6x^3 - 12x^2 + 9x - 4x^2 + 8x - 6$
$= 6x^3 - 16x^2 + 17x - 6$

28.
$$\begin{array}{r|l} & 5x^2 - x + 2 \\ 5x+1 & 25x^3 + 0x^2 + 9x + 2 \\ & \underline{25x^3 + 5x^2} \\ & -5x^2 + 9x \\ & \underline{-5x^2 - x} \\ & 10x + 2 \\ & \underline{10x + 2} \\ & 0 \end{array}$$

29. $8x^3 - 27 = (2x)^3 - 3^3$
$= (2x - 3)((2x)^2 + (2x)(3) + (3)^2)$
$= (2x - 3)(4x^2 + 6x + 9)$

30. $x^3 + 2x^2 - 4x - 8 = x^2(x + 2) - 4(x + 2)$
$= (x + 2)(x^2 - 4)$
$= (x + 2)(x + 2)(x - 2)$

31. $2x^3 + 15x^2 - 8x = x(2x^2 + 15x - 8)$
$= x(2x - 1)(x + 8)$

32. $x^2 + 15x + 54 = 0$
$(x + 9)(x + 6) = 0$
$x + 9 = 0$ $\quad x + 6 = 0$
$x = -9$ $\quad x = -6$

33. $\dfrac{9x^3 - x}{3x^2 - 8x - 3} = \dfrac{x(9x^2 - 1)}{(3x + 1)(x - 3)}$
$= \dfrac{x(3x + 1)(3x - 1)}{(3x + 1)(x - 3)}$
$= \dfrac{x(3x - 1)}{(x - 3)}$

34. $\dfrac{x^2 - 9}{2x^2 + 7x + 3} \div \dfrac{x^2 - 3x}{2x^2 + 11x + 5}$
$= \dfrac{(x + 3)(x - 3)}{(2x + 1)(x + 3)} \cdot \dfrac{(2x + 1)(x + 5)}{x(x - 3)}$
$= \dfrac{x + 5}{x}$

35. $\dfrac{3x}{x + 5} - \dfrac{2}{x^2 + 7x + 10}$
$= \dfrac{3x(x + 2)}{(x + 5)(x + 2)} - \dfrac{2}{(x + 5)(x + 2)}$
$= \dfrac{3x(x + 2) - 2}{(x + 5)(x + 2)}$
$= \dfrac{3x^2 + 6x - 2}{(x + 5)(x + 2)}$

36. $\dfrac{\frac{3}{2x} - 1}{\frac{5}{2} - \frac{1}{x}} = \dfrac{\frac{3}{2x} - 1}{\frac{5}{2} + \frac{1}{x}} \cdot \dfrac{2x}{2x} = \dfrac{3 - 2x}{5x + 2}$

37.
$$\frac{x-1}{x^2-4}=\frac{2}{x+2}+\frac{4}{x-2}$$
$$\frac{x-1}{(x+2)(x-2)}=\frac{2}{x+2}+\frac{4}{x-2}$$
$$(x+2)(x-2)\left(\frac{x-1}{(x+2)(x-1)}\right)=(x+2)(x-2)\cdot\frac{2}{x+2}+(x+2)(x-2)\cdot\frac{4}{x-2}$$
$$x-1=2(x-2)+4(x+2)$$
$$x-1=2x-4+4x+8$$
$$5x=-5$$
$$x=-1$$

38. $16^{3/2}=\sqrt{16^3}=\sqrt{16\cdot16\cdot16}=4\cdot4\cdot4=64$

39. $\sqrt{44a^4b^7c}=\sqrt{4(a^2)^2(b^3)^2\cdot11bc}=2a^2b^3\sqrt{11bc}$

40.
$$5\sqrt{2}-3\sqrt{50}+4\sqrt{98}=5\sqrt{2}-3\sqrt{25\cdot2}+4\sqrt{49\cdot2}$$
$$=5\sqrt{2}-15\sqrt{2}+28\sqrt{2}$$
$$=18\sqrt{2}$$

41.
$$\frac{5}{\sqrt{7}-2}=\frac{5}{\sqrt{7}-2}\cdot\frac{\sqrt{7}+2}{\sqrt{7}+2}$$
$$=\frac{5\sqrt{7}+10}{7-4}$$
$$=\frac{5\left(\sqrt{7}+2\right)}{3}$$

42. $i^3+\sqrt{-25}+\sqrt{-16}=-i+5i+4i=8i$

43.
$$\sqrt{x+7}=x+5$$
$$\left(\sqrt{x+7}\right)^2=(x+5)^2$$
$$x+7=x^2+10x+25$$
$$x^2+9x+18=0$$
$$(x+6)(x+3)=0$$
$x+6=0$ $\quad$ $x+3=0$

$x=-6$ $\quad$ $x=-3$

Check: $\sqrt{-6+7}\stackrel{?}{=}-6+5$
$$\sqrt{1}\stackrel{?}{=}-1$$
$$1\neq-1$$
$$\sqrt{-3+7}\stackrel{?}{=}-3+5$$
$$\sqrt{4}\stackrel{?}{=}-3+5$$
$$2=2$$
$x=-3$ is the solution.

44. $y = kx^2$

$15 = k(2)^2$

$k = \frac{15}{4}$

$y = \frac{15}{4}x^2$

$y = \frac{15}{4}(3)^2 = 33.75$

45. $5x(x+1) = 1+6x$

$5x^2 + 5x = 1+6x$

$5x^2 - x - 1 = 0$, use quadratic formula

$x = \frac{-(-1) \pm \sqrt{(-1)^2 - 4(5)(-1)}}{2(5)}$

$x = \frac{1 \pm \sqrt{21}}{10}$

46. $5x^2 - 9x = -12x$

$5x^2 + 3x = 0$

$x(5x+3) = 0$

$x = 0$ or $5x + 3 = 0$

$x = -\frac{3}{5}$

47. $x^{2/3} + 5x^{1/3} - 14 = 0$, let $x^{1/3} = w$, $x^{2/3} = w^2$

$w^2 + 5w - 14 = 0$

$(w-2)(w+7) = 0$

$w - 2 = 0$ $\quad$ $w + 7 = 0$

$w = 2$ $\quad$ $w = -7$

$x^{1/3} = 2$ $\quad$ $x^{1/3} = -7$

$x = 8$ $\quad$ $x = -343$

48. $3x^2 - 11x - 4 \ge 0$

$3x^2 - 11x - 4 = 0$

$(3x+1)(x-4) = 0$

$3x + 1 = 0$ or $x - 4 = 0$

$x = -\frac{1}{3}$ $\quad$ $x = 4$

Region I: Test $x = -1$

$3(-1)^2 - 11(-1) - 4 = 10 > 0$

Region II: Test $x = 0$

$3(0)^2 - 11(0) - 4 = -4 < 0$

Region III: Test $x = 5$

$3(5)^2 - 11(5) - 4 = 16 > 0$

$x \le -\frac{1}{3}$ or $x \ge 4$

49. $f(x) = -x^2 - 4x + 5$

parabola, opening downward

$f(0) = -0^2 - 4(0) + 5 = 5 \Rightarrow$ y-int: (0, 5)

$-x^2 - 4x + 5 = 0$

$x^2 + 4x - 5 = 0$

$(x+5)(x-1) = 0$

$x = -5, x = 1$

x-int: (−5, 0), (1, 0)

$-\frac{b}{2a} = -\frac{-4}{2(-1)} = -2$

$f(-2) = -(-2)^2 - 4(-2) + 5 = 9$

$V(-2, 9)$

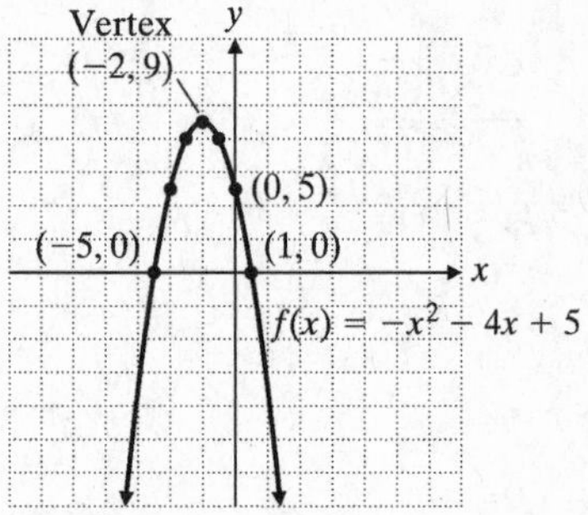

Scale: Each unit = 2

50. $L = 3W + 1$

$A = 52 \text{ cm}^2$

$A = LW = (3W+1)W = 52$

$3W^2 + W - 52 = 0$

$(W-4)(2W+13) = 0$

$W = 4$ or $W = -\frac{13}{3}$ reject, $W > 0$

$L = 3W + 1 = 13$

The width is 4 cm and the length is 13 cm.

51. $x^2 + y^2 + 6x - 4y = -9$

$x^2 + 6x + 9 + y^2 - 4y + 4 = -9 + 9 + 4 = 4$

$(x+3)^2 + (y-2)^2 = 2^2$

$C(-3, 2), r = 2$

52. $\dfrac{x^2}{16}+\dfrac{y^2}{25}=1$

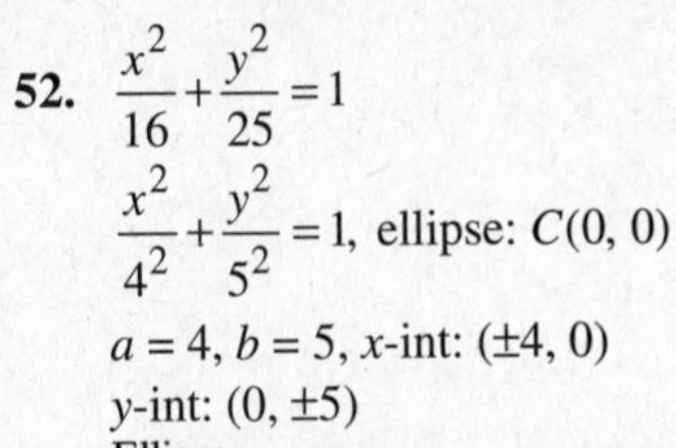

$\dfrac{x^2}{4^2}+\dfrac{y^2}{5^2}=1$, ellipse: $C(0, 0)$

$a = 4$, $b = 5$, x-int: $(\pm 4, 0)$

y-int: $(0, \pm 5)$

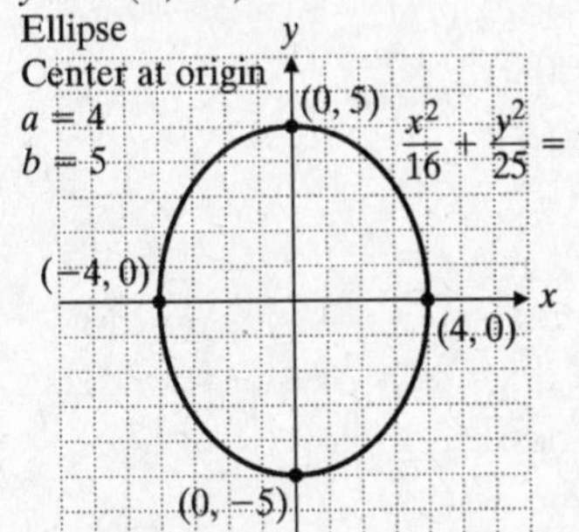

53. $\dfrac{x^2}{4}-\dfrac{y^2}{9}=1$

$\dfrac{x^2}{2^2}-\dfrac{y^2}{3^2}=1$, hyperbola

$C(0, 0)$, $a = 2$, $b = 3$

x-int: $(\pm 2, 0)$

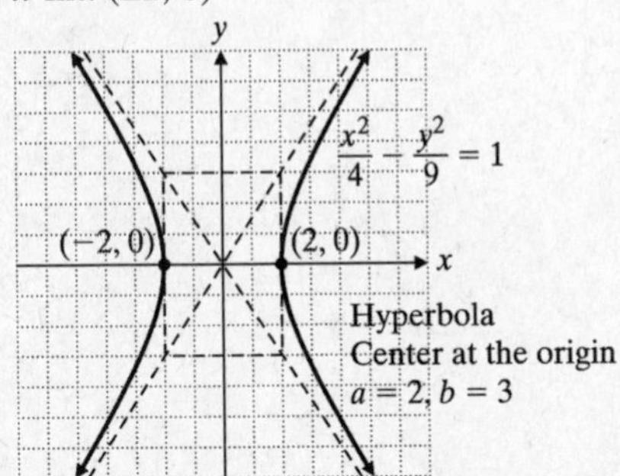

54. $x=(y-3)^2+5$

parabola opening right, $V(5, 3)$

$x=(0-3)^2+5=14$, x-int: $(14, 0)$

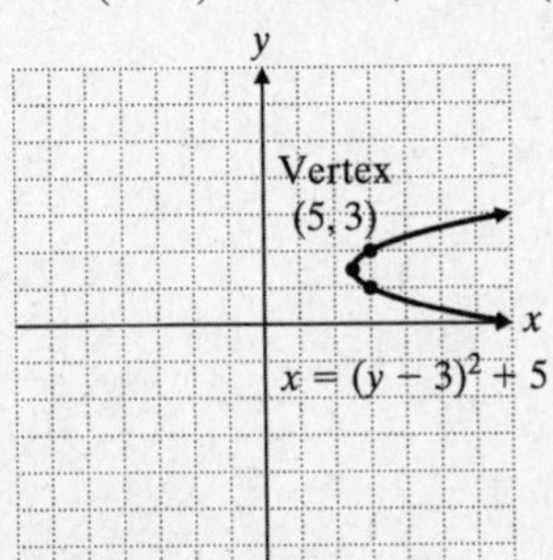

Scale: Each unit = 2

55. $x^2+y^2=16$

$x^2-y=4 \Rightarrow y=x^2-4$

$x^2+(x^2-4)^2=16$

$x^2+x^4-8x^2+16=16$

$x^4-7x^2=0$

$x^2(x^2-7)=0$

$x^2=0 \qquad x^2=7$

$x=0 \qquad x=\pm\sqrt{7}$

$y=x^2-4=0^2-4=-4$

$y=x^2-4=\left(\pm\sqrt{7}\right)^2-4=3$

$(0, -4)$, $\left(\pm\sqrt{7}, 3\right)$ is the solution.

56. $f(x)=3x^2-2x+5$

a. $f(-1)=3(-1)^2-2(-1)+5=3+2+5=10$

b. $f(a)=3a^2-2a+5$

c. $f(a+2)=3(a+2)^2-2(a+2)+5$

$=3a^2+12a+12-2a-4+5$

$=3a^2+10a+13$

57. $f(x)=5x^2-3$, $g(x)=-4x-2$

$f[g(x)]=f(-4x-2)$

$=5(-4x-2)^2-3$

$=5(16x^2+16x+4)-3$

$=80x^2+80x+20-3$

$=80x^2+80x+17$

58. $f(x)=\dfrac{1}{2}x-7$, $f(x)\to y$

$y=\dfrac{1}{2}x-7$, $x \leftrightarrow y$

$x=\dfrac{1}{2}y-7$

$\dfrac{1}{2}y=x+7$

$y=2x+14$

$y \to f^{-1}(x)$

$f^{-1}(x)=2x+14$

59. $M = \left\{(3, 7), (2, 8), \left(7, \frac{1}{2}\right), (-3, 7)\right\}$ is not one-to-one because two pairs, (3, 7) and (–3, 7), have the same second coordinate.

60. $f(x) = 2^{1-x}$

x	$y = f(x) = 2^{1-x}$
–1	4
0	2
1	1

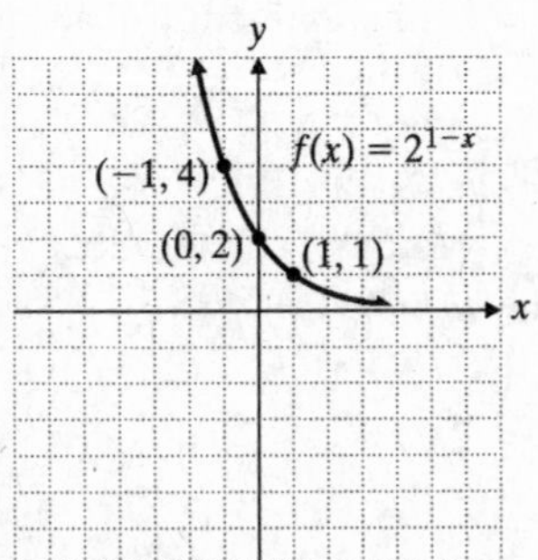

61. $\log_6 1 = x$

$$6^x = 1$$
$$6^x = 6^0$$
$$x = 0$$

62. $\log_4(3x+1) = 3$

$$3x+1 = 4^3 = 64$$
$$3x+1 = 64$$
$$3x = 63$$
$$x = 21$$

63. $\log_{10} 0.01 = y$

$$10^y = 0.01$$
$$10^y = 10^{-2}$$
$$y = -2$$

64.

$$\log_2 6 + \log_2 x = 4 + \log_2(x-5)$$
$$\log_2(6x) = 4 + \log_2(x-5)$$
$$\log_2(6x) - \log_2(x-5) = 4$$
$$\log_2 \frac{6x}{x-5} = 4$$
$$\frac{6x}{x-5} = 2^4 = 16$$
$$6x = 16(x-5) = 16x - 80$$
$$16x - 80 = 6x$$
$$10x = 80$$
$$x = 8$$